INTERNATIONAL UNION OF THEORETICAL AND APPLIED MECHANICS

THEORY OF THIN SHELLS

SECOND SYMPOSIUM, COPENHAGEN
SEPTEMBER 5–9, 1967

EDITOR

F. I. NIORDSON

WITH 86 FIGURES

SPRINGER-VERLAG BERLIN HEIDELBERG GMBH 1969

Professor Dr. FRITHIOF I. NIORDSON, Ph. D.
Department of Applied Mechanics
The Technical University of Denmark,
Copenhagen

Originally published by Springer-Verlag Berlin Heidelberg New York in 1969
Softcover reprint of the hardcover 1st edition 1969

Library of Congress Catalog Card Number 68-26458

Title No. 1530

Preface

During its meeting in Vienna on the 25th of June 1966 the General Assembly of the International Union of Theoretical and Applied Mechanics (IUTAM) decided to sponsor a second international symposium on the theory of thin shells. It was decided to hold this symposium in Copenhagen and the scientific organization was left in the hands of a committee consisting of A. E. Green, W. T. Koiter, F. I. Niordson (chairman), W. Olszak, E. Reissner, and I. N. Vekua.

The main topics of the symposium were announced to be:

"Foundations", "Non-linear theory", "Buckling and post-buckling", and "Inelastic behaviour".

The impetus to the advancement of shell theory from the first IUTAM Symposium on this subject, held in Delft in 1959, has been widely recognized. The purpose of this second Symposium was to take stock of the situation and, by bringing leading scientists in this field together, to open and encourage new studies in this seemingly inexhaustible discipline.

It was felt, in the development following the first Symposium in Delft, that solid foundations of shell theory had been obtained and that this second Symposium could be devoted almost entirely to the buckling and post-buckling behaviour and to the non-linear theory of shells. However, as it proved to be, nearly half the lecturers were mainly concerned with the foundations of shell theory and the discussions indicated that more will be said in this matter. Nevertheless, also the sessions devoted to stability and non-linear behaviour undoubtably contributed greatly to the scientific value of the Symposium.

Participants were invited personally for their active interest in any of the fields to be covered by the symposium and participation was strictly limited.

In all there were 75 participants from 15 countries and 23 papers were presented. A period of 35 minutes was allocated to the presentation of each paper, and a period of 10 minutes was reserved for discussion. All papers were presented in English although this was not a requirement.

The local arrangements of the symposium were in the hands of a committee consisting of T. Brøndum-Nielsen, J. Christoffersen (secretary), F. I. NIORDSON (chairman), and B. HØJLUND RASMUSSEN. Financial support was generously provided by the following Danish companies: The Carlsberg Breweries, Christiani & Nielsen A/S, A/S Hellesens, Højgaard & Schultz A/S, A. Jespersen & søn A/S, Johs. Jørgensen A/S, Larsen & Nielsen Consultor A/S, P. E. Malmstrøm, Monberg & Thorsen A/S, Chr. Ostenfeld & W. Jønson, F. L. Smidth & Co. A/S.

The working sessions of the symposium were held at the Lyngby Campus of the Royal Technical University of Denmark. In the opening session the address of welcome by the Rector of Technical University, Professor Dr. Count E. KNUTH-WINTERFELDT was followed by the addresses of Professor M. ROY, President of IUTAM and the chairman. Afterwards all participant and accompanying persons were received by the Mayor of Copenhagen in the City Hall.

For the successful completion of the Symposium my thanks are due to the devoted work by the assistants, Miss H. BYSKOV, Mrs. K. CHRISTOFFERSEN, Mr. P. FLENSBORG, Mr. A. GUDMANN NIELSEN, Mr. P. MADSEN, Miss P. NIORDSON, and Mr. P. TERNDRUP PEDERSEN. A special thank is due to my secretary Mrs. E. CHRISTENSEN, who has been of great help.

Next, I wish to thank the authors for their cooperation in my editoral work of these Proceedings. Finally, I am greatly indebted to the Springer-Verlag, who has undertaken the difficult task of printing this book and who has eased my work considerably.

Copenhagen, December 1968 **Frithiof I. Niordson**

Scientific Committee

A. E. Green, Oxford
W. T. Koiter, Delft
F. Niordson, Copenhagen (chairman)
W. Olszak, Warsaw
E. Reissner, Cambridge, Mass.
I. N. Vekua, Tbilisi

List of Participants

Ambartsumyan, S. A.,	Erevan, USSR
Argyris, J. H.,	Stuttgart, Germany
Augusti, G.,	Napoli, Italy
Bergholt, K.,	Copenhagen, Denmark
Bouma, A. L.,	Delft, Netherlands
Bourgain, L. M.,	Gif sur Yvette, France
Brøndum-Nielsen, T.,	Copenhagen, Denmark
Budiansky, B.,	Cambridge, Mass., USA
Bychawski, Z.,	Warsaw, Poland
Christoffersen, J.,	Copenhagen, Denmark
Clark, R. A.,	Cleveland, Ohio, USA
Djubek, J.,	Bratislava, CSSR
Duddeck, H.,	Braunschweig, Germany
Duszek, M. K.,	Warsaw, Poland
Eringen, A. C.,	Princeton, N. J., USA
Esslinger, M.,	Braunschweig, Germany
Fischer, G.,	Porz-Wahn, Germany
Flügge, W.,	Stanford, Cal., USA
Gol'denweizer, A. L.,	Moscow, USSR
Green, A. E.,	Oxford, England
Gustafson, A.,	Trondheim, Norway
Günther, W.,	Karlsruhe, Germany
Görtler, H.,	Freiburg, Germany
Hoff, N. J.,	Stanford, Cal., USA
Hutchinson, J. W.,	Cambridge, Mass., USA
Højlund Rasmussen, B.,	Copenhagen, Denmark
John, F.,	New York, N. Y., USA
Kalnins, A.,	Bethlehem, Pa., USA
Kildegaard, A.,	Copenhagen, Denmark

KODNAR, R.,	Bratislava, CSSR
KOITER, W. T.,	Delft, Netherlands
VAN KOTEN, H.,	Delft, Netherlands
KŘUPKA, V.,	Brno, CSSR
LARDNER, T. J.,	Cambridge, Mass., USA
LECKIE, F. A.,	Cambridge, England
LEKKERKERKER, J. G.,	Delft, Netherlands
LEONARD, R. W.,	Hampton, Va., USA
LUKASIEWICZ, S.,	Warsaw, Poland
MASUR, E. F.,	Chicago, Ill., USA
MIKELADZE, M. SH.,	Tbilisi, USSR
MIKHAILOV, G. K.,	Moscow, USSR
MIŞICU, M.,	Bucharest, Rumania
NACHBAR, W.,	San Diego, Cal., USA
NAGHDI, P. M.,	Berkeley, Cal., USA
NASH, W. A.,	Gainesville, Fla., USA
VAN DER NEUT, A.,	Delft, Netherlands
NIELSEN, M. P.,	Aalborg, Denmark
NIORDSON, F.,	Copenhagen, Denmark
NOVÁK, O.,	Praha, CSSR
ODQVIST, F.,	Stockholm, Sweden
OLSZAK, W.,	Warsaw, Poland
POGORELOV, A. V.,	Charkov, USSR
REISSNER, E.,	Cambridge, Mass., USA
RIMROTT, F. P. J.,	Toronto, Ont., Canada
ROBERT, A. J.,	Paris, France
ROSS, E. W.,	Natick, Mass., USA
ROUGEE, P.,	Paris, France
ROY, M.,	Paris, France
RUTTEN, H. S.,	Delft, Netherlands
SANDERS, J. L.,	Cambridge, Mass., USA
SAVIN, G. N.,	Kiev, USSR
SAWCZUK, A.,	Grenoble, France
SCHNELL, W.,	Darmstadt, Germany
SHERBOURNE, A. N.,	Waterloo, Ont., Canada
SEDOV, L. I.,	Moscow, USSR
SEIDE, P.,	Los Angeles, Cal., USA
SIMMONDS, J. G.,	Charlottesville, Va., USA
SINGER, J.,	Haifa, Israel
STEELE, C. R.,	Stanford, Cal., USA
STEIN, E.,	Stuttgart, Germany
TEMPLE, G.,	Oxford, England
THIELEMANN, W. F.,	Braunschweig, Germany
VALID, R.,	Chatillon, France
VEKUA, I. N.,	Tbilisi, USSR
WAN, F. Y.,	Cambridge, Mass., USA
WITTRICK, W. H.,	Birmingham, England

Authors

Prof. S. A. Ambartsumyan, Institute of Mathematics and Mechanics, Academy of Sciences of the Armenian SSR, Barekamutian 24b, Erevan, USSR

Prof. B. Budiansky, Pierce Hall, Harvard University, Cambridge, Mass. 02138, USA

Prof. Z. Bychawski, Institute of Basic Technical Problems, Polish Academy of Sciences, Warsaw, Swietokrzyska 21, Poland

Dr. M. K. Duszek, Institute of Basic Technical Problems, Polish Academy of Sciences, Warsaw, Swietokrzyska 21, Poland

Prof. M. Esslinger, Institut für Flugzeugbau, Deutsche Forschungsanstalt für Luft- und Raumfahrt e. V., 33 Braunschweig-Kralenriede, Bienroder Weg 53, Germany

Prof. A. L. Gol'denweizer, Institute of Mechanical Problems, USSR Academy of Sciences, Moscow A-40, Leningradskoj Prosp. 7, USSR

Prof. A. E. Green, Department of Mathematics, Oxford University, Oxford, England

Prof. F. John, Courant Institute of Mathematical Sciences, New York University, 251 Mercer Street, New York, N. Y. 10012, USA

Mr. A. Kildegaard, Afdelingen for Styrkelære, Danmarks tekniske Højskole, Rigensgade 13, Copenhagen, Denmark

Prof. W. T. Koiter, Laboratorium voor Technische Mechanica, Technische Hogeschool, Mekelweg 2, Delft, Netherlands

Dr. F. A. Leckie, University Engineering Laboratories, Trumpington Street, Cambridge, England

Dr. S. Lukasiewicz, Politechnika Warszawska, Warsaw, Nowowiejska 24, Poland

Prof. M. Sh. Mikeladze, Chair of Structural Mechanics Georgian Polytechnic Institute Tbilisi, Georgia, USSR

Prof. M. Mişicu, Center of Solid Mechanics, Academia R. S. R., Str. Constantin Mille No. 15, Bucharest, Rumania

Prof. W. Nachbar, University of California, San Diego, P. O. Box 109, La Jolla, Cal. 92038, USA

Prof. P. M. Naghdi, Division of Applied Mechanics, University of California, 6141 Etchevery Hall, Berkeley, Cal. 94720, USA

Prof. W. Olszak, Polish Academy of Sciences, Warsaw, Swietokrzyska 21, Poland

Prof. A. V. Pogorelov, Physical-Technical Institute of Low Temperatures, Prosp. Lenin 47, Charkov 86, USSR

Prof. E. Reissner, Department of Mathematics, Massachusetts Institute of Technology, Cambridge, Mass. 02139, USA

Mr. H. S. Rutten, Department of Mathematics, Technological University, Julianalaan 132, Delft, Netherlands

Prof. J. L. Sanders, Pierce Hall, Harvard University, Cambridge, Mass. 02138, USA

Prof. A. Sawczuk, Laboratoires de Mécanique des Fluides, Université de Grenoble, 44–46, Avenue Félix Viallet, Grenoble, France, and The Institute for Basic Technical Research, Warsaw, Poland

Prof. J. G. Simmonds, University of Virginia, Charlottesville, Va., 22901, USA

Prof. J. Singer, Department of Aeronautical Engineering, Technion-Israel Institute of Technology, Technion City, Haifa, Israel

Prof. W. F. Thielemann, Institut für Flugzeugbau, Deutsche Forschungsanstalt für Luft- und Raumfahrt e. V., 33 Braunschweig-Kralenriede, Bienroder Weg 53, Germany

Prof. I. N. Vekua, University of Tbilisi, Prospekt Tchavchavadze 1, Tbilisi, USSR

Contents

Refined Interior Shell Equations

By

F. John

New York University, New York, N.Y., USA

1. Introduction

In a previous paper[1] the author derived a set of approximate interior equations for equilibrium of thin perfectly elastic shells on the basis of a priori estimates, taking advantage of the elliptic character of the 3-dimensional differential equations. I want to explain here the point of view, the significance, and the limitations of those results, and add a refinement which appears to be needed in certain situations.

One feature of the theory presented here is the fact that the equations for solids and the resulting interior shell equations involve only stresses and strains and make no mention of displacements[2]. The main advantage for us is that the conclusions arrived at do not depend on any assumption on the displacements or their derivatives, but hold as soon as the strains (or stresses) are sufficiently small.

2. General 3-Dimensional Equations for Stresses and Strains

The deformation of the solid is considered as a mapping of the points in euclidean 3-space representing the position of particles in the original

Acknowledgement. This paper represents results obtained at the Courant Institute of Mathematical Sciences, New York University, with the Office of Naval Research, Contract ONR-285(46). Reproduction in whole or in part is permitted for any purpose of the United States Government.

[1] "Estimates for the derivatives of the stresses in a thin shell and interior shell equations," Communications on Pure and Applied Mathematics 18, 235—267 (1965).

[2] The full system of 3-dimensional equations for the stresses or strains, as presented here, consists of 9 (dependent) equations for 6 unknown functions. They are "less non-linear" than the equations in terms of displacements, in that the coefficients of the second derivatives of the unknown functions do not contain their first derivatives. Boundary conditions expressing prescribed forces on the boundary also take a simpler form in the system used here.

"unstrained" state onto points representing the position of the same particles in the "strained" equilibrium state. In the unstrained state particles have curvilinear "material" coordinates U^1, U^2, U^3. The element dS of euclidean distance between particles in the unstrained state is given by[1]

$$dS^2 = G_{ik}\, dU^i\, dU^k. \tag{1a}$$

In the strained state a particle has the same curvilinear coordinates U^i; only the element ds of euclidean distance is now a different one:

$$ds^2 = g_{ik}\, dU^i\, dU^k. \tag{1b}$$

We consider the form $G_{ik}\, dU^i\, dU^k$ as known. All identifications of covariant and contravariant components of tensors shall refer to this form. Covariant differentiation with respect to this form is denoted by the symbol ";". The *strain tensor* ε_{ik} is defined by

$$\varepsilon_{ik} = \frac{1}{2}\,(g_{ik} - G_{ik}). \tag{1c}$$

Since the deformation takes place in euclidean space the g_{ik} or ε_{ik} satisfy the *compatibility conditions*, which express the vanishing of the Riemann curvature tensor:

$$\begin{aligned} O &= R_{hijk} \\ &= \varepsilon_{hk;ij} + \varepsilon_{ij;hk} - \varepsilon_{hj;ik} - \varepsilon_{ik;hj} + g_{ab}\,(c^b_{hk} c^a_{ij} - c^b_{hj} c^a_{ik}). \end{aligned} \tag{2a}$$

Here the quantities c^i_{jk} are the components of the *tensor* that represents the difference of the Christoffel symbols with respect to the metrics ds^2 and dS^2; they can be expressed invariantly from the formulae

$$g_{si}\, c^i_{jk} = \varepsilon_{sj;k} + \varepsilon_{sk;j} - \varepsilon_{jk;s}. \tag{2b}$$

For a perfectly elastic material, which is homogeneous and isotropic in the the unstrained state, the constituent equations can be derived from a *strain energy density function*

$$W = W(s_1, s_2, s_3). \tag{3a}$$

Here s_i is the sum of the i-th powers of the *principal strains*, i.e. of the eigenvalues of the matrix with components $\varepsilon_k{}^i = G^{ij} \varepsilon_{jk}$. In terms of the $\varepsilon_k{}^i$ we have

$$s_1 = \varepsilon_i{}^i, \quad s_2 = \varepsilon_j{}^j \varepsilon_i{}^i, \quad s_3 = \varepsilon_j{}^i \varepsilon_k{}^j \varepsilon_i{}^k. \tag{3b}$$

[1] In this paper Roman indices range over the values 1, 2, 3 and Greek indices over the values 1, 2.

We assume that W has derivatives up to order 6 near, $s_1 = s_2 = s_3 = 0$ and an expansion starting with

$$W = 2\mu\left[\frac{p}{2q}(s_1)^2 + \frac{1}{2}s_2 + \frac{\alpha}{3}(s_1)^3 + \beta s_1 s_2 + \frac{\gamma}{3}s_3\right] + \\ + \text{fourth degree terms in } \varepsilon + \dots \tag{3c}$$

Here

$$2\mu = \frac{E}{1+\nu}, \qquad p = \frac{\nu}{1-\nu}, \qquad q = \frac{1-2\nu}{1-\nu} = 1 - p \tag{3d}$$

in terms of Young's modulus E and Poisson's ratio ν, while the dimensionless constants α, β, γ, determine the cubic terms[1].

The *constituent equations* connecting the stress tensor $t_k^{\,i}$ with the strain tensor $\varepsilon_k^{\,i}$ are

$$\begin{aligned} t_k^{\,i} &= (\det g_s^{\,r})^{-1/2}\frac{\partial W}{\partial \varepsilon_i^{\,k}} \\ &= 2\mu\left[\frac{p}{q}\varepsilon_j^{\,j}\delta_k^{\,i} + \varepsilon_k^{\,i} + \left(\alpha - \frac{p}{q}\right)\varepsilon_j^{\,j}\varepsilon_r^{\,r}\delta_k^{\,i} + (2\beta - 1)\varepsilon_j^{\,j}\varepsilon_k^{\,i} + \right. \\ &\qquad \left. + \beta\varepsilon_t^{\,j}\varepsilon_j^{\,t}\delta_k^{\,i} + \gamma\varepsilon_j^{\,i}\varepsilon_k^{\,j} + \dots\right]. \end{aligned} \tag{4a}$$

The *equations of equilibrium* are

$$0 = t^i_{j;i} + c^i_{ik}t_j^{\,k} - c^k_{ji}t_k^{\,i}. \tag{4b}$$

Substituting for the $t_k^{\,i}$ their expressions from (4a) and for the c^i_{jk} their expressions from (2b), we have in (2a) and (4b) a system of essentially 9 partial differential equations for the 6 distinct strain components ε_{ik}.[2] Prescribing the forces acting on the boundary amounts to 3 boundary conditions at each boundary point.

3. Description of the Shell Σ

In the unstrained state the boundary of the region Σ occupied by the elastic material shall consist of two parallel surfaces S_h and S_{-h} a distance $2h$ apart (called the *faces* of the shell) and a remaining boundary portion

[1] Formulae (3c), (3d) constitute our precise definition of the constants E, ν, μ, α, β, γ. This definition is tied to the use of equation (1c) to define the strain. Other definitions of strain will lead to numerically different coefficients for the "cubic" terms in the strain energy. We assume that $0 < \nu < \frac{1}{2}$.

[2] The discrepancy in the numbers of dependent variables and equations is due to the dependencies ("Bianchi identities") existing between the compatibility Eqs. (2a).

(called the *edge*). The faces are the two parallel surfaces of distance h to a certain surface S_0, the *undeformed middle surface*. The shell has the constant thickness $2h$. We choose our curvilinear coordinates U^i in a special way, so that in the unstrained state U^1, U^2 are constant along a normal to S_0 and U^3 measures the (signed) distance from S_0. Then

$$G_{13} = G_{23} = 0, \qquad G_{33} = 1, \tag{5a}$$

$$G_{\alpha\beta} = E_{\alpha\beta} - 2L_{\alpha\beta}U^3 + M_{\alpha\beta}U^3U^3, \tag{5b}$$

where $E_{\alpha\beta}$, $L_{\alpha\beta}$, $M_{\alpha\beta}$ are respectively the coefficients of the 1st, 2nd and 3rd fundamental form of the surface S_0[1].

There are three "typical" lengths we can associate geometrically with the shell Σ in its unstrained state. There is first of all the *thickness* $2h$. The second one, which we shall call R, is associated with the undeformed middle surface S_0. It is defined as the largest number with the property that in every cartesian $X_1X_2X_3$-coordinate system with origin 0 on S_0 and with the X_1X_2-plane tangential to S_0 at 0 the surface S_0 has an equation $Z = f(X, Y)$ for which

$$\left|\left(\frac{\partial^n f}{\partial X_{\alpha_1}\ldots\partial X_{\alpha_n}}\right)_{X_1=X_2=0}\right| \le R^{1-n} \quad \text{for } n = 2, 3, \ldots, 6. \tag{6}$$

In particular R does not exceed any radius of curvature of S_0. The third typical length is the distance D of a point in the unstrained state from the edge. In constrast to h and R the length D *varies* with the point of Σ under consideration.

We assume that in the strained state no forces act on the faces of the shell. This leads to the boundary conditions

$$t_3^{\,i} = 0 \quad \text{for} \quad U^3 = \pm h \quad \text{and} \quad i = 1, 2, 3. \tag{7}$$

No boundary conditions are specified on the edge.

An *equilibrium state* is a set of functions t_{ik}, ε_{ik} satisfying the compatibility conditions (2a), equilibrium Eqs. (4b), constituent Eqs. (4a) and boundary conditions (7) for curvilinear coordinates (U^1, U^2, U^3) corresponding to points of Σ. For a given equilibrium state the *maximum strain* ε is defined as the dimensionless scalar

$$\varepsilon = \sup_{\Sigma} \sqrt{\varepsilon_k^{\,i}\varepsilon_i^{\,k}}. \tag{8}$$

For sufficiently small ε the $\varepsilon_k^{\,i}$ lie in the domain of definition of the energy function W. We assume, of course, that the function W describing the

[1] Here $M_{\alpha\beta} = L_{\alpha\gamma}L_\beta^{\,\gamma}$.

physical properties of the material is known, and is fixed in what follows. Then $W/2\mu$ is a function of the $\varepsilon_k{}^i$ which is independent of any choice of units or of curvilinear coordinates. A positive scalar that only depends on knowledge of the function $W/2\mu$ and not on the particular shell or its deformation will be called a *universal constant*. For two scalar variables a, b the relation

$$a = 0(b)$$

shall indicate that there exists a universal constant K such that

$$|a| \leq Kb.$$

For example, by (4a) there exists a universal constant ε_0 such that

$$\frac{1}{2\mu}\sqrt{t_k{}^i t_i{}^k} = 0\left(\sqrt{\varepsilon_k{}^i \varepsilon_i{}^k}\right) \quad \text{for} \quad \varepsilon \leq \varepsilon_0; \tag{9a}$$

that is the stresses (made dimensionless by dividing by the Lamé-constant 2μ) are of the order of the strains, provided the strains are not too large.

More generally for a scalar b and a tensor $a_{ijk\cdots}$ the symbolic equation

$$a_{ijk\cdots} = 0(b)$$

shall indicate the existence of a universal constant K such that

$$\sqrt{a_{ijk\ldots}\, a^{ijk\ldots}} \leq Kb.$$

Similarly

$$a_{ij\alpha\beta\ldots} = 0(b)$$

stands for

$$\sqrt{a_{ij\alpha\beta\ldots}\, a^{ij\alpha\beta\ldots}} \leq Kb$$

where Greek indices are summed over the values 1, 2, Roman ones over the values 1, 2, 3. For example, by (9a) there exists a universal constant ε_0 such that

$$\frac{1}{2\mu} t_{ik} = 0(\varepsilon) \quad \text{for} \quad \varepsilon < \varepsilon_0. \tag{9b}$$

For a point P of Σ not on the edge of the shell we define the *critical ratio* as the dimension-less coefficient

$$\Theta = \Theta(P) = \text{Max}\left(\frac{h}{D}, \sqrt{\frac{h}{R}}, \sqrt{\varepsilon}\right).$$

The *effective wave length* at P is the quantity

$$\lambda = \lambda(P) = \frac{h}{\Theta} = \text{Min}\left(D, \sqrt{Rh}, \frac{h}{\sqrt{\varepsilon}}\right).$$

Determination of Θ or λ at a point P requires only knowledge of h and R (i.e. of the geometry of the shell in the undeformed state), of D (i.e. of the location of P in Σ in the undeformed state) and of the maximum strain ε of the equilibrium state under consideration. As will be seen smallness of Θ automatically forces a certain behavior on the equilibrium state of Σ that can be identified with "behaving like a thin shell". It is this behavior that is under discussion here. In all that follows we assume that

$$\Theta(P) < \Theta_0,$$

where Θ_0 is a certain universal constant[1]. The smaller Θ, the more the shell behaves like a two-dimensional system.

Results of this type alone are not sufficient to reduce the problem of determining the deformation of a 3-dimensional shell for given boundary conditions at the edge to a two-dimensional problem. No conclusions are presented for points near the edge, where Θ is bound to be large. Moreover no upper bound for Θ can be established without knowledge of a bound on the maximum strain ε, which somehow would have to be obtained from the given boundary conditions[2]. My point of view here is that the estimates and approximations given below, which presuppose information on the size of Θ in order to be meaningful, are comparable to statements made in other parts of continuum mechanics, say, in Hydrodynamics for flows with Reynolds numbers of a certain size. I believe the results to be pertinent for any rigorous mathematical study of the relations between 3-dimensional elasticity and shell theory[3].

---

[1] This excludes from consideration: shells too strongly curved in an unstrained state (those with $\sqrt{h/R} > \Theta_0$) or deformed with too large strains ($\sqrt{\varepsilon} > \Theta_0$) or portions of the shell too close to the edge ($h/D > \Theta_0$).

[2] In principle it is conceivable that one can estimate the maximum strain in terms, say, of bounds on prescribed boundary loads and certain of their derivatives. Such estimates are likely to become very poor for small h and near a state of buckling. In practice moreover the prescribed forces and their derivatives are hardly known with sufficient accuracy.

[3] Such a study doubtlessly is irrelevant for many practical and theoretical purposes, e.g. for the axiomatic approach to shells. Any attempt, however, to derive shell equations from 3-dimensional equations has to be based purely on *mathematical* considerations, if the materials of shells and solids have the same physical properties. The virtue of "rigour" is to bring out clearly the limits, beyond which the arguments become uncertain, and possibly then also the results become factually wrong.

4. Estimates for the Strain Derivatives and Lowest Order Interior Shell Equations

Using standard techniques from the theory of elliptic partial differential equations one derives estimates for the derivatives of the strain components. These estimates have a slightly different form when all differentiations take place in directions tangential to the middle surface ("wave length" $\lambda = h/\Theta$) or when some of the derivatives are taken normal to the middle surface ("wave length" h). Thus for $n = 0, 1, 2, 3, 4$

$$\varepsilon_{i_1 i_2; k_1 \ldots k_n} = 0\left(\frac{\varepsilon}{h^n}\,\Theta^{n-1}\right) = 0\left(\frac{\varepsilon}{h\,\lambda^{n-1}}\right) \tag{10a}$$

$$\varepsilon_{i_1 i_2; \alpha_1 \ldots \alpha_n} = 0\left(\frac{\varepsilon}{h^n}\,\Theta^{n}\right) = 0\left(\frac{\varepsilon}{\lambda^n}\right). \tag{10b}$$

It follows from these inequalities that the *Kirchhoff hypotheses* are valid in the sense that

$$\frac{1}{2\mu}\,t_\alpha{}^3 = 0(\Theta\,\varepsilon), \qquad \frac{1}{2\mu}\,t_3{}^3 = 0(\Theta^2\,\varepsilon) \tag{11a}$$

whereas for the other stress components we have from (9b) only that

$$\frac{1}{2\mu}\,t_\beta{}^\alpha = 0(\varepsilon). \tag{11b}$$

As a consequence of these estimates one can derive two-dimensional *approximate differential equations with precise error estimates*. Various choices are possible for the dependent variables. The deformed middle surface is determined within rigid motions by the knowledge of its first and second fundamental forms, which we denote respectively by

$$e_{\alpha\beta}\,d\,U^\alpha\,d\,U^\beta \quad \text{and} \quad l_{\alpha\beta}\,d\,U^\alpha\,d\,U^\beta.$$

We put

$$e_{\alpha\beta} = E_{\alpha\beta} + 2\,\varepsilon_{\alpha\beta}, \qquad l_{\alpha\beta} = L_{\alpha\beta} + w_{\alpha\beta}.$$

Here $\varepsilon_{\alpha\beta}$ has the same meaning as before (but taken for $U^3 = 0$) and $w_{\alpha\beta}$ measures in some sense changes in curvature due to the deformation of the middle surface. The tensors $\varepsilon_{\alpha\beta}$ and $w_{\alpha\beta}$ are symmetric by definition. Moreover

$$\varepsilon_{\alpha\beta} = 0(\varepsilon), \qquad w_{\alpha\beta} = 0\left(\frac{\varepsilon}{h}\right). \tag{11c}$$

We can also introduce the symmetric stress tensors

$$\tau_{\alpha\beta} = (t_{\alpha\beta})_{U^3=0}, \qquad \sigma_{\alpha\beta} = (t_{\alpha\beta;3})_{U^3=0}$$

that are connected with $\varepsilon_{\alpha\beta}$ and $w_{\alpha\beta}$ by the "approximate constituent shell equations"

$$\frac{1}{2\mu}\,\tau_\beta{}^\alpha = \varepsilon_\beta{}^\alpha + p\varepsilon_\gamma{}^\gamma\delta_\beta{}^\alpha + 0(\varepsilon\Theta^2) \tag{12a}$$

$$\frac{1}{2\mu}\,\sigma_\beta{}^\alpha = -\frac{h^3}{3}\,(w_\beta{}^\alpha + pw_\gamma{}^\gamma\delta_\beta{}^\alpha) + 0(\varepsilon h\Theta^2). \tag{12b}$$

The stress components $t_{\alpha\beta}$ in any point of the shell can be approximated linearly by the formulae

$$t_{\alpha\beta} = \tau_{\alpha\beta} + \frac{3}{h^2}\,\sigma_{\alpha\beta}U^3 + 0(\mu\varepsilon\Theta^2). \tag{12c}$$

The *lowest order interior shell equations* can be written, in analogy to the v. Karman-Föppl equations, as

$$\frac{1}{2\mu}\,\tau_\alpha{}^\beta{}_{|\beta} = 0\left(\frac{\varepsilon}{h}\,\Theta^3\right) \tag{13a}$$

$$w_\alpha{}^\beta{}_{|\beta} - w_\beta{}^\beta{}_{|\alpha} = 0\left(\frac{\varepsilon}{h^2}\,\Theta^3\right) \tag{13b}$$

$$\frac{1}{2\mu}\,\frac{1+p}{1+2p}\,\tau_\alpha{}^\alpha|_\beta{}^\beta + L_\alpha{}^\alpha w_\beta{}^\beta - L_\beta{}^\alpha w_\alpha{}^\beta + \frac{1}{2}\,(w_\alpha{}^\alpha w_\beta{}^\beta - w_\beta{}^\alpha w_\alpha{}^\beta) = 0\left(\frac{\varepsilon}{h^2}\,\Theta^4\right) \tag{13c}$$

$$-\frac{1}{3}\,(1+p)h^2 w_\alpha{}^\alpha|_\beta{}^\beta + \frac{1}{2u}\,\tau_\beta{}^\alpha l_\alpha{}^\beta = 0(\Theta^4). \tag{13d}$$

In terms of the $\varepsilon_{\alpha\beta}$, $w_{\alpha\beta}$ alone we have instead the equations

$$p\varepsilon_\beta{}^\beta{}_{|\alpha} + \varepsilon_\alpha{}^\beta{}_{|\beta} = 0\left(\frac{\varepsilon}{h}\,\Theta^3\right) \tag{14a}$$

$$w_\alpha{}^\beta{}_{|\beta} - w_\beta{}^\beta{}_{|\alpha} = 0\left(\frac{\varepsilon}{h^2}\,\Theta^3\right) \tag{14b}$$

$$(1+p)\varepsilon_\alpha{}^\alpha|_\beta{}^\beta + L_\alpha{}^\alpha w_\beta{}^\beta - L_\beta{}^\alpha w_\alpha{}^\beta + \frac{1}{2}\,(w_\alpha{}^\alpha w_\beta{}^\beta - w_\beta{}^\alpha w_\alpha{}^\beta) = 0\left(\frac{\varepsilon}{h^2}\,\Theta^4\right) \tag{14c}$$

$$-\frac{1}{3}\,(1+p)h^2 w_\alpha{}^\alpha|_\beta{}^\beta + pl_\alpha{}^\alpha\varepsilon_\beta{}^\beta + l_\beta{}^\alpha\varepsilon_\alpha{}^\beta = 0\left(\frac{\varepsilon}{h}\,\Theta^4\right). \tag{14d}$$

Here the symbol "|" indicates covariant differentiation with respect to the first fundamental form $E_{\alpha\beta}\,dU^\alpha\,dU^\beta$ of the undeformed middle surface.

All the results mentioned so far are contained in the paper quoted on p. 1. What follows is new and will be derived elsewhere in more detail.

5. Refined Interior Shell Equations

In certain situations the lowest order interior shell equations (13a, b, c, d) or (14a, b, c, d) appear to be insufficiently precise. It is possible now to derive by the same methods approximate differential equations for the same quantities and of the same orders[1] with *a smaller error term*. It is easiest to work with differential equations for the quantities $\varepsilon_{\alpha\beta}$, $w_{\alpha\beta}$ corresponding to the first and second fundamental form of the deformed middle surface. We replace Eqs. (14b) and (14c) by the precise Codazzi and Gauss equations for a surface in Euclidean space:

$$w_\alpha{}^\beta{}_{|\beta} - w_\beta{}^\beta{}_{|\alpha} - E^{\beta\mu}(C^\lambda_{\mu\beta} l_{\lambda\alpha} - C^\lambda_{\mu\alpha} l_{\lambda\beta}) = 0 \tag{15a}$$

$$\begin{aligned} &\varepsilon_{11|22} + \varepsilon_{22|11} - \varepsilon_{12|21} - \varepsilon_{21|12} + l_{11} l_{22} - l_{12} l_{21} \\ &\quad - \frac{1}{2} e_\gamma{}^\gamma (L_{11} L_{22} - L_{12} L_{21}) + e_{\lambda\mu}(C^\mu_{11} C^\lambda_{22} - C^\mu_{12} C^\lambda_{21}) = \end{aligned} \tag{15b}$$

Here the tensor $C^\lambda_{\mu\beta}$ represents the difference of Christoffel symbols with respect to the metrics $e_{\alpha\beta}\, dU^\alpha\, dU^\beta$ and $E_{\alpha\beta}\, dU^\alpha\, dU^\beta$:

$$e_{\alpha\lambda} C^\lambda_{\mu\beta} = \varepsilon_{\alpha\mu|\beta} + \varepsilon_{\alpha\beta|\mu} - \varepsilon_{\mu\beta|\alpha}.$$

The remaining 3 equations, corresponding to equilibrium conditions are refinements of Eqs. (14a) and (14d). These equations contain not only Poisson's ratio, or equivalently the coefficient $p = \nu/(1-\nu)$, but also the cubic coefficients α, β, γ from the expansion (3c) of the energy function. We introduce the abbreviation

$$A = \alpha q^3 + 3\beta p^2 q - \gamma p^3.$$

Then (14a) is to be replaced by

$$\begin{aligned} &\Big[p\varepsilon_\gamma{}^\gamma \delta_\beta{}^\alpha + \varepsilon_\beta{}^\alpha + \left(A - \frac{p}{2}\right) \varepsilon_\xi{}^\xi \varepsilon_\eta{}^\eta \delta_\beta{}^\alpha + \left(\beta q - \frac{1}{2}\right) \varepsilon_\eta{}^\xi \varepsilon_\xi{}^\eta \delta_\beta{}^\alpha + \\ &\quad + (2\beta q + 2p)\varepsilon_\gamma{}^\gamma \varepsilon_\beta{}^\alpha + (\gamma + 2)\varepsilon_\gamma{}^\alpha \varepsilon_\beta{}^\gamma + \frac{h^2}{6}\Big(-(p + \\ &\quad + 3p^2) L_\xi{}^\xi w_\eta{}^\eta \delta_\beta{}^\alpha - 3p L_\eta{}^\xi w_\xi{}^\eta \delta_\beta{}^\alpha - p L_\beta{}^\alpha w_\gamma{}^\gamma - L_\gamma{}^\alpha w_\beta{}^\gamma - \\ &\quad - L_\beta{}^\gamma w_\gamma{}^\alpha + \left(2A - \frac{3}{2} p - 3p^2\right) w_\xi{}^\xi w_\eta{}^\eta \delta_\beta{}^\alpha + \\ &\quad + \left(2\beta q - 1 - \frac{1}{2} p\right) w_\eta{}^\xi w_\xi{}^\eta \delta_\beta{}^\alpha + ((4\beta q + 3p) w_\gamma{}^\gamma w_\beta{}^\alpha + \\ &\quad + (2\gamma + 5) w_\gamma{}^\alpha w_\beta{}^\gamma)\Big]_{|\alpha} + \frac{h^2}{6}\big((2 + p) L_\alpha{}^\alpha w_\gamma{}^\gamma|_\beta + 2L_\beta{}^\alpha w_\gamma{}^\gamma|_\alpha - \\ &\quad - 2L_\gamma{}^\alpha w_\alpha{}^\gamma|_\beta\big) + 0\left(\frac{\varepsilon}{h}\,\Theta^5\right) \end{aligned} \tag{15c}$$

[1] Higher derivatives that do occur can all be eliminated with the help of the lowest order equations.

and (14d) by

$$
\begin{aligned}
&-\frac{1}{3}(1+p)h^2 w_\alpha{}^\alpha|_\beta{}^\beta + p l_\alpha{}^\alpha \varepsilon_\beta{}^\beta + l_\beta{}^\alpha \varepsilon_\alpha{}^\beta + (A-p^2) l_\alpha{}^\alpha \varepsilon_\beta{}^\beta \varepsilon_\gamma{}^\gamma + \\
&+ (2\beta q - p) l_\beta{}^\alpha \varepsilon_\alpha{}^\beta \varepsilon_\gamma{}^\gamma + \beta q l_\alpha{}^\alpha \varepsilon_\gamma{}^\beta \varepsilon_\beta{}^\gamma + \gamma l_\beta{}^\alpha \varepsilon_\gamma{}^\beta \varepsilon_\alpha{}^\gamma + \\
&+ \frac{h^2}{6}\Big[-\left(\frac{24}{5}p + 2p^2\right) \varepsilon_\alpha{}^\alpha|_\beta{}^\beta L_\gamma{}^\gamma - \Big(4A - 4pq\beta + \frac{14}{5}p - \\
&- 4p^2\Big) \varepsilon_\alpha{}^\alpha|_\beta{}^\beta w_\gamma{}^\gamma - \left(\frac{14}{5} + p\right) \varepsilon_\gamma{}^\alpha|_\beta{}^\beta L_\alpha{}^\gamma - \left(\frac{4}{5} + 5p + 4\beta q\right) \varepsilon_\gamma{}^\alpha|_\beta{}^\beta w_\alpha{}^\gamma + \\
&+ (2+3p) \varepsilon_\alpha{}^\alpha|_\gamma{}^\beta L_\beta{}^\gamma + (-4q\beta + 4\gamma p + 2 + 13p) \varepsilon_\alpha{}^\alpha|_\gamma{}^\beta w_\beta{}^\gamma + \\
&+ \left(2p^2 - \frac{4}{5}p - 4A - 4q\beta\right) w_\alpha{}^\alpha|_\beta{}^\beta \varepsilon_\gamma{}^\gamma - \Big(8\beta q + 4\gamma + 4p + \\
&+ \frac{34}{5}\Big) w_\alpha{}^\alpha|_\gamma{}^\beta \varepsilon_\beta{}^\gamma + \left(2 - \frac{18}{5}p - 6p^2\right) \varepsilon_\alpha{}^\alpha|^\beta L_\gamma{}^\gamma|_\beta + \Big(2 + \frac{12}{5}p + \\
&+ 2p^2 - 8A - 8q^2\beta + 4p\gamma\Big) \varepsilon_\alpha{}^\alpha|^\beta w_\gamma{}^\gamma|_\beta - \left(\frac{8}{5} + 2p\right) \varepsilon_\gamma{}^\alpha|^\beta L_\alpha{}^\gamma|_\beta - \\
&- \left(\frac{58}{5} + 10p + 8q\beta + 4\gamma\right) \varepsilon_\gamma{}^\alpha|^\beta w_\alpha{}^\gamma|_\beta + (2p - 3p^2) L_\alpha{}^\alpha L_\beta{}^\beta w_\gamma{}^\gamma + \\
&+ (2-4p) L_\alpha{}^\alpha L_\gamma{}^\beta w_\beta{}^\gamma - 3p L_\beta{}^\alpha L_\alpha{}^\beta w_\gamma{}^\gamma - 4 L_\beta{}^\alpha L_\gamma{}^\beta w_\alpha{}^\gamma + \\
&+ (2p - 6p^2 + 2A)\, L_\alpha{}^\alpha w_\beta{}^\beta w_\gamma{}^\gamma + (2 - p + 2\beta q) L_\alpha{}^\alpha w_\gamma{}^\beta w_\beta{}^\gamma + \\
&+ (4q\beta - 6p) L_\beta{}^\alpha w_\alpha{}^\beta w_\gamma{}^\gamma + (2\gamma - l)\, L_\beta{}^\alpha w_\gamma{}^\beta w_\alpha{}^\gamma + \\
&+ (2A - 3p^2) w_\alpha{}^\alpha w_\beta{}^\beta w_\gamma{}^\gamma + 6q\beta w_\beta{}^\alpha w_\alpha{}^\beta w_\gamma{}^\gamma + (2\gamma+3) w_\beta{}^\alpha w_\gamma{}^\beta w_\alpha{}^\gamma\Big] \\
&= 0\left(\frac{\varepsilon}{h}\,\Theta^6\right). \qquad (15\text{d})
\end{aligned}
$$

6. Deformations of Plates into Cylindrical Shells

The Eqs. (15a, b, c, d) are admittedly horribly complicated. For the purpose of analysing the type of information that might be extracted from them we consider the simplest of all examples. We take in the unstrained state a flat plate referred to cartesian coordinates:

$$E_{\alpha\beta} = \delta_{\alpha\beta}, \qquad L_{\alpha\beta} = 0, \qquad U^\alpha = X^\alpha.$$

The plate shall be deformed into a *cylinder* with generators parallel to the X_2-axis, the coordinate X_2 staying fixed in the process. Then $\varepsilon_{\alpha\beta}$ and $w_{\alpha\beta}$ will vanish, except for $\alpha = \beta = 1$, and ε_{11}, w_{11} will depend on X_1 only: We write

$$X^1 = U^1 = X, \qquad \varepsilon_{11} = \eta(X), \qquad w_{11} = v(X).$$

Here the curvature of the deformed plate is $v/(1+2\eta)$. To begin with we have from (11c) the estimates

$$\eta = 0(\varepsilon), \qquad v = 0\left(\frac{\varepsilon}{h}\right). \tag{16}$$

The derivative of a function $f(X)$ with respect to X will be denoted by f'. Here the cubic constants from the energy function center only in the combination

$$B = \frac{A + 3\beta + \gamma}{1+p} = \frac{\alpha q^3 + 3\beta q(1+p^2) + \gamma(1-p^3)}{1+p}.$$

The Gauss-Codazzi equations (15a, b) are satisfied identically. Equation (15c) here corresponds to a relation

$$\eta + \left(B + \frac{1}{2}\right)\eta^2 + \frac{h^2}{6}(2B + 4 - 3p)v^2 = c + 0(\varepsilon\Theta^4) \tag{17}$$

where c is a constant[1]. Using (16) we see that

$$c = 0(\varepsilon).$$

We find from (15d) that

$$-\frac{1}{3}h^2 v'' + \eta w + (B-p)\eta^2 w + \frac{h^2}{6}\left(-4B + 2p - \frac{34}{5}\right)\eta v'' + $$
$$+ \frac{h^2}{6}(2B + 3 - 3p)v^3 = 0\left(\frac{\varepsilon}{h}\Theta^6\right).$$

Using here (17) we conclude that, similar to the Elastica,

$$\frac{h^2}{3}\left(v'' + \frac{1}{2}v^3\right) - \gamma v = 0\left(\frac{\varepsilon}{h}\Theta^6\right), \tag{18}$$

where γ is the constant given by

$$\gamma = c - \left(2B + \frac{49}{10}\right)c^2.$$

From the lowest order shell equations, (as also from the classical v. Karman-Föppl plate equations) we would obtain instead only the *linear* differential equation

$$\frac{h^2}{3}v'' - cv = 0\left(\frac{\varepsilon}{h}\Theta^4\right). \tag{19}$$

---

[1] Actually (17) does not follow from (15c) but from an integrated form of that equation, valid for plates, which asserts that

$$\frac{\partial}{\partial X^\alpha}\int_{-h}^{h} t_{\alpha\beta}\, dX^3 = 0.$$

We like to conclude that the solution v of (18), respectively (19) differs little from a suitable solution $\overline{v}$ of the corresponding equations

$$\frac{h^2}{3}\left(\overline{v}'' + \frac{1}{2}\,\overline{v}\right) - \gamma\overline{v} = 0, \quad \text{(Elastica Equation)} \tag{18'}$$

respectively

$$\frac{h^2}{3}\,\overline{v}'' - c\overline{v} = 0, \tag{19'}$$

in which the error term has been neglected. This actually is not necessarily the case, if the length L of the interval I in which we like to approximate v by $\overline{v}$ is too large. Here the behavior of the equations is different in the cases of $c < 0$ (compression) and $c > 0$ (tension).

For $c < 0$ we construct a counter example. Let I be an interval of length

$$L = \frac{2n\pi h}{\sqrt{-3c}} \tag{20a}$$

where n is a positive integer. Consider the function

$$v = \frac{n\varepsilon^3}{hc}\sin\frac{2\pi}{L}(n+1)X. \tag{20b}$$

Here

$$\frac{h^2}{3}\,v'' - cv = \left(2 + \frac{1}{n}\right)\frac{\varepsilon^3}{h} = 0\left(\frac{\varepsilon}{h}\,\Theta^4\right)$$

since $\Theta \leq \sqrt{\varepsilon}$. In the interval I the function v given by (20b) is orthogonal to the general solution

$$\overline{v} = a\sin\frac{2\pi n}{L}X + b\cos\frac{2\pi n}{L}X$$

of (19′). It follows that

$$\operatorname*{Max}_{\mathrm{I}}|v - \overline{v}| \geq \frac{1}{\sqrt{2}}\,\frac{n\varepsilon^3}{hc}$$

for every solution $\overline{v}$ of (19′). For $c \leq n\varepsilon^2$ the maximum of $|v - \overline{v}|$ is then of the order $\frac{\varepsilon}{h}$, which is the maximum order v itself can have in our problems, and thus solutions of the truncated Eq. (19′) do not furnish significant approximations to v. This happens, e. g. with $n = 1$ for very small changes in length, namely for $c = 0(\varepsilon^2)$; it happens under all circumstances for n of the order h/ε, since $c = 0(\varepsilon)$. In these situations one has to have recourse to the solutions $\overline{v}$ of the *refined* Eq. (18′) to obtain significant approximations for v.

The situation is different for $c > 0$, due to the *maximum principle* that in that case is valid for the linear Eq. (19'). Assume that I is an interval of length L, so far removed from the edges that in I

$$\Theta = \mathrm{Max}\left(\frac{h}{D}, \sqrt{\varepsilon}\right),$$

is small. We take for $\bar{v}$ the solution of (19') with the same values as v in the endpoints of I. Then it follows from

$$\frac{h^2}{3}(v'' - \bar{v}'') - c(v - \bar{v}) = 0\left(\frac{\varepsilon}{h}\Theta^4\right)$$

that

$$\mathop{\mathrm{Max}}_{\mathrm{I}} |v - \bar{v}| = 0\left(\frac{\varepsilon}{h}\frac{1}{c}\Theta^4\right),$$

quite independently of the length L of the interval I. Thus $v - \bar{v}$ is small compared to ε/h for fixed $c > 0$ and sufficiently small Θ. Moreover the solution $\bar{v}$ of (19') *decays exponentially* in I, as we move away from the endpoints of I. More precisely in the endpoints $\bar{v}$ agrees with v and hence has values of order $0(\varepsilon/h)$. In the middle third of the interval I we have then

$$\bar{v} = 0\left(\frac{\varepsilon}{h}\frac{\sinh\frac{2}{3}\frac{\sqrt{c}}{h}L}{\sinh\frac{\sqrt{c}}{h}L}\right) = 0\left(\frac{\varepsilon}{h}e^{-\frac{1}{3}\frac{\sqrt{c}}{h}L}\right).$$

For fixed c and small h then, $\bar{v}$ will be small compared to ε/h. Thus for small ε, small h and fixed positive c the solution v will be small compared to ε/h at sufficient distances from the edge. Since w is essentially the curvature of the deformed cylindrical plate, the deformed plate will be nearly flat in the portions away from the edge. *Its behavior is that of a membrane.*

7. Concluding Remarks

The work described here is still in the preliminary stage, and I am quite aware of its fragmentary character. The "lowest order equations" as well as the "refined shell equation" of this report have to be related to the various other shell theories working with displacements, and based either on conservation laws, formal expansions or on an axiomatic approach. Something will have to be said about *boundary conditions* or *boundary layers* near the edges. Also to be investigated remains the question under what circumstances the lowest order equations are sufficient, and when one has to have recourse to the refined equations. Can the refined equations be simplified in the situations where they are needed? Special cases, such as plates and shells of revolution should be discussed. Is the transition from shells to membranes also automatically forced on the shell under certain conditions, just as the Kirchhoff hypotheses?

Discussion

J. G. Simmonds: It seems strange to me that it should be necessary to introduce cubic terms into the 3-dimentional strain energy density in order to obtain refined interior shell equations. I would say that the reason the analysis in your paper in the *Comm. on Pure and Appl. Math.* produces only Donnell-Vlassov type equations is that you obtain all stress estimates via the stress-strain relations. For example, it is well known that if a shell is in a state of near inextensional bending, then the terms involving the transverse shear stress resultants, which are neglected in the tangential force equilibrium equations in the Donnell-Vlasov theory, become important. However, in this case, the membrane stress resultants are *not* calculated from the stress-strain relations, but rather become statically determined. One can already see this illustrated in your example of the elastica. Your arguments seem to imply that only by the introduction of cubic terms in the elastic energy can one replace the inaccurate linear equation (19) by the accurate non-linear equation (18). But note that the coefficient of the single new term, $\frac{1}{2}v^3$, does *not* involve any of the cubic elastic parameters. Therefore, it must be present when these parameters are zero, which implies that it must be possible to derive (18), but only with the error estimate of (19), without the need to consider cubic terms in the strain energy function.

W. T. Koiter: My comment is related to the previous question by Professor Simmonds. I also feel that a set of refined shell equations should be obtainable without introducing cubic terms in the strain energy function. Professor John's theory is essentially based on the larger of the three quantities h/D, $\sqrt{h/R}$ and $\sqrt{\varepsilon}$ as a "small" parameter, and in this connection cubic terms in the strain energy become important in a refined theory. If one assumes, however, the strains to remain "infinitesimal", such that higher order terms in the strains are always negligible, a simpler refined theory appears to be possible, in which the larger of h/D and $\sqrt{h/R}$ occurs as the "small" parameter.

F. John: The cubic terms would actually enter also in the elastica, but the way one usually derives elastica equations involves displacements and static considerations and makes it difficult to compare. They should really be contained in the general equations and should be a consequence of them.

When one writes down the refined equations the cubic terms enter as much as E and ν in all the equations. That seems to be unavoidable. Notice, however, that the cubic coefficients from the energy do not really complicate the form of the resulting shell equations; the combinations of the quadratic coefficients are actually responsible for the unsightly appearance of the equations.

H. S. Rutten: W. Z. Chien derived his intrinsic theory of shells and plates in exactly the same quantities $\varepsilon_\alpha{}^\beta$ and $w_{,}{}^\beta$, as you did. Are the resulting equations similar or are there any essential differences?

F. John: A comparison should be made. However, Chien's notation is complicated. He starts out from integrated forms of the equilibrium equations, and uses formal expansions instead of estimates.

On the Foundations of Generalized Linear Shell Theory

By

E. Reissner

Massachusetts Institute of Technology, Cambridge, Mass., USA

1. Introduction

In speaking of the foundations of shell theory, we mean the problem of rationally deriving a system of two-dimensional equations from a given three-dimensional formulation, subject only to the limitation that all characteristic lengths measured along curves on the middle surface of the shell are large compared to the thickness of the shell.

Much recent insightful work in this direction has been based on the concepts of iteration and parametric expansion in a formal asymptotic sense and the present account will again make use of these concepts.

The starting point of our considerations is the remarkable simplification in form of those parts of (linear) shell theory which are amenable to a completely two-dimensional analysis, upon extension of the classical theory by the introduction of couple stress stress couples turning about the normal to the middle surface of the shell.

On the basis of this simplification, it seemed to be of interest to consider the derivation of a (generalized) two-dimensional shell theory from a suitable version of three-dimensional couple stress elasticity theory. It will be shown in what follows, that the derivation of two-dimensional shell theory via three-dimensional couple stress theory does in fact have particular attractions, even for the case that the actual shell medium is one without capability of supporting couple stress.

The present account begins with a brief restatement of those parts of linear shell theory which can be treated without three-dimensional considerations, following presentations by Günther [6] and in [12]. This

---

Acknowledgement. The work reported on in this paper has been supported by the Office of Naval Research under a contract with the Massachusetts Institute of Technology. Thanks are due to T. J. Lardner and F. Y. M. Wan for stimulating discussions on the subject of this paper.

is followed by a preliminary discussion of two-dimensional stress strain relations, and by a statement of the shell problem within the framework of three-dimensional couple stress elasticity theory, in terms of equilibrium equations and compatibility equations for suitably defined pseudo stresses and strains.

It is next shown that the simple project of eliminating derivatives in thickness direction in the three-dimensional equilibrium and compatibility equations of couple stress theory automatically leads to the correct form of the two-dimensional equilibrium and compatibility equations, in terms of stress resultants and couples and in terms of strain resultants and couples.

The introduction of the results of the elimination process into the three-dimensional stress strain relations reduces the three-dimensional boundary value problem to a system of integro-differential equations. Use of the concept of characteristic lengths along middle surface curves of the shell, together with order of magnitude considerations for classes of stress and strain, along the lines of earlier work by Johnson and the author [7, 11] and by Reiss [9], Green [3, 4] and Gol'denweizer [1, 2] shows the possibility of iterative or parametric expansion type solutions of the integro-differential equations, in such a way that the leading term of the parametric expansion or, equivalently, the first step of the iterative procedure produces a statement of two-dimensional shell theory.

2. Two-Dimensional Shell Theory Including Couple Stress Stress Couples

We assume an orthogonal system of coordinates ξ_1, ξ_2 on the middle surface of the shell with position vector $\boldsymbol{r} = \boldsymbol{r}(\xi_1, \xi_2)$, tangent unit vectors $\boldsymbol{t}_i$, a normal unit vector $\boldsymbol{n} = \boldsymbol{t}_1 \times \boldsymbol{t}_2$, fundamental forms $d\boldsymbol{r} \cdot d\boldsymbol{r} = \alpha_1^2\, d\xi_1^2 + \alpha_2^2\, d\xi_2^2$ and $d\boldsymbol{r} \cdot d\boldsymbol{n} = (\alpha_1^2/R_{11})\,d\xi_1^2 + 2(\alpha_1\alpha_2/R_{12})\,d\xi_1\, d\xi_2 + (\alpha_2^2/R_{22})\,d\xi_2^2$.

We define stress resultant and stress couple vectors $\boldsymbol{N}_i$ and $\boldsymbol{M}_i$ with component representations

$$\boldsymbol{N}_i = \sum N_{ij}\boldsymbol{t}_j + Q_i\boldsymbol{n}, \quad \boldsymbol{M}_i = \boldsymbol{n} \times \sum M_{ij}\boldsymbol{t}_j + P_i\boldsymbol{n} \tag{1}$$

and, analogously, force and moment load intensity vectors $\boldsymbol{p}$ and $\boldsymbol{q}$, which are subject to equilibrium equations of the form

$$(\alpha_2\boldsymbol{N}_1)_{,1} + (\alpha_1\boldsymbol{N}_2)_{,2} + \alpha_1\alpha_2\boldsymbol{p} = 0 \tag{2a}$$

$$(\alpha_2\boldsymbol{M}_1)_{,1} + (\alpha_1\boldsymbol{M}_2)_{,2} + \boldsymbol{r}_{,1} \times (\alpha_2\boldsymbol{N}_1) + \boldsymbol{r}_{,2} \times (\alpha_1\boldsymbol{N}_2) + \alpha_1\alpha_2\boldsymbol{q} = 0. \tag{2b}$$

Corresponding to the definitions of stress resultant and couple vectors, we define strain resultant and couple vectors

$$\boldsymbol{\varepsilon}_i = \sum \varepsilon_{ij} \boldsymbol{t}_j + \gamma_i \boldsymbol{n}, \qquad \boldsymbol{\varkappa}_i = \boldsymbol{n} \times \sum \varkappa_{ij} \boldsymbol{t}_j + \lambda_i \boldsymbol{n} \tag{3}$$

and corresponding to the definitions of force and moment load intensity vectors, we define translational and rotational displacement vectors $\boldsymbol{u}$ and $\boldsymbol{\varphi}$. Use of the principle of virtual work then leads to two-dimensional strain displacement relations

$$\alpha_i \boldsymbol{\varepsilon}_i = \boldsymbol{u}_{,i} + \boldsymbol{r}_{,i} \times \boldsymbol{\varphi}, \qquad \alpha_i \boldsymbol{\varkappa}_{,i} = \boldsymbol{\varphi}_{,i} \tag{4}$$

as logical counterparts of the two-dimensional equilibrium eqs. (2).

An immediate consequence of (4) are two vectorial compatibility equations

$$(\alpha_2 \boldsymbol{\varkappa}_2)_{,1} - (\alpha_1 \boldsymbol{\varkappa}_1)_{,2} = 0 \tag{5a}$$

$$(\alpha_2 \boldsymbol{\varepsilon}_2)_{,1} - (\alpha_1 \boldsymbol{\varepsilon}_1)_{,2} + \boldsymbol{r}_{,1} \times (\alpha_2 \boldsymbol{\varkappa}_2) - \boldsymbol{r}_{,2} \times (\alpha_1 \boldsymbol{\varkappa}_1) = 0. \tag{5b}$$

Equations (5), in conjunction with the homogeneous form of the equilibrium eqs. (3), very simply imply a static-geometric duality which, with the help of the strain displacement relations (4), is readily extended to a duality for displacements and stress functions as well.

The completely two-dimensional eqs. (2) and (5) must be complemented by a system of constitutive equations. Assuming the property of elasticity, and more specifically, the existence of a strain energy function, these constitutive equations will be of the form

$$N_{ij} = \frac{\partial A}{\partial \varepsilon_{ij}}, \quad Q_i = \frac{\partial A}{\partial \gamma_i}, \quad M_{ij} = \frac{\partial A}{\partial \varkappa_{ij}}, \quad P_i = \frac{\partial A}{\partial \lambda_i}. \tag{6}$$

The principal problem of the formulational phase of shell theory (other than non-linear aspects requiring non-linear versions of equilibrium and strain displacement relations) is the determination of the form of the stress potential A, either on the basis of suitable experiments for shell elements, or on the basis of a rational system of stress strain relations for the shell considered as a three-dimensional continuum.

3. Preliminary Consideration of Two-Dimensional Stress Strain Relations

In order to see the nature of the problem of establishing stress strain relations of the form (2.6), it is helpful to consider the form of the strain displacement relations (2.4) upon introducing the component representation (2.3), together with the representations

$$\boldsymbol{u} = \sum u_j \boldsymbol{t}_j + w\boldsymbol{n}, \qquad \boldsymbol{\varphi} = \boldsymbol{n} \times \sum \varphi_j \boldsymbol{t}_j + w\boldsymbol{n} \tag{1}$$

In this way and upon use of appropriate base vector differentiation formulas there result the scalar strain displacement relations

$$\varepsilon_{11} = \frac{u_{1,1}}{\alpha_1} + \frac{\alpha_{1,2} u_2}{\alpha_1 \alpha_2} + \frac{w}{R_{11}}, \qquad \varepsilon_{22} = \frac{u_{2,2}}{\alpha_2} + \frac{\alpha_{2,1} u_1}{\alpha_1 \alpha_2} + \frac{w}{R_{22}}, \tag{2a}$$

$$\varepsilon_{12} = \frac{u_{2,1}}{\alpha_1} - \frac{\alpha_{1,2} u_1}{\alpha_1 \alpha_2} + \frac{w}{R_{12}} - \omega, \qquad \varepsilon_{21} = \frac{u_{1,2}}{\alpha_2} - \frac{\alpha_{2,1} u_2}{\alpha_1 \alpha_2} + \frac{w}{R_{12}} + \omega, \tag{2b}$$

$$\gamma_1 = \varphi_1 + \frac{w_{,1}}{\alpha_1} - \frac{u_1}{R_{11}} - \frac{u_2}{R_{12}}, \qquad \gamma_2 = \varphi_2 + \frac{w_{,2}}{\alpha_2} - \frac{u_2}{R_{22}} - \frac{u_1}{R_{12}}, \tag{2c}$$

$$\varkappa_{11} = \frac{\varphi_{1,1}}{\alpha_1} + \frac{\alpha_{1,2} \varphi_2}{\alpha_1 \alpha_2} + \frac{\omega}{R_{12}}, \qquad \varkappa_{22} = \frac{\varphi_{2,2}}{\alpha_2} + \frac{\alpha_{2,1} \varphi_1}{\alpha_1 \alpha_2} - \frac{\omega}{R_{12}}, \tag{2d}$$

$$\varkappa_{12} = \frac{\varphi_{2,1}}{\alpha_1} - \frac{\alpha_{1,2} \varphi_1}{\alpha_1 \alpha_2} - \frac{\omega}{R_{11}}, \qquad \varkappa_{21} = \frac{\varphi_{1,2}}{\alpha_2} - \frac{\alpha_{2,1} \varphi_2}{\alpha_1 \alpha_2} + \frac{\omega}{R_{22}}, \tag{2e}$$

$$\lambda_1 = \frac{\omega_{,1}}{\alpha_1} - \frac{\varphi_1}{R_{12}} + \frac{\varphi_2}{R_{11}}, \qquad \lambda_2 = \frac{\omega_{,2}}{\alpha_2} + \frac{\varphi_2}{R_{12}} - \frac{\varphi_1}{R_{22}}. \tag{2f}$$

An important property of this system is that, in general, $\varepsilon_{12} \neq \varepsilon_{21}$ and $\varkappa_{12} \neq \varkappa_{21}$. Contact with more usual strain representations may be established by introducing the symmetrical combinations $\varepsilon_{12} + \varepsilon_{21}$ and $\varkappa_{12} + \varkappa_{21} - \varepsilon_{12}/R_{11} - \varepsilon_{21}/R_{22}$.

The existence of a system of stress strain relations of the form (2.6) together with strain displacement relations of the form (2) seems first to have been postulated in 1962 for shell theory without couple stress stress couples, that is, for a function A which does not involve the strain components λ_i [10]. Subsequently, the stress strain relations known as the Flügge-Lurje-Byrne relations as well as the ones known as the Koiter-Sanders relations were written in the form (2.6) [14, 15].

One of the purposes of the present report is to suggest an alternate, and in certain respects simpler system of linear stress strain relations for an isotropic shell, in the form

$$N_{11} = C(\varepsilon_{11} + \nu_N \varepsilon_{22}), \qquad N_{12} = (1 - \nu_N)\, C\varepsilon_{12}, \tag{3a}$$

$$M_{11} = D(\varkappa_{11} + \nu_M \varkappa_{22}), \qquad M_{12} = (1 - \nu_M) D\varkappa_{12} \tag{3b}$$

together with corresponding expressions for N_{21}, N_{22}, M_{12}, M_{22} and together with the relations

$$Q_i = C_Q \gamma_i, \qquad P_i = D_P \lambda_i. \tag{3c}$$

It is not obvious, but may be shown by simple order-of magnitude considerations of relevant scalar compatibility and equilibrium equations that, when $D_P = 0$ and $C_Q = \infty$, the second Eqs. in (3a) and (3b) may be replaced with negligible error by relations of the form

$$M_{12} = M_{21} = \frac{1}{2}(1 - \nu_M) D(\varkappa_{12} + \varkappa_{21}), \tag{4a}$$

$$(1 - \nu_N) C\varepsilon_{12} = (1 - \nu_N) C\varepsilon_{21} = \frac{1}{2}(N_{12} + N_{21}). \tag{4b}$$

As an indication of the relative simplicity of the system (3a, b) the following may be mentioned. Assume that (3a, b) have first been derived so as to be valid for lines-of-curvature coordinates. It then follows from the form of A and from the invariance properties of the ε_{ij} and the $\varkappa_{ij}$ that (3a, b) are equally valid for general orthogonal surface coordinates.

It is not known yet whether any practical advantages in the solution of specific shell problems are associated with the use of the system (3) instead of any of the various other equally accurate systems which are known. Independent of this question, the system (3) seems to recommend itself by the aesthetic advantage of its simple symmetrical appearance.

4. Three-Dimensional Shell Theory Including Couple Stresses

We assume a three-dimensional orthogonal coordinate system ξ_1, ξ_2, ζ with position vector $\boldsymbol{x}(\xi_1, \xi_2, \zeta)$ where the coordinate curves ξ_i are the lines of curvature on the middle surface of the shell and the ζ-curves are the straight lines perpendicular to the middle surface. From $\boldsymbol{x} = \boldsymbol{r} + \zeta\boldsymbol{n}$ follows then $d\boldsymbol{x} \cdot d\boldsymbol{x} = \alpha_1^2(1 + \zeta/R_1)^2\, d\xi_1^2 + \alpha_2^2(1 + \zeta/R_2)^2\, d\xi_2^2 + d\zeta^2$ where R_1 and R_2 are the values of R_{11} and R_{22} for the case that $R_{12} = \infty$.

We define force and couple stress vectors $\boldsymbol{\sigma}_i$ and $\boldsymbol{\tau}_i$ with component representations

$$\boldsymbol{\sigma}_i = \sum \sigma_{ij}\boldsymbol{t}_j + \sigma_{i\zeta}\boldsymbol{n}, \qquad \boldsymbol{\tau}_i = \boldsymbol{n} \times \sum \tau_{ij}\boldsymbol{t}_j + \tau_{i\zeta}\boldsymbol{n}, \tag{1a}$$

$$\boldsymbol{\sigma}_\zeta = \sum \sigma_{\zeta j}\boldsymbol{t}_j + \sigma_{\zeta\zeta}\boldsymbol{n}, \qquad \boldsymbol{\tau}_\zeta = \boldsymbol{n} \times \sum \tau_{\zeta j}\boldsymbol{t}_j + \tau_{\zeta\zeta}\boldsymbol{n} \tag{1b}$$

and analogously, force and couple strain vectors $\boldsymbol{e}_i, \boldsymbol{e}_\zeta, \boldsymbol{k}_i, \boldsymbol{k}_\zeta$.

We assume that the stress vectors are subject to equations of force and moment equilibrium and that the strain vectors depend on translational and rotational displacement vectors $\boldsymbol{v}$ and $\boldsymbol{\psi}$ in such a way as follows through application of the principle of virtual work.

The formulation of equilibrium equations, strain displacement relations and compatibility equations is somewhat simplified by writing these relations in terms of pseudo-stresses and strains, defined as follows

$$\begin{pmatrix}\boldsymbol{\sigma}_1{}^*\\ \boldsymbol{\tau}_1{}^*\end{pmatrix} = \left(1+\frac{\zeta}{R_2}\right)\begin{pmatrix}\boldsymbol{\sigma}_1\\ \boldsymbol{\tau}_1\end{pmatrix}, \qquad \begin{pmatrix}\boldsymbol{e}_i{}^*\\ \boldsymbol{k}_i{}^*\end{pmatrix} = \left(1+\frac{\zeta}{R_i}\right)\begin{pmatrix}\boldsymbol{e}_i\\ \boldsymbol{k}_i\end{pmatrix}, \tag{2a}$$

$$\begin{pmatrix}\boldsymbol{\sigma}_\zeta{}^*\\ \boldsymbol{\tau}_\zeta{}^*\end{pmatrix} = \left(1+\frac{\zeta}{R_1}\right)\left(1+\frac{\zeta}{R_2}\right)\begin{pmatrix}\boldsymbol{\sigma}_\zeta\\ \boldsymbol{\tau}_\zeta\end{pmatrix}. \tag{2b}$$

The basic equations of the theory are then two equations of equilibrium

$$\frac{(\alpha_2\boldsymbol{\sigma}_1{}^*)_{,1} + (\alpha_1\boldsymbol{\sigma}_2{}^*)_{,2}}{\alpha_1\alpha_2} + \boldsymbol{\sigma}^*_{\zeta,\zeta} + \boldsymbol{p}^* = 0, \tag{3a}$$

$$\frac{(\alpha_2\boldsymbol{\tau}_1{}^*)_{,1} + (\alpha_1\boldsymbol{\tau}_2{}^*)_{,2}}{\alpha_1\alpha_2} + \boldsymbol{\tau}^*_{\zeta,\zeta} + \boldsymbol{q}^* + \boldsymbol{n}\times\boldsymbol{\sigma}_\zeta{}^* +$$

$$+\left(1+\frac{\zeta}{R_1}\right)\boldsymbol{t}_1\times\boldsymbol{\sigma}_1{}^* + \left(1+\frac{\zeta}{R_2}\right)\boldsymbol{t}_2\times\boldsymbol{\sigma}_2{}^* = 0 \tag{3b}$$

and six equations of compatibility

$$(\alpha_2\boldsymbol{k}_2{}^*)_{,1} - (\alpha_1\boldsymbol{k}_1{}^*)_{,2} = 0, \tag{4a}$$

$$\frac{(\alpha_2\boldsymbol{e}_2{}^*)_{,1} - (\alpha_1\boldsymbol{e}_1{}^*)_{,2}}{\alpha_1\alpha_2} + \left(1+\frac{\zeta}{R_1}\right)\boldsymbol{t}_1\times\boldsymbol{k}_2{}^* - \left(1+\frac{\zeta}{R_2}\right)\boldsymbol{t}_2\times\boldsymbol{k}_1{}^* = 0, \tag{4b}$$

$$\frac{\boldsymbol{k}_{\zeta,i}}{\alpha_i} - \boldsymbol{k}^*_{i,\zeta} = 0, \qquad i = 1, 2, \tag{4c}$$

$$\frac{\boldsymbol{e}_{\zeta,i}}{\alpha_i} - \boldsymbol{e}^*_{i,\zeta} + \left(1+\frac{\zeta}{R_i}\right)\boldsymbol{t}_i\times\boldsymbol{k}_\zeta - \boldsymbol{n}\times\boldsymbol{k}_i{}^* = 0, \qquad i = 1, 2. \tag{4d}$$

Equations (4) are a consequence of strain displacement relations of the form

$$\boldsymbol{k}_i{}^* = \frac{\boldsymbol{\psi}_{,i}}{\alpha_i}, \qquad \boldsymbol{k}_\zeta = \boldsymbol{\psi}_{,\zeta} \tag{5a, b}$$

and

$$\boldsymbol{e}_i{}^* = \frac{\boldsymbol{v}_{,i}}{\alpha_i} + \left(1+\frac{\zeta}{R_i}\right)\boldsymbol{t}_i\times\boldsymbol{\psi}, \qquad \boldsymbol{e}_\zeta = \boldsymbol{v}_{,\zeta} + \boldsymbol{n}\times\boldsymbol{\psi}. \tag{5c, d}$$

The above, except for the introduction of pseudo-stresses and strains in place of the actual quantities, coincides with Günther's version of couple stress theory [5, 15].

The system (3) and (4) is complemented by stress strain relations of the form $\sigma_{ij} = \partial A/\partial e_{ij}$, $\tau_{ij} = \partial A/\partial k_{ij}$, etc. [15], and by boundary conditions for the faces $\zeta = \pm(1/2)\,h(\xi_1, \xi_2)$ of the shell.

Insofar as the face boundary conditions are concerned, assume for simplicity's sake that the shell is of constant thickness and without face surface tractions, that is

$$\zeta = \pm \frac{1}{2} h; \qquad \sigma_\zeta^* = 0, \qquad \boldsymbol{\tau}_\zeta^* = 0. \tag{6}$$

Stress resultants and couples, as introduced in sec. 2, may be considered as given in terms of σ_i^* and τ_i^* as follows:

$$\boldsymbol{N}_i = \int_{-h/2}^{h/2} \sigma_i^* \, d\zeta, \qquad \boldsymbol{M}_i = \int_{-h/2}^{h/2} (\boldsymbol{\tau}_i^* + \zeta \boldsymbol{n} \times \sigma_i^*) \, d\zeta \tag{7}$$

It is evident from (7) that couple stresses, if they exist, contribute to all stress couple components and not just to the normal components P_i.

5. Integro-Differential Equation Formulation of the Three-Dimensional Problem

In this section we reduce the three-dimensional boundary value problem by systematic elimination of ζ-derivatives.

Beginning with the equilibrium equations (3) we write

$$\sigma_\zeta^* = \int_{-h/2}^{h/2} \frac{1}{2} \operatorname{sig}(\eta - \zeta) \, [R(\sigma_i^*) + \boldsymbol{p}^*] \, d\eta \tag{1a}$$

and

$$\boldsymbol{\tau}_\zeta^* = \int_{-h/2}^{h/2} \frac{1}{2} \operatorname{sig}(\eta - \zeta) \Big[R(\boldsymbol{\tau}_i^*) + \boldsymbol{q}^* + \\ + \sum \left(1 + \frac{\eta}{R_j}\right) \boldsymbol{t}_j \times \sigma_j^* + \boldsymbol{n} \times \sigma_\zeta^* \Big] \, d\eta . \tag{1b}$$

In this $\operatorname{sig}(\eta - \zeta) \equiv (\eta - \zeta)/|\eta - \zeta|$ and $R(\sigma_i^*) \equiv [(\alpha_2 \sigma_1^*)_{,1} + (\alpha_1 \sigma_2^*)_{,2}]/ /\alpha_1 \alpha_2$, with a corresponding definition of $R(\boldsymbol{\tau}_i^*)$. The verification of (1) and (2) depends on the differentiation formula $(\operatorname{sig} \zeta)_{,\zeta} = 2\delta(\zeta)$, δ being the customary delta function.

Introduction of (1a, b) into the face boundary conditions (4.6) leaves the relations

$$\int_{-h/2}^{h/2} [R(\sigma_i^*) + \boldsymbol{p}^*)] \, d\eta = 0, \tag{2a}$$

$$\int_{-h/2}^{h/2} \Big[R(\boldsymbol{\tau}_i^*) + \boldsymbol{q}^* + \sum \left(1 + \frac{\eta}{R_j}\right) \boldsymbol{t}_j \times \sigma_j^* + \boldsymbol{n} \times \sigma_\zeta^* \Big] \, d\eta = 0, \tag{2b}$$

which, with the help of Eqs. (4.7) for stress resultants and couples, may readily be seen to be equivalent to the two-dimensional equilibrium equations (2.2), written in the form

$$\left[\alpha_2 \int_{-h/2}^{h/2} \boldsymbol{\sigma}_1^* \, d\zeta\right]_{,1} + \left[\alpha_1 \int_{-h/2}^{h/2} \boldsymbol{\sigma}_2^* \, d\zeta\right]_{,2} + \alpha_1 \alpha_2 \int_{-h/2}^{h/2} \boldsymbol{p}^* \, d\zeta = 0 \tag{3a}$$

and

$$\left[\alpha_2 \int_{-h/2}^{h/2} (\boldsymbol{\tau}_1^* + \zeta \boldsymbol{n} \times \boldsymbol{\sigma}_1^*) \, d\zeta\right]_{,1} + \left[\alpha_1 \int_{-h/2}^{h/2} (\boldsymbol{\tau}_2 + \zeta \boldsymbol{n} \times \boldsymbol{\sigma}_2^*) \, d\zeta\right]_{,2} +$$

$$+ \alpha_1 \alpha_2 \int_{-h/2}^{h/2} (\boldsymbol{t}_1 \times \boldsymbol{\sigma}_1^* + \boldsymbol{t}_2 \times \boldsymbol{\sigma}_2^*) \, d\zeta + \alpha_1 \alpha_2 \int_{-h/2}^{h/2} (\boldsymbol{q}^* + \zeta \boldsymbol{n} \times \boldsymbol{p}^*) \, d\zeta = 0 . \tag{3b}$$

We next deduce from the compatibility equations (4.4c) and (4.4d) the relations

$$\boldsymbol{k}_i^* = \boldsymbol{\varkappa}_i + \int_0^{\zeta} \frac{\boldsymbol{k}_{\zeta,i}}{\alpha_i} \, d\eta , \tag{4a}$$

$$\boldsymbol{e}_i^* = \boldsymbol{\varepsilon}_i - \zeta \boldsymbol{n} \times \boldsymbol{\varkappa}_i + \boldsymbol{t}_i \times \int_0^{\zeta} \boldsymbol{k}_\zeta \, d\eta + \int_0^{\zeta} \left[\frac{\boldsymbol{e}_{\zeta,i}}{\alpha_i} + \eta \, \frac{(\boldsymbol{n} \times \boldsymbol{k}_\zeta)_{,i}}{\alpha_i} - \zeta \, \frac{\boldsymbol{n} \times \boldsymbol{k}_{\zeta,i}}{\alpha_i}\right] d\eta . \tag{4b}$$

In this use has been made of the differentiation formula $\boldsymbol{n}_{,i} = \alpha_i \boldsymbol{t}_i / R_i$ and $\boldsymbol{\varepsilon}_i$ and $\boldsymbol{\varkappa}_i$ are as yet undetermined functions of ξ_1 and ξ_2.

Introduction of (4a) into equation (4.4a) leads to a two-dimensional compatibility equation for $\boldsymbol{\varkappa}_1$ and $\boldsymbol{\varkappa}_2$, of the form

$$(\alpha_2 \boldsymbol{\varkappa}_2)_{,1} - (\alpha_1 \boldsymbol{\varkappa}_1)_{,2} = 0 . \tag{5a}$$

Introduction of (4b) into equation (4.4b) gives a second two-dimensional compatibility equation, of the form

$$\frac{(\alpha_2 \boldsymbol{\varepsilon}_2)_{,1} - (\alpha, \boldsymbol{\varepsilon}_1)_{,2}}{\alpha_1 \alpha_2} + (\boldsymbol{t}_1 \times \boldsymbol{\varkappa}_2 - \boldsymbol{t}_2 \times \boldsymbol{\varkappa}_1) = 0 . \tag{5b}$$

It is very remarkable that Eqs. (5) agree completely, without any approximations whatever having been made, with the two-dimensional compatibility equations (2.5), upon identifying the quantities $\boldsymbol{\varepsilon}_i$ and $\boldsymbol{\varkappa}_i$ in (5) with the corresponding quantities in (2.5).

Equations (1) to (5) are all the formal consequences which may be derived from the three-dimensional equilibrium differential equations, together with the face boundary conditions, and from the three-dimensional compatibility equations. We note that these consequences include besides the system of two-dimensional equilibrium equations and compatibility equations previously deduced by purely two-dimensional considerations also expressions for the stresses $\boldsymbol{\sigma}_\zeta^*$, and $\boldsymbol{\tau}_\zeta^*$ in terms of the stresses $\boldsymbol{\sigma}_i^*$ and $\boldsymbol{\tau}_i^*$ and their derivatives, and expressions for the strains $\boldsymbol{e}_i^*$ and $\boldsymbol{k}_i^*$ in terms of the strains $\boldsymbol{e}_\zeta$ and $\boldsymbol{k}_\zeta$ and their derivatives.

There remains the problem of introducing the foregoing consequences into a system of three-dimensional stress strain relations, in such a way that a systematic rational derivation of two-dimensional stress strain relations may be accomplished. To bring out the essence of the procedure as compactly as possible we limit ourselves to a consideration of the following special system of stress strain relations

$$e_{11} = \frac{\sigma_{11} - \nu\sigma_{22}}{E} - \frac{\nu_\zeta \sigma_{\zeta\zeta}}{E_\zeta}, \qquad e_{12} = \frac{(1+\nu)\sigma_{12}}{E}, \quad \text{etc.} \tag{6a}$$

$$e_{i\zeta} = \frac{\sigma_{i\zeta}}{G}, \qquad e_{\zeta i} = \frac{\sigma_{\zeta i}}{G}, \qquad e_{\zeta\zeta} = \frac{\sigma_{\zeta\zeta} - \nu_\zeta\sigma_{11} - \nu_\zeta\sigma_{22}}{E_\zeta}, \tag{6b}$$

$$k_{ij} = \frac{\tau_{ij}}{h^2 F}, \qquad k_{i\zeta} = \frac{\tau_{i\zeta}}{h^2 H}, \qquad k_{\zeta i} = \frac{\tau_{\zeta i}}{h^2 H_\zeta}, \qquad k_{\zeta\zeta} = \frac{\tau_{\zeta\zeta}}{h^2 F_\zeta}. \tag{7}$$

Introduction of pseudo stress and strain components changes (6) and (7) into

$$e_{11}^* = \frac{1+\zeta/R_1}{1+\zeta/R_2}\frac{\sigma_{11}^*}{E} - \nu\frac{\sigma_{22}^*}{E} - \frac{\nu_\zeta}{1+\zeta/R_2}\frac{\sigma_{\zeta\zeta}^*}{E_\zeta}, \quad \text{etc.} \tag{8a}$$

$$e_{12}^* = \frac{1+\zeta/R_1}{1+\zeta/R_2}\frac{(1+\nu)\sigma_{12}^*}{E}, \qquad e_{1\zeta}^* = \frac{1+\zeta/R_1}{1+\zeta/R_2}\frac{\sigma_{1\zeta}^*}{G}, \tag{8b}$$

$$\left(1+\frac{\zeta}{R_1}\right)\left(1+\frac{\zeta}{R_2}\right) e_{\zeta i} = \frac{\sigma_{\zeta i}^*}{G}, \tag{8c}$$

$$\left(1+\frac{\zeta}{R_1}\right)\left(1+\frac{\zeta}{R_2}\right) e_{\zeta\zeta} = \frac{\sigma_{\zeta\zeta}^*}{E_\zeta} - \left(1+\frac{\zeta}{R_1}\right)\frac{\nu_\zeta\sigma_{11}^*}{E_\zeta} - \left(1+\frac{\zeta}{R_2}\right)\frac{\nu_\zeta\sigma_{22}^*}{E_\zeta}, \tag{8d}$$

$$k_{11}^* = \frac{1+\zeta/R_1}{1+\zeta/R_2}\frac{\tau_{11}^*}{h^2 F}, \qquad k_{12}^* = \frac{1+\zeta/R_1}{1+\zeta/R_2}\frac{\tau_{12}^*}{h^2 F}, \tag{9a}$$

$$k_{1\zeta}^* = \frac{1+\zeta/R_1}{1+\zeta/R_2}\frac{\tau_{1\zeta}^*}{h^2 H}, \qquad \left(1+\frac{\zeta}{R_1}\right)\left(1+\frac{\zeta}{R_2}\right) k_{\zeta i} = \frac{\tau_{\zeta i}^*}{h^2 H_\zeta}, \tag{9b}$$

$$\left(1+\frac{\zeta}{R_1}\right)\left(1+\frac{\zeta}{R_2}\right) k_{\zeta\zeta} = \frac{\tau_{\zeta\zeta}^*}{h^2 F_\zeta}. \tag{9c}$$

We now take the quantities e_{ij}^*, k_{ij}^*, etc. in accordance with Eqs. (4a) and (4b) as follows

$$k_{ij}^* = \varkappa_{ij} + \int_0^\zeta T_{ij}(k_{\zeta l}, k_{\zeta\zeta})\, d\eta, \tag{10a}$$

$$k_{i\zeta}^* = \lambda_i + \int_0^\zeta T_i(k_{\zeta l}, k_{\zeta\zeta})\, d\eta, \tag{10b}$$

$$e_{ii}^* = \varepsilon_{ii} + \zeta\varkappa_{ii} + \int_0^\zeta S_{ii}(e_{\zeta l}, e_{\zeta\zeta}\, k_{\zeta l}, k_{\zeta\zeta})\, d\eta, \tag{11a}$$

$$e_{12}^* = \varepsilon_{12} + \zeta\varkappa_{12} - \int_0^\zeta k_{\zeta\zeta}\, d\eta + \int_0^\zeta S_{12}\, d\eta, \tag{11b}$$

$$e_{21}^* = \varepsilon_{21} + \zeta\varkappa_{21} + \int_0^\zeta k_{\zeta\zeta}\, d\eta + \int_0^\zeta S_{21}\, d\eta, \tag{11c}$$

$$e_{i\zeta}^* = \gamma_i + \int_0^\zeta k_{\zeta i}\, d\eta + \int_0^\zeta S_i\, d\eta. \tag{11d}$$

Next we take the quantities $\sigma_{\zeta i}^*$ and $\sigma_{\zeta\zeta}^*$ in accordance with (1a) in the form

$$(\sigma_{\zeta i}^*, \sigma_{\zeta\zeta}^*) = \frac{1}{2}\int_{-h/2}^{h/2} \operatorname{sig}(\eta - \zeta)\,(R_i, R_\zeta)\,(\sigma_{ij}^*, \sigma_{i\zeta}^*)\, d\eta \tag{12}$$

where for simplicity's sake we assume for what follows that $\boldsymbol{p}^* = 0$.

In writing corresponding expressions for $\tau_{\zeta i}^*$ and $\tau_{\zeta\zeta}^*$ in accordance with (1b) it is of importance to separate different terms in the following manner

$$\tau_{\zeta i}^* = -\frac{1}{2}\int_{-h/2}^{h/2} \operatorname{sig}(\eta - \zeta)\, \sigma_{i\zeta}^*\, d\eta + \\ + \frac{1}{2}\int_{-h/2}^{h/2} \operatorname{sig}(\eta - \zeta)\left[\frac{\eta}{R_i}\,\sigma_{i\zeta}^* - \sigma_{\zeta i}^* + R_i(\tau_{ij}^*, \tau_{i\zeta}^*)\right] d\eta, \tag{13a}$$

$$\tau_{\zeta\zeta}^* = \frac{1}{2}\int_{-h/2}^{h/2} \operatorname{sig}(\eta - \zeta)\,(\sigma_{12}^* - \sigma_{21}^*)\, d\eta + \\ + \frac{1}{2}\int_{-h/2}^{h/2} \operatorname{sig}(\eta - \zeta)\left[\left(\frac{\sigma_{12}^*}{R_1} - \frac{\sigma_{21}^*}{R_2}\right)\eta + R_\zeta(\tau_{ij}^*, \tau_{i\zeta}^*)\right] d\eta. \tag{13b}$$

Introduction of Eqs. (10) to (13) into Eqs. (8) and (9) leaves the following system of eighteen equations

$$\varepsilon_{11} + \zeta\varkappa_{11} + \int_0^\zeta S_{11}\,d\eta = \frac{1+\zeta/R_1}{1+\zeta/R_2}\frac{\sigma_{11}^*}{E} - \nu\frac{\sigma_{22}^*}{E} - \frac{\nu_\zeta/E_\zeta}{1+\zeta/R_2}\int_0^\zeta \frac{\operatorname{sig}(\eta-\zeta)}{2} R_\zeta(\sigma)\,d\eta, \tag{14a}$$

$$\varepsilon_{12} + \zeta\varkappa_{12} - \int_0^\zeta k_{\zeta\zeta}\,d\eta + \int_0^\zeta S_{12}\,d\eta = \frac{1+\zeta/R_1}{1+\zeta/R_2}\frac{(1+\nu)\sigma_{12}^*}{E}, \tag{14b}$$

$$\gamma_1 + \int_0^\zeta k_{\zeta 1}\,d\eta + \int_0^\zeta S_1\,d\eta = \frac{1+\zeta/R_1}{1+\zeta/R_2}\frac{\sigma_{1\zeta}^*}{G}, \tag{14c}$$

$$\left(1+\frac{\zeta}{R_1}\right)\left(1+\frac{\zeta}{R_2}\right) e_{\zeta i} = \frac{1}{G}\int_{-h/2}^{h/2} \frac{\operatorname{sig}(\eta-\zeta)}{2} R_i(\sigma)\,d\eta, \quad i = 1, 2, \tag{14d}$$

$$\left(1+\frac{\zeta}{R_1}\right)\left(1+\frac{\zeta}{R_2}\right) e_{\zeta\zeta} = \frac{1}{E_\zeta}\int_{-h/2}^{h/2} \frac{\operatorname{sig}(\eta-\zeta)}{2} R_\zeta(\sigma)\,d\eta - \left(1+\frac{\zeta}{R_1}\right)\frac{\nu_\zeta\sigma_{11}^*}{E_\zeta} - \left(1+\frac{\zeta}{R_2}\right)\frac{\nu_\zeta\sigma_{22}^*}{E_\zeta} \tag{14e}$$

with three corresponding equations, and

$$\varkappa_{11} + \int_0^\zeta T_{11}\,d\eta = \frac{1+\zeta/R_1}{1+\zeta/R_2}\frac{\tau_{11}^*}{h^2 F}, \tag{15a}$$

$$\lambda_1 + \int_0^\zeta T_1\,d\eta = \frac{1+\zeta/R_1}{1+\zeta/R_2}\frac{\tau_{1\zeta}^*}{h^2 H}, \tag{15b}$$

$$\left(1+\frac{\zeta}{R_1}\right)\left(1+\frac{\zeta}{R_2}\right) k_{\zeta i} = -\frac{1}{h^2 H_\zeta}\int_{-h/2}^{h/2} \frac{\operatorname{sig}(\eta-\zeta)}{2}\sigma_{i\zeta}^*\,d\eta + \frac{1}{h^2 H_\zeta}\int_{-h/2}^{h/2} \frac{\operatorname{sig}(\eta-\zeta)}{2}\left[\frac{\eta}{R_i}\sigma_{i\zeta}^* - \sigma_{\zeta i}^* + R_i(\tau)\right] d\eta, \tag{15c}$$

$$\left(1+\frac{\zeta}{R_1}\right)\left(1+\frac{\zeta}{R_2}\right) k_{\zeta\zeta} = \frac{1}{h^2 F_\zeta}\int_{-h/2}^{h/2} \frac{\operatorname{sig}(\eta-\zeta)}{2}(\sigma_{12}^* - \sigma_{21}^*)\,d\eta + \frac{1}{h^2 F_\zeta}\int_{-h/2}^{h/2} \frac{\operatorname{sig}(\eta-\zeta)}{2}\left[\left(\frac{\sigma_{12}^*}{R_1} - \frac{\sigma_{21}^*}{R_2}\right)\eta + R_\zeta(\tau)\right] d\eta \tag{15d}$$

with four corresponding equations.

The system (14) and (15), comprising altogether eighteen equations, is to be considered as a system of integral equations for the determination of the ζ-dependence of the eighteen quantities σ^*_{ij}, $\sigma^*_{i\zeta}$, τ^*_{ij}, $\tau^*_{i\zeta}$, $e_{\zeta i}$, $e_{\zeta\zeta}$ and $k_{\zeta i}$ and $k_{\zeta\zeta}$.

Beyond this, (14) and (15) together with the six scalar equations in (3) and the six scalar equations in (5), may be considered as a system of thirty integro-differential equations for the eighteen quantities above, together with the twelve quantities ε_{ij}, γ_i, $\varkappa_{ij}$, λ_i which by definition are independent of ζ. The importance of reducing the three-dimensional boundary value problem to this form lies in the fact that in this way the problem is concentrated in those of its aspects which represent the essential difficulty of the step from three dimensions to two dimensions, that is, the step from three-dimensional to two-dimensional stress strain relations.

6. Order-of-Magnitude Relations and Iterative and Parametric Expansion Procedures

Use of appropriate order of magnitude considerations in conjunction with the system of integro-differential equations of the three-dimensional shell problem leads to recognition of the possibility of iterative or parametric expansion procedures for this system of equations, and to the further possibility of defining two-dimensional shell theory as the contents of the first step in the iterative or parametric expansion procedure.

Let σ and τ be representative force and couple stress measures such that

$$(\sigma^*_{ij}, \sigma^*_{i\zeta}) = O(\sigma), \qquad (\tau^*_{ij}, \tau^*_{i\zeta}) = O(\tau). \tag{1}$$

Let L be the smallest characteristic length along curves on the middle surface of the shell, in the sense that in Eqs. (5.14) and (5.15)

$$\int_{-h/2}^{h/2} \operatorname{sig}(\zeta - \eta)\, R(\sigma)\, d\eta = O(h\sigma/L), \tag{2a}$$

$$\int_{-h/2}^{h/2} \operatorname{sig}(\zeta - \eta)\, R(\tau)\, d\eta = O(h\tau/L). \tag{2b}$$

Equations (2a) in conjunction with (5.12) imply as order of magnitude relations for transverse force stress components

$$(\sigma^*_{\zeta i}, \sigma^*_{\zeta\zeta}) = O(h\sigma/L). \tag{3a}$$

Corresponding order of magnitude relations for transverse couple stress components follow from Eqs. (1), (2b) and (5.13) in the less simple form

$$(\tau^*_{\zeta i}, \tau^*_{\zeta\zeta}) = O\big(\mathrm{Max}\,(h\tau/L, h\sigma)\big). \tag{3b}$$

Equations (5.13) also imply a refined version of (3b), namely

$$\tau^*_{\zeta i} = \int_{-h/2}^{h/2} \frac{1}{2} \operatorname{sig}(\zeta - \eta)\, \sigma^*_{i\zeta}\, d\eta + O\big(\operatorname{Max}(h\tau/L, h^2\sigma/L)\big), \tag{4a}$$

$$\tau^*_{\zeta\zeta} = \int_{-h/2}^{h/2} \frac{1}{2} \operatorname{sig}(\zeta - \eta)\, (\sigma^*_{12} - \sigma^*_{21})\, d\eta + O\big(\operatorname{Max}(h\tau/L, h^2\sigma/L)\big) \tag{4b}$$

provided we assume, as we shall do, that the curvature radii R_i are subject to the restriction

$$\frac{1}{|R_i|} = O\left(\frac{1}{L}\right). \tag{5}$$

Corresponding considerations of force and couple strain components are as follows. We introduce force and couple strain measures ε and $\varkappa$ in such a way that

$$(\varepsilon_{ij}, \gamma_i) = O(\varepsilon), \qquad (\varkappa_{ij}, \lambda_i) = O(\varkappa). \tag{6}$$

We then assume that ε and $\varkappa$ also indicate the order of magnitude of the transverse strain vectors $\boldsymbol{e}_\zeta$ and $\boldsymbol{k}_\zeta$, that is, we assume

$$(e_{\zeta i}, e_{\zeta\zeta}) = O(\varepsilon), \qquad (k_{\zeta i}, k_{\zeta\zeta}) = O(\varkappa). \tag{7}$$

Having (6) and (7), we obtain from Eqs. (5.4) and (5.10) the following orders of magnitude for the various strain integrals in (5.14) and (5.15)

$$\int_0^\zeta T\, d\eta = O(h\varkappa/L), \tag{8a}$$

$$\int_0^\zeta S\, d\eta = O\big(\operatorname{Max}(h\varepsilon/L, h^2\varkappa/L)\big). \tag{8b}$$

Equations (1) to (8) make it apparent that the stress strain relations (5.14) and (5.15) may be solved by means of expansions

$$\varepsilon_{ij} = \varepsilon\left[\hat{\varepsilon}^{(0)}_{ij} + \frac{h}{L}\,\hat{\varepsilon}^{(1)}_{ij} + \ldots\right], \qquad \varkappa_{ij} = \varkappa\left[\hat{\varkappa}^{(0)}_{ij} + \frac{h}{L}\,\hat{\varkappa}^{(1)}_{ij} + \ldots\right]. \tag{9}$$

$$\frac{\sigma^*_{ij}}{E} = \varepsilon\left[\hat{\sigma}^{(0)}_{ij} + \ldots\right] + h\varkappa\left[\tilde{\sigma}^{(0)}_{ij} + \ldots\right], \tag{10a}$$

$$\frac{\tau^*_{ij}}{h^2 F} = \varkappa\left[\hat{\tau}^{(0)}_{ij} + \frac{h}{L}\,\hat{\tau}^{(1)}_{ij} + \ldots\right] + \frac{\varepsilon}{L}\left[\tilde{\tau}^{(1)}_{ij} + \ldots\right]. \tag{10b}$$

together with corresponding expansions for γ_1, λ_i, $k_{\zeta i}$, $k_{\zeta\zeta}$, $e_{\zeta i}$, $e_{\zeta\zeta}$, and that these same expansions may also be introduced into the two-dimensional equilibrium and compatibility equations (5.3) and (5.5).

Alternately, we may use the order of magnitude relations (1) to (8) in order to solve the stress strain relations (5.14) and (5.15) by an iterative procedure, omitting in the first round all terms in these equations which are of relative order h/L.

7. Shell Theory without Couple Stresses

In considering shell theory without couple stresses, we may retain the formulation as in Eqs. (4.1) to (4.7), *thereby retaining the advantage of the generalized formulation, insofar as equilibrium and compatibility are concerned.* We introduce the stipulation of vanishing couple stresses by way of the strain energy function A being independent of k_{ij}, etc. This means that $F = H = H_\zeta = F_\zeta = 0$ in (5.7), and these equations, together with (5.9) and (5.15), disappear from the integral equation formulation of the stress strain relation system. Instead, (5.6), (5.8) and (5.14) are supplemented by the relations

$$\sigma_{12}^* - \sigma_{21}^* = \zeta\left(\frac{\sigma_{21}^*}{R_2} - \frac{\sigma_{12}^*}{R_1}\right), \qquad \sigma_{i\zeta}^* = \sigma_{\zeta i}^* - \frac{\zeta}{R_i}\,\sigma_{i\zeta}^* \tag{1}$$

which are to be used in conjunction with (5.14).

The first step of an iterative procedure for the solution of the system (5.14), supplemented by (1), then leads to the relations

$$E(\varepsilon_{11} + \zeta\varkappa_{11}) = \sigma_{11}^* - \nu\sigma_{22}^*, \qquad E(\varepsilon_{22} + \zeta\varkappa_{22}) = \sigma_{22}^* - \nu\sigma_{11}^*, \tag{2a}$$

$$E\left(\varepsilon_{12} + \zeta\varkappa_{12} - \int_0^\zeta k_{\zeta\zeta}\,d\eta\right) = (1+\nu)\,\sigma_{12}^*, \tag{2b}$$

$$E\left(\varepsilon_{12} - \zeta\varkappa_{21} + \int_0^\zeta k_{\zeta\zeta}\,d\eta\right) = (1+\nu)\,\sigma_{21}^*, \tag{2c}$$

$$G\left(\gamma_i + \int_0^\zeta k_{\zeta i}\,d\eta\right) = \sigma_{i\zeta}^*, \tag{2d}$$

$$e_{\zeta i} = 0, \qquad E_\zeta e_{\zeta\zeta} = -\nu_\zeta(\sigma_{11}^* + \sigma_{22}^*), \tag{2e}$$

together with

$$\sigma_{12}^* - \sigma_{21}^* = O\left(\frac{h}{L}\,\sigma\right), \qquad \sigma_{i\zeta}^* = O\left(\frac{h}{L}\,\sigma\right). \tag{3a, b}$$

The solution of (2a) and (2e) is self evident and leads to expected portions of a system of two-dimensional shell stress strain relations.

The solution of (2b) and (2c) together with (3a) is effected as follows. We attempt a solution by postulating

$$k_{\zeta\zeta} = 0 \tag{4}$$

and then show that (4) is consistent with all other requirements for this solution.

From (2b, c) and (4) follows, for $i \neq j$,

$$N_{ij} = \frac{Eh}{1+\nu}\,\varepsilon_{ij} \qquad M_{ij} = \frac{Eh^3}{12(1+\nu)}\,\varkappa_{ij}\,. \tag{5a, b}$$

To see that the remaining condition (3a) is satisfied, we write

$$(1+\nu)\,(\sigma_{12}^* - \sigma_{21}^*) = E\,[\varepsilon_{12} - \varepsilon_{21} + \zeta(\varkappa_{12} - \varkappa_{21})]\,. \tag{6}$$

In this compatibility implies

$$\varkappa_{12} - \varkappa_{21} = \frac{\varepsilon_{21}}{R_1} - \frac{\varepsilon_{12}}{R_2} = \frac{1+\nu}{Eh}\left(\frac{N_{21}}{R_1} - \frac{N_{12}}{R_2}\right) = O\left(\frac{\sigma}{EL}\right) \tag{7a}$$

and equilibrium implies

$$\varepsilon_{12} - \varepsilon_{21} = \frac{1+\nu}{Eh}\,(N_{12} - N_{21}) = \frac{1+\nu}{Eh}\left(\frac{M_{21}}{R_2} - \frac{M_{12}}{R_1}\right) = O\left(\frac{h\sigma}{EL}\right) \tag{7b}$$

showing that (6) is in fact consistent with (3a).

The remaining reduction of (2d) consists in noting that the stipulations.

$$\gamma_i = 0, \qquad k_{\zeta i} = 0 \tag{8a, b}$$

are consistent with (2d) and (3b).

References

1. Gol'denweizer, A. L.: Prikl. mat. mekh. **27**, 593–608 (1963).
2. Gol'denweizer, A. L.: Proc. 11th Intern. Congr. Appl. Mech. (Munich 1964), 306–311 (1966).
3. Green, A. E.: Proc. Roy. Soc. **A 266**, 143–160 (1962).
4. Green, A. E.: Proc. Roy. Soc. **A 269**, 481–491 (1963).
5. Günther, W.: Abh. Braunschweigische Wiss. Ges. **10**, 195–213 (1958).
6. Günther, W.: Ingenieur-Archiv **30**, 160–186 (1961).
7. Johnson, M. W., and E. Reissner: J. Math. and Phys. **37**, 374–392 (1958).
8. Koiter, W. T.: Proc. IUTAM Symposium on Shells (Delft 1959) 12–33 (1960).
9. Reiss, E. L.: Comm. Pure and Appl. Math. **13**, 531–550 (1960).
10. Reissner, E.: J. Eng. Mech. Div. ASCE 88, 23–57 (1962).
11. Reissner, E.: J. Math and Phys. **42**, 263–277 (1963).

12. REISSNER, E.: Proc. 11th Intern. Congr. Appl. Mech. (Munich 1964) 20–30 (1966).
13. REISSNER, E., and F. Y. M. WAN: J. Appl. Math. and Phys. (ZAMP) **17**, 676 to 681 (1966).
14. REISSNER, E., and F. Y. M. WAN: The Folke ODQUIST Volume, 487–500 (1967).
15. REISSNER, E., and F. Y. M. WAN: Proc. IUTAM Symp. on the Generalized Cosserat Continuum (Freudenstadt-Stuttgart 1967), Berlin/Heidelberg/New York: Springer 1968.

Discussion

W. T. KOITER: Professor REISSNER has given a most interesting reduction of three-dimensional couple stress theory to a two-dimensional theory of shells. In so far as the "ordinary" theory of shells is concerned, the initial introduction of couple stresses appears to be an artifice, however, and one is inclined to believe that a similar justification of shell theory should be obtainable without an appeal to couple stress theory.

E. REISSNER: The thought in considering the derivation of 2D-shell theory from 3D-elasticity theory, as done here, was that there ought to be advantages to a derivation of a two-dimensional theory in which forces and moments are of comparable importance from a three-dimensional theory for which the same is the case, rather than from a three-dimensional theory which deals with forces only. Mathematically, the advantages of the present approach seem to depend on the fact that here all three-dimensional equilibrium and compatibility equations are first-order differential equations while in the three-dimensional theory without couple stresses the equilibrium equations are a combination of first and zeroth order differential equations while the compatibility equations are second order differential equations. Having this system of first order equations of couple stress theory, it would seem possible to rewrite the equations of the theory without couple stresses in an analogous fashion, through use of appropriate auxiliary variables, and proceed from there to a derivation of two-dimensional shell theory.

H. S. RUTTEN: In particular, the stress-strain relations you established are limited to the case in which the curvatures are of the order of magnitude of the inverse of the smallest wave length L. If, for instance, the wave length L is significantly smaller than the curvature radii R_i, what are the consequences in your theory?

E. REISSNER: The statement made in Eq. (5.6) amounted to saying that the smallest characteristic length L should not be large compared to the smallest radius of curvature of the middle surface of the shell, for Eq. (5.4) to be valid. Going beyond this, it appears that the procedure proposed in the paper should be appropriate in all cases for which the shell thickness h is small compared to the smaller of the two quantities L and $R_{\min}$.

N. J. HOFF: Referring to Professor KOITER's comment I am willing to admit that couple stresses may appear somewhat artificial. But thin-shell theory is also somewhat artificial. If we are willing to consider a thin shell as a curved surface of infinitesimal wall thickness yet capable of carrying moment resultants, we might as well call the moment resultants couple stresses.

Problems in the Rigorous Deduction of the Theory of Thin Elastic Shells

By

A. L. Gol'denweizer

USSR Academy of Sciences, Moscow, USSR

This investigation is based on the well-known concept of a partition of the state of stress into an interior state of stress and a boundary-layer; here we shall deal mainly with the interior state of stress.

Integrals of the differential equations of the theory of elasticity corresponding to the interior state of stress may be developed with the use of an interative process, which will be defined as the basic process. The first approximation in this process is equivalent to a modified Love theory.

By the modified Love theory we mean a theory where some stress resultants $T^{\alpha\beta}$, N^{α} and stress couples $M^{\alpha\beta}$ are considered instead of the stresses, and where the displacements v^{α}, w and the strains $\varepsilon_{\alpha\beta}$, $\mu_{\alpha\beta}$ of the middle surface are considered instead of the displacements and strains of an elastic medium. The above mentioned variables satisfy the following equations:

Equations of equilibrium

$$\nabla_{\lambda} T^{\alpha\lambda} - b_{\lambda}{}^{\alpha} N^{\lambda} + X^{\alpha} = 0; \qquad b_{\lambda\varrho} T^{\lambda\varrho} + \nabla_{\lambda} N^{\lambda} - x = 0;$$
$$\nabla_{\lambda} M^{\lambda\alpha} - N^{\alpha} + c^{\varrho\alpha} Y_{\varrho} = 0.$$

Strain-displacement relations

$$\varepsilon_{\alpha\beta} = \frac{1}{2}\left(\nabla_{\alpha} v_{\beta} + \nabla_{\beta} v_{\alpha} + 2 b_{\alpha\beta} w\right); \qquad \mu_{\alpha\beta} = \nabla_{\beta}\gamma_{\alpha} + \frac{1}{2} c_{\lambda\alpha} b_{\beta}{}^{\lambda} c^{\gamma\varrho} \nabla_{\gamma} v_{\varrho}$$
$$(\gamma_{\alpha} = \nabla_{\alpha} w - b_{\alpha}{}^{\lambda} v_{\lambda}).$$

Stress-strain relations

$$2Eh\,\varepsilon_{\alpha\beta} = P_{\alpha\beta\lambda\varrho} T^{\lambda\varrho} + Q_{\alpha\beta\lambda\varrho} M^{\lambda\varrho} + E_{\alpha\beta},$$
$$\frac{2Eh^3}{3}\mu_{\alpha\beta} = R_{\alpha\beta\lambda\varrho} M^{\lambda\varrho} + \frac{h^2}{3} S_{\alpha\beta\lambda\varrho} T^{\lambda\varrho} + F_{\alpha\beta}.$$

This theory reduces to Love's theory when the tensors E and F are taken equal to zero and when the proper choice of P, Q, R and S is made. For example, in the simplest case it is assumed that

$$P_{\alpha\beta\lambda\varrho} = R_{\alpha\beta\lambda\varrho} = (1+\sigma)\, a_{\alpha\lambda} a_{\beta\varrho} - \sigma a_{\alpha\beta} a_{\lambda\varrho},$$
$$Q_{\alpha\beta\lambda\varrho} = S_{\alpha\beta\lambda\varrho} = 0.$$

The modified Love theory differs from the theory described above in that the P, Q, R and S tensors may have another meaning and that in this theory E and F are functions of the applied loading.

It seems important to consider the following problem: Develop a modified Love theory, which corresponds to such a first approximation of the basic iterative process, which yields the least possible asymptotic error.

Starting with the solution of this problem, we introduce for an elastic medium a curvilinear coordinate system (x^1, x^2, x^3):

$$\boldsymbol{R} = \boldsymbol{r}(x^1, x^2) + x^3 \boldsymbol{n}$$

where g_{ij} is the metric tensor and where $g = |g_{ij}|$. (Latin indices take the values 1, 2, 3).

Let σ^{ij} be the stress tensor, γ_{ij} the strain tensor, and $a_{\alpha\beta}$ and $b_{\alpha\beta}$ the first and second fundamental tensors of the middle surface $x^3 = 0$. We define the asymmetric stress tensor τ^{ij} by the relations

$$\sqrt{\frac{g}{a}}\, \sigma_\beta{}^i = (a_{\lambda\beta} - x^3 b_{\lambda\beta})\, \tau^{i\lambda}; \qquad \sqrt{\frac{g}{a}}\, \sigma^{i3} = \tau^{i3}$$
$$a = |a_{\alpha\beta}|$$

(Greek indices take the values 1, 2).

Then the boundary value problem for the theory of elastisity corresponding to the shell theory problem is the problem of integrating the system of equations including the

Equations of equilibrium

$$\nabla_\alpha \tau^{\alpha\beta} - b_\alpha{}^\beta \tau^{\alpha 3} + \frac{\partial \tau^{3\beta}}{\partial x^3} = 0$$

$$\nabla_\alpha \tau^{\alpha 3} + b_{\alpha\beta} \tau^{\alpha\beta} + \frac{\partial \tau^{33}}{\partial x^3} = 0; \qquad \tau^{3\alpha} = \tau^{\alpha 3} - x^3 b_\beta{}^\alpha \tau^{\beta 3}.$$

Strain-displacement relations

$$2\gamma_{\alpha\beta} = \nabla_\beta U_\alpha + \nabla_\alpha U_\beta + 2 b_{\alpha\beta} W - x^3 [b_\beta{}^\lambda (\nabla_\alpha U_\lambda + b_{\lambda\alpha} W) + $$
$$+ b_\alpha{}^\lambda (\nabla_\beta U_\lambda + b_{\lambda\beta} W)]$$

$$2\gamma_{\alpha 3} = -\nabla_\alpha W + b_\alpha{}^\lambda U_\lambda + \frac{\partial U_\alpha}{\partial x^3} - x^3 b_\alpha{}^\lambda \frac{\partial U_\lambda}{\partial x^3}; \qquad \gamma_{33} = -\frac{\partial W}{\partial x^3}.$$

Stress-strain relations

$$E\sqrt{\frac{g}{a}}\,\gamma_{33} = \tau^{33} - \sigma(a_{\lambda\varrho} - x^3 b_{\lambda\varrho})\,\tau^{\lambda\varrho}; \qquad E\sqrt{\frac{g}{a}}\,\gamma_{\alpha 3} = (1+\sigma)\,g_{\alpha\lambda}\tau^{\lambda 3}$$

$$E\sqrt{\frac{g}{a}}\,\gamma_{\alpha\beta} = (1+\sigma)\,g_{\alpha\lambda}(a_{\varrho\beta} - x^3 b_{\varrho\beta})\,\tau^{\lambda\varrho} - \sigma g_{\alpha\beta}\left[(a_{\lambda\varrho} - x^3 b_{\lambda\varrho})\,\tau^{\lambda\varrho} + \tau^{33}\right].$$

To these equations one should add conditions for the surfaces, i.e. (for $x^3 = \pm h$)

$$\tau^{33} = \sqrt{\frac{g}{a}}\left(\pm\frac{1}{2}\overset{*}{x} + \frac{1}{2}\overset{*}{m}\right); \qquad \tau^{\alpha 3} = \sqrt{\frac{g}{a}}\left(\pm\frac{1}{2}\overset{*}{X}_\varrho + \frac{1}{2}\overset{*}{Y}_\varrho\right)a^{\alpha\varrho}$$

and also conditions for the lateral surfaces (the type of the latter contions is not important in the following derivations).

The desired theory may be developed with the use of the following conditional hypotheses:

Hypothesis 1. The stress and the displacement distributions over the coordinate x^3 are of the form

$$\tau^{\alpha\beta} = \overset{0}{\tau}{}^{\alpha\beta} + \frac{x^3}{h}\overset{1}{\tau}{}^{\alpha\beta}; \qquad \tau^{\alpha 3} = \overset{0}{\tau}{}^{\alpha 3} + \frac{x^3}{h}\overset{1}{\tau}{}^{\alpha 3} + \left(\frac{x^3}{h}\right)^2\overset{2}{\tau}{}^{\alpha 3}$$

$$\tau^{33} = \overset{0}{\tau}{}^{33} + \frac{x^3}{h}\overset{1}{\tau}{}^{33} + \left(\frac{x^3}{h}\right)^2\overset{2}{\tau}{}^{33} + \left(\frac{x^3}{h}\right)^3\overset{3}{\tau}{}^{33}$$

$$U_\alpha = \overset{0}{U}_\alpha + \frac{x^3}{h}\overset{1}{U}_\alpha; \qquad W = \overset{0}{W} + \frac{x^3}{h}\overset{1}{W}$$

Hypothesis 2. The terms τ^{33} and $\tau^{\lambda 3}$ in the first and second stress-strain relation may be neglected.

Hypothesis 3. The stress τ^{33} is of minor importance in the third elasticity equation. With the use of the first condition for the free surfaces this stress may be expressed in terms of the load in the following way:

$$\tau^{33} = \frac{\overset{*}{m}}{2} + \frac{x^3}{h}\,\frac{\overset{*}{x}}{2}.$$

These hypotheses are said to be conditional, as there is no necessity to take them. One can develop the desired theory rigorously proceeding from the given definition of a theory.

These hypotheses were formulated with the sole purpose of making clear the mechanical sense of the results obtained.

According to the hypotheses made above, the equations of the theory of elasticity together with given conditions on the surfaces lead to three groups of equations.

Group 1 consists of a system of differential equations for the fundamental unknowns in the theory of shells, namely

$$\overset{0}{\tau}{}^{\alpha\beta},\ \overset{1}{\tau}{}^{\alpha\beta},\ \overset{0}{\tau}{}^{\alpha 3} + \frac{1}{3}\overset{2}{\tau}{}^{\alpha 3},\ \overset{0}{U}_{\alpha},\ \overset{0}{W},\ \overset{1}{U}_{\alpha}.$$

Group 2 permits direct calculation of all other unknowns in the right-hand sides of the equations contained in hypothesis I, namely

$$\overset{0}{\tau}{}^{33},\ \overset{1}{\tau}{}^{33},\ \overset{2}{\tau}{}^{33},\ \overset{3}{\tau}{}^{33},\ \overset{0}{\tau}{}^{\alpha 3} + \overset{2}{\tau}{}^{\alpha 3},\ \overset{1}{\tau}{}^{\alpha 3},\ \overset{1}{W}.$$

Group 3 consists of superfluous equations which must be ignored (Hypothesis I expresses the possibility of ignoring such equations).

The connection between the unknowns in Group I and the unknowns in Love's theory is expressed by the following relations

$$T^{\beta\alpha} = \int_{-h}^{+h} \tau^{\alpha\beta}\, dx^3; \qquad M^{\beta\alpha} = \int_{-h}^{+h} \tau^{\alpha\beta}\, x^3 dx^3, \qquad N^{\alpha} = \int_{-h}^{+h} \tau^{\alpha 3}\, dx^3$$

$$\overset{0}{U}_{\alpha} = v_{\alpha} \qquad \overset{0}{W} = w.$$

The connection between the quantities $\overset{*}{X}{}^{\alpha}$, $\overset{*}{Y}{}^{\alpha}$, $\overset{*}{x}$, $\overset{*}{m}$ which describe the load on the surfaces and the quantities X^{α}, Y_{λ}, x which describe the surface load on the middle surface $x^3 = 0$ are expressed by the following relations

$$\overset{*}{X}{}^{\alpha} - h(b_{\varrho}{}^{\lambda} + b_{\lambda}{}^{\lambda} a^{\alpha}{}_{\varrho})\, \overset{*}{Y}{}^{\varrho} = X^{\alpha}, \qquad h\,\overset{*}{Y}{}^{\alpha} = c^{\varrho\alpha}\, Y_{\varrho}$$

$$\overset{*}{x} - h b_{\lambda}{}^{\lambda} \overset{*}{m} = -x.$$

With the use of these formulae one can show that the theory presented is equivalent to a modified Love theory in which the following relations must be introduced into the stress-strain relations:

$$P_{\alpha\beta\lambda\varrho} = R_{\alpha\beta\lambda\varrho} = (1 + \sigma)\, a_{\alpha\lambda} a_{\beta\varrho} - \sigma a_{\alpha\beta} a_{\varrho\lambda}; \qquad Q_{\alpha\beta\lambda\varrho} = 0$$

$$S_{\alpha\beta\lambda\varrho} = b_{\beta}{}^{\gamma} P_{\alpha\gamma\lambda\varrho} + b_{\alpha}{}^{\gamma} P_{\gamma\beta\lambda\varrho} + 2 b_{\gamma}{}^{\gamma} P_{\alpha\beta\lambda\varrho} - 2\sigma a_{\alpha\beta} b_{\lambda\varrho} - L_{\alpha\beta\lambda\varrho} - L_{\beta\alpha\lambda\varrho}$$

$$L_{\alpha\beta\lambda\varrho} = (1 + \sigma)\,(a_{\alpha\lambda} b_{\varrho\beta} + 2 b_{\alpha\lambda} a_{\varrho\beta}) - \sigma(a_{\alpha\beta} b_{\lambda\varrho} - 2 a_{\lambda\varrho} b_{\alpha\beta})$$

$$E_{\alpha\beta} = -\sigma h a_{\alpha\beta} \overset{*}{m}; \qquad F_{\alpha\beta} = -\frac{\sigma h^2}{3}\, a_{\alpha\beta} \overset{*}{x}.$$

There is one conditionality in the last statement. The first equation of equilibrium in the theory presented may be obtained only after certain transformations based upon the assumption that N^λ can be expressed in terms of $\overset{*}{Y}{}^\lambda$ by the use of the formulae $N^\lambda = h\,\overset{*}{Y}{}^\lambda$ which is the result of simplification of the third equation of equilibrium.

The hypotheses of the theory presented can be confirmed and corresponding asymptotic errors can be estimated for the case when the stress strain field is represented as

$$\begin{gathered}\tau^{\alpha\beta} = \varkappa^{q+d} \sum \varkappa^{-s}\, \tau^{\alpha\beta}_{(s)}; \qquad \tau^{\alpha 3} = \varkappa^{p+d} \sum \varkappa^{-s}\, \tau^{\alpha 3}_{(s)} \\ \tau^{33} = \varkappa^{c+d} \sum \varkappa^{-s}\, \tau^{33} \\ U_\alpha = \varkappa^{q-p+d} \sum \varkappa^{-s}\, U_\alpha{}^{(s)}; \qquad W = \varkappa^{q-c+d} \sum \varkappa^{-s}\, W^{(s)} \\ c = 0 \quad \text{if} \quad q \geq 2p \qquad c = 2p - q \quad \text{if} \quad q < 2p.\end{gathered} \tag{*}$$

Here d is a value which must be chosen depending on the problem stated; p and q are relatively prime integers ($p < q$) defined by the relations

$$\varkappa^{-q} = h_*{}^{-1}; \qquad \varkappa^{p} = h_*{}^{-t}$$

(h_* is the relative semi-thickness, t is the index of variation).

Let us perform, in the differential equations of the theory of elasticity and in the conditions at the surfaces, a scaling typical for asymptotic methods, by substituting in the equations obtained expansions of the unknowns and by equating coefficients of terms with the same powers of $\varkappa$. Then we obtain a recurrent system of equations for the coefficients in the expansions.

After investigating the structure of this system one can easily obtain the formal asymptotic error of the theory presented

$$\varepsilon = O(\varkappa^{-2q+2p}) = O(h_*{}^{2-2t})$$

while for the Love theory it is

$$\varepsilon = O(\varkappa^{-q+c}) = \begin{cases} O(h_*{}^{1}) & t \leq \dfrac{1}{2} \\ O(h_*{}^{2-2t}) & t > \dfrac{1}{2}. \end{cases}$$

By our definition the formal asymptotic error of a certain approximate theory is of the order $O(x^{-s})$ if the influence of the neglected terms in expansions (*) begins to manifest itself beginning with the number s.

This justification and the resulting error estimate is based on the formulae (*). It establishes certain asymptotic properties of the state of stress which may be not valid. The class of admissible problems is restricted by two requirements.

Requirement 1: Solutions of the boundary value problems connected with the iterative process resulting from (*) should exist.

Requirement 2: The iterative processes under consideration should converge asymptotically.

Consideration of the first requirement is closely connected with the results of a qualitative analysis of the differential equations of two-dimensional shell theory as small parameter equations. In the following, we shall use the appropriate terms established in the author's book[1].

In the first approximation iterative processes resulting from (*) correspond to the basic state of stress and the state of stress with a large index of variation, the latter including non-degenerated edge effects as special cases. Therefore, one condition for the validity of the theory presented is that in the framework of this theory the unknown state of stress must either be that with a large index of variation or may be subdivided into a basic state of stress and a non-degenerated edge effect (of course, the superposition of such states of stress is permissible).

The class of such problems is rather extensive but does not include all cases. Shells with middle surfaces having certain peculiar features, for example, those which degenerate into plates or touch a plane along a closed curve, such as torus, or shells which have a vertex, such as the conical shell excepted. Furthermore, the condition is not satisfied for the case when the unknown state of stress include degenerated edge effects, as the case would be in a long, open cylindrical shell. And, finally, the infinite shells may also be exceptions. In such shells the stress resultants and the stress couples of the basic state of stress may grow infinitely although the true state of stress is bounded; such is the case, for example, in the semi-infinite cylindrical closed shell, subject to an edge load in equilibrium.

This list of singular cases may be not complete. Besides, it must be kept in mind, that cases when the shell is closed to being singular should also be treated as singular ones. Then one can get a formal solution with the help of a subdivision of the state of stress into a basic one and an edge effect but this solution will be very approximate. A study of the convergence of the basic iterative processes would reveal all the phenomena related.

Thus, it is shown, that small modifications to the Love theory yields a possibility for developing a two-dimensional shell theory, which for

[1] Gol'denweizer, A. L. "Theory of Elastic Thin Shells", Pergamon Press 1961.

$t < \frac{1}{2}$ is considerably more accurate than Love's theory. This theory is intended for the interior state of stress only, but it is not universal even in this case. It seems to the author that any other convenient two-dimensional theory will have a bounded region of applicability. The evaluation of the framework of the two-dimensional theory is a difficult mathematical problem, which implies the study of boundary-value problems for the two-dimensional shell theory under consideration and an estimation of the remainder terms in the iterative process used.

The range of applicability of the two-dimensional theory presented, may certainly be wider than the range where the described method of the substantiation of this theory is valid. For example, it is possible to prove, that the theory presented is applicable also to shallow shells. Alternatively, this theory has no chance for success in the case of long cylindrical shells.

In conclusion I would like to point out that the given asymptotic estimate of errors is inherently based on the concept of the "index of variation", which is widely used in the soviet literature on shell theory (in the literature of other countries an equivalent concept of the typical lenght of the deformation pattern is used).

Both of these concepts are real only for problems that are homogeneous over a variation or may be easily broken into a number of problems homogeneous over a variation. A problems, homogeneous over a variation is a problem concerned with the construction of a state whose variation is uniform everywhere.

If the problem is broken into a number of homogeneous ones, then it is necessary to estimate the errors for each of the problems separately. Particularly, if the state of stress under consideration is of the type "basic state of stress plus edge effect" then it is necessary for the basic state of stress to take $t = 0$ (if $t = 0$ for the external actions) and for the edge effect to take $t = \frac{1}{2}$ (if it is the simple one). As a result we would get $\varepsilon = O(h_*^2)$ for the basic state of stress and $\varepsilon = O(h_*^1)$ for the simple edge effect.

In a very general case it can be shown that the division of statical problems into homogeneous problems is roughly speaking equivalent to the same kind of division of the external actions. However, for dynamical problems this matter is yet to be investigated.

Discussion

E. Reissner: I would like to ask Professor Gol'denweizer whether his talk means that he is pessimistic about a universal shell theory so-called, i.e. a two-dimensional theory which is valid to first-order at least for *all* shell problems, as long as the characteristic lengths are large in comparison with the thickness of the shell.

A. L. GOL'DENWEIZER: Yes, I belive that so many different problems in shell theory can be encountered, that if we do not restrict the domain of analysis, no two-dimensional theory could cover them all.

E. REISSNER: I think it would then be of great interest to have specificly described a problem, where one would *think* that two-dimensional shell theory would apply and it could then be shown that for some reason it would *not* apply.

A. L. GOL'DENWEIZER: I think there is a misunderstanding here. I believe that there always will be a two-dimensional theory to cover any specific problem, but that different problems might require different theories.

Shells in the Light of Generalized Continua

By ·

A. E. Green

Oxford University, Oxford, England

and

P. M. Naghdi

University of California, Berkeley, Cal., USA

Abstract. Some aspects of recent developments in the nonlinear theories of deformable surfaces (as two dimensional generalized continua), including the Cosserat surface, are outlined with a view toward their application to the (three dimensional) elastic shells. A fairly detailed development of a linear theory of an elastic Cosserat surface is presented and some features of its constitutive equations are discussed.

1. Introduction

The past few years have witnessed a revival of interest in the formulation of general theories of oriented media and generalized continua. In particular, we call attention to recent complete theories of deformable surfaces — as two dimensional generalized continua embedded in a Euclidean 3-space — in which, in addition to the ordinary displacement vector of a point of the surface, other kinematical ingredients are also admitted. Some of these developments bear directly on the foundations of the classical theories of shells and plates. Moreover, there is already ample evidence in special cases of these theories to suggest that consideration of shells (and plates) in the light of generalized continua is likely to be fruitful.

Among recent contributions of particular interest, we mention the theory of a Cosserat surface by Green, Naghdi and Wainwright [1], a related work on directed surfaces by Cohen and DeSilva [2] and a theory of deformable surfaces with simple force multipoles by Balaban, Green and Naghdi [3]. The nonlinear theories in [1,3] are quite general and are valid for any material; the development in [2] is concerned only

Acknowledgement. The work of one of us (P. M. N.) was supported by U. S. Office of Naval Research under Contract Nonr-222(69) with the University of California, Berkeley.

with a nonlinear (isothermal) theory for elastic surfaces. A recent paper by GREEN, LAWS and NAGHDI [4], among others, considers the question of the relationship between the nonlinear theory of an elastic Cosserat surface and a theory of shells derived from the three dimensional theory of classical continuum mechanics. Also several other aspects of the Cosserat surface, including the linear theory of an elastic Cosserat plate, have been further discussed by GREEN and NAGHDI [5—8]. For historical background on the theories of oriented media and generalized continua, we refer the reader to [1, 3, 6, 8] where other related references on the subject may be found.

It was noted above that in the construction of the theories of deformable surfaces, within the framework of two dimensional generalized continua, other basic kinematical ingredients besides the ordinary (monopolar) displacement of the surface are also admitted. The additional kinematical ingredient in the theory of a Cosserat surface is a single director (i.e., a deformable vector) assigned to every point of the surface; the directors not necessarily lying along the normals to surface. In the theory of deformable surfaces with simple force multipoles, however, the additional kinematical ingredients are suitable second and higher order displacement gradients.

The development of the theory of thin shells, usually confined to elastic shells with infinitesimal deformations, in the past has been pursued mainly along two lines. One method of approach, including sometimes the use of variational methods, begins by integrating the three dimensional equations of classical linear elasticity across the thickness of the shell and strives to supply a systematic derivation of an approximate system of equations of the bending theory of shells in terms of stress-resultants and stress-couples; these two dimensional equations are then valid on the middle surface of the shell. The second approach also begins with the three dimensional equations of (classical) linear elasticity and by expanding all variables in powers of a parameter (e.g., the ratio of shell thickness to a minimum radius of curvature of the middle surface), utilizes a method of asymptotic integration in order to arrive at successive approximations of the two dimensional 'interior' and 'boundary layer' problems of shell theory. That each of these methods have their own shortcomings is evident, in part, from the continual efforts (in the last decade) of a number of investigators to rigorize and systematize the derivation of the equations of the classical shell theory. However, it should be noted that the developments referred to above have served a useful purpose and have also shed considerable light on the subject.

In the light of the theory of a deformable surface in [1], if a shell is regarded as a Cosserat surface (defined in Sec. 2), then we have a general nonlinear theory [1] which is based on dynamical and thermodynamical principles of continuum mechanics. Under certain assumptions and with reference to an elastic Cosserat surface, a restricted theory results which corresponds to the classical elastic shell theory [5, 6]. When the equations of the special theory are linearized, they reduce to a form equivalent to those derived from various points of view in [9—11][1].

By way of additional background, we recall that using the three dimensional (nonlinear) theory of classical continuum mechanics as a starting point and by employing an exact representation for expansion (in powers of the thickness coordinate) of the position vector, in addition to other considerations, in [4] Green, Laws and Naghdi have included the development of a two dimensional theory for a shell (or plate) from the three dimensional theory[2]. In entirely another development, using the three dimensional linear theory of an elastic continuum which admits a director as a starting point and by employing a method of asymptotic expansion, Green and Naghdi [8] have indicated successive approximations for a director theory of plates[3].

Given the basic postulates of Green, Naghdi and Wainwright [1], the theory of a Cosserat surface may be regarded as an exact theory of deformable surfaces which is also applicable to shells. However, Green, Laws and Naghdi [4] have also suggested a method of approximation for elastic shells which enables them to recover (from their two dimensional theory referred to above) a theory of the same form as that of an elastic Cosserat surface included in [1]. In addition it is worth noting that Green and Naghdi, via a method of asymptotic expansion, have shown in [8] that the equations of the linear theory of an elastic Cosserat plate [7] may be obtained as a first approximation to an exact linear three dimensional director theory for plates mentioned above.

In the present paper, after some background information in Sec. 2, we collect in Sec. 3 the basic equations of the nonlinear theory of an elastic Cosserat surface [1] and briefly indicate their relationships to those of the approximate theory of (the three dimensional) elastic shells in [4]. Next, we discuss a fairly detailed development of the linearized theory of an elastic Cosserat surface in which the directors are initially coincident with the unit normals to the surface; the material

---

[1] A number of additional references are cited in [10].

[2] The content of [4] is more than that indicated above. It deals with nonlinear thermodynamical theories of rods, plates and shells. In obtaining the one dimensional theory for rods and the two dimensional theory for shells and plates (from the three dimensional theory) an approximation is used for the expansion of the temperature. In the isothermal case, however, the theories are exact.

[3] Ref. [8] contains other developments, including that of the linear micropolar theory of plates.

is assumed to be homogeneous and to posess isotropy with a center of symmetry. Finally, in Sec. 5, we give an account of the theory of a deformable surface with simple force dipole contained in [3]. After a brief outline of this theory for an elastic surface, we discuss its main features in comparison with those of the theory of a Cosserat surface discussed in Sec. 3.

The theory presented in Sec. 4, apart from linearization, is exact and is valid for non-isothermal deformation. The basic field equations are recast in terms of variables which are particularly convenient when the reduction is sought to the equations of the classical shell theory. The linear constitutive equations deduced in Sec. 4 are among the simplest that can be provided for a general linear theory of a Cosserat surface. In the isothermal case, these constitutive relations contain nine coefficients and some of them are determined from comparison of solutions in the linear theory of a Cosserat surface with corresponding known exact solutions in the three dimensional (classical) theory of elasticity.

2. Preliminaries. General Background

Let points of a three dimensional continuum, in a Euclidean 3-space, be defined by a general convected coordinate system x^i $(i = 1, 2, 3)$. Let $\boldsymbol{r}^*$ be the position vector, relative to a fixed origin, of a typical particle x^i at time t. Then

$$\boldsymbol{r}^* = \boldsymbol{r}^*(x^1, x^2, x^3, t), \qquad \boldsymbol{g}_i = \frac{\partial \boldsymbol{r}^*}{\partial x^i}, \qquad [\boldsymbol{g}_1 \boldsymbol{g}_2 \boldsymbol{g}_3] > 0,$$
$$g_{ij} = \boldsymbol{g}_i \cdot \boldsymbol{g}_j, \qquad \boldsymbol{g}^i \cdot \boldsymbol{g}_j = \delta_j^i, \qquad g^{ij} = \boldsymbol{g}^i \cdot \boldsymbol{g}^j, \tag{2.1}$$

where $\boldsymbol{g}_i$ and $\boldsymbol{g}^i$ are the covariant and the contravariant base vectors at points of the continuum at time t, g_{ij} is the metric tensor, g^{ij} its conjugate and δ_j^i is the Kronecker symbol in 3-space.

The parametric equation $x^3 = 0$ defines a surface s in space at time t. If $\boldsymbol{r}$ is the position vector (relative to the same fixed origin) of any point of s, then

$$\boldsymbol{r} = \boldsymbol{r}(x^1, x^2, t) = \boldsymbol{r}^*(x^1, x^2, 0, t). \tag{2.2}$$

Apart from suitable regularity and smoothness assumptions which we omit here, a shell may be defined as follows: A continuum bounded by (1) the surfaces

$$x^3 = \bar{\alpha}, \qquad x^3 = \bar{\beta} \qquad (\bar{\alpha} < 0 < \bar{\beta}), \tag{2.3}$$

with s lying entirely between them; and (2) a surface

$$f(x^1, x^2) = 0 \tag{2.4}$$

which is such that $x^3 =$ const. are closed curves on this surface. The middle surface is arbitrarily situated with respect to the boundary surfaces $x^3 = \bar{\alpha}$ and $x^3 = \bar{\beta}$; however, a reference state may be chosen in which the initial middle surface is midway between the surfaces defined in (2.3).

It is convenient to use the notation $\boldsymbol{a}_\alpha$ $(\alpha = 1, 2)$ for the base vectors along the x^α-curves on s $(x^3 = 0)$ and $\boldsymbol{a}_3$ (a function of x^α and t) for the unit normal to s. The base vectors $\boldsymbol{a}_i$, their reciprocals $\boldsymbol{a}^i$ and the corresponding surface metric tensors satisfy the relations

$$\begin{aligned} \boldsymbol{a}_\alpha &= \boldsymbol{r}_{,\alpha}, \qquad \boldsymbol{a}_\alpha \cdot \boldsymbol{a}_\beta = a_{\alpha\beta}, \qquad a = \det(a_{\alpha\beta}), \\ \boldsymbol{a}^\alpha \cdot \boldsymbol{a}_\beta &= \delta_\beta{}^\alpha, \qquad a^{\alpha\gamma} a_{\gamma\beta} = \delta_\beta{}^\alpha, \qquad \boldsymbol{a}^\alpha = a^{\alpha\gamma} \boldsymbol{a}_\gamma, \end{aligned} \tag{2.5}$$

and

$$\boldsymbol{a}_\alpha \cdot \boldsymbol{a}_3 = 0, \qquad \boldsymbol{a}_3 \cdot \boldsymbol{a}_3 = 1, \qquad \boldsymbol{a}_3 = \boldsymbol{a}^3, \tag{2.6}$$

where a comma stands for partial differentiation with respect to x^α and $\delta_\beta{}^\alpha$ is the Kronecker symbol in 2-space. Denoting the second fundamental form of s by $b_{\alpha\beta}$, we also have

$$\boldsymbol{a}_{\alpha|\beta} = b_{\alpha\beta} \boldsymbol{a}_3, \qquad \boldsymbol{a}_{3,\beta} = -b_\beta{}^\alpha \boldsymbol{a}_\alpha, \qquad b_{\alpha\beta|\gamma} = b_{\alpha\gamma|\beta}. \tag{2.7}$$

In the above equations and throughout the paper, all Latin indices have the range 1, 2, 3, Greek indices have the range 1, 2 only and a vertical line stands for covariant differentiation with respect to the surface metric $a_{\alpha\beta}$. We denote the initial value of $\boldsymbol{r}$ at time $t = 0$ by $\boldsymbol{R}$ and refer to the (initial) undeformed surface by $\mathscr{S}$. Also, we designate the initial values of $\boldsymbol{a}_i$, $a_{\alpha\beta}$ and $b_{\alpha\beta}$ by $\boldsymbol{A}_i$, $A_{\alpha\beta}$ and $B_{\alpha\beta}$, respectively, and note that results similar to (2.5) to (2.7) hold also for the surface $\mathscr{S}$.

Let a deformable vector $\boldsymbol{d}$ be assigned to every point of s. The deformable vector $\boldsymbol{d}$ at $\boldsymbol{r}$, called a *director*, is not necessarily along the normal to s and possesses the property that it remains invariant in length under rigid body motions; the initial director at $\boldsymbol{R}$ will be denoted by $\boldsymbol{D}$. A surface s with deformable directors $\boldsymbol{d}$ is called a *Cosserat surface* and its motion is characterized by

$$\boldsymbol{r} = \boldsymbol{r}(x^\alpha, t), \qquad \boldsymbol{d} = \boldsymbol{d}(x^\alpha, t), \qquad [\boldsymbol{a}_1 \boldsymbol{a}_2 \boldsymbol{d}] > 0. \tag{2.8}$$

Let $\boldsymbol{v}$ and $\boldsymbol{w}$ (both three dimensional vector fields) denote, respectively, the velocity of a typical particle of s at time t, and the director velocity at time t. Then

$$\boldsymbol{v} = \dot{\boldsymbol{r}}, \qquad \boldsymbol{w} = \dot{\boldsymbol{d}}, \tag{2.9}$$

where a superposed dot denotes the material derivative with respect to t, holding x^α fixed.

Let σ be an arbitrary area of s bounded by a closed curve c lying in the surface, let ν be the outward unit normal to c and let ϱ stand for the mass density per unit area of s. If $\boldsymbol{N}$ is a three dimensional vector field and if, for all arbitrary velocity fields $\boldsymbol{v}$, the scalar $\boldsymbol{N}\cdot\boldsymbol{v}$ represents a rate of work per unit length of c, then $\boldsymbol{N}$ is called a curve force vector (or simply a force vector) per unit length of c. Similarly, if $\boldsymbol{M}$ is a three dimensional vector field and if, for all arbitrary director velocity $\boldsymbol{w}$, the scalar $\boldsymbol{M}\cdot\boldsymbol{w}$ is a rate of work per unit length, then $\boldsymbol{M}$ is called the director force vector per unit length of c. Also, let $\boldsymbol{F}$ and $\boldsymbol{L}$ stand, respectively, for the three dimensional fields of assigned surface force and assigned surface director force, per unit mass of s at time t, such that for all arbitrary $\boldsymbol{v}$ and $\boldsymbol{w}$, the scalars $\boldsymbol{F}\cdot\boldsymbol{v}$ and $\boldsymbol{L}\cdot\boldsymbol{w}$ represent rate of work per unit area of s. Further, let r be the heat supply function per unit mass per unit time, h the flux of heat across c per unit length per unit time, U the internal energy function per unit mass, S the entropy per unit mass and $T(>0)$ the temperature. Then, an equation for balance of energy may be stated in the form

$$\frac{D}{Dt}\int_\sigma \varrho E\,d\sigma - \int_\sigma \varrho\{r + \boldsymbol{F}\cdot\boldsymbol{v} + \boldsymbol{L}\cdot\boldsymbol{w}\}\,d\sigma = \int_c [\boldsymbol{N}\cdot\boldsymbol{v} + \boldsymbol{M}\cdot\boldsymbol{w} - h]\,dc, \tag{2.10}$$

where D/Dt stands for the material time derivative, $E = U + K$ is the sum of internal and kinetic energies per unit mass with

$$K = \frac{1}{2}(\boldsymbol{v}\cdot\boldsymbol{v} + \alpha\boldsymbol{w}\cdot\boldsymbol{w}) \tag{2.11}$$

and with the coefficient α dependent only on the surface coordinates x^α.

With the above background and using the balance of energy (2.10) and an entropy production inequality (not recorded here), together with invariance conditions under superposed rigid body motions, a general nonlinear theory for a Cosserat surface has been developed in [1] which is valid for any Cosserat surface. However, here and in the remainder of the paper we limit the discussion to an elastic Cosserat surface and quote freely from [1], as well as several other related recent papers [4, 7].

Let $\boldsymbol{n}^\alpha$, $\boldsymbol{m}^\alpha$ be the physical force and director force vectors across the x^α-curves, respectively. Put

$$\boldsymbol{N}^\alpha = \boldsymbol{n}^\alpha(a^{\alpha\alpha})^{1/2}, \qquad \boldsymbol{M}^\alpha = \boldsymbol{m}^\alpha(a^{\alpha\alpha})^{1/2}. \tag{2.12}$$

Then, if $\boldsymbol{\nu} = \nu_\alpha\boldsymbol{a}^\alpha$ is the unit normal to c, it can be shown that

$$\boldsymbol{N} = \boldsymbol{N}^\alpha\nu_\alpha \tag{2.13}$$

for any deformable surface. Also, for an elastic Cosserat surface, we have

$$\boldsymbol{M} = \boldsymbol{M}^{\alpha}\nu_{\alpha}. \tag{2.14}$$

It follows from (2.13) and (2.14) that

$$\boldsymbol{N}^{\alpha} = N^{\alpha i}\boldsymbol{a}_i, \qquad \boldsymbol{M}^{\alpha} = M^{\alpha i}\boldsymbol{a}_i, \tag{2.15}$$

and if we put

$$\boldsymbol{N} = N^{i}\boldsymbol{a}_i, \qquad \boldsymbol{M} = M^{i}\boldsymbol{a}_i, \tag{2.16}$$

we deduce

$$N^{i} = N^{\alpha i}\nu_{\alpha}, \qquad M^{i} = M^{\alpha i}\nu_{\alpha}. \tag{2.17}$$

The notations $N^{\alpha i}$ and $M^{\alpha i}$ are in agreement with conventional notations for the components of membrance force and bending couple vectors in shell theory, but differ from those used previously in [1, 4, 7]. Also, let h^{α} be the value of the flux of heat across the x^{α}-curves. Then, provided c is not a curve of discontinuity, we can show that for any deformable surface

$$h = \boldsymbol{q}\cdot\boldsymbol{\nu} = q^{\alpha}\nu_{\alpha}, \tag{2.18}$$

where

$$\boldsymbol{q} = q^{\alpha}\boldsymbol{a}_{\alpha}, \qquad q^{\alpha} = h^{\alpha}(a^{\alpha\alpha})^{1/2}. \tag{2.19}$$

3. Cosserat Surface and Three Dimensional Elastic Shells

We collect in this Section the basic equations of the nonlinear theory of an elastic Cosserat surface as given in [1] allowing for the change of notations for $N^{\alpha i}$, $M^{\alpha i}$ mentioned earlier. These equations of a Cosserat surface are applicable to elastic shells and may be brought into correspondence with a two dimensional theory of shells derived from the three dimensional equations of classical continuum mechanics [4]. Further reference to this will be made at the end of this Section.

The kinematic variables for a Cosserat surface (defined in Sec. 2) may be taken in the form

$$2e_{\alpha\beta} = a_{\alpha\beta} - A_{\alpha\beta}, \qquad \varkappa_{i\alpha} = \lambda_{i\alpha} - \Lambda_{i\alpha}, \qquad \delta_i = d_i - D_i, \tag{3.1}$$

where

$$\begin{gathered} \lambda_{i\alpha} = \boldsymbol{a}_i\cdot\boldsymbol{d}_{,\alpha}, \qquad \Lambda_{i\alpha} = \boldsymbol{A}_i\cdot\boldsymbol{D}_{,\alpha}, \qquad \boldsymbol{d} = d_i\boldsymbol{a}^i, \\ \lambda_{\beta\alpha} = d_{\beta|\alpha} - b_{\alpha\beta}d_3, \qquad \lambda^{\beta}_{\cdot\alpha} = a^{\beta\gamma}\lambda_{\gamma\alpha}, \\ \lambda_{3\alpha} = d_{3,\alpha} + b_{\alpha}^{\beta}d_{\beta}, \qquad \lambda^{3}_{\cdot\alpha} = \lambda_{3\alpha}, \end{gathered} \tag{3.2}$$

together with corresponding expressions for $\Lambda_{\beta\alpha}$ and $\Lambda_{3\alpha}$. We also note that the material derivatives of the quantities in (3.1) are

$$2\dot{e}_{\alpha\beta} = \dot{a}_{\alpha\beta}, \qquad \dot{\varkappa}_{i\alpha} = \dot{\lambda}_{i\alpha}, \qquad \dot{\delta}_i = \dot{d}_i. \tag{3.3}$$

The equation for conservation of mass is

$$\frac{D}{Dt}(\varrho a^{1/2}) = 0. \tag{3.4}$$

The equations of motion are

$$\begin{aligned} N^{\alpha\beta}{}_{|\alpha} - b_\alpha{}^\beta N^{\alpha 3} + \varrho F^\beta &= \varrho c^\beta, \\ N^{\alpha 3}{}_{|\alpha} + b_{\alpha\beta} N^{\alpha\beta} + \varrho F^3 &= \varrho c^3, \end{aligned} \tag{3.5}$$

and

$$\begin{aligned} M^{\alpha\beta}{}_{|\alpha} - b_\alpha{}^\beta M^{\alpha 3} + \varrho \bar{L}^\beta &= m^\beta, \\ M^{\alpha 3}{}_{|\alpha} + b_{\alpha\beta} M^{\alpha\beta} + \varrho \bar{L}^3 &= m^3, \end{aligned} \tag{3.6}$$

together with the restrictions

$$\begin{aligned} &\varepsilon_{\beta\alpha}[N^{\alpha\beta} + m^\beta d^\alpha + M^{\gamma\beta}\lambda^{\alpha}_{\cdot\gamma}] = 0, \\ &N^{\alpha 3} + (m^3 d^\alpha - m^\alpha d^3) + M^{\gamma 3}\lambda^{\alpha}_{\cdot\gamma} - M^{\gamma\alpha}\lambda^{3}_{\cdot\gamma} = 0, \end{aligned} \tag{3.7}$$

where $\varepsilon_{\alpha\beta} = a^{1/2}\bar{e}_{\alpha\beta}$, $\bar{e}_{\alpha\beta}$ is a permutation symbol with non-zero components $\bar{e}_{12} = -\bar{e}_{21} = 1$, $c^i = (c^\beta, c^3)$ are the components of the acceleration, $\bar{L}^i = \bar{\boldsymbol{L}} \cdot \boldsymbol{a}^i$ are the components of $\bar{\boldsymbol{L}}$, the difference of the assigned director force per unit mass and the inertia terms due to the director displacement $\boldsymbol{d}$, and the quantities m_i for which in general constitutive equations are required are defined through (3.6).

The (local) energy equation for an elastic Cosserat surface may be expressed as

$$\begin{aligned} &\varrho r - q^\alpha{}_{|\alpha} - \varrho(T\dot{S} + \dot{T}S + \dot{A}) \\ &\qquad + N'^{\alpha\beta}\dot{e}_{\alpha\beta} + M^{\alpha i}\dot{\varkappa}_{i\alpha} + m^i\dot{\delta}_i = 0, \end{aligned} \tag{3.8}$$

where $A = U - TS$ is the Helmholtz free energy function per unit mass and

$$N'^{\alpha\beta} = N'^{\beta\alpha} = N^{\alpha\beta} - m^\alpha d^\beta - M^{\gamma\alpha}\lambda^{\beta}_{\cdot\gamma}. \tag{3.9}$$

Assuming that A, S, $N'^{\alpha\beta}$, $M^{\alpha i}$, m^i, q^α are all functions of[1] T, $e_{\alpha\beta}$, $\varkappa_{i\alpha}$, δ_i, (as well as the initial values $A_{\alpha\beta}$, $\Lambda_{i\alpha}$, D_i), and that q^α depends in addi-

[1] The dependence of the constitutive equations on the particular variables $e_{\alpha\beta}$, $\varkappa_{i\alpha}$, δ_i is assumed for later convenience. Alternatively, the kinematic variables in the constituve equations of the nonlinear theory may be taken as $e_{\alpha\beta}$, $\lambda_{i\alpha}$, d_i together with the initial values $A_{\alpha\beta}$, $\Lambda_{i\alpha}$, D_i.

tion on $T_{,\alpha}$, we can then deduce from the combination of the energy equation (3.8) and a local entropy production inequality (not recorded here) the results

$$S = -\frac{\partial A}{\partial T} \tag{3.10}$$

$$N'^{\alpha\beta} = \varrho \frac{\partial A}{\partial e_{\alpha\beta}}, \qquad M^{\alpha i} = \varrho \frac{\partial A}{\partial \varkappa_{i\alpha}}, \qquad m^i = \varrho \frac{\partial A}{\partial \delta_i}, \tag{3.11}$$

and the inequality

$$-q^\alpha T_{,\alpha} \geqq 0. \tag{3.12}$$

The partial derivative $\frac{\partial A}{\partial e_{\alpha\beta}}$ in (3.10) is understood to have the symmetric form $\frac{1}{2}\left[\frac{\partial A}{\partial e_{\alpha\beta}} + \frac{\partial A}{\partial e_{\beta\alpha}}\right]$. A similar remark holds for all derivatives with respect to a symmetric second order tensor encountered in the rest of the paper. In view of (3.10) and (3.11), the energy equation reduces to

$$\varrho r - q^\alpha{}_{|\alpha} - \varrho T \dot{S} = 0. \tag{3.13}$$

The nonlinear constitutive equations (3.9) and (3.10) are valid for an elastic Cosserat surface which is anisotropic in some preferred state, usually taken to be the (initial) undeformed state. Detailed construction of the constitutive equations for an isotropic material with a center of symmetry in terms of joint invariants of the kinematic variables is discussed in [1].

We also recall here that under the conditions

$$D_\alpha = 0, \quad D_3 = 1, \quad d_\alpha = 0, \tag{3.14}$$

together with $L^3 = 0$ and the neglect of the director inertia in this direction and the assumption that the free energy is independent of $\lambda_{3\alpha} = d_{3,\alpha}$, a restricted theory can be deduced from the above equations which corresponds to one form of the classical theory of shells. The detailed development of this restricted theory of an elastic Cosserat surface (in which d_3 remains variable) can be found in [5, 6] and will not be given here.

In order to give an idea of the relationship between the theory of the elastic Cosserat surface characterized by the Eqs. (3.1) to (3.13) and the two dimensional theory of shells developed by Green, Laws and Naghdi [4], we consider a shell as a three dimensional continuum defined in Sec. 2 and recall the formulae

$$\boldsymbol{t} = \frac{n_i \boldsymbol{T}_i}{g^{1/2}}, \qquad \boldsymbol{T}_i = g^{1/2} \tau^{ij} \boldsymbol{g}_j, \qquad \boldsymbol{n} = n_i \boldsymbol{g}^i = n^i \boldsymbol{g}_i, \tag{3.15}$$

where $\boldsymbol{t}$ is the stress vector across a surface whose outward unit normal is $\boldsymbol{n}$, τ^{ij} is the symmetric contravariant stress tensor, $\boldsymbol{g}_i$, $\boldsymbol{g}^i$, g_{ij} are defined in (2.1) and $g = \det(g_{ij})$. If we admit the Maclaurin series representation

$$\boldsymbol{r}^* = \boldsymbol{r}(x^1, x^2, t) + \sum_{N=1}^{\infty} (x^3)^N \boldsymbol{d}_N \tag{3.16}$$

and assume that (3.16) is continuously differentiable, then by $(2.1)_2$ we have

$$\begin{aligned} \boldsymbol{g}_\alpha &= \boldsymbol{a}_\alpha + \sum_{N=1}^{\infty} (x^3)^N \frac{\partial \boldsymbol{d}_N}{\partial x^\alpha}, \qquad \boldsymbol{a}_\alpha = \frac{\partial \boldsymbol{r}}{\partial x^\alpha}, \\ \boldsymbol{g}_3 &= \sum_{N=1}^{\infty} N (x^3)^{N-1} \boldsymbol{d}_N, \end{aligned} \tag{3.17}$$

where $\boldsymbol{d}_N$ are vector functions of x^1, x^2 and t. The vectors $\boldsymbol{d}_N$ possess the property that they remain invariant in length when the shell is subjected to superposed rigid body motions and hence they may be called directors. Let ϱ^* be the mass density of the shell. Then, we define a mass per unit area by the relation

$$\varrho a^{1/2} = \int_{\alpha}^{\beta} \varrho^* g^{1/2} \, dx^3. \tag{3.18}$$

We also introduce the definitions

$$\begin{aligned} \boldsymbol{N}^\alpha a^{1/2} &= \int_{\alpha}^{\beta} \boldsymbol{T}_\alpha \, dx^3, \qquad \boldsymbol{N} = \boldsymbol{N}^\alpha \nu_\alpha, \\ \boldsymbol{M}^{\alpha N} &= \int_{\alpha}^{\beta} \boldsymbol{T}_\alpha (x^3)^N \, dx^3, \qquad \boldsymbol{M}^N = \boldsymbol{M}^{\alpha N} \nu_\alpha. \end{aligned} \tag{3.19}$$

With the foregoing background, a very brief outline of the theory of shells developed in [4] may be stated as follows. Starting with the three dimensional nonlinear theory of classical continuum mechanics and using the representation (3.16), definitions of the type (3.18) and (3.19), as well as a number of other considerations which include those for the thermodynamic variables, the three dimensional theory is reduced to a two dimensional theory[1] for an elastic shell (or plate). This two dimensional theory, however, consists of an infinite system of equations in an infinite number of unknowns. By introducing the approximation

$$\boldsymbol{g}_\alpha = \boldsymbol{a}_\alpha + x^3 \frac{\partial \boldsymbol{d}_1}{\partial x^\alpha}, \tag{3.20}$$

[1] In the development of the two dimensional theory an approximation is made for the temperature, but the rest of the theory is exact.

so that only the first two terms in the expansion $(3.17)_1$ are retained (but keeping the exact form for $\boldsymbol{g}_3$) and also neglecting certain body and inertia forces, an approximate theory of shells is obtained in [4] which is of the same form as that of the Cosserat surface. More specifically, if we put

$$\boldsymbol{d} = \boldsymbol{d}_1, \qquad \boldsymbol{M}^{\alpha 1} = \boldsymbol{M}^{\alpha} = M^{\alpha i}\boldsymbol{a}_i, \tag{3.21}$$

and if we identify ϱ in (3.18) with the mass density of the Cosserat surface, $\boldsymbol{N}$ in $(3.19)_2$ with that in (2.13), etc., then the theory of an elastic Cosserat surface discussed earlier in this Section can be brought into correspondence with the approximate theory of shells in [4].

4. The Linear Theory of an Elastic Cosserat Surface

In this section we confine attention to a particular form of the linearized theory of an elastic Cosserat surface in which the directors are initially coincident with the unit normals to the surface, i.e., when $\boldsymbol{D} = \boldsymbol{A}_3$ or equivalently

$$D_\alpha = 0, \quad D_3 = 1,$$

so that

$$\Lambda_{\beta\alpha} = -B_{\beta\alpha}, \quad \Lambda_{3\alpha} = 0. \tag{4.1}$$

Let

$$\boldsymbol{u} = \boldsymbol{r} - \boldsymbol{R}, \qquad \boldsymbol{u} = u^i \boldsymbol{A}_i, \qquad \boldsymbol{v} = \dot{\boldsymbol{u}}, \qquad \bar{\boldsymbol{\delta}} = \boldsymbol{d} - \boldsymbol{D}, \qquad \bar{\boldsymbol{\delta}} = \bar{\delta}_i \boldsymbol{A}^i, \qquad \boldsymbol{w} = \dot{\bar{\boldsymbol{\delta}}}. \tag{4.2}$$

The linear theory under discussion is such that $\boldsymbol{u}$, $\bar{\boldsymbol{\delta}}$, as well as their surface and time derivatives, are all infinitesimals of the first order. We also assume that the kinematical quantities and their derivatives, the curve and director forces (expressed in suitable non-dimensional form) and their derivatives, $(T - T_0)/T_0$ and $(S - S_0)/S_0$ (with T_0, S_0 as a standard temperature and entropy of the initial undeformed surface), are all infinitesimals of the first order.

Thus, under the condition (4.1) and when $\boldsymbol{u}$ and $\bar{\boldsymbol{\delta}}$ are infinitesimals of the first order, kinematic variables reduce to[1].

$$e_{\alpha\gamma} = \frac{1}{2}(u_{\alpha|\gamma} + u_{\gamma|\alpha}) - B_{\alpha\gamma} u_3 = e_{\gamma\alpha}, \tag{4.3}$$

[1] These results follow from Sec. 6 of Ref. [1] when D_i are specified by (4.1). Some aspects of the developments in this Section, with particular reference to a cylindrical Cosserat surface, are discussed in [12].

and

$$\begin{aligned} \varkappa_{\gamma\alpha} &= (\bar{\delta}_{\gamma|\alpha} - B_{\alpha\gamma}\bar{\delta}_3) - B_{\alpha\nu}(u^\nu{}_{|\gamma} - B_\gamma{}^\nu u_3), \\ \varkappa_{3\alpha} &= (\bar{\delta}_{3,\alpha} + B_\alpha{}^\gamma\bar{\delta}_\gamma) - B_{\gamma\alpha}\beta^\gamma, \end{aligned} \tag{4.4}$$

where

$$\begin{aligned} \delta_\alpha = d_\alpha = \bar{\delta}_\alpha - \beta_\alpha, \qquad \delta_3 = \bar{\delta}_3 = d_3 - 1, \\ \beta_\alpha = -(u_{3,\alpha} + B_\alpha{}^\gamma u_\gamma), \end{aligned} \tag{4.5}$$

and in (4.3) and (4.4) and throughout this Section a vertical line stands for covariant differentiation with respect to $A_{\alpha\beta}$ and a comma refers to partial differentiation with respect to the coordinates of the (initial) undeformed surface. With the help of (4.5), the kinematic variables in (4.4) may be put in a more revealing form, namely

$$\varkappa_{\alpha\gamma} = \varrho_{\alpha\gamma} - B_{\alpha\gamma}\delta_3, \qquad \varkappa_{3\alpha} = \varrho_{3\alpha} + B_\alpha{}^\gamma\delta_\gamma, \tag{4.6}$$

where

$$\begin{aligned} \varrho_{3\alpha} &= \delta_{3,\alpha}, \\ \varrho_{\alpha\gamma} &= \delta_{\alpha|\gamma} - [u_{3|\alpha\gamma} + B^\beta_{\alpha|\gamma}u_\beta + B_\gamma{}^\beta u_{\beta|\alpha} + B_\alpha{}^\beta u_{\beta|\gamma} - B_{\alpha\nu}B_\gamma{}^\nu u_3]. \end{aligned} \tag{4.7}$$

For future reference we also note that

$$\begin{aligned} \varrho_{(\alpha\gamma)} = \frac{1}{2}(\delta_{\alpha|\gamma} + \delta_{\gamma|\alpha}) - (u_{3|\alpha\gamma} + B^\beta_{\alpha|\gamma}u_\beta + B_\gamma{}^\beta u_{\beta|\alpha} \\ + B_\alpha{}^\beta u_{\beta|\gamma} - B_{\alpha\nu}B_\gamma{}^\nu u_3), \\ \varrho_{[\alpha\gamma]} = \frac{1}{2}(\delta_{\alpha|\gamma} - \delta_{\gamma|\alpha}), \end{aligned} \tag{4.8}$$

where the usual notation for symmetry and anti-symmetry in α, γ is used in (4.8).

To the order of approximation considered, in all of the basic field equations of Secs. 2 and 3, e. g., (2.12) to (2.19) and (3.5) to (3.9), all variables are now referred to the initial undeformed surface, covariant differentiation is with respect to the metric of the initial surface and $b_{\alpha\beta}$, $\lambda_{i\alpha}$, d_i in these equations must be replaced with $B_{\alpha\beta}$, $\Lambda_{i\alpha}$, D_i, respectively. In particular, Eqs. (2.15) and (2.16) now become

$$\begin{aligned} \boldsymbol{N} &= N^i\boldsymbol{A}_i = N^{\alpha i}\boldsymbol{A}_i\nu_\alpha \\ \boldsymbol{M} &= M^i\boldsymbol{A}_i = M^{\alpha i}\boldsymbol{A}_i\nu_\alpha \end{aligned} \tag{4.9}$$

where $\nu_\alpha = \boldsymbol{v}\cdot\boldsymbol{A}_\alpha$. For future convenience we record here the equations of motion of the linear theory which are given by

$$\begin{aligned} N^{\alpha\beta}{}_{|\alpha} - B_\alpha{}^\beta N^{\alpha 3} + \varrho_0 F^\beta &= \varrho_0 c^\beta, \\ N^{\alpha 3}{}_{|\alpha} + B_{\alpha\beta}N^{\alpha\beta} + \varrho_0 F^3 &= \varrho_0 c^3, \end{aligned} \tag{4.10}$$

and

$$M^{\alpha\beta}{}_{|\alpha} - B_\alpha{}^\beta M^{\alpha 3} + \varrho_0 \bar{L}^\beta = m^\beta,$$
$$M^{\alpha 3}{}_{|\alpha} + B_{\alpha\beta} M^{\alpha\beta} + \varrho_0 \bar{L}^3 = m^3, \tag{4.11}$$

together with

$$\varepsilon_{\beta\alpha}[N^{\alpha\beta} - M^{\gamma\beta} B_\gamma{}^\alpha] = 0, \qquad N^{\alpha 3} - m^\alpha - M^{\gamma 3} B_\gamma{}^\alpha = 0, \tag{4.12}$$

where ϱ_0 is the mass density of the initial surface. Also, the linearized energy equation becomes

$$\varrho_0 r - Q^\alpha{}_{|\alpha} - \varrho_0(T\dot{S} + \dot{T}S + \dot{A})$$
$$+ N'^{\alpha\beta}\dot{e}_{\alpha\beta} + M^{\alpha i}\dot{\varkappa}_{i\alpha} + m^i\dot{\delta}_i = 0, \tag{4.13}$$

where in (4.13) T is the temperature difference from an initial constant temperature T_0, Q^α are the components of the heat flux per unit length (in the undeformed surface) per unit time and $N'^{\alpha\beta}$ is now given by

$$N'^{\alpha\beta} = N'^{\beta\alpha} = N^{\alpha\beta} + M^{\gamma\alpha} B_\gamma{}^\beta. \tag{4.14}$$

From (4.7) we have

$$\dot{\varkappa}_{\alpha\beta} = \dot{\varrho}_{\alpha\beta} - B_{\alpha\beta}\dot{\delta}_3, \qquad \dot{\varkappa}_{3\alpha} = \dot{\varrho}_{3\alpha} + B_\alpha{}^\gamma \dot{\delta}_\gamma. \tag{4.15}$$

Using (4.15), (4.13) may be put in the form

$$\varrho_0 r - Q^\alpha{}_{|\alpha} - \varrho_0(T\dot{S} + \dot{T}S + \dot{A})$$
$$+ N'^{\alpha\beta}\dot{e}_{\alpha\beta} + M^{\alpha i}\dot{\varrho}_{i\alpha} + V^i\dot{\delta}_i = 0, \tag{4.16}$$

where $N'^{\alpha\beta}$ is given by (4.14) and

$$V^\alpha = m^\alpha + B_\gamma{}^\alpha M^{\gamma 3}, \qquad V^3 = m^3 - B_{\alpha\gamma} M^{\alpha\gamma}. \tag{4.17}$$

In view of the form of the energy equation (4.16), we now assume that the Helmholtz free energy is a function of T, $e_{\alpha\beta}$, $\varrho_{i\alpha}$ and δ_i, namely

$$A = A^*(e_{\alpha\beta}, \varrho_{i\alpha}, \delta_i, T). \tag{4.18}$$

We also assume that S, $N'^{\alpha\beta}$, $M^{\alpha i}$, V^i, Q^α are functions of the same variables[1] and that Q^α depends in addition on $T_{,\alpha}$. Then, as in (3.10) to (3.12), from the combination of the energy equation (4.16) and a local

[1] In all these constitutive assumptions, including (4.18), the dependence on the metric tensor $A_{\alpha\beta}$ is understood. In general A^* (as well as other constitutive functions) depends also on $B_{\alpha\beta}$, although this is not exhibited explicitly in (4.18).

entropy production inequality we can deduce the results $S = -\frac{\partial A^*}{\partial T}$,

$$N'^{\alpha\beta} = \varrho_0 \frac{\partial A^*}{\partial e_{\alpha\beta}}, \qquad M^{\alpha i} = \varrho_0 \frac{\partial A^*}{\partial \varrho_{i\alpha}}, \qquad V^i = \varrho_0 \frac{\partial A^*}{\partial \delta_i}, \tag{4.19}$$

as well as an inequality of the form (3.12) with q^α replaced by Q^α. In terms of the variables V^i defined by (4.17), the Eqs. (4.11) may be expressed as

$$M^{\alpha\beta}{}_{|\alpha} + \varrho_0 \bar{L}^\beta = V^\beta, \qquad M^{\alpha 3}{}_{|\alpha} + \varrho_0 \bar{L}^3 = V^3, \tag{4.20}$$

and $(4.12)_2$ becomes

$$N^{\alpha 3} = V^\alpha. \tag{4.21}$$

Hence, when the directors are initially coincident with the unit normals to the surface, the equations of motion of the linear theory are (4.10), (4.20), $(4.12)_1$ and (4.21).

For infinitesimal deformation of a Cosserat surface, which is initially homogeneous and isotropic with a center of symmetry, an explicit form for the free energy as a quadratic function of temperature and the kinematic variables $e_{\alpha\beta}$, δ_i and $\varkappa_{i\alpha}$ (instead of $\varrho_{i\alpha}$) has been given by GREEN, NAGHDI and WAINWRIGHT [1, Sec. 6]. For simplicity, in [1], attention was restricted to the values of the free energy which did not depend explicitly on D_i. However, since D_i as specified by (4.1) may be regarded as a surface vector ($D_\alpha = 0$) and a scalar ($D_3 = 1$), it is clear that D_i does not occur explicitly in the free energy. Also, as already noted in (4.1), in this case

$$\Lambda_{3\alpha} = 0 \quad \text{and} \quad \Lambda_{\beta\alpha} = -B_{\beta\alpha}.$$

In [1], it was assumed that the free energy does not depend explicitly on D_i and $\Lambda_{i\alpha}$. Here, in view of (4.1), we assume that $\varrho_0 A^*$ does not depend explicitly on $B_\beta{}^\alpha$ and limit further discussion to the case in which[1]

$$\begin{aligned} 2\varrho_0 A^* = {} & [\alpha_1 A^{\alpha\beta} A^{\gamma\delta} + \alpha_2 (A^{\alpha\gamma} A^{\beta\delta} + A^{\alpha\delta} A^{\beta\gamma})] e_{\alpha\beta} e_{\gamma\delta} \\ & + \alpha_3 A^{\alpha\beta} \delta_\alpha \delta_\beta + \alpha_4 (\delta_3)^2 + 2\alpha_4' \delta_3 T + \alpha_4'' T^2 \\ & + [\alpha_5 A^{\alpha\beta} A^{\gamma\delta} + \alpha_6 A^{\alpha\gamma} A^{\beta\delta} + \alpha_7 A^{\alpha\delta} A^{\beta\gamma}] \varrho_{\alpha\beta} \varrho_{\gamma\delta} \\ & + \alpha_8 A^{\alpha\beta} \varrho_{3\alpha} \varrho_{3\beta} + 2\alpha_9 A^{\alpha\beta} e_{\alpha\beta} \delta_3 \\ & + 2\alpha_9' A^{\alpha\beta} e_{\alpha\beta} T, \end{aligned} \tag{4.22}$$

where the coefficients $\alpha_1, \alpha_2, \ldots, \alpha_9$ are constants. With the use of (4.22) and the linearized version of (3.11), the constitutive relations for

---

[1] This form of free energy, for isothermal deformations, is similar to that used in the linear theory of a Cosserat plate [7].

an isotropic Cosserat surface (aside from an expression for entropy) are of the form

$$N'^{\alpha\beta} = [\alpha_1 A^{\alpha\beta} A^{\gamma\delta} + 2\alpha_2 A^{\alpha\gamma} A^{\beta\delta}] e_{\gamma\delta} + \alpha_9 A^{\alpha\beta} \delta_3 + \alpha_9' A^{\alpha\beta} T, \quad (4.23)$$

$$\begin{aligned} M^{(\alpha\beta)} &= [\alpha_5 A^{\alpha\beta} A^{\gamma\delta} + (\alpha_6 + \alpha_7) A^{\alpha\gamma} A^{\beta\delta}] \varrho_{(\gamma\delta)}, \\ M^{[\alpha\beta]} &= (\alpha_6 - \alpha_7) A^{\alpha\delta} A^{\beta\gamma} \varrho_{[\gamma\delta]}, \end{aligned} \quad (4.24)$$

$$M^{\alpha 3} = \alpha_8 A^{\alpha\gamma} \varrho_{3\gamma}, \quad (4.25)$$

$$V^\alpha = \alpha_3 A^{\alpha\gamma} \delta_\gamma, \qquad V^3 = \alpha_4 \delta_3 + \alpha_9 A^{\alpha\beta} e_{\alpha\beta} + \alpha_4' T. \quad (4.26)$$

Apart from linearization and the assumed form of the free energy in (4.22), the Eqs. (4.3) and (4.6) to (4.26) are exact and contain a contribution which corresponds to the effect of[1] "transverse shear deformation". In the case of an initially flat surface, these equations reduce to the equations of the linear theory of a Cosserat plate discussed in [7] and, in fact, separate into those for purely extensional and those for purely flexural motions. The above basic field equations and constitutive relations of the linear theory of an elastic Cosserat surface include several features which are ordinarily absent (or are accounted for only approximately) in the existing derivations of the linear theory of shells. Among these we mention the presence of (i) the components $\bar{\delta}_\alpha$ which represent the surface components of rotation of the normal $\boldsymbol{A}_3 (= \boldsymbol{D})$; (ii) the normal component δ_3 which may be regarded as the extension of the normal $\boldsymbol{A}_3 (= \boldsymbol{D})$; (iii) a normal couple ($M^{\alpha 3}$) and a normal force (V^3), each per unit length; (iv) constitutive relation for the anti-symmetric bending couple $M^{[\alpha\beta]}$; (v) constitutive relations for V^α which may be regarded as representing the effect of "transverse shear deformation"; and (vi) nine constitutive coefficients for the isothermal theory. It is perhaps of interest to indicate here the reduction of (4.23)—(4.26) to one form of the equations of the classical theory of shells corresponding to the restricted theory of a Cosserat surface mentioned in Sec. 3 [immediately after (3.14)]. Under the conditions (3.14), we must have $\delta_\alpha \to 0$ and $\alpha_3 \to \infty$ in such a way that V^α remains finite and undetermined by a constitutive equation. Also, $\alpha_8 \to 0$ and $V^3 = 0$. The latter determines δ_3 in terms of $e_{\alpha\beta} A^{\alpha\beta}$ and hence $N'^{\alpha\beta}$. Moreover, since $\delta_\alpha = 0$, the kinematic variable $\varrho_{\alpha\beta}$ becomes symmetric in α, β and $M^{\alpha\beta}$ can be expressed entirely in terms of $\varrho_{\alpha\beta}$ with $\alpha_6 = \alpha_7$.

---

[1] The effect of "transverse shear deformation" in the existing derivations of shell theory (from the three dimensional equations of linear elasticity) is accounted for only approximately. In this connection, the equations of this Section may be compared to those given in [13—15].

For convenience, in the rest of this Section we limit the discussion to the equilibrium of a Cosserat surface with isothermal deformation. Equations (4.23) to (4.26), in the absence of the coefficients α_4' and α_9', are among the simplest constitutive relations that can be provided for a general linear (isothermal) theory of a Cosserat surface and contain nine coefficients. Some of these coefficients may be identified by comparison of certain solutions of the above theory with corresponding known exact solutions of the three dimensional (classical) linear elasticity. For example, two of the coefficients $[\alpha_5, (\alpha_6 + \alpha_7)]$ may be identified by comparing the solution for pure bending of a Cosserat plate with a corresponding exact solution in the (three dimensional) linear theory of isotropic elasticity [7][1]. Also, two of the coefficients $[\alpha_2, (\alpha_6 + \alpha_7)]$ may be identified by comparison of the solution for torsion of a cylindrical Cosserat surface with a corresponding exact solution in the Saint-Venant theory of torsion for a hollow circular cylinder [12][1]. However, not all coefficients may be identified by such comparisons as the constitutive relations of the theory of a Cosserat surface (and plate) contain such coefficients as $(\alpha_6 - \alpha_7)$ which have no counterpart in (classical) three dimensional linear elasticity.

It follows from the above discussion that for an isotropic Cosserat surface, we may set

$$\begin{gathered} 2\alpha_2 = (1 - \nu) C, \\ \alpha_5 = \nu D, \quad (\alpha_6 + \alpha_7) = (1 - \nu) D, \end{gathered} \tag{4.27}$$

where $C = \dfrac{Eh}{1 - \nu^2}$, $D = \dfrac{Eh^3}{12(1 - \nu^2)}$, h is the thickness of the shell (or plate) in the three dimensional solution, E is Young's modulus and ν is Poisson's ratio. A further identification of the constitutive coefficients may be effected by considering the equations of the extensional theory of a Cosserat plate[2] in the light of known results from the (classical) three dimensional theory. If V^3 is regarded as an average value of τ_{33} and δ_3 as an average value of e_{33}, then reasonable values of the coefficients α_1, α_4 and α_9 are

$$\alpha_1 = \frac{\nu(1 - \nu)}{1 - 2\nu} C = \alpha_9, \qquad \alpha_4 = \frac{(1 - \nu)^2}{1 - 2\nu} C, \tag{4.28}$$

and these are consistent with $(4.27)_1$. The determination of the remaining constitutive coefficients in (4.23)—(4.26), namely α_3, α_8 and $(\alpha_6 - \alpha_7)$ requires additional consideration. At present, it appears that some light on this aspect of the subject may be shed by investigating some parti-

---

[1] In making such comparisons we should put $(\alpha_6 - \alpha_7) = 0$.

[2] Recall that with A^* in the form (4.22), the equations of an initially flat surface separate into those for purely flexural and purely extensional deformations.

cularly relevant solutions within the scope of the theory of Cosserat surface, e. g., the stress concentration problem of an infinite Cosserat plate studied in [7].

5. A Deformable Surface with Simple Force Dipole

We conclude the present paper with some remarks concerning a theory of deformable surface with simple force multipoles [3]. For simplicity, however, we limit the discussion to an elastic surface with simple force dipole.

In the theory of a deformable surface with simple force dipole, the basic kinematical ingredients are the velocity $\boldsymbol{v}$ and the velocity gradient $\boldsymbol{v}_{|\alpha}$ (in contrast to $\boldsymbol{v}$ and the director velocity $\boldsymbol{w}$ of the Cosserat surface in Secs. 2—3). We again begin the development of the theory with an energy balance similar in form to (2.10) but with some modifications. In particular, the terms involving $\boldsymbol{w}$ in the surface and line integrals, namely $\boldsymbol{L} \cdot \boldsymbol{w}$ and $\boldsymbol{M} \cdot \boldsymbol{w}$, are now replaced by $\boldsymbol{F}^\alpha \cdot \boldsymbol{v}_{|\alpha}$ and $\boldsymbol{T}^\alpha \cdot \boldsymbol{v}_{|\alpha}$, respectively. In these (scalar) rate of work terms, $\boldsymbol{T}^\alpha$ is a simple dipolar force per unit length and $\boldsymbol{F}^\alpha$ is an assigned simple dipolar force per unit mass defined in a manner similar to $\boldsymbol{M}$ and $\boldsymbol{L}$ in Sec. 2. When referred to the base vectors $\boldsymbol{a}_i$, $\boldsymbol{T}^\alpha$ and $\boldsymbol{F}^\alpha$ can be expressed as

$$\boldsymbol{T}^\alpha = T^{\alpha i}\boldsymbol{a}_i , \qquad \boldsymbol{F}^\alpha = F^{\alpha i}\boldsymbol{a}_i . \tag{5.1}$$

Also in (2.11), $\alpha \boldsymbol{w} \cdot \boldsymbol{w}$ should be replaced by $y^{\alpha\beta}\boldsymbol{v}_{|\alpha} \cdot \boldsymbol{v}_{|\beta}$, $y^{\alpha\beta}$ being a function of surface coordinates only.

It should be clear that Eqs. $(2.12)_1$, (2.13), $(2.15)_1$, $(2.16)_1$, $(2.17)_1$ and (2.18) still hold for a deformable surface with simple force dipole. But, instead of (2.14) $(2.15)_2$ and (2.17), we now have[1]

$$\boldsymbol{T}^\alpha = \boldsymbol{T}^{\alpha\eta}\nu_\eta , \qquad \boldsymbol{T}^{\alpha\eta} = \boldsymbol{t}'^{\alpha\eta}(a^{\eta\eta})^{1/2} , \tag{5.2}$$

$$\boldsymbol{T}^{\alpha\eta} = T^{\alpha\eta i}\boldsymbol{a}_i , \qquad T^{\alpha i} = T^{\alpha\eta i}\nu_\eta , \tag{5.3}$$

where $\boldsymbol{t}'^{\alpha\eta}$ is the (physical) force dipole vector across the x^α-curves on s.

The equations of motion (3.5) still hold and the remaining basic equations for an elastic surface with simple force dipole are

$$\begin{aligned} G^{\alpha\lambda} &= N^{\alpha\lambda} + T^{\alpha\eta\lambda}{}_{|\eta} - b_\eta{}^\lambda T^{\alpha\eta 3} + \varrho \bar{F}^{\alpha\lambda} , \\ G^{\alpha 3} &= N^{\alpha 3} + T^{\alpha\eta 3}{}_{|\eta} + b_{\eta\lambda} T^{\alpha\eta\lambda} + \varrho \bar{F}^{\alpha 3} , \end{aligned} \tag{5.4}$$

[1] Previously in [3], instead of $\boldsymbol{F}^\alpha$ and $\boldsymbol{T}^\alpha$, the notations $\boldsymbol{f}^\alpha$ and $\boldsymbol{n}^\alpha$ were employed. Also, the order of indices in $T^{\alpha\eta i}$ differs from the corresponding quantity in [3], namely $N^{i\eta\alpha}$, but is in agreement with the more usual notation in shell theory.

and

$$G^{[\alpha\lambda]} = 0, \qquad G^{\lambda 3} - b_{\alpha\beta} T^{\alpha\beta\lambda} = 0, \tag{5.5}$$

where $\overline{F}^{\alpha i} = \overline{\boldsymbol{F}}^{\alpha} \cdot \boldsymbol{a}_i$ are the components of $\overline{\boldsymbol{F}}^{\alpha} = \boldsymbol{F}^{\alpha} - y^{\alpha\beta} (\dot{\overline{\boldsymbol{v}_{|\beta}}})$ (the difference of the assigned dipolar force and the inertia terms due to the velocity gradient), and the quantities $G^{\alpha i}$ (for which in general constitutive equations are required) are defined through (5.4). The local energy equation [corresponding to (3.8)] is

$$\varrho r - q^{\alpha}{}_{|\alpha} - \varrho (T\dot{S} + \dot{T}S + \dot{A}) + G^{(\alpha\beta)} \dot{e}_{\alpha\beta} + T^{(\alpha\beta) i} \eta_{i(\alpha\beta)} = 0, \tag{5.6}$$

where $\eta_{3(\alpha\beta)} = \dot{b}_{\alpha\beta}$ and $\eta_{\lambda(\alpha\beta)}$ may be expressed in terms of $(\dot{\overline{A_{\alpha\beta|\lambda}}})$. After introducing suitable constitutive assumptions and by combining (5.6) with an entropy production inequality, we can then deduce constitutive equations for S, $G^{(\alpha\beta)}$ and $T^{(\alpha\beta)i}$ in terms of A. We do not, however, record these results here and refer the reader to [3, Sec. 7].

In the above theory, the rate of work terms $\boldsymbol{T}^{\alpha} \cdot \boldsymbol{v}_{|\alpha}$ and $\boldsymbol{F}^{\alpha} \cdot \boldsymbol{v}_{|\alpha}$ in the balance of energy may be written in the form

$$\boldsymbol{T}^{\alpha} \cdot \boldsymbol{v}_{|\alpha} = (T^{\alpha i} \boldsymbol{a}_i) \cdot (v_{j;\alpha} \boldsymbol{a}^j) = T^{\alpha j} v_{j;\alpha} \tag{5.7}$$

with a similar expression for $\boldsymbol{F}^{\alpha} \cdot \boldsymbol{v}_{|\alpha}$, where $\boldsymbol{v} = v_i \boldsymbol{a}^i$ and the notation $v_{j;\alpha}$ stands for

$$\begin{aligned} v_{\lambda;\alpha} &= \boldsymbol{v}_{|\alpha} \cdot \boldsymbol{a}_\lambda = v_{\lambda|\alpha} - b_{\alpha\lambda} v_3, \\ v_{3;\alpha} &= \boldsymbol{v}_{|\alpha} \cdot \boldsymbol{a}_3 = v_{3,\alpha} + b_{\alpha}{}^{\lambda} v_\lambda. \end{aligned} \tag{5.8}$$

If instead of full velocity gradient $\boldsymbol{v}_{|\alpha}$ we only admit the component $v_{3;\alpha}$ [defined by $(5.8)_2$] as a basic kinematical ingredient, then the original balance of energy will contain $T^{\alpha 3} v_{3;\alpha}$ and $F^{\alpha 3} v_{3;\alpha}$ instead of $T^{\alpha j} v_{j;\alpha}$ and $F^{\alpha j} v_{j;\alpha}$. Thus, in the absence of $T^{\alpha\lambda}$ and $F^{\alpha\lambda}$ and keeping (5.1)—(5.3) in mind, the equations of motion (5.4) and (5.5) simplify and will not include terms involving $T^{\alpha\eta\lambda}$ and $\overline{F}^{\alpha\lambda}$. Similarly, on the left-hand side of the energy equation (5.6), in place of $T^{(\alpha\beta)i} \eta_{i(\alpha\beta)}$ we will have $T^{(\alpha\beta)3} \dot{b}_{\alpha\beta}$. If we identify $T^{(\alpha\beta)3}$ with $M^{(\alpha\beta)}$, the resulting theory would then be similar to the restricted theory of a Cosserat surface mentioned in Sec. 3. Both the theory of a Cosserat surface and the theory of a deformable surface with full simple force dipoles contain features not present in the restricted theory of a Cosserat surface or the similar theory of a deformable surface with only v_i and $v_{3;\alpha}$ as kinematic ingredients. The additional features of the two theories are, however, somewhat different in character.

References

1. Green, A. E., Naghdi, P. M., and W. L. Wainwrihgt: Arch. Rat. Mech. Anal. **20**, 287 (1965).
2. Cohen, H., and C. N. DeSilva: J. Math. Phys. **7**, 960 (1966).

3. Balaban, M. M., Green, A. E., and P. M. Naghdi: J. Math. Phys. 8, 1026 (1967).
4. Green, A. E., Laws, N., and P. M. Naghdi: Proc. Cambridge Philos. Soc. **64** (1968) to appear.
5. Green, A. E., and P. M. Naghdi: Quart. J. Mech. Appl. Math. **21**, 135 (1968).
6. Green, A. E., and P. M. Naghdi: Proc. IUTAM Symp. on the "Generalized Cosserat continuum and the theory of dislocations with applications", (Freudenstadt and Stuttgart, August 1967), Berlin/Heidelberg/New York: Springer, to appear.
7. Green, A. E., and P. M. Naghdi: Proc. Cambridge Philos. Soc. **63**, 537 and 922 (1967).
8. Green, A. E., and P. M. Naghdi: Quart. J. Mech. Appl. Math. **20**, 183 (1967).
9. Koiter, W. T.: Proceedings of the IUTAM Symposium on the Theory of Thin Elastic Shells (Delft 1959), North-Holland 1960, 12.
10. Green, A. E., and P. M. Naghdi: Quart. J. Mech. Appl. Math. **18**, 257 (1965).
11. Naghdi, P. M.: Proceedings of the Eleventh Intern. Congress Appl. Mech. (Munich 1964), Berlin/Heidelberg/New York: Springer 1966, p. 262.
12. Naghdi, P. M.: Proceedings — Symposium on the theory of shells, L. H. Donnell Anniversary Volume, (Houston 1966), McCutchan 1967, 25.
13. Green, A. E., and W. Zerna: Quart. J. Mech. Appl. Math. **3**, 1 (1950).
14. Reissner, E.: J. Math. and Phys. **31**, 109 (1952).
15. Naghdi, P. M.: Quart. Appl. Math. **14**, 369 (1957).

Discussion

M. Roy: I don't intend to discuss in any way the foundations of this very interesting theoretical work and would only ask two questions for my information.

These two questions refer to the equation of energy.

1. In the energy equation the term h relates to the heat flux through the line which constitutes the boundary of the surface. How is the heat flux through the surface itself taken account of?

2. Which is the connection between the director (vector $\boldsymbol{d}$) and the physical state of the two-dimensional continuum, as this state determines the internal energy, entropy, etc. of the medium concerned?

P. M. Naghdi: 1. There is a term in the energy equation allowing for contributions of heat sources distributed over the surface.

2. The kinematical variables involve $\boldsymbol{d}$ and its derivatives and the internal energy (or the Helmoltz free energy) are functions of kinematic variables and entropy (or temperature) as well as the first and the second fundamental forms of the initial surface.

F. John: In embedding your theory as you did at the end in a three-dimensional one, do you compare the differential equations eventually or do you have a way of identifying the expressions for energy?

P. M. Naghdi: In making a comparison of the theory of Cosserat surface with the three-dimensional theory for a plate we compare solutions for a flat plate and torsion of a cylindrical surface which are exact in both theories. In addition we compare extensional stress-strain relations with averages of stress-strain relations in three dimensions. The differential equations of motion or equlibrium are exact.

W. T. KOITER: The interesting presentation of the paper by Professors GREEN and NAGHDI raises the question whether this theory contributes to a better understanding of shell theory or, perhaps, conversely that its merits lie more specifically in attaching a more definite meaning to the theory of Cosserat surfaces, by interpreting it in terms of shell theory. I am inclined to take the second view, but I agree that there is considerable room here for divergent opinions.

P. M. NAGHDI: The theory of Cosserat surface is exact, but shell theory derived from the three-dimensional equations is approximate. It may be a matter of taste, but we prefer to regard an exact theory as more fundamental. The Cosserat theory for shells is on a comparable footing with any exact three-dimensional continuum theory.

H. S. RUTTEN: I am wondering if it makes sense to introduce explicit assumptions regarding the constitutive formulae governing the deformations and forces and moments within the theory of Cosserat media independently of the classical continuum theory; and, to compare thereafter the ultimate results to show the correspondance in specific cases.

P. M. NAGHDI: Assumptions about the energy function can certainly be made independently of classical continuum mechanics and results can then be compared in specific cases. This has already been done to some extent for a plate.

On one Version of the Consistent Theory of Elastic Shells

By

I. N. Vekua

University of Tbilisi, Tbilisi, USSR

I

The lack of the proper agreement between physical and kinematic assumptions forming the basis of the classical moment theory of elastic shells explains a certain incompleteness of the corresponding shell equilibrium differential equations. According to the basic physical assumption, the stress resultants and couples are to be defined instead of stresses distributed continuously inside the shell. This assumption leads to the five independent physical boundary conditions — the tangent normal and shearing forces, the transverse shearing forces, the bending and twisting moments are to be arbitrarily given on the boundary beforehand. But the classical shell theory equations disagree with the boundary conditions mentioned above.

The equilibrium equations for stress resultants and couples do not allow to define all sought quantities. Therefore one has to consider some kinematic model of deformation to derive the corresponding shell equilibrium equations. But great difficulties are encountered at an attempt to formulate the optimal kinematic model of strain which would agree with the basic physical assumption. The optimality requirement means that the system of the shell equilibrium equations has to have a unique solution, satisfying the arbitrarily given five physical boundary conditions mentioned above.

The conventional kinematic Kirchhoff-Love assumptions used widely in classical treatieses do not satisfy this requirement. The corresponding elliptic system of equations is of the eighth order and it secures the realization of only four out of five physical boundary conditions. Attempts to substantiate the reduction of the five conditions to four by means of mechanical considerations cannot be thought as logically faultless. That is why the further perfection of the classical moment theory is important and proceeds very intensively today. (E. Reissner, A. E. Green,

W. I. Koiter, F. Johns, O. K. Friedrichs, P. N. Naghdi, A. L. Gol'denweizer).

Speaking on the shortcomings of the classical moment theory and the necessity of its perfection we do not mean that it should be rejected completely. One should take into account the fact that many conclusions of the classical moment theory are sufficiently well confirmed in practice. Therefore it is not desirable to go far away from the conventional constructions. At first we are to try to improve the classical theory making the necessary corrections without any essential analytical complications.

The shell theory is an applied science and its main object consists in elaboration of several approximate methods to solve different problems of elastic shells. Classifying the problems of the theory of elasticity by means of various physical, kinematic or geometric factors one must construct the corresponding version of the shell theory. But each version of the theory must satisfy two main requirements. The first one consists in security of inner consistency of the theory and the second necessary requirement is that the theory must have the sufficiently wide range of practical applications. One must also try to make analytical difficulties of the theory as small as possible. The extraordinary complexity of the mathematical theory depreciates to a great extent its practical significance. But this does not mean that one should refuse from the construction of the consistent mathematical theory only because of its complexity. It is better to have a complicated theory than no theory at all the more so that a strong mathematical theory may be sometimes simplified maintaining the requirement of inner consistency. I venture to say that the theory which will be proposed below satisfies all these requirements.

II

One of the conventional methods of shell theory constructions going back to Poisson and Cauchy is the method of expansion of the sought solutions in the power series over the distance between the point and the middle surface. Having taken this method in a modified form as the basis, the author developed a special approach to study elastic shell problems [1, 2, 3, 4]. This approach does not use the assumptions of the classical shell theory. The sought functions are represented approximately as polynomials of some power N on the coordinate x^3 where $|x^3|$ is the distance of the considered point to the middle surface ($-h \leq x^3 \leq h$, $2h$ is the shell thickness). The functions to be determined may be expressed in the following form

$$f(x^1, x^2, x^3) = \sum_{k=0}^{N} f_k(x^1, x^2) P_k\left(\frac{x^3}{h}\right)$$

where P_k are Legendre's polynomials. The non-negative integer N is called the order of approximation. To each value of N corresponds its own version of the consistent ellastic shell theory which agrees with the natural (for a considered version) boundary conditions [4].

The version corresponding to the approximations of the order of $N = 1$ is the closest to the classical shell theory. Therefore below we discuss just this particular case of the general theory.

III

In case $N = 1$ the stress forces on a cross section Σ_l with the unit normal $\boldsymbol{l}$ are distributed according to the formula

$$\boldsymbol{P}_{(l)} = \frac{1}{2h} \boldsymbol{T}_{(l)} + \frac{3x^3}{2h^3} \boldsymbol{S}_{(l)} \tag{1}$$

where

$$\boldsymbol{T}_{(l)} = T^{\alpha\beta} l_\alpha \boldsymbol{r}_\beta + T^\alpha l_\alpha \boldsymbol{n}, \qquad \boldsymbol{S}_{(l)} = S^{\alpha\beta} l_\alpha \boldsymbol{r}_\beta + S^\alpha l_\alpha \boldsymbol{n}. \tag{2}$$

Here $\boldsymbol{r}_1, \boldsymbol{r}_2$ are the basic vectors of the coordinate system on the middle surface F, $l_\alpha = \boldsymbol{l}\,\boldsymbol{r}_\alpha$ $(\alpha = 1, 2)$, $\boldsymbol{n}$ is the unit normal to the surface F (see Fig. 1), $T^{\alpha\beta}$, $S^{\alpha\beta}$ and T^α, S^α are contravariant tensors of the 2nd and 1st order respectively belonging to the middle surface.

Formula (1) may be also written in the form

$$\boldsymbol{P}_{(l)} = \frac{1}{2h} \boldsymbol{T}_{(l)} + \frac{3x^3}{2h^3} \boldsymbol{M}_{(l)} \times \boldsymbol{n} + \frac{2x^3}{h^2} Q_{(l)} \boldsymbol{n} \tag{3}$$

where

$$\boldsymbol{M}_{(l)} = \boldsymbol{n} \times \boldsymbol{S}_{(l)}, \qquad Q_{(l)} = \frac{3}{4h} \boldsymbol{n} \boldsymbol{S}_{(l)}. \tag{4}$$

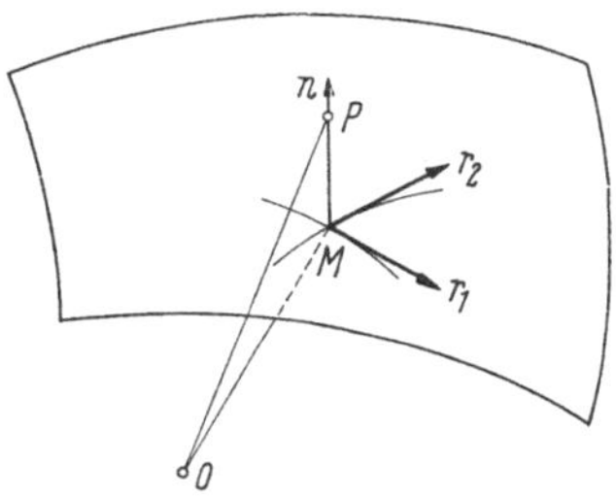

Fig. 1. The space reference frame x^1, x^2, x^3 normally connected with the middle surface.

The suggested model of the stress distribution on Σ_l differs from classical one due to the presence of the third term in the right side of (3). This term expresses a self-equilibrated system of forces acting on Σ_l. For shells having essential rigidity the mechanical effect of their action is approximately the same as the action of the transverse couple $(Q_{(l)}\boldsymbol{n}, -Q_{(l)}\boldsymbol{n})$. The colinear transverse forces of this couple are applied to the points situated symmetrically with respect to the middle surface. The action of these couples causes extensions or contractions of the transverse fibres of the shell; in the former case $Q_{(l)} > 0$ and it seems to split the shell promoting formation of longitudinal internal cracks. Therefore one can call $Q_{(l)}$ the cracking or splitting force. Further we shall use this term in the both cases $Q_{(l)} \geq 0$ or $Q_{(l)} < 0$.

Therefore according to the new theory the stress resultants $\boldsymbol{T}_{(l)}$, the stress couples $\boldsymbol{M}_{(l)}$ and also the self-equilibrated transverse couples $(Q_{(l)}\boldsymbol{n}, -Q_{(l)}\boldsymbol{n})$ act on any cross section Σ_l (see Fig. 2).

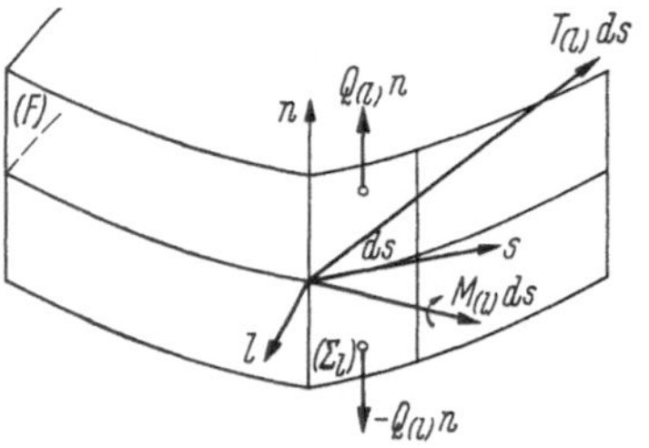

Fig. 2. The distribution of stress resultants and couples on the cross section.

IV

Further we shall limit ourselves with the consideration of thin or shallow shells with constant thickness. In this case one can assume that

$$T^{\alpha\beta} = T^{\beta\alpha} = \int_{-h}^{h} P^{\alpha\beta} dx^3, \qquad T^{\alpha} = \int_{-h}^{h} P^{\alpha 3} dx^3 \tag{5}$$

$$S^{\alpha\beta} = S^{\beta\alpha} = \int_{-h}^{h} P^{\alpha\beta} x^3 \, dx^3, \qquad S^{\alpha} = \int_{-h}^{h} P^{\alpha 3} x^3 \, dx^3 \tag{6}$$

where P^{ik} are the contravariant components of the stress tensor with respect to the space reference frame x^1, x^2, x^3. Therefore $T^{\alpha\beta}$ and $S^{\alpha\beta}$ are symmetric tensors belonging to the middle surface.

Using the system of the continuous medium equilibrium equations one can easily prove that the contravariant stress resultant tensors $T^{\alpha\beta}$, T^{α}, $S^{\alpha\beta}$ and S^{α} satisfy the system of equations [4]

$$\begin{aligned}
&\nabla^{\alpha} T^{\alpha\beta} - b_{\alpha}^{\ \beta} T^{\alpha} + X^{\beta} = 0 \quad (\beta = 1, 2)\\
&\nabla_{\alpha} T^{\alpha} + b_{\alpha\beta} T^{\alpha\beta} + X = 0\\
&\nabla_{\alpha} S^{\alpha\beta} - b_{\alpha}^{\ \beta} S^{\alpha} - T^{\beta} = 0 \quad (\beta = 1, 2)\\
&\nabla_{\alpha} S^{\alpha} + b_{\alpha\beta} S^{\alpha\beta} - T = 0
\end{aligned} \tag{7}$$

where

$$T = \int_{-h}^{h} P^{33} \, dx^3. \tag{8}$$

The symbols of the covariant derivatives are constructed with the help of the metric tensor of the middle surface, $b_{\alpha\beta}$ are the coefficients of the second fundamental quadratic form of the middle surface F and X^α, X are the components of the reduced external load

$$\boldsymbol{X} = \boldsymbol{X}_+ + \boldsymbol{X}_- + \int_{-h}^{h} \boldsymbol{F} dx^3 = X^\alpha \boldsymbol{r}_\alpha + X\boldsymbol{n} \tag{9}$$

where $\boldsymbol{X}_+$ and $\boldsymbol{X}_-$ are the forces acting on the face surfaces $x^3 = \pm h$ of the shell and $\boldsymbol{F}$ is the mass force. In addition we assume that the moments of the external forces equal to zero.

The quantity T given by formula (8) may be called the transverse force. In the classical shell theory this quantity is neglected assuming that $T = 0$ identically. But we shall not neglect it to construct the more complete theory of shells.

V

Our further considerations will be based on the kinematic model of deformation according to which the displacement field is given by the formula

$$\boldsymbol{U}(x^1, x^2, x^3) = \boldsymbol{u}(x^1, x^2) + x^3 \boldsymbol{v}(x^1, x^2) \tag{10}$$

where $\boldsymbol{u}$ and $\boldsymbol{v}$ ate the vector fields on the middle surface. For the strain tensor components we shall obtain the expressions [4]

$$e_{\alpha\beta} = \varepsilon_{\alpha\beta} + x^3 \eta_{\alpha\beta}, \qquad e_{\alpha 3} = \varepsilon_\alpha + x^3 \eta_\alpha, \qquad e_{33} = v \tag{11}$$

where

$$\varepsilon_{\alpha\beta} = \frac{1}{2} (\nabla_\alpha u_\beta + \nabla_\beta u_\alpha) - b_{\alpha\beta} u$$

$$\varepsilon_\alpha = \frac{1}{2} (\nabla_\alpha u + b_\alpha^{\ \beta} u_\beta + v_\alpha$$

$$\eta_{\alpha\beta} = \frac{1}{2} (\nabla_\alpha v_\beta + \Delta_\beta b_\alpha) - b_{\alpha\beta} v$$

$$\eta_\alpha = \frac{1}{2} (\nabla_\alpha v + b_\alpha^{\ \beta} v_\beta), \tag{12}$$

Here $u_\alpha = \boldsymbol{u}\boldsymbol{r}_\alpha$, $u = \boldsymbol{u}\boldsymbol{n}$, $v_\alpha = \boldsymbol{v}\boldsymbol{r}_\alpha$, $v = \boldsymbol{v}\boldsymbol{n}$; u_α, v_β are the covariant vectors and u, v are scalars belonging to the middle surface.

The expression $u(x^1, x^2) + x^3 v(x^1, x^2)$ is the normal displacement of the point (x^1, x^2, x^3). Therefore $(u(x^1, x^2)$ is the normal deflection of the point (x^1, x^2) of the middle surface and $v(x^1, x^2)$ is the elongation of the transverse shell fibre passing through this point. Formulae (11) reveal the kinematic meanings of tensors $\varepsilon_{\alpha\beta}$, ε_α, $\eta_{\alpha\beta}$, η_α.

The cubical extension is expressed by the formula

$$\Theta = e_i^i = \varepsilon_\nu^\nu + v + x^3 \eta_\nu^\nu \tag{13}$$

where

$$\varepsilon_\nu^\nu = \nabla_\nu u^\nu - 2Hu = \frac{1}{\sqrt{a}} \frac{\partial \sqrt{a}\, u^\nu}{\partial x^\nu} - 2Hu$$

$$\eta_\nu^\nu = \nabla_\nu v^\nu - 2Hv = \frac{1}{\sqrt{a}} \frac{\partial \sqrt{a}\, v^\nu}{\partial x^\nu} - 2Hv. \tag{14}$$

Here H is the mean curvature of the middle surface, a is the discriminant of the first fundamental (metric) quadratic form of the middle surface ($a = a_{11} a_{22} - a_{12}^2 > 0$, $a_{\alpha\beta} = \boldsymbol{r}_\alpha \boldsymbol{r}_\beta$). Formula (13) reveals the geometrical (kinematic) meanings of the invariants ε_ν^ν and η_ν^ν.

The rotation of an element of the surface $x^3 = \text{const}$ is expressed by the formula

$$\omega = \omega_0 + x^3 \varrho, \qquad \omega_0 = \frac{1}{2} (\nabla_1 u_2 - \nabla_2 u_1), \qquad \dot{\varrho} = \frac{1}{2} (\nabla_1 v_2 - \nabla_2 v_1). \tag{15}$$

It is obvious that $\nabla_\alpha u_\beta = \varepsilon_{\alpha\beta} + \omega_{\alpha\beta}$ where $\omega_{\alpha\beta} = \frac{1}{2} (\nabla_\alpha u_\beta - \nabla_\beta u_\alpha)$. These formulae reveal the kinematic meanings of the covariant derivatives of the covariant vector u_α. Similarly we can find the kinematic meanings of $\nabla_\alpha v_\beta$. The quantity ϱ is the twist with respect to the middle surface of a cross prismatic element of the shell.

VI

For homogeneous isotropic elastic shells, according to Hooke's law, one can derive the following stress-strain relations:

$$\begin{aligned} T^{\alpha\beta} &= 2h[\lambda(\varepsilon_\nu^\nu + v)\, a^{\alpha\beta} + 2\mu \varepsilon^{\alpha\beta}] \\ T^\alpha &= 4h\mu\varepsilon^\alpha \\ T &= 2h[\lambda\varepsilon_\nu^\nu + (\lambda + 2\mu) v] \\ S^{\alpha\beta} &= \frac{2h^3}{3} [\lambda(\eta_\nu^\nu)\, a^{\alpha\beta} + 2\mu\eta^{\alpha\beta}] \\ S^\alpha &= \frac{4h^3}{3} \mu\eta^\alpha \end{aligned} \tag{16}$$

where λ and μ are Lame's constants,

$$\lambda = \frac{E\sigma}{(1+\sigma)(1-2\sigma)}, \qquad \mu = \frac{E}{2(1+\sigma)} \tag{17}$$

Here E is Young's modulus and σ is Poisson's ratio.

Substitution of these relations into Eqs. (7) leads to the elliptic system of equations of the 12th order. We have mentioned above that the classical shell theory based on Kirchhoff-Love's assumptions leads to the elliptic system of the 8th order.

VII

For any solution of the above-obtained system of equations one can prove the following integral identity [4] (see Fig. 3)

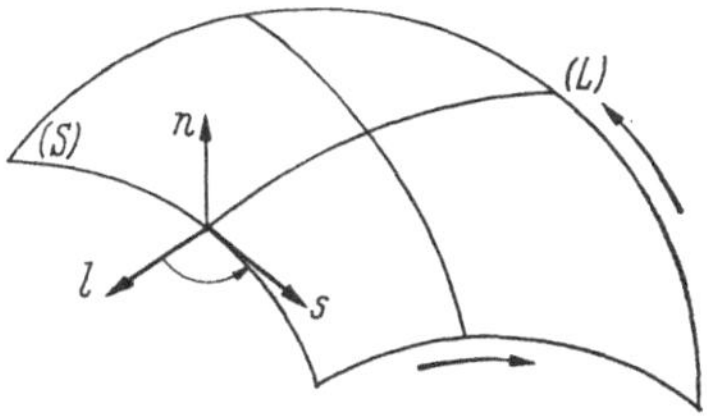

Fig. 3. The middle surface and its boundary with the positive direction of rounds.

$$\int_L (\boldsymbol{u} \boldsymbol{T}_{(l)} + \boldsymbol{v} \boldsymbol{S}_{(l)})\, ds = - \iint_F (X u + X^\alpha u_\alpha)\, dF + I(\boldsymbol{u}, \boldsymbol{v}) \qquad (18)$$

where L is the boundary of the shell, $\boldsymbol{l}$ is the unit tangent normal to L and

$$I(\boldsymbol{u}, \boldsymbol{v}) = \iint_F \{2h[\lambda(\varepsilon_\nu^\nu + v)^2 + 2\mu \varepsilon^{\alpha\beta} \varepsilon_{\alpha\beta} + 4\mu \varepsilon^\alpha \varepsilon_\alpha + 2\mu v^2] + \\ + \frac{2h^3}{3} [\lambda(\eta_\nu^\nu)^2 + 2\mu \eta^{\alpha\beta} \eta_{\alpha\beta} + 4\mu \eta^\alpha \eta_\alpha]\}\, dF. \qquad (19)$$

The non-negative integral functional $I(\boldsymbol{u}, \boldsymbol{v})$ may be considered as the potential energy of an elastic shell (for the case of the approximation of the order of $N = 1$).

It is easy to prove that the elastic shell equilibrium equations obtained above are Euler equations of the variational problem for the functional

$$I(\boldsymbol{u}, \boldsymbol{v}) - \iint_F (X u + X^\alpha u_\alpha)\, dF. \qquad (20)$$

If $I(\boldsymbol{u}, \boldsymbol{v}) = 0$ then

$$\varepsilon_{\alpha\beta} = 0, \qquad \varepsilon_\alpha = 0, \qquad \eta_{\alpha\beta} = 0, \qquad \eta_\alpha = 0, \qquad v = 0. \qquad (21)$$

To find the general solution of Eqs. (21) we ought to consider three different cases:

1. The middle surface F is not a minimal one, i.e. the mean curvature H does not vanish identically.

2. The surface F is a minimal one, but it does not coincide with the plane, i.e. $H = 0$, but the principal curvatures k_1 and k_2 do not vanish identically ($k_1 = -k_2 \neq 0$).

3. The surface F is a plane.

A comparatively simple investigation shows that the general solution of the set of Eqs. (21) is expressed by the following formulae [4].

In the case 1°

$$\boldsymbol{u} = \boldsymbol{u}_0, \qquad \boldsymbol{v} = 0 \tag{22}$$

where $\boldsymbol{u}_0$ denotes (here and below) an arbitrary constant vector field on F. If we add to the Eqs. (21) the requirement that $\boldsymbol{u}$ vanishes at some fixed point M_0 belonging to F or L then we will find that $\boldsymbol{u} = \boldsymbol{v} = 0$ everywhere.

In the case 2°

$$\boldsymbol{u} = c_0 \int\limits_{M_0 M} M \times dr + \boldsymbol{u}_0, \qquad \boldsymbol{v} = 0 \tag{23}$$

where c_0 is an arbitrary constant; the curvilinear integral does not depend on the way of integration if the middle surface F is single connected. If we add to the Eqs. (21) the requirements that $\boldsymbol{u}$ and $\partial \boldsymbol{u}/\partial \boldsymbol{s}$ vanish at some fixed point M_0 belonging to F or L then we will find that $\boldsymbol{u} = \boldsymbol{v} = 0$ everywhere.

In the case 3°

$$\boldsymbol{u} = c_0 \boldsymbol{n} \times \boldsymbol{r} + (\boldsymbol{v}_0 \boldsymbol{r})\, \boldsymbol{n} + \boldsymbol{u}_0, \qquad \boldsymbol{v} = \boldsymbol{v}_0. \tag{24}$$

If we add to the Eqs. (21) the requirements that $\boldsymbol{u}$, $\partial \boldsymbol{u}/\partial \boldsymbol{s}$ and $\boldsymbol{v}$ vanish at some fixed point M_0 belonging to F or L then we will obtain that $\boldsymbol{u} = \boldsymbol{v} = 0$ everywhere.

One can easily find that

$$\begin{aligned} \boldsymbol{u} \boldsymbol{T}_{(l)} + \boldsymbol{v} \boldsymbol{S}_{(l)} &= u_{(l)} T_{(ll)} + u_{(s)} T_{(ls)} + u T_{(l)} + \\ &+ v_{(l)} M_{(ls)} - v_{(s)} M_{(ll)} + \frac{4h}{3} v Q_{(l)} \end{aligned} \tag{25}$$

where $T_{(ll)} = T^{\alpha\beta} l_\alpha l_\beta$ is tangent normal force (acting on Σ_l), $T_{(ls)} = T^{\alpha\beta} l_\alpha s_\beta$ is the tangent shearing force, $T_{(l)} = T^\alpha l_\alpha$ is the transverse shearing force, $M_{(ls)} = S^{\alpha\beta} l_\alpha l_\beta$ is the bending moment, $M_{(ll)} = -S^{\alpha\beta} l_\alpha s_\beta$ is the twisting moment and $Q_{(l)} = \dfrac{3}{4h} S^\alpha l_\alpha$ is the cracking force; $u_{(l)}$, $u_{(s)}$, u, $v_{(l)}$, $v_{(s)}$, v are the projections of the vectors $\boldsymbol{u}$ and $\boldsymbol{v}$ on $\boldsymbol{l}$, $\boldsymbol{s}$, $\boldsymbol{n}$ respectively; $\boldsymbol{s} = \boldsymbol{n} \times \boldsymbol{l}$ is the unit tangent vector to L (see fig. 3).

VIII

The fundamental identity (18) allows to formulate a number of boundary problems for the system of the elastic shell equilibrium equations obtained above. For example, taking the boundary values of $T_{(ll)}$, $T_{(ls)}$, $T_{(l)}$, $M_{(ls)}$, $M_{(ll)}$, $Q_{(l)}$ or $u_{(l)}$, $u_{(s)}$, u, $v_{(l)}$, $v_{(s)}$, v we shall have the mutually conjugated boundary conditions of physical and kinematic type, the fundamental boundary value problems I and II, respectively.

One can also consider different boundary value problems of the mixed type [4].

For the above-mentioned boundary problems one can prove the theorems of uniqueness using formulae (22), (23) and (24), as from the corresponding homogeneous boundary conditions, according to (25) and (18), follows that $I(\boldsymbol{u}, \boldsymbol{v}) = 0$. The additional conditions securing the vanishing of $\boldsymbol{u}$, $\boldsymbol{v}$ are mostly fulfilled due to the boundary conditions. But sometimes it is necessary to add these requirements to the initial boundary conditions, for example, we have such a case when the fundamental problem I is under consideration.

For proving the existence theorems the method of the theory of integral equations may be used [4].

It should be emphasized that boundary conditions which follow from the consideration of the integral identity (18) are natural for physical and kinematic models of an elastic shell described above. The above-indicated optimal compatibility of these boundary conditions with the system of Eqs. (7) and (16) shows a complete agreement of these models.

It is worth noting that the above-obtained results may be easily generalized for elastic anisotropic shells and also for shells of varying thickness.

IX

The constructed theory secures the realization of all the five physical boundary conditions of the classical moment theory, but according to the earlier suggested model (1) the sixth boundary condition should be added to them, i.e. boundary values of the transverse couples. In the classical treatment of the problem this never occurs as it is assumed that $Q_{(l)} = 0$ everywhere. This fact causes, naturally, some analytical complication; the classical theory leads to the elliptic systems of equations of the 8th order and the new one to the elliptic system of the 12th order. But no principal difficulties arise. It is sufficient to note that the system of Eqs. (7) and (16) is integrated in the explicit form for a plate and spherical shells of constant thickness [4]. Their solutions, the same as in the classical case [7, 8], are expressed by means of arbitrary analytic functions of one complex variable. The difference is that the general solution formulae contain four analytic functions in the classical case while in ours they have six of them. (See the Sections XI, XII).

It is worth noting that for the sufficiently rigid shell the addition of the above-mentioned sixth boundary condition (for the transverse couples) to the five classical physical boundary conditions does not essentially change the picture of the stress distribution in the distance from the boundary. It follows immediately from Saint-Venant's principle since the transverse couples compose the self-equilibrated system of forces acting along the boundary. This remark does not mean, however, that one can neglect $Q_{(l)}$ assuming that $Q_{(l)} = 0$ inside the shell, everywhere. If even $Q_{(l)} = 0$ along the boundary, it does not mean that $Q_{(l)} = 0$ everywhere inside the shell. Sometimes $Q_{(l)}$ may attain in some points the considerable values and that causes local deformations promoting the formation of the longitudinal internal cracks and protrusions of the shell.

From (4) and (16) it follows that

$$Q_{(l)} = \frac{1}{2} h^2 \mu \left[\frac{dv}{dl} + b_{\beta}{}^{\alpha} l_{\alpha} v^{\beta} \right]. \tag{26}$$

This formula shows that $Q_{(l)}$ depends essentially on v and one cannot assume that $v = 0$ everywhere, in general. In some points v may have the considerable values and the transverse fibres of the shell passing through those points may be reptured. As a result of this phenomenon, we will have the formation of the longitudinal cracks and protrusions of the shell.

X

At the consideration of the general theory there is no need to use any special coordinates. But quite a different situation takes place at an attempt of actual integration of the system of Eqs. (7) and (16). Then it is important to reach the maximum simplification of equations by means of a special choice of coordinates. Sometimes it is very convenient to use the isometric coordinates on the middle surface [4], i.e.

$$ds^2 = \Lambda (dx^2 + dy^2) = \Lambda \, dz \, d\bar{z}$$

$$(a_{11} = a_{22} = \Lambda, \qquad a_{12} = 0, \qquad z = x + iy, \qquad \bar{z} = x - iy) \tag{27}$$

The transition from one system of isometric coordinates to another is made by means of the conformal transformations of the 1st or 2nd order. Due to such properties of isometric coordinates, wide possibilities are open for the application of the theory of functions of a complex variable, in particular, of conformal transformation, to problems of the shell theory. Due to this fact one can always assume, for instance, that the boundaries of the shell are the isometric coordinate lines. Having that in mind it is more convenient to write down the corresponding equations in the complex form.

Simple calculations show that the system of Eqs. (7) and (16) with respect to the isometric coordinates may be written in the following form

$$\frac{1}{\Lambda}\frac{\partial}{\partial z}(T_{11}-T_{22}+2iT_{12})+\frac{\partial}{\partial \bar z}(T_1{}^1+T_2{}^2)-HT_+-H_*\overline{T}_++X_+=0$$

$$\frac{1}{\Lambda}\left(\frac{\partial T_+}{\partial z}+\frac{\partial \overline{T}_+}{\partial \bar z}\right)+H(T_1{}^1+T_2{}^2)+Re[\overline{H}_*(T_1{}^1-T_2{}^2+2iT_2{}^1)]+X=0$$

$$\frac{1}{\Lambda}\frac{\partial}{\partial z}(S_{11}-S_{22}+2iS_{12})+\frac{\partial}{\partial \bar z}(S_1{}^1+S_2{}^2)-HS_+-H_*\overline{S}_+-T_+=0$$

$$\frac{1}{\Lambda}\left(\frac{\partial S_+}{\partial z}+\frac{\partial \overline{S}_+}{\partial \bar z}\right)+H(S_1{}^1+S_2{}^2)+Re[\overline{H}_*(S_1{}^1-S_2{}^2+2iS_2{}^1)]-T=0 \tag{28}$$

where

$$T_1{}^1-T_1{}^2+2iT_2{}^1=8h\mu\left[\frac{\partial}{\partial \bar z}(u^1+iu^2)-H_*u\right]$$

$$T_1{}^1+T_2{}^2=4h[(\lambda+\mu)\,\Theta_1-2(u+\mu)\,Hu+\lambda v]$$

$$S_1{}^1-S_2{}^2+2iS_2{}^1=\frac{8h^3}{3}\mu\left[\frac{\partial}{\partial \bar z}(v^1+iv^2)-H_*v\right]$$

$$S_1{}^1+S_2{}^2=\frac{4h^3}{3}(\lambda+\mu)[\Theta_2-2Hv]$$

$$T_+=T_1+iT_2=2h\mu\left[2\frac{\partial u}{\partial \bar z}+H(u_1+iu_2)+H_*(u_1-iu_2)+v_1+iv_2\right]$$

$$S_+=S_1+iS_2=\frac{2h^3}{3}\mu\left[2\frac{\partial V}{\partial \bar z}+H(v_1+iv_2)+H_*(v_1-iv_2)\right]$$

$$T=2h[\lambda\Theta_1-2\lambda Hu+(\lambda+2\mu)\,v]. \tag{29}$$

Here

$$\left.\begin{aligned}\Theta_1&=\nabla_\alpha u^\alpha=\frac{1}{\Lambda}\left(\frac{\partial(u_1+iu_2)}{\partial z}+\frac{\partial(u_1-iu_2)}{\partial \bar z}\right)\\ \Theta_2&=\nabla_\alpha v^\alpha=\frac{1}{\Lambda}\left(\frac{\partial(v_1+iv_2)}{\partial z}+\frac{\partial(v_1-iv_2)}{\partial \bar z}\right)\end{aligned}\right\} \tag{30}$$

$$H=\frac{1}{2}(b_1{}^1+b_2{}^2),\qquad H_*=\frac{1}{2}(b_1{}^1-b_2{}^2+2ib_2{}^1). \tag{31}$$

One should bear in mind that the operations of rising and lowering of indices are made according to the formulae

$$u_\alpha=\Lambda u^\alpha,\qquad T_{\beta\alpha}=\Lambda T_\beta{}^\alpha\qquad \text{etc.} \tag{32}$$

The physical components of the covariant vector u_α are expressed by the formulae

$$u_{(\alpha)}=\frac{1}{\sqrt{\Lambda}}u_\alpha\,.$$

The physical components of the symmetric tensor $T^{\alpha\beta}$ of the 2nd order are equal to the components of the mixed type,

$$T_{(\alpha\beta)} = T_{(\beta\alpha)} = T_\beta^{\ \alpha}.$$

Therefore it is easy to write the system of Eqs. (28) and (29) with the help of the physical components of the sought tensors.

It is easy to see that $|H_*|^2 = H^2 - K = \frac{1}{4}(k_1 - k_2)^2$. Therefore $H_* = 0$ for spherical shells.

For the shell the middle surface of which is a minimal surface $H = 0$, i.e. $k_1 = -k_2 = k > 0$. Besides one can choose the isometric coordinate in such a way that [13]

$$I = ds^2 = \frac{1}{k}(dx^2 + dy^2), \qquad \mathrm{II} = dx^2 - dy^2.$$

Therefore in this case we have

$$\Lambda = \frac{1}{k}, \qquad H_* = k, \qquad H = 0.$$

The advantage of Eqs. (28) and (29) is that the more conventional partial derivatives are used instead of covariant ones while equations maintain rather a compact form. Another advantage of isometric coordinates is that they reveal a new important aspect of the shell theory connecting it with the theory of conformal invariant differential forms.

One can easily prove the formulae

$$(u_{(s)} - iu_{(l)})\,ds = (u_1 + iu_1)\,d\bar{z}, \qquad (v_{(s)} - iv_{(l)})\,ds = (v_1 + iv_2)\,d\bar{z}$$
$$(T_{(s)} - iT_{(l)})\,ds = T_+\,d\bar{z}, \qquad (Q_{(s)} - iQ_{(l)})\,ds = \frac{3}{4h}S_+\,d\bar{z}$$
$$T_{(ll)} + iT_{(ls)} = \frac{1}{2}T_\alpha^{\ \alpha} - \frac{1}{2}(T_1^{\ 1} - T_2^{\ 2} + 2iT_2^{\ 1})\frac{d\bar{z}}{dz}$$
$$M_{(ls)} - iM_{(ll)} = \frac{1}{2}S_\alpha^{\ \alpha} - \frac{1}{2}(S_1^{\ 1} - S_2^{\ 2} + 2iS_2^{\ 1}\frac{d\bar{z}}{dz} \qquad (33)$$

where x and y ($z = x + iy$) are isometric coordinates on the middle surface. Since $T_\alpha^{\ \alpha}$, $S_\alpha^{\ \alpha}$ and $ds^2 = \Lambda\,dz\,d\bar{z}$ are invariants, one can conclude from formulae (33) that the quantities

$$(u_1 + iu_2)\,d\bar{z}, \qquad (v_1 + iv_2)\,d\bar{z}, \qquad T_+d\bar{z}, \qquad S_+d\bar{z}$$
$$(T_{11} - T_{22} + 2iT_{12})\,d\bar{z}^2, \qquad (S_{11} - S_{22} + 2iS_{12})\,d\bar{z}^2 \qquad (34)$$

are conformal invariants. Due to this fact the shell theory problems may be connected with the theory of the invariant differential forms on the Riemannian surfaces.

XI

Let us consider, as an example, the case of an elastic plate of constant thickness. In this case we shall have the equations ($A = 1$, $b_{\alpha\beta} = 0$, $h = \text{const}$)

$$\mu \Delta u_1 + (\lambda + \mu) \frac{\partial \Theta_1}{\partial x} + \lambda \frac{\partial v}{\partial x} + \frac{1}{2h} X_1 = 0$$

$$\mu \Delta u_2 + (\lambda + \mu) \frac{\partial \Theta_1}{\partial y} + \lambda \frac{\partial v}{\partial y} + \frac{1}{2h} X_2 = 0$$

$$\mu \Delta v - 3(\lambda + 2\mu) h^{-2} v - 3\lambda h^{-2} \Theta_1 = 0$$

$$\left(\Theta_1 = \frac{\partial u_1}{\partial x} + \frac{\partial u_2}{\partial y}\right). \tag{35}$$

$$\mu \Delta v_1 + (\lambda + \mu) \frac{\partial \Theta_2}{\partial x} - 3\mu h^{-2} \left(\frac{\partial u}{\partial x} + v_1\right) = 0$$

$$\mu \Delta v_2 + (\lambda + \mu) \frac{\partial \Theta_2}{\partial y} - 3\mu h^{-2} \left(\frac{\partial u}{\partial y} + v_2\right) = 0$$

$$\mu \Delta u + \mu \Theta_2 + \frac{1}{2h} X = 0.$$

$$\left(\Theta_2 = \frac{\partial v_1}{\partial x} + \frac{\partial v_2}{\partial y}\right). \tag{36}$$

Thus the elastic plate equilibrium equations are splitting into two independent systems of Eqs. (35) and (36), each of them contains three sought functions u_1, u_2, v and v_1, v_2, u respectively.

The corresponding stress-strain states of the plate can be called (T) and (M) states. The state (T) defines the tangent components of the stress resultants and the transverse couples. As to the state (M) it defines the bending and twisting moments and the transverse shearing forces. If the boundary conditions do not hook each other, i.e. they are prescribed separately for u_1, u_2, v and v_1, v_2, u then one can study the states (T) and (M) independently.

For the state (M) of a plate the similar theory was constructed by E. Reissner as well (see, for example, [12], chap. VII, sec. 7.7). But Reissner's equations do not coincide with ours since P^{33} seems to be actually neglected in the stress-strain relations.

If we assume that $X_1 = X_2 = 0$ then all solutions of Eqs. (35) and (36) my bae expressed by formulae [I, 4]

$$u_1 + i u_2 = 2 \frac{\partial}{\partial \bar{z}} (U + i U) \tag{37}$$

$$v = -\frac{\sigma}{1-\sigma} \Delta U - \frac{(1-2\sigma) h^2}{6\sigma} \Delta \Delta U \tag{38}$$

$$v_1 + i v_2 = 2 \frac{\partial}{\partial \bar{z}} \left(V + i \frac{2h^2}{3} \varrho \right) \tag{39}$$

$$u = - V + \frac{B(1+\sigma)}{Eh} \Delta V \tag{40}$$

where

$$U = - \frac{\sigma h^2}{6} x + \frac{1-\sigma}{2(1+\sigma)} \left[z \overline{f(z)} + \bar{z} f(z) - \frac{1}{2} \left(f_0(z) + \overline{f_0(z)} \right) \right] \tag{41}$$

$$U_* = \frac{i}{1+\sigma} \left[z \overline{f(z)} - \bar{z} f(z) \right]. \tag{42}$$

Here f and f_0 are arbitrary analytic functions of $z = x + iy$, χ, ϱ and V are arbitrary real-valued solutions of equations

$$\Delta \chi - \varkappa_1^2 \chi = 0, \qquad \Delta \varrho - \varkappa_2^2 \varrho = 0$$

$$\left(\varkappa_1^2 = \frac{6}{(1-\sigma) h^2}, \qquad \varkappa_2^2 = \frac{3}{h^2} \right) \tag{43}$$

$$\Delta \Delta V = - \frac{1}{B} X \tag{44}$$

where

$$B = \frac{2}{3} (\lambda + 2\mu) h^3 = \frac{2(1-\sigma) E h^3}{3(1+\sigma)(1-2\sigma)}. \tag{45}$$

It follows from (37)—(44)

$$v = \chi - \frac{2\sigma}{1+\sigma} \left[f'(z) + \overline{f'(z)} \right] \tag{46}$$

$$\Theta_1 = - \frac{\sigma}{1+\sigma} \chi + \frac{2(1-\sigma)}{1+\sigma} \left[f'(z) + \overline{f'(z)} \right] \tag{47}$$

$$\omega_0 = \frac{1}{2} \left(\frac{\partial u_2}{\partial x} - \frac{\partial u_1}{\partial y} \right) = \frac{2}{i(1+\sigma)} \left[f'(z) - \overline{f'(z)} \right] \tag{48}$$

$$\chi = \frac{1-\sigma}{1-2\sigma} \left[(1-\sigma) v + \sigma \Theta_1 \right] = \frac{1-\sigma^2}{2Eh} T \tag{49}$$

$$f'(z) = \frac{1-\sigma}{4(1-2\sigma)} \left[\sigma v + (1-\sigma) \Theta_1 \right] + i \omega_0 \tag{50}$$

$$\Theta_2 = \Delta V = \frac{\partial v_1}{\partial x} + \frac{\partial v_2}{\partial y} \tag{51}$$

$$\varrho = \frac{1}{2} \left(\frac{\partial v_2}{\partial x} - \frac{\partial v_1}{\partial y} \right) \tag{52}$$

$$V = - u + \frac{B(1+\sigma)}{Eh} \Theta_2. \tag{53}$$

The stress resultants and stress couples may be expressed by the formulae

$$(T_{(ll)} + i T_{(ls)})\, dz = \frac{2Eh}{1+\sigma} d\left[f(z) + z\overline{f'(z)} + \overline{f_0'(z)} - \frac{\sigma h^2}{3}\frac{\partial \chi}{\partial \bar{z}}\right] \tag{54}$$

$$Q_{(l)} = \frac{Eh^2}{4(1+\sigma)}\frac{d}{dl}\left[\varkappa - \frac{2\sigma}{1+\sigma}\left(f'(z) + \overline{f'(z)}\right)\right] \tag{55}$$

$$(M_{(ls)} - i M_{(ll)})\, dz = B\left[\left(\Delta V - i\frac{1-2\sigma}{1-\sigma}\varrho\right) dz - \frac{1-2\sigma}{1-\sigma} d(v_1 + i v_2)\right] \tag{56}$$

$$T_{(l)} = B\left[\frac{d\Delta V}{dl} - \frac{1-2\sigma}{1-\sigma}\frac{d\varrho}{ds}\right] \tag{57}$$

Formulae (46)—(57) reveal the physical and kinematic meanings of functions occuring in formulae (37)—(42).

It follows from (40) and (44) that

$$\Delta \Delta u = \frac{1}{B}\left(X - \frac{B(1+\sigma)}{Eh}\Delta X\right) \tag{58}$$

$$V = -u - \frac{B(1+\sigma)}{Eh}\Delta u - \frac{B(1+\sigma)^2}{(Eh)^2} X. \tag{59}$$

The general solution of Eq. (44) may be expressed by the formula

$$V = w + V_0 \tag{60}$$

where w is an arbitrary biharmonic function,

$$\Delta \Delta w = 0$$

and V_0 is a particular solution of (44). It is easy to show that V_0 may be expressed by the double integral

$$V_0 = -\frac{1}{8\pi B}\iint |\zeta - z|^2 \ln|\zeta - z|\, X(\xi, \eta)\, d\xi\, d\eta.$$

Here the integral may be taken over any domain which contains the middle plane of the plate itself or coincides with it. If X depends only on $r = \sqrt{x^2 + y^2}$ then V_0 can be expressed by a simple integral. In particular, if X is a polinomial of r, for instance,

$$X = a_0 + a_1 r + \cdots + a_m r^m \qquad (a_i \text{ are constants})$$

then

$$V_0 = -\frac{1}{B}\sum_{k=0}^{m} a_k \frac{r^{k+4}}{(k+2)^2\,(k+4)^2}. \tag{61}$$

The function V may be expressed by the formula

$$V = \bar{z} g(z) + z \overline{g(z)} + g_0(z) + \overline{g_0(z)} + V_0 \tag{62}$$

where g and $\bar{g}_0$ are arbitrary analytic functions of $z = x + iy$. Then formulae (56) and (57) may be written in the form

$$(M_{(ls)} - i M^*_{(ll)})\, dz = \frac{2B(1-2\sigma)}{1-\sigma}\, d \left[\frac{3-2\sigma}{1-2\sigma}\, g - z\bar{g}' - \bar{g}_0' - i\, \frac{2h^2}{3} \frac{\partial \varrho}{\partial \bar{z}}\right] + \\ + B\left[\Delta V_0\, dz - \frac{2(1-2\sigma)}{1-\sigma}\, d\, \frac{\partial V_0}{\partial \bar{z}}\right] \tag{63}$$

$$T_{(l)} = B\, \frac{d}{ds}\left[\frac{4}{i}\, (g' - \bar{g}') - \frac{1-2\sigma}{1-\sigma}\, \varrho\right] + B\, \frac{d\Delta V_0}{dl} \tag{64}$$

where $M^*_{(ll)}$ denotes a modified twisting moment,

$$M^*_{(ll)} = M_{(ll)} - \int^{s} \left(T_{(l)} - B\, \frac{d\Delta V_0}{dl}\right) ds. \tag{65}$$

For the components of the vectors $\boldsymbol{u}$ and $\boldsymbol{v}$ we have the formulae

$$u_1 + i u_2 = -\frac{\sigma h^2}{3} \frac{\partial \chi}{\partial \bar{z}} + \frac{3-\sigma}{1+\sigma}\, f(z) - z\overline{f'(z)} - \overline{f_0'(z)} \tag{66}$$

$$v = \chi - \frac{2\sigma}{1+\sigma}\, [f'(z) + \overline{f'(z)}] \tag{67}$$

$$v_1 + i v_2 = 2(g + z\bar{g}' + \bar{g}_0') + i\, \frac{4h^2}{3} \frac{\partial \varrho}{\partial \bar{z}} + 2\, \frac{\partial V_0}{\partial \bar{z}} \tag{68}$$

$$u = \frac{4B(1+\sigma)}{Eh}\, (g' + \bar{g}') - \bar{z}g - z\bar{g} - g_0 - \bar{g}_0 - V_0 + \frac{B(1+\sigma)}{Eh}\, \Delta V_0. \tag{69}$$

It is known that the functions χ and ϱ satisfying Eq. (44) may be expressed by formulae (78)

$$\chi = Re\left[f_1(z) - \int^{z} f_1(t)\, \frac{\partial}{\partial t}\, I_0\left(\varkappa_1 \sqrt{\bar{z}(z-t)}\right) dt\right]$$

$$\varrho = Re\left[g_1(z) - \int^{z} g_1(t)\, \frac{\partial}{\partial t}\, I_0\left(\varkappa_2 \sqrt{\bar{z}(z-t)}\right) dt\right] \tag{70}$$

where f_1 and g_1 are arbitrary analytic functions of $z = x + iy$.

Thus the above-obtained formulae expressing the solutions of the system of eqs. (35) and (36) contain the six arbitrary analytic functions f, f_0, f_1, g, g_0, g_1 of $z = x + iy$.

Due to this fact in the shell theory may be used the methods, developed for solution of the plane elasticity problems, which are based on application of the theory of analytic function of one complex variable [9, 7, 8].

By the suitable coice of above mentioned six analytic functions one can satisfy the six independently given boundary conditions. This confirms the situation pointed out for shells of arbitrary shapes given in Sec. VIII.

The results, established above, reveal many similarities with the plane elasticity theory. But one can also see some discrepancies stipulated by the presence of functions χ and ϱ in the formulae expressing the stress resultants and displacement fields of the plate. If we assume that χ or ϱ or both of them vanish identically, we shall obtain the simpler formulae much more similar to those of the classical plane elasticity theory. But such a simplification is not always expedient since one cannot satisfy all the natural boundary conditions with the help of the simplified formulae corresponding to the assumptions $\chi = 0$ or $\varrho = 0$. In spite of this it is expedient to consider these cases separately since they have some peculiarities.

XII

Let us consider at first the case $\chi = 0$. It means that, according to formula (49), the transverse force $T = 0$. This assumption is widely used in the classical shell theory. Setting $\chi = 0$ the formulae (37), (46), (54) and (55) become

$$u_1 + i u_2 = \frac{3-\sigma}{1+\sigma} f(z) - z\overline{f'(z)} - \overline{f_0'(z)} \tag{71}$$

$$v = -\frac{\sigma}{1-\sigma}\Theta_1 = -\frac{2\sigma}{1+\sigma}\left[f'(z) + \overline{f'(z)}\right] \tag{72}$$

$$(T_{(ll)} + i T_{(ls)})\, dz = \frac{2Eh}{1+\sigma}\, d\left[f(z) + z\overline{f'(z)} + \overline{f_0'(z)}\right]. \tag{73}$$

$$Q_{(l)} = \frac{Eh^2}{4(1+\sigma)}\frac{dv}{dl} = -\frac{\sigma E h^2}{2(1+\sigma)^2}\frac{d}{dl}\left[f'(z) + \overline{f'(z)}\right]. \tag{74}$$

Instead of Eqs. (35) we obtain the system of equations (here is supposed that $X_1 = X_2 = 0$)

$$\begin{aligned} \mu\,\Delta u_1 + (\lambda^* + \mu)\frac{\partial\Theta_1}{\partial x} &= 0 \\ \mu\,\Delta u_2 + (\lambda^* + \mu)\frac{\partial\Theta_2}{\partial y} &= 0 \end{aligned} \qquad \left(\Theta_1 = \frac{\partial u_1}{\partial x} + \frac{\partial u_2}{\partial y}\right) \tag{75}$$

where

$$\lambda^* = \lambda \frac{1 - 2\sigma}{1 - \sigma} = \frac{E\sigma}{1 - \sigma^2}. \tag{76}$$

The third Eq. (35) is reduced to the Laplace equation and is fulfilled due to formula (72). Therefore it is impossible to consider this equation as an independent one.

The system of Eqs. (75) for the tangent displacement field on the middle plane ($x^3 = 0$) is the same which we have for the case of the so-called generalized plane stress state of the elastic body [9, 12]. Formula (71) gives a general representation of solutions of Eqs. (75). Formula (72) shows that the elongations of the transverse fibres of a plate depend only on the tangent displacements field on the middle plane ($x^3 = 0$). Therefore the function v is completely defined if the boundary values of u_1 and u_2 are given beforehand, i.e. the presence of the tangent displacement field brings the elongations of the transverse fibres about. Besides, it follows from formulae (72) that the function v may attain maxima on the boundary (if v is not constant). The neighbourhoods of the boundary points where v has a positive maximum, are the places where the ruptures of the transverse fibres and therefore the appearance of longitudinal cracks may be expected.

Formula (74) shows that the cracking forces $Q_{(l)}$ completely depend on the distribution of the tangent stress resultants. Therefore if the values of $T_{(ll)}$ and $T_{(ls)}$ are given beforehand on the boundary, the field of the transverse couples is defined uniquely. In addition, the assumption $\chi = 0$ (or $T = 0$) leaves the formulae for v_1, v_2, u unaltered. Due to this fact, the simplified formulae corresponding to the assumption $\chi = 0$ allow to satisfy the five classical physical or kinematic boundary conditions. For instance, setting $\chi = 0$ and prescribing beforehand the boundary values of $T_{(ll)}$, $T_{(ls)}$, $T_{(l)}$, $M_{(ls)}$, $M_{(ll)}$ or u_1, u_2, v_1, v_2, u the corresponding displacements, stress resultants and couples together with the quantities v and $Q_{(l)}$ will be defined uniquely.

Thus in the case of a plate of constant thickness, formulae (71), (72), (39) and (40) define the kinematic model of deformation compatible with the five physical boundary conditions, corresponding to the classical physical basic assumption. But the field of the transverse couples and the elongations of the transverse fibres do not vanish identically in general. These fields vanish identically if $T_{(ll)} = 0$, $T_{(ls)} = 0$ or $u_1 = 0$, $u_2 = 0$ on the boundary everywhere.

If we assume $v = 0$ then according to formula (67), we obtain

$$\chi = 0, \qquad f'(z) + \overline{f'(z)} = 0. \tag{77}$$

Therefore $f(z) = ic_0 z + c' + ic''$ where c', c'' and c_0 are real constants. Then formulae (71)—(74) become

$$u_1 + iu_2 = \frac{4i}{1+\sigma} c_0 z + \frac{3-\sigma}{1+\sigma}(c' + ic'') - \overline{f_0'(z)}$$

$$v = 0$$

$$(T_{(ll)} + iT_{(ls)})\, dz = \frac{2Eh}{1+\sigma} d\overline{f_0'(z)}$$

$$Q_{(l)} = 0\,. \tag{78}$$

If one neglects the rigid translations and rotations of the shell the problem will be reduced to the determination of one analytic function $f_0''(z)$. Therefore one can give beforehand only one physical or kinematic condition on the boundary. For example, one can give beforehand only the boundary values of $T_{(ll)}$, or $u_{(l)}$.

It is worth noting that in the case of a plate the assumption $v = 0$ makes no influence either on the formulae defining the normal deflection u of the middle plane ($x^3 = 0$) or on the field of the secondary tangent displacements v_1, v_2 on the parallel planes $x^3 =$ const or the fields of the stress couples and the transverse shearing forces. Therefore the assumption $v = 0$ is compatible with the four independently given boundary conditions. For example, one can arbitrarily give the values of $T_{(ll)}$, $M_{(ls)}$, $M_{(ll)}$, $T_{(l)}$ or $u_{(l)}$, $v_{(l)}$, $v_{(s)}$, u on the boundary. Then the corresponding stress-strain state of the plate will be defined uniquely.

XIII

Assume now that $\varrho = 0$. It means that the elementary prismatic cross sections of the shell are rigid with respect to twisting deformations. This assumption does not change formulae (37), (38), (54) and (55), expressing the tangent displacement field u_1, u_2 on the middle plane and the elongations of the transverse fibres v and the corresponding tangent stress resultants and transverse couples. Formulae (39), (57) and (59) become

$$v_1 + iv_2 = 2\frac{\partial V}{\partial \bar{z}} \tag{79}$$

$$(M_{(ls)} - iM_{(ll)})\, dz = -\frac{B}{1-\sigma}\left[\Delta V\, dz + 2(1-2\sigma)\frac{\partial^2 V}{\partial \bar{z}^2}\, d\bar{z}\right] \tag{80}$$

$$T_{(l)} = B\frac{d\Delta V}{dl} \tag{81}$$

where V is the same function as before, i.e. it is an arbitrary solution of Eq. (44). Therefore the assumption $\varrho = 0$ has no influence on the values of the normal deflections of the middle plane, as it is clear from formula (40). By means of (63) and (64), formulae (80) and (81) may be written in the form

$$(M_{(ls)} - i M^*_{(ll)})\, dz = \frac{2B(1-2\sigma)}{1-\sigma}\, d \left[\frac{3-2\sigma}{1-2\sigma}\, g - z\bar{g}' - \bar{g}_0'\right] + \\ + B \left[\Delta V_0\, dz - \frac{2(1-2\sigma)}{1-\sigma}\, d\, \frac{\partial V_0}{\partial \bar{z}}\right] \tag{82}$$

$$T_{(l)} = B \frac{d}{ds} \left[\frac{4}{i}\, (g' - \bar{g}') + B\, \frac{d \Delta V_0}{dl}\right] \tag{83}$$

where $M^*_{(ll)}$ is the reduced twisting moment given by formuly (65). Formulae (79), (82) and (83) show that one can arbitrarily give beforehand the boundary values of v_1, v_2 or u, $\frac{\partial u}{\partial r}$ or two of three quantities $M_{(ls)}$, $M^*_{(ll)}$, $T_{(l)}$.

Thus the assumption $\varrho = 0$ leads to the formulae which allow to satisfy the five independent boundary conditions. For example, one can independently give the boundary values of $T_{(ll)}$, $T_{(ls)}$, $Q_{(l)}$ and $M_{(ls)}$, $M^*_{(ll)}$ (or $T_{(l)}$) or u_1, u_2, v and v_1, v_2 $\left(\text{or } u, \frac{\partial u}{\partial r}\right)$ and define uniquely the strain-stress state of the shell.

XIV

If we assume that $\chi = 0$, $\varrho = 0$ simultaneously we obtain the formulae

$$u_1 + i u_2 = \frac{3-\sigma}{1-\sigma}\, f(z) - z\, \overline{f'(z)} - \overline{f_0'(z)} \tag{84}$$

$$v = -\frac{\sigma}{1-\sigma}\, \Theta_1 = -\frac{2\sigma}{1+\sigma}\, [f'(z) + \overline{f'(z)}] \tag{85}$$

$$v_1 + i v_2 = 2\, \frac{\partial V}{\partial \bar{z}} \tag{86}$$

$$u = -V + \frac{B(1+\sigma)}{Eh}\, \Delta V \tag{87}$$

$$(T_{(ll)} + i T_{(ls)})\, dz = \frac{2Eh}{1+\sigma}\, d\, [f(z) + z\, \overline{f'(z)} + \overline{f_0'(z)}] \tag{88}$$

$$Q_{(l)} = -\frac{\sigma E h^2}{2(1+\sigma)^2}\, \frac{d}{dl}\, [f'(z) + \overline{f'(z)}] \tag{89}$$

$$(M_{(ls)} - i M^*_{(ll)})\, dz = \frac{2B(1-2\sigma)}{1-\sigma}\, d\left[\frac{3-2\sigma}{1-2\sigma}\, g - z\bar{g}' - \bar{g}_0'\right] +$$

$$+ B\left[\Delta V_0\, dz - \frac{2(1-2\sigma)}{1-\sigma}\, d\, \frac{\partial V_0}{\partial \bar{z}}\right] \tag{90}$$

$$T_{(l)} = B \frac{d}{ds}\left[\frac{4}{i}\,(g' - \bar{g}')\right] + B \frac{d\Delta V_0}{dl} \tag{91}$$

where

$$V = \bar{z} g + z\bar{g} + g_0 + \bar{g}_0 + V_0. \tag{62}$$

These formulae allow to satisfy the four boundary conditions corresponding to Kirchhoff-Love's theory. For instance, one can give on the boundary the values of $T_{(ll)}$, $T_{(ls)}$, $M_{(ls)}$, $M^*_{(ll)}$ (or $T_{(l)}$) at will. But the formulae expressing the displacement field and stress resultants and couples do not coincide completely with corresponding formulae of Kirchhoff-Love's theory. We have some discrepancies due to the presence of the elongations of the transverse fibres of the shell and the transverse couples which do not take part in the classical considerations. The action of the transverse couples may sometimes lead to the appearance of internal longitudinal cracks and warping of the face surface of the shell. This phenomenon, which is excluded from consideration by Kirchhoff-Love's assumptions, cannot be explained by means of the classical shell theory though it may be frequently observed in practice.

XV

Let us consider now the plate equilibrium problems corresponding to the case when $u_1 = u_2 = v = 0$. These requirements will be realized if $X_1 = X_2 = 0$ and the boundary conditions $u_1 = 0$, $u_2 = 0$, $v = 0$ are fulfilled.

Let us assume also that the stress-strain picture has a polar symmetry with respect to the origin of coordinates ($x = 0$, $y = 0$). It means that $\varrho = 0$ and therefore according to (40), (79), (80) and (81) we obtain

$$u = -V(r) + \frac{B(1+\sigma)}{Eh} \frac{1}{r} \frac{d}{dr}\, r\, \frac{dV(r)}{dr} \tag{92}$$

$$v_{(r)} = \frac{dV}{dr}, \qquad v_{(\varphi)} = 0 \tag{93}$$

$$M_{(ls)} - i M_{(ll)} = B\left[\frac{1}{r}\frac{d}{dr}\, r\, \frac{dV}{dr} - \frac{1-2\sigma}{1-\sigma}\left(\frac{1}{r}\frac{dV}{dr} + \frac{d}{dr}\left(\frac{1}{r}\frac{dV}{dr}\right) z\, \frac{dr}{dz}\right)\right] \tag{94}$$

$$T_{(r)} = -\frac{1}{2} X r, \qquad T_{(\varphi)} = 0 \tag{95}$$

where V depends only on $r = \sqrt{x^2 + y^2}$ and therefore satisfies the equation

$$\left(\frac{1}{r}\frac{d}{dr}r\frac{d}{dr}\right)^2 V = -\frac{1}{B}X. \tag{96}$$

It is clear that X depends only on r as well. Setting $X = \text{const}$ we obtain

$$V = a + br^2 - \frac{X}{64B}r^4 \qquad (a \text{ and } b \text{ are real constants}). \tag{97}$$

It follows from (94) that

$$M_{(rr)} = 0, \qquad M_{(\varphi\varphi)} = 0 \tag{98}$$

$$M_{(r\varphi)} = B\left[\frac{d^2V}{dr^2} + \frac{\sigma}{1-\sigma}\frac{1}{r}\frac{dV}{dr}\right] \tag{99}$$

$$M_{(\varphi r)} = -B\left[\frac{\sigma}{1-\sigma}\frac{d^2V}{dr^2} + \frac{1}{r}\frac{dV}{dr}\right]. \tag{100}$$

Let us consider now some boundary value problems for the circular plate ($0 \leq r \leq c$, $c > 0$) the solutions of which may be effectively constructed with the help of formulae (92)—(100). The obtained results may be compared with classical ones to reveal the similarities and discrepancies.

Let us consider firstly the circular plate ($0 \leq r \leq c$) with the rigid edge, that means that

$$u|_{r=c} = 0, \qquad v_{(r)}|_{r=c} = 0. \tag{101}$$

It is easy to find that the corresponding function V is

$$V = -\frac{Xc^4}{64B}\left[\left(1 - \frac{r^2}{c^2}\right)^2 + \frac{16(1-\sigma)}{3(1-2\sigma)}\left(\frac{h}{c}\right)^2\right] \tag{102}$$

and according to (92), (93), (99) and (100) we obtain

$$u = \frac{Xc^4}{64B}\left[\left(1 - \frac{r^2}{c^2}\right)^2 + \frac{32(1-\sigma)}{3(1-2\sigma)}\left(\frac{h}{c}\right)^2\left(1 - \frac{r^2}{c^2}\right)\right] \tag{103}$$

$$v_{(r)} = \frac{Xc^2}{16B}r\left(1 - \frac{r^2}{c^2}\right) \tag{104}$$

$$M_{(r\varphi)} = \frac{Xc^2}{1-\sigma}\left[1 - (3 - 2\sigma)\frac{r^2}{c^2}\right],$$

$$M_{(\varphi r)} = -\frac{Xc^2}{1-\sigma}\left[1 - (1 + 2\sigma)\frac{r^2}{c^2}\right]. \tag{105}$$

If we solve the same problem using the classical bending equations for a plate [10, 11],

$$\Delta\,\Delta\bar{u} = \frac{X}{D}, \qquad \bar{v}_{(r)} = -\frac{d\bar{u}}{dr} \qquad \left(D = \frac{2Eh^3}{3(1-\sigma^2)} = \frac{1-2\sigma}{(1-\sigma)^2}B\right), \tag{106}$$

then we will obtain

$$\bar{u} = \frac{Xc^4}{64D}\left(1 - \frac{r^2}{c^2}\right)^2 \tag{107}$$

$$\bar{v}_{(r)} = \frac{Xc^2}{16D}\,r\left(1 - \frac{r^2}{c^2}\right) = \frac{(1-\sigma)^2}{1-2\sigma}\,v_{(r)}. \tag{108}$$

The comparison of formulae (107) and (108) with (103) and (104) shows that for the both cases the picture of radial displacements on the planes $x^3 =$ const are the same qualitatively but there exist the quantitative differences. For instance, setting $\sigma = 0{,}3$ we obtain that $v_{(r)} = 0{,}82\,\bar{v}_{(r)}$. It means that the new more precise plate bending theory leads to 18% decrease for the radial displacements on the planes $x^3 =$ const (in comparison with the classical values). We have the different picture for the normal deflections. Denoting by u_0 and $\bar{u}_0$ the maximum deflections corresponding to (103) and (107) we will obtain (at the point $x = y = 0$)

$$u_0 = \frac{1-2\sigma}{(1-\sigma)^2}\left[1 + \frac{32(1-\sigma)}{3(1-2\sigma)}\left(\frac{h}{c}\right)^2\right]\bar{u}_0. \tag{109}$$

Setting $\sigma = 0{,}3$ we will have

$$u_0 = 0.82\left[1 + 18.66\left(\frac{h}{c}\right)^2\right]\bar{u}_0. \tag{110}$$

It follows from (103) and (107) that on the boundary we have the equalities

$$\left.\frac{du}{dr}\right|_{r=c} = -\frac{(1+\sigma)Xc}{2Eh}, \qquad \left.\frac{d\bar{u}}{dr}\right|_{r=c} = 0. \tag{111}$$

Eqs. (109), (110) and (111) shows that there exist essential discrepancies between u and $\bar{u}$.

The corresponding radial displacements $U_{(r)} = x^3 v_{(r)}$ and $\bar{U}_{(r)} = x^3\bar{v}_{(r)}$ reach the maximum values on the face surfaces $x^3 = \pm h$ along the circumference $r = c/\sqrt{3}$, i.e.

$$U^0_{(r)} = \frac{1-2\sigma}{(1-\sigma)^2}\,\bar{U}^0_{(r)} = \pm\frac{Xhc^3}{24\sqrt{3}B}. \tag{112}$$

Therefore within the zone near the circumference $r = c/\sqrt{3}$ the appearance of the transverse cracks may be expected if $U^0_{(r)}$ exceeds some critical value.

For the plate with the clamped edge the boundary conditions have the form

$$u|_{r=c} = 0, \qquad \left.\frac{du}{dr}\right|_{r=c} = 0. \tag{113}$$

Setting, as before, $X = \text{const}$ the classical solution of this problem is given by formulae (107) and (108). It means that the classical bending theory of plates makes no difference between the different boundary value problems (101) and (113). Actually the corresponding deformations do not coincide and there exist essential discrepancies between them. This will be revealed if we solve problem (113) with the help of the new more precise bending theory of plates. According to formulae (92), (93) and (94) the sought functions have the form

$$u' = u - \frac{(1+\sigma)Xc^2}{4Eh}\left(1 - \frac{r^2}{c^2}\right) = \frac{(1-\sigma)^2}{1-2\sigma}\,\overline{u} \tag{114}$$

$$v'_{(r)} = v_{(r)} - \frac{(1+\sigma)X}{2Eh}\,r = \frac{(1-\sigma)^2}{1-2\sigma}\,\overline{v}_{(r)} - \frac{(1+\sigma)X}{2Eh} \tag{115}$$

where u, $v_{(r)}$, $\overline{u}$ and $\overline{v}_{(r)}$ are defined by formulae (103), (104), (107) and (108) correspondingly. The quantitative and qualitative similarities and discrepancies between the classical and new solutions of the problem (113) and between the solutions of the problems (101) and (113) may be revealed by means of these relations. It is clear that we have here not only similarities but essential discrepancies between the corresponding deformation pictures. About 18% decrease we have for the normal deflections (in comparison with the classical values). Approximately the same picture we have for the radial displacements near the centre of the disc, but near the boundary the discrepancies are more essential as one may see from the equalities

$$v'_{(r)}|_{r=c} = -\frac{(1+\sigma)Xc}{2Eh}, \qquad \overline{v}_{(r)}|_{r=c} = 0. \tag{116}$$

Apparently, the picture for the radial displacements of the clamped plate, given by formula (115) is nearer to the observed phenomenon than the picture corresponding to classical formula (108).

The radial displacements $U'_{(r)} = x^3 v'_{(r)}$ reach the maxima on the face surfaces $x^3 = \pm h$ along the circumference

$$r = \frac{c}{\sqrt{3}}\left[1 - \frac{16(1-\sigma)}{3(1-2\sigma)}\left(\frac{h}{c}\right)^2\right]^{1/2}. \tag{117}$$

Therefore for the clamped plate the zone where the transverse cracks may be expected is nearer to the centre of the disc than for the plate with the rigid boundary. Between the corresponding maxima of the radial displacements there exists the relation of the form

$$U'^{0}_{(r)} = U^{0}_{(r)}\left[1 + \frac{16(1-\sigma)}{3(1-2\sigma)}\left(\frac{h}{c}\right)^{2}\right] + 0\left[\left(\frac{h}{c}\right)^{4}\right]. \tag{118}$$

Finally we shall consider the circular plate leaning along the edge on the rigid support. In this case the boundary conditions can be written in the form

$$u|_{r=c} = 0, \qquad M_{(r\varphi)}|_{r=c} = 0. \tag{119}$$

According to formulae (92), (97) and (99) we will obtain (we assume again that $X = \text{const}$)

$$u = \frac{Xc^4}{64B}\left[\left(1 - \frac{r^2}{c^2}\right)^2 + 4(1+\sigma)\left(1 + \frac{8}{3(1-\sigma)}\left(\frac{h}{c}\right)^2\right)\left(1 - \frac{r^2}{c^2}\right)\right] \tag{120}$$

$$v_{(r)} = \frac{Xc^2}{16B}\, r\left(3 - 2\sigma - \frac{r^2}{c^2}\right) \tag{121}$$

$$M_{(r\varphi)} = \frac{(3-2\sigma)Xc^2}{16(1-\sigma)}\left(1 - \frac{r^2}{c^2}\right) \tag{122}$$

$$M_{(\varphi r)} = -\frac{(3-2\sigma)Xc^2}{16(1-\sigma)}\left(1 - \frac{1+2\sigma}{3-2\sigma}\frac{r^2}{c^2}\right) \tag{123}$$

$$T_{(r)} = -\frac{1}{2}Xr. \tag{124}$$

We do not dwell upon the question on relations of these formulae with the classical ones. It is not difficult to discuss this question.

References

1. Vekua, I. N.: On one method of approximate computation of prismatic shells. Trudi Tbilisskogo Matematich. Instituta 21 (1955).
2. Vekua, I. N.: New methods in mathematical shell theory. Proceedings of the Eleventh International Congress of Applied Mechanisc. Munich (Germany): Springer-Verlag 1964, 47–58.
3. Vekua, I. N.: Theory of thin and shallow elastic shells with variable thickness. Applications of the theory of functions in continuum mechanics. Proceedings of the International Symposium, Tbilisi, 1963. Isdatelstvo „Nauka", Moskow: 1965, 410–431.
4. Vekua, I. N.: Theory of thin shallow shells with variable thickness. Trudi Tbilisk. Matematich. Instituta, 30 (1965), 5–102.
5. Vekua, I. N.: Generalized analytic functions, Moskow: Fizmatgiz 1959.

6. VEKUA, I. N.: Verallgemeinerte analytische Functionen, Berlin: Akademie-Verlag 1963.
7. VEKUA, I. N.: New methods for solving elliptic equations, Moskow: Gostekhizdat 1948.
8. VEKUA, I. N.: New methods for solving elliptic equations, Amsterdam: North-Holland 1967.
9. MUSKHELISHVILI, N. I.: Some fundamental problems of mathematical theory of elasticity, Moskow 1966.
10. GEKKELER, I. V.: Statics of an elastic body (translated from German) Leningrad-Moskow: ONTI 1934.
11. TIMOSHENKO, S. P., and S. VOINOVSKI-CRIEGER: Plates and shells (translated from English), Moskow: Fizmatgiz 1963.
12. GREEN, A. E., and W. ZERNA: Theoretical elasticity, Oxford: The Clarendon press. 1954.
13. VEKUA, I. N.: Foundations of tensor analysis, Tbilisi University Press 1967.

Discussion

W. T. KOITER: My question is what relation exists between Professor VEKUA's theory, when applied to flat plates, and Professor REISSNER's theory of bending of plates with account of transverse shear deformation.

I. N. VEKUA: My additional equation is of the form $\Delta\chi - 3h^{-2}\chi = 0$, whereas the corresponding equation of Professor REISSNER is $\Delta\chi - \frac{5}{2}h^{-2}\chi = 0$.

The Purposes and Problems of the Technical Theory of Thin Plastic Shells

By

M. Sh. Mikeladze

Georgian Polytechnic Institute, Tbilisi, USSR

At present a great deal of attention is devoted to designing thin shells by means of the methods of the theory of plasticity. However, the overwhelming majority of papers investigating this problem deal with cases of axial symmetry. In connection with this it is natural that we observe a tendency to broaden the circle of problems under research by including therein general cases which are of interest from the practical engineering viewpoint. On the other hand studying such problems involves a great deal of difficulties of mathematical nature caused not only by the amount of calculation, but also by the essence of the task. This circumstance determined to a great extent the ways and tendencies of further investigation and development of the applied theory of plastic shells, where the evaluation of the load carrying capacity of structures as well as the questions connected with optimum design of the latter, are of prime importance.

Resorting to simplifying assumptions of statical, geometrical and physical nature permits research of a sufficiently large number of problems, the essence of the methods of solution of which, contributes, in a way, to the formation of the technical theory of plastic shells.

The present communication gives a concise survey of some results recently obtained by the author in the given field.

I. On the Load Carrying Capacity of Shells [1—7]

Bearing in mind, chiefly, the shallow shells as well as the circular cylindrical ones, the principal statical and geometrical relations may be written as follows

$$\frac{1}{A_1}\frac{\partial T_1}{\partial \alpha_1} + \frac{1}{A_2}\frac{\partial S}{\partial \alpha_2} + q_1 = 0, \qquad \frac{1}{A_1}\frac{\partial S}{\partial \alpha_1} + \frac{1}{A_2}\frac{\partial T_2}{\partial \alpha_2} + q_2 = 0,$$

$$\frac{1}{A_1}\frac{\partial Q_1}{\partial \alpha_1} + \frac{1}{A_2}\frac{\partial Q_2}{\partial \alpha_2} - \frac{T_1}{R_1} - \frac{T_2}{R_2} + q_n = 0,$$

$$\frac{1}{A_1}\frac{\partial M_1}{\partial \alpha_1} + \frac{1}{A_2}\frac{\partial H}{\partial \alpha_2} = Q_1, \qquad \frac{1}{A_1}\frac{\partial H}{\partial \alpha_1} + \frac{1}{A_2}\frac{\partial M_2}{\partial \alpha_2} = Q_2; \tag{1.1}$$

$$\varepsilon_1 = \frac{1}{A_1}\frac{\partial U}{\partial \alpha_1} + \frac{W}{R_1}, \qquad \varepsilon_2 = \frac{1}{A_2}\frac{\partial V}{\partial \alpha_2} + \frac{W}{R_2},$$

$$\omega = \frac{1}{A_1}\frac{\partial V}{\partial \alpha_1} + \frac{1}{A_2}\frac{\partial U}{\partial \alpha_2}, \qquad \varkappa_1 = -\frac{1}{A_1{}^2}\frac{\partial^2 W}{\partial \alpha_1{}^2},$$

$$\varkappa_2 = -\frac{1}{A_2{}^2}\frac{\partial^2 W}{\partial \alpha_2{}^2}, \qquad \chi = -\frac{1}{A_1 A_2}\frac{\partial^2 W}{\partial \alpha_1 \partial \alpha_2}; \tag{1.2}$$

where A_1, A_2 are coefficients of the first quadratic form;
α_1, α_2 are coordinate lines; R_1, R_2 are radii of curvature;
q_1, q_2, q_n are components of external load;
T_1, T_2 are the normal forces; S is the shearing force;
Q_1, Q_2 are the transverse shearing forces;
M_1, M_2 are the bending moments; H is the twisting moment;
U, V, W are the projections of the displacement velocity vector of the arbitrary point on the middle surface respectively;
ε_1, ε_2, ω, $\varkappa_1$, $\varkappa_2$, χ are the rates of membrane strain and curvature.

In order to obtain the basic physical relations, the material of the shells is assumed to be an anisotropic, rigid-plastic medium, whose principal directions of anisotropy coincide with the coordinate lines on the middle surface. For the sake of further simplification of the rigid-plastic design scheme, not only the elastic, but also the elastic-plastic parts of the shell are considered to be rigid. Such an assumption together with the traditional one of the rigidity of the normal element results in satisfaction of the yield condition only in some integral form. As to the yield condition itself, it is a somewhat modified, well-known Hill's condition.

Thus, finally, the yield condition and the associated flow rule for shells can be expressed as follows:

$$\mathscr{F}(T_1, T_2, S, M_1, M_2, H) \equiv \frac{1}{h^2}\left(\frac{T_1{}^2}{\sigma_{s1}^2} - \frac{T_1 T_2}{\sigma_{s1}\sigma_{s2}} + \frac{T_2{}^2}{\sigma_{s2}^2} + \frac{S^2}{\tau_s{}^2}\right)$$

$$+ \frac{12}{h^4}\left(\frac{M_1{}^2}{\sigma_{s1}^2} - \frac{M_1 M_2}{\sigma_{s1}\sigma_{s2}} + \frac{M_2{}^2}{\sigma_{s2}^2} + \frac{H^2}{\tau_s{}^2}\right) = 1, \tag{1.3}$$

$$\frac{\varepsilon_1}{\partial\mathscr{F}/\partial T_1} = \frac{\varepsilon_2}{\partial\mathscr{F}/\partial T_2} = \frac{\omega}{\partial\mathscr{F}/\partial S} = \frac{\varkappa_1}{\partial\mathscr{F}/\partial M_1} = \frac{\varkappa_2}{\partial\mathscr{F}/\partial M_2} = \frac{\chi}{\partial\mathscr{F}/\partial H} = \frac{1}{2}D, \tag{1.4}$$

where h is the thickness of the shell; σ_{s1}, σ_{s2}, τ_s are the yield limits in simple tension and shear; D is the rate of dissipation of mechanical energy per unit area of the middle surface.

$$D = 2h \left[\frac{1}{3} (\sigma_{s1}^2 \varepsilon_1^2 + \sigma_{s1} \sigma_{s2} \varepsilon_1 \varepsilon_2 + \sigma_{s2}^2 \varepsilon_2^2) + \frac{\tau_s^2 \omega^2}{4} \right.$$
$$\left. + \frac{h^2}{36} (\sigma_{s1}^2 \varkappa_1^2 + \sigma_{s1} \sigma_{s2} \varkappa_1 \varkappa_2 + \sigma_{s2}^2 \varkappa_2^2) + \frac{\tau_s^2 h^2}{48} \chi^2 \right]^{1/2}. \quad (1.5)$$

Relations (1.1)—(1.5) enable us to estimate the load carrying capacity of the shell with the aid of the well-known limit design theorems.

Along with this we have:

1. The lower bound can be obtained by means of successive solution of two relatively simpler problems: the plane problem and the problem of limit equilibrium of a certain flat plate; it can be performed if the yield condition (1.3) is written as follows:

$$\frac{1}{h^2} \left(\frac{T_1^2}{\sigma_{s1}^2} - \frac{T_1 T_2}{\sigma_{s1} \sigma_{s2}} + \frac{T_2^2}{\sigma_{s2}^2} + \frac{S^2}{\tau_s^2} \right) = \alpha,$$

$$\frac{12}{h^4} \left(\frac{M_1^2}{\sigma_{s1}^2} - \frac{M_1 M_2}{\sigma_{s1} \sigma_{s2}} + \frac{M_2^2}{\sigma_{s2}^2} + \frac{H^2}{\tau_s^2} \right) = 1 - \alpha,$$

where α is a constant, limited by inequalities $0 < \alpha < 1$. The best choice for constant α will be the value maximizing the lower bound.

2. In order to obtain the upper bound of the load carrying capacity of the shell, it appears expedient to approximate the uniform shell by a sandwich one, assuming after Nakamura [7], that only one layer of the shell is in the plastic state; the advantage of such an approximation is that the evaluation of the power of work done by the internal forces is, in fact, reduced to one function — the rate of the normal displacement w.

3. As regards simply supported shells, we can obtain estimation data permitting us to compare the load carrying capacity of a shell with that of a flat plate. So, for instance, for a simply supported circular cylindrical shell ($A_1 = A_2 = 1$, $1/R_1 = 0$, $R_2 = R$) subjected only to uniform normal load ($q_n = q$) the corresponding estimations are:

$$q > \frac{2h\sigma_{s2}}{\sqrt{3}\,R}, \qquad q > q_{Pl}, \qquad q < \frac{2h\sigma_{s2}}{\sqrt{3}\,R} + q_{Pl},$$

where q_{Pl} denotes the load carrying capacity of a flat plate which is obtained by the development of the cylinder.

Estimations of the load carrying capacity of structures have been obtained for other cases as well: for the anisotropic shallow shell with the elliptic boundary; for the shallow shell of revolution and also for circular plates subjected to extension and bending, where the former may be caused by the action of centrifugal forces.

II. Semi-Momentless Theory of Cylindrical Shells and Shells Slightly Different from Them [8—10]

The theory is based on simplifying assumptions of statical nature, according to which the twisting moment, the longitudinal bending moment and the corresponding transverse shearing force are equal to zero. The absence of these two moments and the transverse shearing force leads to such a plastic state which is charakterized by the symmetric flow with respect to the middle surface and the proportional relation between the normal forces:

$$T_2 = \frac{H}{H+F} T_1, \tag{2.1}$$

where H and F are the known parameters, involved into the Hill's yield condition for an orthotropic body.

This condition, expressed through the forces and the moment, can approximately be written as follows:

$$\frac{GH+HF+FG}{(H+F)h^2} T_1^2 + \frac{2N}{h^2} S^2 + \frac{16(H+F)}{h^4} M_2^2 = 1, \tag{2.2}$$

where

$$2G = \sigma_{sn}^{-2} + \sigma_{s1}^{-2} - \sigma_{s2}^{-2}, \qquad 2H = \sigma_{s1}^{-2} + \sigma_{s2}^{-2} - \sigma_{sn}^{-2},$$
$$2F = \sigma_{s2}^{-2} + \sigma_{sn}^{-2} - \sigma_{s1}^{-2}, \qquad 2N = \tau_s^{-2},$$

and σ_{sn} denotes the yield limit in tension in the direction of the normal.

The maximum deviation of approximate relation (2.2) from the exact one is about 9% [1, 11].

Relations (2.1), (2.2) and the equilibrium equations with the corresponding boundary conditions permit us to find the unknown forces and the moment as well as the thickness of the shell.

The problem of determination of forces and the moment is a linear and statically determinate one. All the non-linearity, typical for the theory of plasticity in general, is represented here by Eq. (2.2), which is biquadratic with respect to the unknown thickness of the shell h.

Thus, the problem is to find such a law of the change of the shell's thickness when all the structure, as a whole, is completely transformed into the plastic state under the given system of forces.

In the case of circular cylindrical shell ($R_2 = a$) the governing equation with respect to the normal force T_2 is:

$$\frac{\partial^2 T_2}{\partial \Theta^2} = \frac{a^2(H+F)}{H} \frac{\partial^2 T_2}{\partial z^2} - T_2 - a q_n - a \frac{\partial q_2}{\partial \Theta} + a^2 \frac{\partial q_1}{\partial z},$$

where z and Θ are the cylindrical coordinates of a point on the middle surface ($z = \alpha_1$, $\Theta = \alpha_2/a$).

If the butts of the shell are free and the lateral edges are simply supported, then the boundary conditions of the problem will be:

$$T_2(\Theta, 0) = T_2(\Theta, l) = 0, \qquad \left.\frac{\partial T_2}{\partial \Theta}\right|_{\Theta=0} = -aq_2, \qquad T_2(o, z) = 0,$$

where l is the lenght of the shell.

When setting the boundary conditions, both the structure itself and the acting load are to be assumed symmetric with respect to plane $\Theta = 0$.

Investigation of some other cases occuring in engineering also leads to the same mathematical problem.

Finally, we must note that from the mathematical point of view, design of plastic shells according to the semi-momentless theory is reduced to the solution of the mixed problem for the equation of the hyperbolical type. The uniqueness and the continuous dependence of the solution on the initial conditions is proved in the well-known book by I. G. Petrovsky, dedicated to lectures on equations with partial derivatives [12].

III. The Theory of Shells of Uniform Strength [1—4, 13]

The term "shells of uniform strength" will be applied to such shells the transition of which into the plastic state is instantaneous. Depending on the design scheme the state of the shell preceding the newly occured plastic state, may be elastic or rigid. In conformity with this we have two ways of investigation of the problem, which essentially differ very little from one another and lead to an identical system of governing equations in the case of isotropic incompressible material.

Uniform strength of a structure can be achieved in many ways, particularly, by finding such a law of changes of shell thickness h which will ensure instantaneous occurence of the plastic state. The corresponding governing system of equations consists of the equilibrium equations (1.1), the yield condition (1.3) and the three conditions of compatibility, which, according to (1.2), become:

$$\frac{\varkappa_1}{R_2} + \frac{\varkappa_2}{R_1} + \frac{1}{A_1^2}\frac{\partial^2 \varepsilon_2}{\partial \alpha_1^2} - \frac{1}{A_1 A_2}\frac{\partial^2 \omega}{\partial \alpha_1 \partial \alpha_2} + \frac{1}{A_2^2}\frac{\partial^2 \varepsilon_1}{\partial \alpha_2^2} = 0,$$

$$\frac{1}{A_2}\frac{\partial \varkappa_1}{\partial \alpha_2} = \frac{1}{A_1}\frac{\partial \chi}{\partial \alpha_1}, \qquad \frac{1}{A_1}\frac{\partial \varkappa_2}{\partial \alpha_1} = \frac{1}{A_2}\frac{\partial \chi}{\partial \alpha_2}.$$

In these relations $\varkappa_1$, $\varkappa_2$, χ, ε_1, ε_2, ω must be expressed through moments and forces after Hooke's law, or else, in case of rigid-plastic design scheme — according to the flow rule (1.4) assuming the ratio D/h to be constant.

Analysis of the corresponding relations readily reveals identity of the two problems concerning the uniform strength and the minimum weight (the minimum volume, to speak more precisely).

Another approach to the problem of uniform strength may be adopted for shallow shells, namely: for the given load and thickness of the structure, its rational middle surface should be sought, the governing system of equations remaining the same. Only some of the equations (those of equilibrium and compatibility of deformations) undergo some slight changes, as we are not always sure that the selected axes α_1 and α_2 coincide with the lines of principal curvatures.

As is known, the shape of the middle surface of a shell or the law of changes of its thickness are pretty often determined proceeding from considerations of aerodynamic, technologic or aesthetic character. In such cases it is natural to seek other ways, ensuring uniform strength of the structure. In this context radioactive irradiation of the shell along its middle surface and with different dozes of the irradiation appears to be one of such possible ways.

It is of common knowledge [14, 15] that radioactive irradiation of solid bodies brings about changes of their elastic and, especially, plastic characteristics. The yield limit of material at relatively moderate temperatures increases to a great extent depending upon the doze of irradiation. The latter is characterized by the quantity of neutrons passing through 1 cm^2 of the body's surface during the test time t and is measured in nvt units (n — is the quantity of neutrons in 1 cm^3 of the current, and v — is its mean velocity). Taking due account of the stable nature of the mechanical properties acquired in the process of the irradiation, it is expedient to impart to the material such an "optimum non-homogeneity" which would ensure uniform strength of the structure.

Within the range of the theory described in this communication, optimum design of a shell is reduced to determination of the corresponding law of irradiation dozing.

Here we should assume that:

a) the changes of the mechanical properties in each point of the body caused by the irradiation depend solely on its doze in this point [15];

b) owing to an inconsiderable thickness of the shell we may neglect the changes of the mechanical properties in the direction of the normal towards the middle surface.

The governing system of equations for the problem under investigation, along with the equilibrium equations, Hooke's law, conditions of plasticity and compatibility of deformations involve, likewise, the two following relations:

$$E = f_1(N), \qquad \sigma_s = f_2(N),$$

which testify to the dependence of the elasticity modulus E and the yield limit σ_s on the doze of irradiation N.

These relations are assumed to be given as a series of curves or tables.

Since the obtained governing system of equations is non-linear and looks too bulky, it is worthwhile to simplify its solution by means of the successive approximations technique, which is based on the assumptions of physical nature. Namely, considering, the relatively insignificant changes of the elastic properties brought about by irradiation, we deem it possible to neglect, in the first approximation, the change of the shell's modulus of elasticity. Then the forces and the moments can be determined through solution of the classic linear problem. Further, by means of the averaged yield condition (1.3) and the empirical relation $\sigma_s = f_2(N)$ we determine the optimum irradiation law N which in its turn allows us to find the changes in the modulus of elasticity of the shell in the first approximation.

Solution of the problem in the second approximation is actually quite ordinary. The forces and the moments are determined, as in the first approximation, from the linear problem, the difference being that now the elasticity modulus of the shell material is a variable.

This remark holds true for any further approximation as well. However, if we pursue purely practical purposes, then the necessity of such approximations (after the second one) exists no longer.

Finally, we ought to mention that for the optimum design of thin plates and shells it is natural to require satisfaction of some rigidity conditions as well. This approach complicates the task of optimum design on the whole to a considerable extent, and at present is rather difficult to be expressed mathematically. However, some success in this line can be made even now, if we confine ourselves to considering an axial-symmetrical problem for elastic circular plates and shells of revolution, for which the law of changes of the variable thickness is given accurate to two parameters. These latter are selected so that the yield condition should be satisfied along the most stressed points of the structure, and that the maximum deflection should be equal to the admissible value. From the mathematical point of view the solution of the boundary problem for the differential equation or a system of such equations the variable coefficients of which include the unknown parameters is necessary here. For this purpose it is expedient to employ in this case the technique suggested by the author in 1953 [16], which reduces the process of numerical solution of the system of differential equations to using some simple recursion relations.

R. M. Tskhvedadze, under the author's supervision, considered the simplest case: an annular circular isotropic plate of a variable thickness is subjected to a uniform load. The inner edge of the plate is free, the outer one is simply supported. The thickness of the plate varies following the exponential law: $h = \beta e^{\gamma r}$, where β and γ are the unknown parameters. The latter are determined according to the conditions on the inner idge. The problem is relatively simplified, as we know in advance at which values of the radius the deflections and the stresses will attain their maximum.

References

1. Mikeladze, M. Sh.: Statics of anisotropic plastic shells, Tbilisi: Georgian SSR Acad. Sc. Press 1963. (in Russian).
2. Mikeladze, M. Sh.: Theory of shallow anisotropic plastic shells. Dokl. Akad. Nauk SSSR, Vol. 160, N. 4, 1965.
3. Mikeladze, M. Sh.: Shallow anisotropic plastic shells. Nuclear Structural Engineering, Amsterdam: North Holland 1, 1965, pp. 414—418.
4. Mikeladze, M. Sh.: Technical theory of thin plastic shells. Soobshcheniya Akad. Nauk Gruz. SSR (Bulletin of the Acad. Sc. of the Georgian SSR), Vol. XL, N. 3, 1965.
5. Mikeladze, M. Sh.: Estimation of the load carrying capacity of shallow shells of revolution and of circular plates in tension and bending. Prikladnaya Mekhanika (Applied Mechanics), Vol. 1, N. 2, 1965. (in Russian).
6. Mikeladze, M. Sh.: Some problems of the theory of spherical plastic shells. Rev. Roum. Sci. Techn. — Méc. Appl., t. 11, N. 6, pp. 1283—1295, Bucarest, 1966.
7. Nakamura, T.: Limit analysis of non-symmetric sandwich shells. IASS Symposium on Non-Classical Shell Problems, Amsterdam: North-Holland 1964.
8. Mikeladze, M. Sh.: Semi-momentless theory of plastic cylindrical shells. Dokl. Akad. Nauk SSSR, Vol. 154, N. 2, 1964.
9. Mikeladze, M. Sh.: Basic equations of the semi-momentless theory of cylindrical orthotropic plastic shells. Prikladnaya Mekhanika (Applied Mechanics), Vol. 1, N. 1, 1965 (in Russian).
10. Mikeladze, M. Sh.: Semi-momentless theory of thin plastic shells slightly different from cylindrical ones. J. Mech. Phys. Solids, **14**, 89—94, 1966.
11. Mikeladze, M. Sh.: On the plastic flow of anisotropic shells. Izvestiya Akad. Nauk SSSR, Otdeleniye Tekhnicheskikh Nauk, N. 8, 1955 (in Russian).
12. Petrovsky, I. G.: Lectures on equations with partial derivatives. Moscow 1950 (in Russian).
13. Mikeladze, M. Sh.: Analysis of the weight and the strength of rigid-plastic orthotropic shells. Arch. Mech. Stos., Warszawa **11**, 1, 1959 (in Russian).
14. Ilyushin, A. A., and Lensky, V. S.: Strength of materials, Moscow: Fizmatgiz 1959 (in Russian).
15. Lensky, V. S.: Influence of radioactive irradiation on mechanical properties of solid bodies. Inzhenerni Sbornik, Vol. XXVIII, USSR Acad. Sc. Press 1960 (in Russian).
16. Mikeladze, M. Sh.: Numerical solution of system of differential equations. Application of the technique to design of shells of revolution. Prikladnaya Matematika i Mekhanika (Applied Mathematics and Mechanics), Vol. XVII, N. 3, 1953.

Foundations and Basic Equations of Shell Theory
A Survey of Recent Progress

By

W. T. Koiter

Technische Hogeschool, Delft, Netherlands

Abstract. This survey aims at a discussion of progress in the past ten years, both with reference to the foundations of shell theory and to the resulting basic equations.

The foundations have been strengthened in two ways. Several investigators have explored a possible justification of the classical shell equations as a first approximation of asymptotic character. Other investigations aim at concrete estimates of the errors involved in shell theory, starting from the threedimensional theory of elasticity. Both approaches can claim a considerable measure of success. Some open questions at the time of writing this paper, will be discussed.

Many attempts have been made at a simplification of the basic equations without loss in accuracy. Such simplifications have already been known for a long time for spherical shells. Considerable success has now also been achieved for circular cylindrical shells, and some progress has also been made for more general shell shapes. Some remaining open questions will again be discussed, and it may be hoped that a partial answer is contained in other papers at the present symposium.

1. Introduction

Any twodimensional theory of thin shells is necessarily of an approximate character. An exact twodimensional theory of shells cannot exist, because the actual body we have to deal with, thin as it may be, is always threedimensional. We may perhaps illustrate this point in a somewhat facetious way: even if it appears to be the fashion for ladies to be as thin as possible, fortunately, in our view, they remain essentially threedimensional.

Since the theory we have to deal with is approximate in character, we feel that extreme rigour in its development is hardly desirable. Extreme rigour in the analysis of physical problems, we are inclined to believe, may easily lead to rigor mortis. Likewise, too much insistence on a systematic approach seems inadvisable to us. Flexible bodies like thin shells require a flexible approach. We shall have occasion to illustrate this view later on.

A complete and adequate set of equations for the linear theory of thin shells was developed by A. E. H. Love nearly eighty years ago, and this analysis forms a substantial part of his famous treatise on the theory of elasticity [1]. In spite of certain shortcomings in his analysis, such as a lack of symmetry in his expressions for the changes of curvature, the enormous effort spent on the further development of shell theory in the present century has not led to a significant improvement of Love's general equations. It is true that a better understanding has now been achieved of the nature of the approximations, that the derivation has been simplified considerably, and that the theory is now easily written in an invariant form, but the basic contents of Love's approximate equations remain essentially unchanged to the present day. At the risk of appearing facetious, we are inclined to quote a popular Beatle-song as an appropriate slogan for the linear theory of shells: all you need is Love, Love is all you need.

In the present paper we shall discuss two aspects of shell theory, the strengthening of the foundations and possible simplifications of the general equations without loss in accuracy, in particular with reference to the advances achieved in the past ten years.

2. Foundations of Shell Theory

It seems to us that two main approaches have to be distinguished in the strengthening of the foundations of shell theory in the past ten years, viz. the asymptotic approach and the approach based on a priori estimates of the errors involved. We shall leave aside a third approach, in which the shell is considered ab initio as a twodimensional body, viz. a Cosserat surface [2, 3]. Although the latter approach may be quite effective in providing us with a model of shell behaviour, it cannot contribute, in our view, to a strengthening of the foundations of shell theory because it ignores all threedimensional effects which are responsible for the approximate character of shell theory.

2.1. Asymptotic Approximations

The basic idea of asymptotic approximations as a justification of twodimensional shell equations is to expand the solution of the threedimensional equations of elasticity theory with respect to some small parameter related to the thickness of the shell [4—18]. An immediate difficulty is here that a variety of asymptotic expansions appears to be possible, and different types of asymptotic expansion have to be used in different problems, depending on geometry, boundary conditions and surface loads. The resulting shell theories are therefore formulated by a number of different sets of equations where one has to select the appropriate set for application in a particular problem. Moreover, it often happens that one set of equations applies to one part of the shell region and another set to another part of the surface, and their solutions then have to be matched in a transition zone.

This fundamental difficulty in systematic asymptotic expansions in shell theory, however, is not as serious in practice as it seems to be in principle. In the solution of a practical problem, formulated in terms of Love's single set of basic equations, it is usually possible, and from a practical point of view virtually necessary, to apply certain simplifications to these equations pertinent to the problem at hand. Here again it is often necessary to distinguish between several regions of the shell and to match the solutions of the different sets of simplified equations in adjoining regions. The various possible simplifications of Love's equations apparently correspond to the different sets of shell equations obtained from the various possible asymptotic expansions, and the latter expansions therefore provide in a certain sense a justification, on the basis of the threedimensional theory of elasticity, of the classical shell equations.

The occurrence of various types of asymptotic expansions in shell theory appears to be a consequence of the *systematic* application of an asymptotic procedure. Less systematic, but apparently more effective, is Reissner's ingenious idea [6—8] to assume the stress distribution according to the classical shell equations as a first approximation, and to show subsequently that the corrections on these stresses resulting from the threedimensional theory of elasticity are of a smaller order of magnitude. This analysis has established in a sense the validity of Love's shell equations, or any equivalent more modern set of equations, as a valid asymptotic approximation to the threedimensional theory.

2.2. A Priori Error Estimates

An inherent drawback of asymptotic methods is that no information is obtained on the actual magnitude of the errors made in the analysis. In this respect the second main approach to the reinforcement of the foundations of shell theory by means of a priori error estimates seems to be more powerful. Such estimates have been obtained by John [19] from the theory of elliptic partial differential equations. Starting from the threedimensional equations of the nonlinear theory of elasticity for a homogeneous isotropic solid, it is shown first that the state of stress, in the absence of surface loads, is indeed approximately plane. This proof is obtained in the form of concrete estimates for the magnitude of the transverse shear stresses $t_{\alpha 3}$ and the transverse normal stress t_{33}

$$t_{\alpha 3} = O(E\varepsilon\Theta), \qquad t_{33} = O(E\varepsilon\Theta^2), \tag{1}$$

where E is Young's modulus, ε is the maximum principal strain in the shell, and Θ is defined by

$$\Theta = \mathrm{Max}\left(\frac{h}{D}, \sqrt{\frac{h}{R}}, \sqrt{\varepsilon}\right), \tag{2}$$

where h is the shell thickness, D is the distance from the point on the middle surface under consideration to the edge of the shell, and R is effectively the smallest principal radius of curvature of the undeformed middle surface. The estimates (1) are supplemented by similar estimates for all derivatives of stresses in the shell.

These results are applied in [19] to derive a system of twodimensional equations, called by JOHN the lowest order interior equations of shell theory. We shall write these equations in a somewhat different form, which is equivalent to John's equations but is more familiar to workers in shell theory. The equations in question are formulated in terms of the tensor of stress resultants $n_\lambda{}^\varkappa$ and the tensor of changes of curvature $\varrho_\lambda{}^\varkappa$, representing the difference between the second fundamental tensors of the deformed middle surface and the undeformed middle surface,

$$\frac{1}{Eh}\, n_\varkappa{}^\varkappa|_\lambda{}^\lambda + b_\varkappa{}^\varkappa \varrho_\lambda{}^\lambda - b_\lambda{}^\varkappa \varrho_\varkappa{}^\lambda + \frac{1}{2}(\varrho_\varkappa{}^\varkappa \varrho_\lambda{}^\lambda - \varrho_\lambda{}^\varkappa \varrho_\varkappa{}^\lambda) = 0\left(\varepsilon\frac{\Theta^4}{h^2}\right), \tag{3}$$

$$\varrho^{\varkappa}_{\lambda|\mu} - \varrho^{\varkappa}_{\mu|\lambda} = 0\left(\varepsilon\frac{\Theta^3}{h^2}\right), \tag{4}$$

$$\frac{Eh^3}{12(1-\nu^2)}\, \varrho_\varkappa{}^\varkappa|_\lambda{}^\lambda - (b_\lambda{}^\varkappa + \varrho_\lambda{}^\varkappa)\, n_\varkappa{}^\lambda = 0(E\varepsilon\Theta^4), \tag{5}$$

$$n^{\lambda}_{\varkappa|\lambda} = 0(E\varepsilon\Theta^3), \tag{6}$$

where E and ν are Young's modulus and Poisson's ratio, h is the (constant) shell thickness, $b_\lambda{}^\varkappa$ is the undeformed second fundamental tensor, and raising and lowering of indices and covariant differentiation are defined with respect to the undeformed metric.

We emphasize that John's derivation of Eqs. (3)—(6) is completely rigorous and systematic, and it was a great shock to us to see that their linearized version does *not* reduce to any of the complete shell theories based on the classical Kirchhoff-Love assumptions. For example, instead of (6) we have for the equations of equilibrium in the tangential plane in the "best" linear theory of shells [23—26]

$$\left[n_\varkappa{}^\lambda + \frac{1}{2} b_\varkappa{}^\mu m_\mu{}^\lambda - \frac{1}{2} b_\mu{}^\lambda m_\varkappa{}^\mu\right]_{|\lambda} + b_\varkappa{}^\mu m_\mu{}^\lambda|_\lambda = 0, \tag{7}$$

where $m_\lambda^\varkappa$ is the tensor of stress couples. In many problems the bending terms in (7) may be ignored, and the equation then reduces to John's equation (6), but it is also known that the bending terms in (7) cannot be omitted in shell problems where the bending stresses are much larger than the membrane stresses (e. g. [25]).

Since we have complete confidence in any consistent classical linear theory of shells, and in view of the rigorous derivation of John's equation (6), it should evidently be possible to explain the discrepancy between Eqs. (6) and (7). The explanation is indeed quite simple. The right-hand member in (6) indicates the error made, if it is replaced by zero, and ε is the *maximum* principal strain occurring in the shell. If the bending strains are much larger than the membrane strains, each *separate* term in the left-hand member of (6) may be equally small as the error term, and in such cases (6) loses all meaning as an approximate equation. On the other hand, the classical shell equation (7) remains meaningful in these cases. The moral of this story is apparently a confirmation of our introductory remarks that too much insistence on rigour and systematics may lead us astray. In a difficult physical problem like shell theory there is both room and a need for intuition and imagination in addition to mathematical analysis.

The analysis based on a priori estimates has been elaborated further by John in his paper at the present conference [20]. The refined interior shell equations are complicated by the appearance of terms depending upon the cubic terms in the energy expressions, and even the linearized version of these refined equations seems to be more complicated than the linear equations of classical shell theory. A detailed comparison, however, would require more time and reflection than was available for the completion of the present paper. Further applications of John's a priori estimates have been made along different lines by C. B. Sensenig [21, 22]. Here again it is far from easy to draw a direct comparison with the predictions of classical shell theory.

The fundamental advantage of John's approach to shell theory, as outlined in the first paragraph of the present section, is that it yields concrete estimates of the errors made in the equations. It should perhaps be emphasized that these error estimates, although a marked improvement over earlier analyses, do not constitute the last word. It would be even more important to have bounds on the errors of the *solutions* of the approximate equations, as compared with the solutions of the three-dimensional equations. It would seem that this question is at present still a completely open problem, and it underlines the scope for (approximate) variational methods to be discussed presently.

2.3. Variational Derivation

In a previous paper [24] we have shown, on the assumptions that the state of stress in a thin shell is approximately plane and that the strains are small everywhere, that the elastic energy per unit area of the middle surface may be written as the sum of extensional and flexural strain energies. Since the assumptions made in [24] have now been proved rigorously by John's estimates (1) in the absense of surface loads, the approximate decomposition of elastic energy in extensional energy and flexural energy may now be considered to be a proven theorem, and the relative error of the approximate energy expression is of order Θ^2, where Θ is John's parameter (2).

The undeformed middle surface of the shell is characterized by its first and second fundamental tensors $a_{\alpha\beta}$ and $b_{\alpha\beta}$, the deformed middle surface by the similar tensors $\bar{a}_{\alpha\beta}$ and $\bar{b}_{\alpha\beta}$. The deformation of the middle surface may then be described by the middle surface strain tensor $\gamma_{\alpha\beta}$ and the tensor of changes of curvature $\bar{\varrho}_{\alpha\beta}$, defined by

$$2\gamma_{\alpha\beta} = \bar{a}_{\alpha\beta} - a_{\alpha\beta}, \qquad \bar{\varrho}_{\alpha\beta} = \bar{b}_{\alpha\beta} - b_{\alpha\beta}. \tag{8}$$

For some purposes it is more convenient to employ a modified tensor of changes of curvature $\varrho_{\alpha\beta}$, defined by

$$\varrho_{\alpha\beta} = \bar{\varrho}_{\alpha\beta} - \frac{1}{2}(b_\alpha^{\ \varkappa}\gamma_{\varkappa\beta} + b_\beta^{\ \varkappa}\gamma_{\varkappa\alpha}). \tag{9}$$

Three equations between the six measures of strain $\gamma_{\alpha\beta}$ and $\bar{\varrho}_{\alpha\beta}$ are obtained from the equations of GAUSS and CODAZZI. If we consider virtual deformations of the shell which obey the Kirchhoff-Love assumptions, these virtual deformations are specified completely by variations of $\gamma_{\alpha\beta}$ and $\bar{\varrho}_{\alpha\beta}$, and the internal virtual work per unit area of the middle surface is described by

$$\bar{n}^{\alpha\beta}\delta\gamma_{\alpha\beta} + \bar{m}^{\alpha\beta}\delta\bar{\varrho}_{\alpha\beta}, \tag{10}$$

where $\bar{n}^{\alpha\beta}$ and $\bar{m}^{\alpha\beta}$ are the symmetric tensors of stress resultants and stress couples. The principle of virtual work then yields three equations of equilibrium to be satisfied by the stress resultants and stress couples.

We emphasize that the equations of compatibility, obtained from Gauss's and Codazzi's equations, and the equations of equilibrium are fully exact. This set of six equations involving twelve unknowns, however, is incomplete. It has to be supplemented by suitable constitutive equations between the stress resultants and stress couples on the one hand, and the middle surface strains and changes of curvature on the other hand. Here it is that the approximate nature of shell theory appears. Since the elastic energy is (approximately) the sum of stretching and bending energies, we may express the stress resultants in terms of the middle surface strains and the stress couples in terms of the changes of curvature.

A detailed discussion of the resulting complicated equations is given in [27], and it will not be reproduced here. A certain simplification is achieved, if the further assumption of small strains in the middle surface is made, in the sense that no distinction need be made between covariant differentiation with respect to the undeformed and the deformed metric. This assumption does not always apply, for example not in buckling problems with mainly extensional deformation in the buckling mode, but it is appropriate for most problems. Employing the modified tensor of changes of curvature (9), the simplified equations are [27, p. 34].

$$\varepsilon^{\alpha\beta}\varepsilon^{\lambda\mu}\left[\gamma_{\alpha\mu|\beta\lambda} + b_{\alpha\mu}\varrho_{\beta\lambda} + \frac{1}{2}\varrho_{\alpha\mu}\varrho_{\beta\lambda}\right] = 0, \tag{A}$$

$$\varepsilon^{\alpha\beta}\varepsilon^{\lambda\mu}\left[\left(\varrho_{\beta\lambda} + \frac{1}{2}b_\beta^{\ \varkappa}\gamma_{\varkappa\lambda} + \frac{1}{2}b_\lambda^{\ \varkappa}\gamma_{\varkappa\beta}\right)\Big|_\mu - b_\lambda^{\ \varkappa}(\gamma_{\varkappa\beta|\mu} + \gamma_{\varkappa\mu|\beta} - \gamma_{\beta\mu|\varkappa})\right] = 0, \tag{B}$$

$$m^{\alpha\beta}|_{\alpha\beta} - \left(\frac{1}{2}b_\alpha^{\ \varkappa}\varrho_{\varkappa\beta} + \frac{1}{2}b_\beta^{\ \varkappa}\varrho_{\varkappa\alpha} + \varrho_\alpha^{\ \varkappa}\varrho_{\varkappa\beta}\right)m^{\alpha\beta} - (b_{\alpha\beta} + \varrho_{\alpha\beta})n^{\alpha\beta} = 0, \tag{C}$$

$$\left[n^{\beta\alpha} + \frac{1}{2}b_\varkappa^{\ \alpha}m^{\beta\varkappa} - \frac{1}{2}b_\varkappa^{\ \beta}m^{\alpha\varkappa} + \varrho_\varkappa^{\ \alpha}m^{\beta\varkappa}\right]\Big|_\beta + (b_\varkappa^{\ \alpha} + \varrho_\varkappa^{\ \alpha})m^{\beta\varkappa}|_\beta = 0, \tag{D}$$

$$\gamma_{\alpha\beta} = \frac{1}{Eh}\left[(1+\nu)n_{\alpha\beta} - \nu n_\varkappa^{\ \varkappa}a_{\alpha\beta}\right], \quad m^{\alpha\beta} = \frac{Eh^3}{12(1-\nu^2)}\left[(1-\nu)\varrho^{\alpha\beta} + \nu\varrho_\varkappa^{\ \varkappa}a^{\alpha\beta}\right], \tag{E}$$

where $\varepsilon^{\alpha\beta}$ is the contravariant alternating tensor.

It is perhaps of some interest to note here that these equations are also capable of describing the elastica problem of cylindrical bending of a flat plate, as discussed in John's paper [20]. In this problem only one significant Cartesian coordinate x^1 occurs, perpendicular to the generators, and differentiations with respect to this coordinate will be denoted by primes. Eqs. (A) and (B) become identities, and the significant quantities in the remaining equations are $m^{11} = m_{11} = m$, $n^{11} = n_{11} = n$, and $\varrho^{11} = \varrho_{11} = \varrho$. Eqs. (C), (D) and (E) reduce to

$$m'' - \varrho^2 m - \varrho n = 0, \quad (n + \varrho m)' + \varrho m' = 0, \quad m = \frac{Eh^3}{12(1-\nu^2)}\varrho. \tag{11}$$

Eliminating m and n, this system may be integrated once, and the resulting equation

$$\frac{Eh^3}{12(1-\nu^2)}\left(\varrho'' + \frac{1}{2}\varrho^3\right) - A\varrho = 0, \tag{12}$$

where A is a constant of integration, coincides with John's similar equation. We omit its further reduction to the standard form of the elastica equation.

3. Simplifications of General Shell Equations

One of the best known versions of simplified shell equations is associated with the names of DONNELL, MARGUERRE, MUSHTARI and WLASSOW, and it is often referred to as shallow shell theory. Unfortunately, this theory rests on a number of simplifying assumptions with reference to the deformation pattern which are not always applicable, and which are even sometimes not easily capable of an a priori verification. A modified and apparently less restrictive theory of a similar type was developed by LIBAI [28] and in [27]. It may be based on the two assumptions that the orders of magnitude of bending and membrane strains are not too widely different (such that John's lowest order interior shell equations constitute a valid approximation), and that the Gaussian curvature K of the middle surface is small in the sense that $|K|L^2$ is small in comparison with unity, where L is the (minimum) wave length of the deformation pattern on the middle surface. Even if this so-called theory of quasi-shallow shells seems to be more securely founded than the classical theory of shallow shells, it remains a simplification which is incapable of dealing with all types of shell problems. For this reason we shall ignore in the sequel the theories of shallow and quasi-shallow shells, and we restrict our attention to some special shell shapes for which a considerable simplification, applicable to all types of problems, is possible without any significant loss in accuracy.

3.1. Spherical Shells

A significant simplification of the equations has been achieved by VAN DER NEUT [29] in his analysis of the stability of thin spherical shells subjected to external pressure. His analysis has been applied by HAVERS [30] to the classical linear theory of shells, and it is also reproduced in a slightly different form in Wlassow's treatise [31]. Van der Neut's simplification is the result of writing the tangential displacement components in terms of two invariant functions φ and ψ. In tensor notation his substitution reads

$$u_\alpha = \varphi_{,\alpha} + \varepsilon_{\alpha\lambda} a^{\lambda\mu} \psi_{,\mu}. \tag{13}$$

The simplification achieved is due to the fact that the elastic strain energy splits up into two parts, one of them depending only on the invariant ψ, and the other one depending only on the invariant φ and the normal deflection w. Hence we obtain one differential equation for ψ, and a pair of simultaneous equations for φ and w, instead of the three simultaneous equations for the displacement components in general shell theory.

A second type of simplified equations for spherical shells was developed by WLASSOW [31]. This development is based on the fact that the equations of equilibrium in the tangential plane may be solved in terms of a stress function, in view of the circumstance that the Gaussian curvature is a constant. Wlassow's alternative simplified equations thus consist of a pair of simultaneous equations for a stress function and the normal deflection. Similar equations have also been obtained by a slightly different argument in [32], and recent work by LUKASIEWICZ [33] and BARTA [34] indicates that they also apply to *nearly* spherical shells.

3.2. Circular Cylindrical Shells

A substantial improvement of Donnell's equations was suggested by MORLEY [35]. Starting from Donnell's eighth-order homogeneous equation for the normal deflection, in the absence of surface loads,

$$\Delta^4 w + 12(1-\nu^2)\frac{R^2}{h^2}\frac{\partial^4 w}{\partial \xi^4} = 0, \tag{14}$$

where ξ is the non-dimensional axial coordinate, and Δ is the non-dimensional Laplacian operator, he showed that the slightly modified equation

$$\Delta^2(\Delta+1)^2 w + 12(1-\nu^2)\frac{R^2}{h^2}\frac{\partial^4 w}{\partial \xi^4} = 0, \tag{15}$$

is equally accurate as Fluegge's much more complicated equations [36]. Morley's numerical substantiation of his Eq. (15) did not leave anything to desire, but his derivation was not quite convincing. A significant improvement in this respect was achieved by SIMMONDS [37], and here again recent work by LUKASIEWICZ [33] and BARTA [34] points at the approximate validity of similar simplified equations in the case of *nearly* circular cylindrical shells.

It may be worthwhile to mention here a different derivation of equations equivalent to those of MORLEY and SIMMONDS of slightly greater generality. This derivation bears some resemblance to the analysis by LUKASIEWICZ [33], but it appears to be founded more securely. The starting-point is the following observation. The bending terms in (D) and the membrane terms in (B) cannot be significant at the same time, and these terms can therefore always be ignored in at least one of these sets of equations. These equations can now be solved in terms of a stress function F and a curvature function W, both in the case of a spherical shell and a circular cylindrical shell, because the second fundamental tensor is covariantly constant in both cases. This method of solution even applies to the full nonlinear equations (B) and (D). The remaining Eqs. (A) and (C) are thus reduced to two sinultaneous fourth-order equations for F and W.

Finally, it seems useful to note that equally simple *displacement* equations of equilibrium can be derived for circular cylindrical shells by adding suitable negligible terms to the elastic energy per unit area of the middle surface in the spirit of [24]. Eliminating the tangential displacement components, Morley's equation (15) is recovered. It seems a minor advantage of this displacement method that all types of boundary conditions are easily formulated.

3.3. Cylindrical Shells of Arbitrary Cross-Section

In the case of Poisson's ratio ν equal to zero, Novozhilov has shown that the general equations of linear shell theory may be written in complex form, thus reducing the order of the system from eight to four [38], and this reduction involves no further approximations. A similar reduction of the equations for non-zero values of Poisson's ratio in [38] involves certain approximations, the accuracy of which is not easily assessed. In the case of cylindrical shells of arbitrary cross-section, however, Novozhilov's resulting equations seem to be entirely accurate. In fact, in contradistinction to shells of completely arbitrary shape, to be discussed in the next section, we are not aware of any problem for general cylindrical shells in which Novozhilov's simplified equations fail to yield the correct solution. In the absence of evidence to the contrary, it seems therefore that Novozhilov's simplified equations in complex form may be applied with confidence to all problems of cylindrical shells.

3.4. Shells of Arbitrary Shape

We have already mentioned Novozhilov's well-known reduction of the general equations of linear shell theory to a complex form [38], where certain approximations have to be introduced in order to achieve such a reduction for non-zero values of Poisson's ratio. It is not at all easy to assess the accuracy of these additional approximations, in spite of Novozhilov's seemingly entirely plausible argument. In fact, as we have noted in a previous paper [25], this argument cannot apply to all possible situations. In the case of nearly inextensional bending of helicoidal shells the solution of Novozhilov's equations cannot reduce to Cohen's complete solution of the general shell equations for loads independent of the helix angle [39, 40], and the error in Novozhilov's solution is *finite*, that is independent of the ratio h/R.

A somewhat different reduction of the general equations to a complex form has been discussed by NAGHDI [41]. Without any further approximations NAGHDI also obtains a set of equations of the fourth order, but these equations contain the unknowns and their complex conjugates simultaneously, and the effective order of these equations is therefore again eight. Here again it is only possible by means of more or less plausible further approximations to reduce the order of the system to four.

Finally, an entirely new approach is to be discussed by SANDERS in his paper at the present conference [42], and it is to be hoped that his analysis will result in a set of equations valid for all possible problems of linear shell theory.

References

1. LOVE, A. E. H.: The mathematical theory of elasticity, 4th ed., Cambridge: Cambridge University Press 1927.
2. COHEN, H., and C. N. DESILVA: Nonlinear theory of elastic surfaces. J. Math. Phys. **7**, 246–253 (1966).
3. GREEN, A. E., and P. M. NAGHDI: Shells in the light of generalized continua. Proceedings Second IUTAM Symposium on the theory of Thin Shells, Copenhagen, September 1967 (in this book, pp. 39).
4. JOHNSON, M. W., and E. REISSNER: On the foundations of the theory of thin elastic shells. J. Math. and Phys. **37**, 374–392 (1958).
5. REISSNER, E.: On some problems in shell theory. Proc. First Symp. Naval Structural Mechanics, 74–114. Oxford: Pergamon Press 1960.
6. REISSNER, E.: On the derivation of the theory of elastic shells. J. Math. and Phys. **42**, 263–277 (1963).
7. REISSNER, E.: On the foundations of the theory of elastic shells. Proc. 11th Int. Congr. Appl. Mech., Munich 1964, 20–30, Berlin/Heidelberg/New York: Springer 1966.
8. REISSNER, E.: Foundations of generalized shell theory. Proc. Second IUTAM Symposium on the Theory of Thin Shells, Copenhagen, September 1967 (in this book, pp. 15).
9. REISS, E. L.: A theory for the small rotationally symmetric deformations of cylindrical shells. Comm. Pure and Appl. Math. **13**, 531–550 (1960).
10. REISS, E. L.: On the theory of cylindrical shells. Quart. J. Mech. and Appl. Math., **15**, 325–338 (1962).
11. GREEN, A. E.: On the linear theory of thin elastic shells. Proc. Roy. Soc. London **A 266**, 143–160 (1962).
12. GREEN, A. E., and P. M. NAGHDI: Some remarks on the linear theory of shells. Quart. J. Mech. and Appl. Math., **18**, 257–276 (1965).
13. GOL'DENWEIZER, A. L.: Derivation of an approximate theory of bending of a plate by the method of asymptotic integration of the equations of the theory of elasticity. Prikl. Mat. Mekh. **26**, 668–686 (1962). English translation, 1000–1025.
14. GOL'DENWEIZER, A. L.: Derivation of an approximate theory of shells by means of asymptotic integration of the equations of the theory of elasticity. Prikl. Mat. Mekh. **27**, 593–608 (1963). English translation, 903–924.

15. Gol'denweizer, A. L.: The principles of reducing threedimensional problems of elasticity to twodimensional problems of the theory of plates and shells. Proc. 11th Int. Congr. Appl. Mech., Munich 1964, 306—311, Berlin/Heidelberg/New York: Springer 1966.
16. Gol'denweizer, A. L.: Problems in the rigorous deduction of the theory of thin elastic shells. Proc. Second. IUTAM Symp. Theory of Thin Shells, Copenhagen, September 1967 (in this book, pp. 31).
17. Cicala, P.: Systematic approximation approach to linear shell theory. Torino: Levrotto and Bella 1965.
18. Rutten, H. S.: Asymptotic approximation in the threedimensional theory of thin and thick elastic shells. Proc. Second IUTAM Symp. Theory of Thin Shells, Copenhagen, September 1967 (in this book, pp. 115).
19. John, F.: Estimates for the derivatives of the stresses in a thin shell and interior shell equations. Comm. Pure and Appl. Math. **18**, 235—267 (1965).
20. John, F.: Refined interior shell equations. Proc. Second IUTAM Symp. on the Theory of Thin Elastic Shells, Copenhagen, September 1967 (in this book, pp. 1).
21. Sensenig, C. B.: A nonlinear shell theory compared with the classical threedimensional theory of elasticity. Report IMM 349, Courant Institute of Mathematical Sciences, New York University 1966.
22. Sensenig, C. B.: A shell theory compared with the exact threedimensional theory of elasticity. Submitted for publication in Int. Journ. Eng. Science 1967.
23. Sanders, J. L.: An improved first-approximation theory for thin shells. NASA Report 24 (1959).
24. Koiter, W. T.: A consistent first approximation in the general theory of thin elastic shells. Proc. IUTAM Symp. Theory of Thin Elastic Shells, Delft, August 1959. Amsterdam: North-Holland 1960 pp. 12—33.
25. Koiter, W. T.: A systematic simplification of the general equations in the linear theory of thin shells. Proc. Kon. Ned. Ak. Wet. B **64**, 612—619 (1961).
26. Budiansky, B., and J. L. Sanders: On the "best" first-order linear shell theory. Progress in Applied Mechanics (Prager Anniversary Volume), 129—140, New York: Macmillan 1963.
27. Koiter, W. T.: On the nonlinear theory of thin elastic shells. Proc. Kon. Ned. Ak. Wet. **B 69**, 1—54 (1966).
28. Libai, A.: On the nonlinear elastokinetics of shells and beams. Journ. Aerospace Sci. **29**, 1190—1195 (1962).
29. van der Neut, A.: The elastic stability of the thin-walled spherical shell (in Dutch). Thesis Delft, Amsterdam: H. J. Paris 1932.
30. Havers, A.: Asymptotische Biegetheorie der unbelasteten Kugelschalen. Ingenieur Archiv **6**, 282—312 (1935).
31. Wlassow, W. S.: Allgemeine Schalentheorie und ihre Anwendung in der Technik. Berlin: Akademie Verlag 1958.
32. Koiter, W. T.: A spherical shell under point loads at its poles. Progress in Applied Mechanics (Prager Anniversary Volume), 155—169, New York: Macmillan 1963.
33. Lukasiewicz, S.: The equations of the theory of non-shallow thin shells. Arch. Budowy Maszyn **12**, 431—443 (1965).
34. Barta, T. A.: An engineering theory of thin elastic shells, Part 1, Linear theory. Unpublished report of Civil Engineering Dept., University College, London 1966.
35. Morley, L. S. D.: An improvement on Donnell's approximation for thinwalled circular cylinders. Quart. Journ. Mech. Appl. Math. **12**, 89—99 (1959).

36. Flügge, W.: Stresses in shells. Berlin/Heidelberg/New York: Springer 1960 (corrected reprint 1962).
37. Simmonds, J. G.: A set of simple, accurate equations for circular cylindrical elastic shells. Int. J. Solids and Structures 2, 525—541 (1966).
38. Novozhilov, V. V.: The theory of thin shells. Translated from the Russian (1951), ed. by P. G. Lowe. Groningen: Noordhoff 1959.
39. Cohen, J. W.: On stress calculations in helicoidal shells and propeller blades. Thesis, Delft. Delft: Waltman 1955.
40. Cohen, J. W.: The inadequacy of the classical stress-strain relations for the right helicoidal shell. Proc. IUTAM Symposium on the Theory of Thin Elastic Shells, Delft, August 1959, Amsterdam: North-Holland 1960 pp. 415—433.
41. Naghdi, P. M.: On the differential equations of the linear theory of shells. Proc. 11th Int. Congr. Appl. Mech., Munich 1964, Berlin/Heidelberg/New York: Springer 1966 pp. 262—269.
42. Sanders, J. L.: On the shell equations in complex form. Proc. Second IUTAM Symposium on the Theory of Thin Shells, Copenhagen, September 1967 (in this book, pp. 135).

Discussion

P. M. Naghdi: With regard to the development of Professor John leading to a simpler Eq. (6), it seems that an assumption has already been made about the nature of the solution. You mentioned simplifacations pertaining not only to the geometry of the shell, i. e., the smallness of h/R, but also other parameters such as ε. I believe you mentioned approximations involving a wave length which seems to imply some knowledge of the solution.

W. T. Koiter: No, I don't think it really implies knowledge of the solution, but it implies allowance for various possibilities.

P. M. Naghdi: I think it implies some assumption about class of solutions.

W. T. Koiter: I think I would like to have Professor John's opinion on this.

F. John: No assumption about the wave length of the solution is made. For the linear case you would not even need the $\sqrt{\varepsilon}$, but you would need the other two quantities in the definition of Θ.

B. Budiansky: You mentioned that the term analogous to $N^{\alpha\beta}$ in Professor John's work was not in fact the stress resultant but the stress at the middle surface. Now, of course, the stress at the middle surface is a symmetric tensor, whereas the stress resultant is not and I notice that some of the differences between your equations and Professor John's consist of terms involving products of $b_{\alpha\beta}$ and $M^{\alpha\beta}$, which are very much like the terms one introduces to symmetrize the stress resultant tensor. Is it possible that this is responsible for the lack of agreement between the two sets of equations?

W. T. Koiter: That is only part of the explanation because the last term in (7) is the transverse shear force multiplied by the curvature tensor, which also comes in essentially, if the bending stresses dominate over the membrane stresses. What Professor Budiansky refers to is of course contained in the expression within brackets in (7), which is an asymmetric tensor of stress resultants, and $n_\lambda{}^k$ is its symmetric part. The symmetric part is the significant one physically — from my point of view — because the symmetric tensor of stress resultants is introduced through the principle of virtual work.

A Theory of Thin Elastic Shells with Local Structural Effects

By

M. Mișicu

Academia R. S. R., Bucharest, Rumania

The theory of thin shells with an enough fine reticular or alveolar structure can be developed in the frame of the continuum theory, [1—7], of two, [8—10], or three dimensional bodies, [11—15]. The approximation depends on the accuracy of the equivalence between the structural element and the adopted mechanical model of the continuum. While the equivalence was studied only in particular cases, concerning lattices, or composite materials, the theory of generalized continua was more deeply elaborated in the last years. Taking into account this fact, it appears that only the correlation of both directions can more succesfully answer to a practical purpose. Since the equivalence problem shall not be discussed here, being needed additional accumulated results, we restrict ourselves to the case of a shell constituted by a Cosserat material.

Without dealing properly with the equivalence problem, by assuming the vanishing of a linear function of distortions, the theory appears endowed with features compatible with the description of a large class of thin structures. In this sense, the aim of the present analysis is to stimulate subsequent more elaborated developments and to revaluate the classical methods of practical interest including field-correspondence, complex formulations and potential representation, [16—23], of differential conditions[1].

---

[1] In the same direction, previously were published some studies concerning non-constitutive or different particular models, [11, 13]. In a recent work [14], the case of vanishing transversal strain is analysed. More recently, in the paper [15], M. Mișicu and V. Nicolae consider the case of vanishing transversal distortion which allow a systematic treatment of the static-geometric analogy and a complex representation. The classical theory of Love-Kirchoff type can be obtained only by imposing kinematic restrictions. Here, owing to a more general assumption, we need only the vanishing of asymmetry moduli.

The need of cheking up the validity of a first order theory, from the point of view of a continuum shell theory also should be emphasized. In this sense we mention the results of E. REISSNER and F. Y. M. WAN.

Let

$$T\,da = \sum_i t_i\,da_i\,, \qquad M\,da = \sum_i m_i\,da_i \tag{1}$$

and t_i, m_i be the resulting stress and couple-stress on the surface da, normal to n and let be da_i the surface elements in the $\Theta_i = \text{const.}$ planes. We denote the base vectors by g_i and the corresponding metric form by $g_{ij} = g_i g_j$. According to the relations

$$da_i = (g g^{ii})^{1/2} dA_i\,, \quad g = |g_{ij}|\,, \quad dA_i = d\Theta^j\, d\Theta^k, \quad i \neq j \neq k \neq i\,,$$
$$n = g^i n_i\,, \quad n\,da = g^i\,da/(g^{ii})^{1/2}, \tag{2}$$

(1) can be reformulated under the form of the balance equations

$$T = \sum_i T_i\,dA_i\,, \qquad M = \sum_i M_i\,dA_i\,. \tag{3}$$

T_i and M_i stand for

$$T_i = (g g^{ii})^{1/2} t_i = g^{1/2} \sigma^{ji} g_j\,, \qquad M_i = (g g^{ii})^{1/2} m_i = g^{1/2} \mu^{ji} g_j\,. \tag{4}$$

Accordingly, the balance equations reduce as follows

$$T_{i,i} + \varrho(F - f) = 0\,, \qquad M_{i,i} + g_i \times T_i + \varrho(H - h) = 0\,. \tag{5}$$

$F = F^i g_i$, H, f and h represent mass and inertial forces and couples in the material component. Further, using the stress components in the shell space,

$$\Sigma^{\beta i} = \gamma_\alpha{}^\beta \sigma^{\alpha i}, \quad \Sigma^{3i} = \sigma^{3i}, \quad M^{\beta i} = \gamma_\alpha{}^\beta \mu^{\alpha i}, \quad M^{3i} = \mu^{3i},$$
$$(\gamma_\alpha{}^\beta = \delta_\alpha{}^\beta - b_\alpha{}^\beta \Theta_3; \quad \alpha, \beta = 1, 2)\,. \tag{6}$$

$b_\alpha{}^\beta$ being the coefficients of the second fundamental form of the shell median surface, the Eq. (5) can be reformulated under the form

$$(m\Sigma^{\beta\alpha})_{|\alpha} + (m\Sigma^{\beta 3})_{,3} - m\Sigma^{3\alpha} b_\alpha{}^\beta + \varrho m(F^\beta - f^\beta) = 0\,,$$
$$(m\Sigma^{3\alpha})_{|\alpha} + (m\Sigma^{33})_{,3} + m\Sigma^{\beta\alpha} b_{\beta\alpha} + \varrho m(F^3 - f^3) = 0\,,$$
$$(m M^{\beta\alpha})_{|\alpha} + (m M^{\beta 3})_{,3} - m M^{3\alpha} b_\alpha{}^\beta +$$
$$+ m \Sigma_{\delta\gamma} a^{\delta\beta} (\gamma_\alpha{}^\gamma \Sigma^{3\alpha} - \Sigma^{\gamma 3}) + \varrho m(H^\beta - h^\beta) = 0\,,$$
$$(m M^{3\alpha})_{|\alpha} + (m M^{33})_{,3} + m M^{\beta\alpha} b_{\beta\alpha} +$$
$$+ m \gamma_\alpha{}^\gamma \varepsilon_{\beta\gamma} \Sigma^{\gamma\alpha} + \varrho m(H^3 - h^3) = 0\,. \tag{7}$$

$(\ )_{|\alpha}$ stands for the covariant derivative on the surface, $\left(m = (g)^{1/2}\right)$.

After the integration over the shell height (h) and using the integrals

$$(n, m, s)^{ij} = \int_{-h/2}^{h/2} m(\Sigma, \Sigma\Theta_3, M)^{ij}\, d\Theta_3$$

$$(P_n, Q_n)^i = \int_{-h/2}^{h/2} m\big((F - f)\,\Theta_3^n, \quad (H - h)\,\Theta_3^n\big)^i\, d\Theta_3. \tag{8}$$

(7) can be reformulated under the form

$$n^{\alpha\beta}{}_{|\beta} - b_\beta{}^\alpha n^{3\beta} + P_0{}^\alpha = 0, \qquad n^{3\beta}{}_{|\beta} + b_{\alpha\beta} n^{\alpha\beta} + P_0{}^3 = 0, \tag{9}$$

$$m^{\alpha\beta}{}_{|\beta} - b_\beta{}^\alpha m^{3\beta} - n^{\alpha 3} + P_1 = 0, \qquad m^{3\beta}{}_{|\beta} + b_{\alpha\beta} m^{\alpha\beta} - n^{33} + P_1{}^3 = 0, \tag{10}$$

$$s^{\alpha\beta}{}_{|\beta} - b_\beta{}^\alpha s^{3\beta} + \varepsilon^{\alpha}_{.\beta}(n^{3\beta} - n^{\beta 3} - b_\gamma{}^\beta m^{3\gamma}) + Q_0{}^\alpha = 0,$$

$$b_{\alpha\beta} s^{\alpha\beta} + s^{3\beta}{}_{|\beta} - \varepsilon_{\alpha\beta}(n^{\alpha\beta} - b_\gamma{}^\beta m^{\alpha\gamma}) + Q_0{}^3 = 0, \qquad (\alpha, \beta = 1, 2) \tag{11}$$

which is used in the theory and the applications based on the hypothesis of Reissner type, [12, 23], also used here, and the assumptions concerning the vanishing of the transverse strain.

2. The variant theory exposed here resort to the simplified system

$$\varepsilon_{\alpha\beta}(n^{\alpha\beta} - b_\gamma{}^\beta r^{\alpha\gamma} - r^{3\alpha\beta}) = Q_0{}^3$$

$$n^{\alpha\beta}{}_{|\beta} - b_\beta{}^\alpha(r^{\beta\gamma}{}_{|\gamma} + b_\gamma{}^\delta \varepsilon_{\delta.}^{\beta} \varepsilon_{\varphi.}^{\gamma} \cdot r^{3\varphi} - P_1{}^\beta - \varepsilon_\gamma{}^\beta Q_0{}^\gamma) + P_0{}^\alpha = 0$$

$$r^{\alpha\beta}{}_{|\alpha\beta} + b_\gamma{}^\beta \varepsilon_{\beta.}^{\alpha} \varepsilon_{\lambda.}^{\gamma} \cdot r^{3\lambda}{}_{|\alpha} + b_{\alpha\beta} n^{\alpha\beta} + P^{\alpha}_{1|\alpha} + \varepsilon^{\alpha}_{.\gamma} Q^{\gamma}_{0|\alpha} + P_0{}^3 = 0. \tag{12}$$

Here we have denoted the total resulting couples

$$r^{\alpha\beta} = m^{\alpha\beta} + \varepsilon^{\alpha}_{.\gamma} s^{\gamma\beta}, \qquad r^{3\alpha} = \varepsilon^{\alpha}_{.\gamma} \cdot s^{3\gamma}. \tag{13}$$

(12) results from (9) after elimination of the sum $n^{\beta 3} + b_\gamma{}^\beta m^{3\gamma}$ according to $(10)_1$ so that $(11)_2$ becomes

$$r^{\alpha\beta}{}_{|\beta} + b_\gamma{}^\beta \varepsilon_{\beta.}^{\alpha} \varepsilon_{\delta.}^{\gamma} r^{3\delta} + P_1{}^\alpha + \varepsilon^{\alpha}_{.\beta} Q_0{}^\beta = n^{3\alpha}. \tag{14}$$

Hence, substituting $n^{3\alpha}$ given by (14) and $(9)_2$ respectively in $(9)_1$ and (14) we obtain (12).

As in the symmetrical case $(10)_2$ shall not be used. The products of curvatures ($b\,.\,.\,b\,.\,.$) shall be neglected.

The solution of (11), in view of (12) and (14) can be represented under the form

$$n^{\alpha\beta} = \varepsilon^{\beta\gamma}(N^\alpha{}_{|\gamma} - b_\gamma{}^\alpha N^3) + n_0{}^{\alpha\beta}, \quad n^{3\alpha} = \varepsilon^{\alpha\gamma}(N^3{}_{|\gamma} + b_{\gamma\lambda} N^\lambda) + n_0{}^\alpha,$$

$$r^{\alpha\beta} = \varepsilon^{\alpha}_{.\gamma}\big(\varepsilon^{\beta\delta}(R^\gamma{}_{|\delta} - b_\delta{}^\gamma R^3) + a^{\beta\gamma} N^3\big) + r_0{}^{\alpha\beta},$$

$$r^{3\alpha} = -(R^{3\alpha} + b_\beta{}^\alpha R^\beta) + \varepsilon^{\beta\alpha} N_\beta + r_0{}^{3\alpha}. \tag{15}$$

The functions with zero indices correspond to mass and inertial forces or couples.

The kinematic field of displacements $u_\alpha(\Theta^i) = v_\alpha(\Theta^\beta) + \Theta^3\beta(\Theta^\beta)$, $u_3(\Theta^i) = v_3(\Theta^\alpha)$ (see [12, 23]) and of the local rotations $\overline{\omega}^i$ lead to the following expressions of the strain, of the constrained and relative rotations, [11, 15]

$$e_{\alpha\beta} = u_{(\alpha;\beta)}|_{\Theta_3=0} = (v_{\alpha|\beta} + v_{\beta|\alpha} - 2b_{\alpha\beta}v_3)/2$$
$$e_{\alpha 3} = (\beta_\alpha + \gamma_{3\alpha})/2, \qquad \omega_{\alpha\beta} = u_{[\beta;\alpha]}|_{\Theta_3=0} = v_{[\beta|\alpha]}$$
$$\omega_{3\alpha} = u_{[\alpha;3]}|_{\Theta_3=0} = (\beta_\alpha - v_{3|\alpha} - b_\alpha{}^\beta v_\beta)/2, \qquad \Omega^\alpha = \overline{\omega}^\alpha - \omega^\alpha. \tag{16}$$

Here $\varepsilon^{\alpha\beta} = \varepsilon^{\alpha\beta 3}|_{\Theta_3=0}$ and $\omega^\alpha = -\varepsilon^{\alpha\beta}\omega_{3\beta}$, $\omega^3 = \varepsilon^{\alpha\beta}\omega_{\alpha\beta}/2$. Accordingly, the components of the distortion $\beta_{ij} = (u_{i;j} + \varepsilon_{ijk}\overline{\omega}^k)_{\Theta_3=0}$ and the twist curvature tensor $\varkappa_{ij} = \overline{\omega}_{i;j}|_{\Theta_3=0}$ take the form

$$\beta_{\alpha\beta} = \gamma_{\alpha\beta} + \varepsilon_{\alpha\beta}\overline{\omega}^3, \qquad \beta_{3\beta} = \gamma_{3\beta} + \varepsilon_{\beta\alpha}\overline{\omega}^\alpha,$$
$$\beta_{\alpha 3} = e_{3\alpha} - \varepsilon_{\alpha\beta}\Omega^\beta, \qquad \varkappa_{\alpha\beta} = \overline{\omega}_{\alpha|\beta} - b_{\alpha\beta}\overline{\omega}_3, \qquad \varkappa_{3\beta} = \overline{\omega}_{3|\beta} + b_\beta{}^\alpha\overline{\omega}_\beta, \tag{17}$$

where

$$\gamma_{\alpha\beta} = v_{\alpha|\beta} - b_{\alpha\beta}v_3, \qquad \gamma_{3\beta} = v_{3|\beta} + b_\beta{}^\alpha v_\alpha. \tag{18}$$

Further we shall use the following notations

$$\varrho^{\alpha\beta} = \varepsilon^{\alpha\gamma}\varepsilon^{\beta\delta}\beta_{\gamma\delta}, \qquad \varrho^{3\alpha} = -\beta^{3\alpha}, \qquad \nu^{\alpha\beta} = \varepsilon^{\beta\gamma}\varkappa^{\alpha}_{.\gamma}, \qquad \nu^{3\alpha} = \varepsilon^{\alpha\gamma}\varkappa^{3}_{.\gamma}. \tag{19}$$

The compatibility conditions

$$\varepsilon^{\alpha\beta}(\varkappa_{\alpha\beta} + b_\alpha{}^\gamma\gamma_{\gamma\beta}) = 0, \qquad \varepsilon^{\alpha\delta}\varepsilon^{\beta\gamma}(\gamma_{\alpha\beta|\delta\gamma} - b_{\delta\gamma}\varkappa_{\beta\alpha}) = 0,$$
$$\varepsilon^{\beta\gamma}(\varepsilon^{\Theta\alpha}\varkappa_{\alpha\beta|\gamma} + b_\lambda{}^\Theta\varepsilon^{\lambda\alpha}\gamma_{\alpha\beta|\gamma}) = 0 \tag{20}$$

where $\varkappa_{\alpha\beta} = -\gamma_{3\alpha|\beta}$, according to the notations (19) transform as follows

$$\varepsilon_{\alpha\beta}(\nu^{\alpha\beta} - b_\gamma{}^\beta\varrho^{\alpha\gamma} - \varrho^{3\alpha|\beta}) = 0, \qquad \nu^{\alpha\beta}{}_{|\beta} - b_\beta{}^\alpha\varrho^{\beta\gamma}{}_{|\gamma} = 0,$$
$$\varrho^{\alpha\beta}{}_{|\alpha\beta} + b_\gamma{}^\beta\varepsilon_\beta{}^\alpha\varepsilon^{\lambda\gamma}\varrho^{3}_{.\lambda|\alpha} + b_{\alpha\beta}\nu^{\alpha\beta} = 0. \tag{21}$$

We consider also the additional condition[1]

$$\varrho^{\alpha\beta}{}_{|\beta} + b_\gamma{}^\alpha\varepsilon_\alpha{}^\beta\varepsilon^{\gamma\lambda}\varrho^3{}_\lambda = \nu^{3\alpha} \tag{22}$$

These conditions shall be correlated with the following fundamental assumptions

$$\beta_{\alpha 3} + K\beta_{3\alpha} = 0, \qquad \beta_{33} = e_{33} = 0. \tag{23}$$

[1] Starting from the conditions $\varepsilon_{\alpha\beta}v^{\alpha|\beta} = 0$, $\varepsilon_{\alpha\beta}\overline{\omega}^{\alpha|\beta} = 0$, one can obtain the equations analogous with (9), $(12)_{1,2}$ (see [15]).

For different values of K, (23) describes various internal states of distortion. According to the constitutive equations (26), for $K = -(\mu - \alpha)/(\mu + \alpha)$, the stresses on the lateral faces of the shell vanish identically. Hence (23) correspond, in the frame of an approximate theory to a continuous model of a shell with free lateral faces.

The condition $(23)_2$ is identically satisfied according to the assumed kinematic field of displacements. The condition $(23)_1$ implies the relation

$$\beta_\alpha = \varepsilon_{\alpha\gamma}\overline{\omega}^\gamma - K\beta_{3\alpha} \tag{24}$$

or, after derivation

$$k_{\alpha\beta} = \beta_{\alpha|\beta} = \varepsilon_{\alpha\gamma}(\varkappa^\gamma_{.\beta} - K\varepsilon^{\gamma\lambda}\varrho^3{}_{\lambda|\beta}) + k'_{\alpha\beta} \tag{25}$$

where $k'_{\alpha\beta} = \varepsilon_{\alpha\gamma} b_\beta{}^\gamma \overline{\omega}^3$. We observe that $k'_{\alpha\beta|\gamma} = \varepsilon_{\alpha\delta} b_\beta{}^\delta \varkappa^3_{.\gamma}$ depends on the product of the twist-curvature tensor in the median surface. Hence, assuming that these terms are negligible, we have $k_{\alpha\beta|\gamma} \cong k'_{\alpha\beta|\gamma}$. Since the constitutive equations of isotropic Cosserat materials [6, 11] are

$$\sigma^{ij} = E^{ijkl}(\mu + \alpha, \mu - \alpha, \lambda)\beta_{kl}, \quad \mu^{ij} = E^{ijkl}(M + A, M - A, \lambda)\varkappa_{kl} \tag{26}$$

where

$$E^{ijkl}(\xi) = E^{ijkl}(\xi_1, \xi_2, \xi_3) = \xi_1 g^{ik} g^{jl} + \xi_2 g^{il} g^{jk} + \xi_3 g^{ij} g^{kl} \tag{27}$$

it is a matter of simple calculations to eliminate σ^{33} and μ^{33}

$$\begin{aligned} \sigma^{\alpha\beta} &= E^{\alpha\beta\gamma\delta}\left(\mu + \alpha, \mu - \alpha, \frac{2\nu}{1-\nu}\right)\beta_{\gamma\delta} + \frac{\nu}{1-\nu} g^{\alpha\beta}\sigma^{33}, \\ \mu^{\alpha\beta} &= E^{\alpha\beta\gamma\delta}\left(M + A, M - A, \frac{2N}{1-N}\right)\varkappa_{\gamma\delta} + \frac{N}{1-N} g^{\alpha\beta}\mu^{33} \end{aligned} \tag{28}$$

where

$$\nu = \lambda/2(\lambda + \mu), \qquad N = \Lambda/2(\Lambda + M). \tag{29}$$

The integrated constitutive equations for resulting forces and couples are

$$\begin{aligned} n^{\alpha\beta} &= E^{\alpha\beta\gamma\delta}(n)\beta_{\gamma\delta}, \qquad n^{3\alpha} = n\varrho^{3\alpha}, \\ r^{\alpha\beta} &= E^{\alpha\beta\gamma\delta}(r)\nu_{\gamma\delta} + E^{\alpha\beta\gamma\delta}(p)\varrho^3_{.\gamma|\delta} + m'^{\alpha\beta}, \qquad r^{3\alpha} = r\nu^{3\alpha} \end{aligned} \tag{30}$$

where

$$\begin{aligned} &n_1 = -(\mu - \alpha)h, \quad n_2 = -(\mu + \alpha)h, \quad n_3 = 2\mu h/(1 - \nu) \\ &n = -h(\mu + \alpha)\big(1 - (\mu - \alpha)K/(\mu + \alpha)\big) \\ &r_1 = -s_1 - 2MNh/(1 - N), \quad r_2 = -s_2 + 2Mh/(1 - N), \\ &r_3 = -s_3 - (M + A)h, \\ &r = h(M + A), \quad s_i = n_i h^2/12, \quad p_i = s_i K, \quad m'^{\alpha\beta} = E^{\alpha\beta\gamma\delta}(s)k'_{\gamma\delta}. \end{aligned} \tag{31}$$

The inverse equations can be readily obtained by using the multiplication law of the operators E . . .

$$E^{\alpha\beta\gamma\delta}(x)\,E_{\gamma\delta\cdot\cdot}^{\ \ \ \lambda\mu}(y) = E^{\alpha\beta\lambda\mu}(z) \tag{32}$$

where

$$z_1 = x_1 y_1 + x_2 y_2, \qquad z_2 = x_1 y_2 + x_2 y_1,$$
$$z_3 = (x_1 + x_2 + x_3) y_3 + (y_1 + y_2 + y_3) x_3 \tag{33}$$

(33) allows, for instance, the determination of the arguments

$$y_1 = (z_1 x_1 - z_2 x_2)/(x_1^2 - x_2^2), \qquad y_2 = (z_2 x_1 - z_1 x_2)/(x_1^2 - x_2^2),$$
$$y_3 = (z_3(x_1 + x_2) - (z_1 + z_2) x_3)/(x_1 + x_2)\,(x_1 + x_2 + x_3) \tag{34}$$

if (x) and (z) are known.

We deduce

$$\varrho^{\alpha\beta} = E^{\alpha\beta\gamma\delta}(\varrho)\,n_{\gamma\delta}, \qquad \varrho^{3\alpha} = \varrho\, n^{3\alpha},$$
$$\nu^{\alpha\beta} = E^{\alpha\beta\gamma\delta}(\nu)\,\varkappa_{\gamma\delta} + E^{\alpha\beta\gamma\delta}(\pi)\,n^{3}_{\cdot\gamma|\delta} + \nu'^{\alpha\beta}, \qquad \nu^{3\alpha} = r^{3\alpha}/r \tag{35}$$

where, according to (33) and (34)

$$\varrho_1 = -(\mu - \alpha)/4\alpha\mu h, \qquad \varrho_2 = -(\mu + \alpha)/4\alpha\mu h, \qquad \varrho_3 = 1/2(1 + \nu)\mu h,$$
$$\varrho = 1/n,$$
$$\nu_1 = r_1/(r_1^2 - r_2^2), \qquad \nu_2 = -r_2/(r_1^2 - r_2^2), \qquad \nu_3 = -r_3/(r_1 + r_2 + 2r_3),$$
$$\pi_1 = -\varrho(\nu_1 p_1 + \nu_2 p_2), \qquad \pi_2 = -\varrho(\nu_1 p_2 + \nu_2 p_1),$$
$$\pi_3 = -\varrho\big((\nu_1 + \nu_2 + \nu_3) p_3 + (p_1 + p_2 + p_3)\nu_3\big), \qquad \nu'^{\alpha\beta} = E^{\alpha\beta\gamma\delta}(-r)\,m'_{\gamma\delta}. \tag{36}$$

The Eqs. (12), (14), (15) and (21), (22), (19) can be deduced respectively one from the other if we substitute the functions $(n^{i\alpha}, r^{i\alpha}, N^i, R^i)$ respectively by $(\nu^{i\alpha}, \varrho^{i\alpha}, \overline{\omega}^i, v^i)$ and conversely. This fact enables us to use the complex functions (see [15, 24])

$$\mathscr{N}^{i\alpha} = n^{i\alpha} + i\varkappa\nu^{i\alpha}, \qquad \mathscr{R}^{i\alpha} = r^{i\alpha} + i\varkappa\varrho^{i\alpha},$$
$$\mathscr{N}^{i} = N^{i} + i\varkappa\overline{\omega}^{i}, \qquad \mathscr{R}^{i} = R^{i} + i\varkappa v^{i} \tag{37}$$

$\varkappa$ being constant. $(37)_{1,2}$ satisfy the equations (12) in which we replace $n^{i\alpha}$ and $r^{i\alpha}$ by $\mathscr{N}^{i\alpha}$ and $\mathscr{R}^{i\alpha}$. Thus, we obtain the representations (15) in which we perform the substitution $n^{i\alpha}, r^{i\alpha} \to \mathscr{N}^{i\alpha}, \mathscr{R}^{i\alpha}$.

According to this result we eliminate N_β from $(15)_3$ reformulated under a complex form, (see [15])

$$\mathcal{N}_\beta = \varepsilon_{\beta\alpha}(\mathcal{R}^{3\alpha} + \mathcal{R}^{3|\alpha} + b_\gamma{}^\alpha \mathcal{R}^\gamma - r_0{}^{3\alpha}) \tag{38}$$

so that

$$\begin{aligned}
\mathcal{N}^{\alpha\beta} &= \varepsilon^{\beta\delta}[\varepsilon^{\alpha\gamma}(\mathcal{R}^3_{.\gamma} + \mathcal{R}^3_{|\gamma} + b_{\gamma\varphi}\mathcal{R}^\varphi - v^3_{0\cdot\gamma})_{|\delta} - b_\delta{}^\alpha \mathcal{N}^3] + n_0{}^{\alpha\beta}\\
\mathcal{N}^{3\alpha} &= \varepsilon^{\alpha\gamma}[\mathcal{N}^3_{|\gamma} + b_{\gamma\lambda}\varepsilon^{\lambda\mu}(\mathcal{R}^3_{.\mu} + \mathcal{R}^3_{|\mu} + b_{\mu\nu}\mathcal{R}^\nu - r^3_{0\cdot\mu})] + n_0{}^{3\alpha}\\
\mathcal{R}^{\alpha\beta} &= \varepsilon^\alpha_{.\gamma}[\varepsilon^{\beta\delta}(\mathcal{R}^\gamma_{|\delta} - b_\delta{}^\gamma \mathcal{R}^3) + a^{\beta\gamma}\mathcal{N}^3] + r_0{}^{\alpha\beta}.
\end{aligned} \tag{39}$$

The constitutive equations (30) and (35) can also be written under a complex form

$$\begin{aligned}
\mathcal{N}^{\alpha\beta} &= E^{\alpha\beta\gamma\delta}(N^-)\mathcal{R}_{\gamma\delta} + E^{\alpha\beta\gamma\delta}(N^+)\overline{\mathcal{R}}_{\gamma\delta} + E^{\alpha\beta\gamma\delta}\left(\frac{\pi}{2}\right)(\mathcal{R}^3_{.\gamma} - \overline{\mathcal{R}}^3_{.\gamma})_{|\delta} + i\varkappa v'^{\alpha\beta}\\
\mathcal{N}^{3\alpha} &= \frac{i}{2}\left[\left(\frac{\varkappa}{r} - \frac{n}{\varkappa}\right)\mathcal{R}^{3\alpha} + \left(\frac{\varkappa}{r} + \frac{n}{\varkappa}\right)\overline{\mathcal{R}}^{3\alpha}\right]
\end{aligned} \tag{40}$$

or

$$\begin{aligned}
\mathcal{R}^{\alpha\beta} &= E^{\alpha\beta\gamma\delta}(R^-)\mathcal{N}_{\gamma\delta} + E^{\alpha\beta\gamma\delta}(R^+)\overline{\mathcal{N}}_{\gamma\delta} + \frac{1}{n}E^{\alpha\beta\gamma\delta}\left(\frac{p}{2}\right)(\mathcal{N}^3_{.\gamma} + \overline{\mathcal{N}}^3_{.\gamma})_{|\delta}\\
\mathcal{R}^{3\alpha} &= \frac{i}{2}\left[\left(\frac{\varkappa}{n} - \frac{r}{\varkappa}\right)\mathcal{N}^{3\alpha} + \left(\frac{\varkappa}{n} + \frac{r}{\varkappa}\right)\overline{\mathcal{N}}^{3\alpha}\right]
\end{aligned} \tag{41}$$

where

$$\begin{aligned}
N_i^- &= \frac{i}{2}\left(v_i\varkappa - \frac{n_i}{\varkappa}\right), \qquad N_i^+ = \frac{i}{2}\left(v_i\varkappa + \frac{n_i}{\varkappa}\right),\\
R_i^- &= \frac{i}{2}\left(\varrho_i\varkappa - \frac{r_i}{\varkappa}\right), \qquad R_i^+ = \frac{i}{2}\left(\varrho_i\varkappa + \frac{r_i}{\varkappa}\right).
\end{aligned} \tag{42}$$

The substitution of (39) in (41) leads to a differential system of 6 equations for 6 unknown functions $\mathcal{R}^\alpha, \mathcal{R}^3, \mathcal{R}^{3\alpha}, \mathcal{N}^3$ which simplify sensibly in some particular cases [11].

We observe that the functions $n'^{i\alpha}$, $v'^{i\alpha}$ can be neglected in the balance equations if we assume that the products of curvatures are negligible. The constant $\varkappa$ can be chosen so that $R_3^- = 0$. In this case we obtain

$$K = \lambda^*\left[1 + \frac{(1-\nu)(M+A)}{6\mu h^2}\right]^{1/2}, \qquad \lambda^* = \mu h^2[(1+\nu)/6(1-\nu)]^{1/2}. \tag{43}$$

The results include the principles of field-correspondences which, in the frame of the symmetric theory were firstly attempted in [16] and enounced in [18, 19]. In the case of shallow shells, for $K = 0$, from the substitution of (40) in the balance equations (14) and $(12)_3$ we derive the equations

$$R_3^- \Delta R^{3|\alpha} + \varepsilon^{\beta\alpha}[\mathcal{N}^3 - l_-^2 \Delta \mathcal{N}^3 - l_+^2 + \Delta \overline{\mathcal{N}^3}]_{|\beta} + \varepsilon^{\alpha}_{.\beta} Q_0^{\beta} + P_0^{\alpha} = 0$$

$$R_3^- \Delta \Delta R^3 + \varepsilon^{\alpha}_{.\beta} Q^{\beta}_{0|\alpha} + P_0^3 = 0$$

$$a^{\pm} = \frac{i}{2}\left(\frac{\varkappa}{n} \pm \frac{r}{\varkappa}\right), \qquad l_-^2 = -(a^- R_2^- + a^+ R_2^+),$$

$$l^2_+ = -(a^- R_2^+ + a^+ R_2^-). \tag{44}$$

The asymmetry and bending effects expressed by the solutions of (44) superpose over the effect defined by $\mathcal{R}^{\alpha}$, functions which can be determined in a simple way from $(39)_3$ and (41), since $\mathcal{N}^{i\alpha}$ can be expressed in terms including $\mathcal{R}^3$, $\mathcal{N}^3$, using $(39)_2$ and $(41)_2$.

In the case of shallow shells, for $K = 1$, the general solution for spherical shells was previously obtained[1] [11].

References

1. Cosserat, E. and F.: C. R. Acad. Sci. **146**, 169—172 (1908).
2. Günther, W.: Abh. Braunschweigische Wiss. Ges. **10**, 195—213 (1958).
3. Schäfer, H.: Abh. Braunschweigische Wiss. Ges. **7**, 107—213 (1955).
4. Kuvshinski, E. V., and E. C. Aero: Fiz. Tverd. Tela **2**, 7, 1399—1409 (1960).
5. Toupin, R. A.: Arch. for Rat. Mech. Anal. **1**, 5, 385—414 (1962).
6. Palmov, V. A.: Priklad. Mat. Mah. **28**, 401—409 (1964).
7. Koiter, W. T.: Koninkl. Nederl. Akad. van Wetenschappen. Proc. Ser. B, 67, 1 (1964) p. 30—44.
8. Ericksen, J. L., and C. Truesdell: Arch. Rat. Mech. Anal. **1**, 4, 295—323 (1958).
9. Günther, W.: Ing. Archiv **30**, 160—186 (1961).
10. Green, A. E., P. M. Naghdi, and W. L. Wainwright: Arch. for Rat. Mech. Anal. **20**, 4, 287—309 (1965).
11. Mişicu, M.: The mechanics of deformable bodies, Bucarest: Academy of Rumanian Republic 1967.

---

[1] In this case, if we set $n^{\beta\alpha} = \varepsilon^{\alpha\gamma}(\varepsilon^{\beta\delta}\Phi_{|\gamma\delta} - \psi|^{\beta}_{\gamma}) - B_{\beta\alpha}$, $b_{\alpha\beta} \cong z_{|\alpha\beta}$, $r^{3\alpha} = \psi^{|\alpha}$, $F = v_3 + i\varkappa\Phi$, $\varkappa = 1/Eh\lambda^*$, $\lambda^* = h/(12(1-\nu^2))^{1/2}$, $q = p \cdot 12(1-\nu^2)/E$, $\mathcal{L}^2 = \eta(1/\mu + 1/\alpha)$, p being the transverse distributed pressure on the shell, we obtain the differential equation

$$(1 - \mathcal{L}^2\Delta)\,(a^{\alpha\beta}a^{\gamma\delta}F_{|\alpha\beta\gamma\delta} + i\varepsilon^{\alpha\varrho}\varepsilon^{\beta\gamma}b_{\alpha\beta}F_{|\varrho\gamma} - q) \cong 0$$

which splits in a bending and an antisymmetric effect equation. In the case of spheric shells with axial symmmetry of stress state, we have

$$F = A_1 + A_2 \log r + A_3 I_0(\varepsilon r i^{1/2}) + A_4 K_0(\varepsilon r i^{1/2}) + A_5 I_0(r/\mathcal{L}) + A_6 K_0(r/\mathcal{L}),$$

$$\varepsilon = (\beta/\lambda^*)^{1/2}, \quad \beta = z_{|11}, \quad n = (x^2 + y^2)^{1/2} = r_0/\beta.$$

For arbitrar values of K, it seems more convenient to elaborate direct solving methods.

12. Reissner, E.: Proc. 2nd Int. Symp. Thin Shells Copenhagen 1967 (in this book, pp. 15—30).
13. Gevorkian, G.: Priklad. Meh. **2**, 7, 74—79 (1966).
14. Babici, D. V.: Priklad. Meh. **3**, 4, 39—45 (1967).
15. Mişicu, M., and V. Nicolae: Rev. Roum. Sci. Techn. Ser. Mec. Appl. **2**, 339 to 345 (1968).
16. Odqvist, F.: Compt. Rend. Acad. Sci. **205**, 271—272 (1937).
17. Novojilov, V. V.: The theory of thin shells, Leningrad: Sudpromghiz 1962.
18. Lurie, A. I.: Priklad. Mat. Meh. **4**, 7 (1940).
19. Gol'denweizer, A. L.: Priklad. Mat. Meh. **4**, 2 (1940).
20. Savin, G. A.: Stress concentration around holes, Berlin: 1958.
21. Vekua, N. I.: New solving methods of elliptic equations, Moscow-Leningrad 1958.
22. Koiter, W. T.: First. Int. Symp. thin shells, Delft: North Holland 1960.
23. Naghdi, P. M.: Progress in solid mech. Vol. IV, New York: Wiley 1963.
24. Naghdi, P. M.: Proc. 11th. Int. Congr. Appl. Mech. Munich 1964, Berlin/Heidelberg/New York: Springer 1966.

Asymptotic Approximation in the Three-Dimensional Theory of Thin and Thick Elastic Shells

By

H. S. Rutten

Technische Hogeschool, Delft, Netherlands

Summary. Within the scope of three-dimensional, linear elasticity theory, two-dimensional, consistent systems of basic shell equations and reduced edge conditions are derived. The method of derivation consists of systematic asymptotic expansion of the solutions of the three-dimensional equations, after the equations have properly been adapted to the specific shape of shells and the general smoothness of their solutions.

1. Introduction

During the last ten years it has become customary to distinguish two different trends in the search for the foundations of elastic shell theory. The investigators of both trends aim at the same goal: obtaining an adequate and reliable method of describing the displacements, stresses and strains of elastically deformable bodies, one dimension of which is relatively small compared to the other two dimensions.

The first group of investigators deals primarily with the three-dimensional equations of elasticity. For the typical shape the elastic body assumes in the case of plates and shells, systematic approximations to the solutions of the three-dimensional equations of successive order of approximation are sought after, via several asymptotic expansion methods.

The second group of investigators strives to accomplish the compatibility conditions for the deformations and the macroequilibrium equations in terms of stress resultants and stress couples of a shell element with an adequate constitutive law relating the deformations and integrated actions. The governing mechanical principle sometimes is introduced via the use of variational methods.

The approach adopted in this work ranges under the first category mentioned. Taking advantage of the *analyticity* of the solutions of the three-dimensional, linear equations of elasticity and referring the equations to *middle-surface coordinates,* the three-dimensional equations first are reduced to a system of two-dimensional *interior shell equations.*

The interior shell equations, which contain a small thickness parameter, are uniformly valid in the *interior* of a shell.

Near the edge a three-dimensional edge effect generally occurs. For the analysis of this edge effect in the *edge zone*, which is called the *Kirchhoff edge effect*, it is necessary to revert again to the three-dimensional equations of elasticity. Referring the three-dimensional equations to *edge-curve coordinates* and splitting up the complete edge-zone state in an interior and an excess part, *reduced edge conditions* compatible with the character and the order of the interior shell equations are deduced from the actually prescribed, three-dimensional edge conditions, via appropriate asymptotic expansion of the excess-state solutions.

Finally the solutions of the interior shell equations are examined asymptotically. For the lowest order approximations, asymptotic equations are obtained, which are identical to the classical equations of

bending of flat plates and generalized plane stress in flat membranes,
the shallow shell theory and of
inextensional bending and membrane theory of shells.

In the latter case, which is characterized by slowly varying surface loads and edge conditions, a second edge effect in the asymptotic theory of shells is encountered. This second edge effect, which is called *bending edge effect*, is strongly dependent on the geometrical properties of middle surface and edge curve. Several types of bending edge-effect equations are derived corresponding to different types of edge curves, interalia, non-asymptotic lines, simple asymptotic lines, geodesic and non-geodesic double asymptotic lines, etc.

In the case of inextensional bending, finally, a bending edge effect is possible that infinitely expands into the whole shell if the shell thickness shrinks to zero. This is in complete contrast with the limiting behaviour of an ordinary edge effect. For that reason the name *anti-edge effect* is adopted.

None of the following analyses is typically limited to small shell thicknesses. The results apply to thin, as well as to relatively thick elastic shells.

2. Interior Shell Equations

From the theory of elliptic partial differential equations it is known that the solutions of systems of elliptic partial differential equations of the second order are analytic functions of the independent variables at all interior points of the domain of definition, on condition that

the functionals and coefficients in the equations are analytic functions of their arguments and of the independent variables and that the first three partial derivatives of the solutions exist and are continuous.

The governing equations in terms of the displacement field $w^i(x)$ of the linear three-dimensional theory of elasticity consist of a system of three linear elliptic partial differential equations of the second order.

$$g^{jk} w^i|_{jk} + \frac{1}{1-2\nu} g^{ik} w^j|_{jk} + f^i = 0, \qquad i = 1, 2, 3. \tag{2.1}$$

The vertical stroke denotes co-variant differentiation with respect to the metric tensor g_{ij}. The components f^i represent the body forces.

With regard to flat plates and generally curved shells, it is appropriate to refer the components of geometric and physical quantities to *middle-surface coordinate systems*. On Fig. 1 such systems are defined.

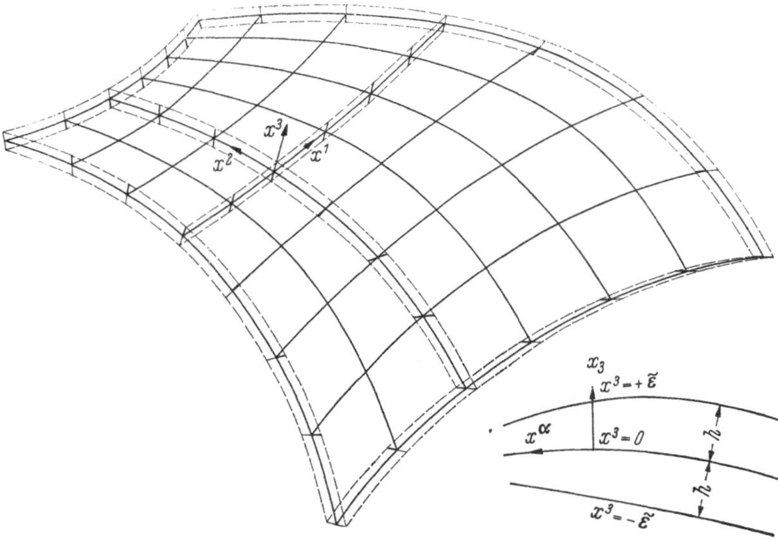

Fig. 1. Middle-surface coordinates.

For the coordinates in Fig. 1 the following relations hold

$$g_{\alpha\beta} = a_{\alpha\beta} - 2\Delta b_{\alpha\beta} x^3 + (\Delta^2 c_{\alpha\beta} + \Delta_{,\alpha}\Delta_{,\beta})(x^3)^2$$

$$g_{\alpha 3} = \Delta\Delta_{,\alpha} x^3$$

$$g_{33} = \Delta^2$$

$$\Delta = \frac{h(x^\alpha)}{h}; \quad \hat{\varepsilon} = \frac{h}{L}$$

where

g_{ij} spatial metric tensor
$a_{\alpha\beta}$ fundamental tensor of middle surface
$b_{\alpha\beta}$ curvature tensor of middle surface $\Big\}\ c_{\alpha\beta} = a_{\alpha\tau} b_\alpha{}^\sigma b_\beta{}^\tau$
$2h$ variable thickness of the shell
h measure of shell thickness
L arbitrary measure of length
$(\ )_{,\alpha}$ partial differentiation with respect to x^α

The continuous variety of points which have equal distances to the upper and the lower face of the shell constitutes the middle surface S. The distances are defined equal, when measured along the normals-to-the-middle surface. The face surfaces ${}_{(+)}S$ and ${}_{(-)}S$ of the shell are made in this manner coordinate surfaces of the middle-surface coordinate system also.

$$\left.\begin{aligned} {}_{(+)}S: x^3 &= +\hat{\varepsilon} \\ {}_{(-)}S: x^3 &= -\hat{\varepsilon}\,. \end{aligned}\right\} \tag{2.2}$$

The analyticity of the solutions $w^i(x)$ of Eqs. (2.1) is guaranteed at the points of the upper and the lower face of the shell as well, if in addition it is assumed that

the face surfaces ${}_{(+)}S$ and ${}_{(-)}S$ of the shell are sufficiently smooth,

the surface loads ${}_{(+)}p^i$ and ${}_{(-)}p^i$ are analytic functions of the coordinates x^α, and that

no quick transitions of displacements, stresses and strains across the thickness of the shell are caused by these surface loads.

As a consequence of the first of the above additional assumptions the middle surface of these shells is a smooth surface also.

Confining, for the time being, the further analysis to problems of plates and shells which comply with the foregoing assumptions, the components of displacements, stresses and strains may be expanded in uniformly convergent Taylor series in the coordinate variable x^3 normal to the middle surface S. E. g.

$$w^i = {}_{[2k]}w^i \quad (x^3)^{2k} + {}_{(2k+1)}w^i \quad (x^3)^{2k+1} + \overset{w}{R}_n{}^i \tag{2.3}$$

Square brackets denote the even and round brackets denote the odd expansion functions with respect to x^3. The lower index and upper exponent k must be summed from $k = 0$ up to and including $k = \frac{1}{2} n$ or $\frac{1}{2}(n-1)$, depending on whether n is even or odd, respectively. Before the expansion in power series in x^3 of the various components is carried through, all components of geometric and physical quantities have been made dimensionless by means of a provisionally arbitrary measure of length L and the shear modulus G.

Insertion of the explicit middle-surface metric and the Taylor-series expansions in the general, three-dimensional equations of elasticity results in the *interior shell equations*. The resulting interior shell equations for shells of uniform thiekness can be summarized as follows

Kinematic Equations

$$ {}_{[00]}E^{\alpha\beta} = (a^{\alpha\varrho}\delta_\sigma{}^\beta + a^{\beta\varrho}\delta_\sigma{}^\alpha)\,{}_{[0]}w^\sigma\big|_{\underset{a}{\varrho}} - 2b^{\alpha\beta}\,{}_{[0]}w^3 $$

$$ \begin{aligned} {}_{(01)}E^{\alpha\beta} = & -(a^{\alpha\varrho}\delta_\sigma{}^\beta + a^{\beta\varrho}\delta_\sigma{}^\alpha)a^{\sigma\tau}\,{}_{[0]}w^3\big|_{\underset{a}{\tau\varrho}} + \\ & + 2\{(b^{\alpha\varrho}\delta_\sigma{}^\beta + b^{\beta\varrho}\delta_\sigma{}^\alpha)\,{}_{[0]}w^\sigma\big|_{\underset{a}{\varrho}} - b^{\alpha\beta}\big|_{\underset{a}{\sigma}}\,{}_{[0]}w^\sigma - 3c^{\alpha\beta}\,{}_{[0]}w^3\} + \\ & + \frac{\nu}{1-\nu} b^{\alpha\beta} a_{\sigma\tau}\,{}_{[00]}E^{\sigma\tau} \end{aligned} $$

Constitutive Equations

$$ {}_{[00]}X^{\alpha\beta} = {}_{[00]}E^{\alpha\beta} + \frac{\nu}{1-\nu} a^{\alpha\beta} a_{\sigma\tau}\,{}_{[00]}E^{\sigma\tau} $$

$$ \begin{aligned} {}_{(01)}X^{\alpha\beta} = & {}_{(01)}E^{\alpha\beta} + \frac{\nu}{1-\nu} a^{\alpha\beta} a_{\sigma\tau}\,{}_{(01)}E^{\sigma\tau} + \\ & + 2\frac{\nu}{1-\nu}(a^{\alpha\beta}b_{\sigma\tau} - b^{\alpha\beta}a_{\sigma\tau})\,{}_{[00]}E^{\sigma\tau} - \frac{\nu}{1-\nu} a^{\alpha\beta} b_{\sigma\tau}\,{}_{[00]}X^{\sigma\tau} \end{aligned} $$

Equilibrium Equations

$$ \begin{aligned} & -\hat{\varepsilon}\,{}_{[00]}X^{\alpha\tau}\big|_{\underset{a}{\tau}} - \frac{1}{3}\hat{\varepsilon}^3\Big[{}_{[02]}X^{\alpha\tau}\big|_{\underset{a}{\tau}} - \{(2b_\sigma{}^\alpha + \delta_\sigma{}^\alpha b_\pi{}^\pi)\,{}_{(01)}X^{\sigma\tau}\}\big|_{\underset{a}{\tau}} + b_\sigma{}^\alpha\big|_{\underset{a}{\tau}}\,{}_{(01)}X^{\sigma\tau} + \\ & + \{(c_\sigma{}^\alpha + 2b_\sigma{}^\alpha b_\pi{}^\pi + \delta_\sigma{}^\alpha K)\,{}_{[00]}X^{\sigma\tau}\}\big|_{\underset{a}{\tau}} - (b_\sigma{}^\alpha\big|_{\underset{a}{\pi}} b_\tau{}^\pi + b_\sigma{}^\alpha\big|_{\underset{a}{\tau}} b_\pi{}^\pi)\,{}_{[00]}X^{\sigma\tau} + \\ & + \frac{3}{2}\frac{\nu}{1-\nu} a^{\alpha\varrho}\{b_{\sigma\tau}\,{}_{(01)}X^{\sigma\tau} - {}_{[00]}X^{\sigma\tau}\big|_{\underset{a}{\tau\sigma}} - (c_{\sigma\tau} + b_{\sigma\tau}b_\pi{}^\pi)\,{}_{[00]}X^{\sigma\tau}\}\big|_{\underset{a}{\varrho}}\Big] + \\ & + \hat{\varepsilon}^5 \ldots = \frac{1}{2}\,{}_{[e]}p^\alpha + \hat{\varepsilon}\Big[{}_{[0]}f^\alpha - \frac{1}{2} b_\sigma{}^\alpha\,{}_{(o)}p^\sigma + \frac{1}{2}\frac{\nu}{1-\nu} a^{\sigma\tau}\,{}_{(o)}p_\alpha{}^3\big|_\sigma\Big] + \hat{\varepsilon}^3 \ldots \\ & \hspace{28em} \alpha = 1, 2 \end{aligned} $$

$$ \begin{aligned} & -\hat{\varepsilon} b_{\sigma\tau}\,{}_{[00]}X^{\sigma\tau} - \frac{1}{3}\hat{\varepsilon}^3\Big[b_{\sigma\tau}\,{}_{[02]}X^{\sigma\tau} + {}_{(01)}X^{\sigma\tau}\big|_{\underset{a}{\tau\sigma}} - (c_{\sigma\tau} + b_{\sigma\tau}b_\pi{}^\pi)\,{}_{(01)}X^{\sigma\tau} + \\ & + \{(b_\sigma{}^\varrho + \delta_\sigma{}^\varrho b_\pi{}^\pi)\,{}_{[00]}X^{\sigma\tau}\}\big|_{\underset{a}{\tau\varrho}} + (c_{\sigma\tau}b_\pi{}^\pi + b_{\sigma\tau}K)\,{}_{[00]}X^{\sigma\tau} + \\ & + \frac{3}{2}\frac{\nu}{1-\nu} b_\varrho{}^\varrho\{b_{\sigma\tau}\,{}_{(01)}X^{\sigma\tau} - {}_{[00]}X^{\sigma\tau}\big|_{\underset{a}{\tau\sigma}} - (c_{\sigma\tau} + b_{\sigma\tau}b_\pi{}^\pi)\,{}_{[00]}X^{\sigma\tau}\}\Big] + \\ & + \hat{\varepsilon}^5 \ldots = \frac{1}{2}\,{}_{[e]}p^3 + \hat{\varepsilon}\Big[{}_{[0]}f^3 + \frac{1}{2}\,{}_{(o)}p^\sigma\big|_{\underset{a}{\sigma}} + \frac{1}{2}\frac{\nu}{1-\nu} b_\sigma{}^\sigma\,{}_{(o)}p^3\Big] + \hat{\varepsilon}^3 \ldots \end{aligned} $$

where

$_{[0]}w^i$ constant part across shell thickness of *displacement* vector
$_{[00]}E^{\alpha\beta}$ constant part across shell thickness of *strain* tensor (membrane strains)
$_{(01)}E^{\alpha\beta}$ linear part across shell thickness of *strain* tensor (flexural strains)
$_{[00]}X^{\alpha\beta}$ constant part across shell thickness of *stress* tensor (membrane stresses)
$_{(01)}X^{\alpha\beta}$ linear part across shell thickness of *stress* tensor (bending stresses)
$\delta_\sigma{}^\alpha$ Kronecker's delta
ν Poisson's ratio
$_{[e]}p^i$ resultants of surface loads $_{(+)}p^i$ and $_{(-)}p^i$
$_{(o)}p^i$ couples of surface loads $_{(+)}p^i$ and $_{(-)}p^i$
$_{[0]}f^i$ constant part across shell thickness of volumetric loads
$\underset{a}{|}$ co-variant differentiation with respect to $a_{\alpha\beta}$

$\hat{\varepsilon} = \frac{h}{L}$; L = arbitrary measure of length

The Taylor series expansions enable the elimination of x^3 from the three-dimensional equations and as the result a series of recurrent relations among the various expansion coefficients is obtained by the aid of which it proves possible to express any and every expansion coefficient in terms of $_{[0]}w^i$ or $_{[00]}E^{\alpha\beta}$ and $_{(01)}E^{\alpha\beta}$ or $_{[00]}X^{\alpha\beta}$ and $_{(01)}X^{\alpha\beta}$.

Finally the dynamic boundary conditions on the upper and the lower face lead to a system of three differential equations for the determination of $_{[0]}w^i$. Above they have been named: Equilibrium equations. It is also via the boundary conditions on the upper and the lower face of the shell that the thickness parameter $\hat{\varepsilon}$ enters the interior shell equations.

Insertion of the explicit middle-surface metric and the Taylor-series expansions in the compatibility conditions of the three-dimensional theory results in a system of three compatibility conditions which must be satisfied by $_{[00]}E^{\alpha\beta}$ and $_{(01)}E^{\alpha\beta}$.

These conditions are

Compatibility conditions

$$e^{\alpha\beta} e^{\lambda\mu} a_{\alpha\sigma} a_{\lambda\tau}\, {}_{[00]}E^{\sigma\tau}\underset{a}{|}_{\beta\mu} + a\,(b_{\sigma\tau} - a_{\sigma\tau} b_\pi{}^\pi)\, {}_{(01)}E^{\sigma\tau} + \frac{3-\nu}{1-\nu}\, a\, K\, a_{\sigma\tau}\, {}_{[00]}E^{\sigma\tau} = 0$$

$$e^{\alpha\beta}\Big\{a_{\pi\sigma} a_{\alpha\tau}\, {}_{(01)}E^{\sigma\tau}\underset{a}{|}_\beta - 2\,(b_{\pi\sigma} a_{\alpha\tau}\, {}_{[00]}E^{\sigma\tau})\underset{a}{|}_\beta - a_{\pi\sigma}\,(b_{\alpha\tau}\, {}_{[00]}E^{\sigma\tau}\underset{a}{|}_\beta - a_{\alpha\tau} b_\beta{}^\varrho\, {}_{[00]}E^{\sigma\tau}\underset{a}{|}_\varrho) + \\ - b_{\beta\sigma} a_{\alpha\tau}\, {}_{[00]}E^{\sigma\tau}\underset{a}{|}_\pi - \frac{\nu}{1-\nu}\, b_{\pi\alpha} a_{\sigma\tau}\, {}_{[00]}E^{\sigma\tau}\underset{a}{|}_\beta\Big\} = 0 \qquad \pi = 1.2$$

where

$e^{\alpha\beta}$ permutation density ($e^{11} = e^{22} = 0,\quad e^{12} = -e^{21} = 1$)
K Gaussian curvature of middle surface

Herewith the alternative dynamic formulation of the interior shell equations is achieved also.

The method adopted for the derivation of the interior shell equations is an extension of the method firstly employed by W. Z. CHIEN in 1944 to derive his intrinsic theory of thin shells and plates.

3. Reduced Edge Conditions and Kirchhoff Edge Effect

The uniform validity of the Taylor series expansions diminishes, in general, in the vicinity of the edge of the shell, near singular lines of the middle surface or of the faces of the shell and also in the neighbourhood of singularities in the boundary conditions on the upper and the lower face. The generic name of *edge zone* of a shell includes all these places. The remainder of the shell, where the Taylor series expansions are uniformly valid, is known as the *interior* of shells.

In the edge zone a three-dimensional edge effect is generally present. This edge effect is, for evidential reasons, called the *Kirchhoff edge effect*. In the edge zone of shells the complete state of displacement, stress and strain consists of an interior and an excess part.

$$w^i = w_I{}^i + \hat{w}^i. \tag{3.1}$$

The interior part $w_I{}^i$ is determined by the interior shell equations and for the determination of the excess part $\hat{w}^i$, it is necessary to revert again to the three-dimensional equations of elasticity.

The analysis of the excess state in the edge zone on the basis of the three-dimensional equations is greatly facilitated by the introduction

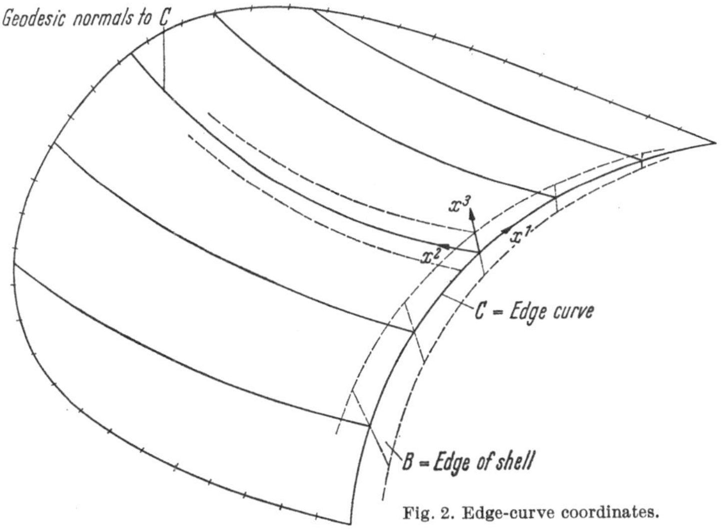

Fig. 2. Edge-curve coordinates.

of *edge-curve coordinates*. Edge-curve coordinates are special middle-surface coordinates, in which, besides the middle surface, the edge of the shell is a coordinate surface also.

The definition of edge-curve coordinates is clarified on fig. 2.

For these coordinates it applies that

$$a_{11} = 1 - 2\varkappa_g x^2 - ({}_{(0)}K - \varkappa_g^2)(x^2)^2 - \frac{1}{3}({}_{(1)}K - 4\varkappa_{g\,(0)}K)(x^2)^3 + \dots$$
$$a_{12} = 0$$
$$a_{22} = 1$$

and

$$b_{11} = \varkappa_n + (\dot{\tau}_g - 2\varkappa_{g\,(0)}H)x^2 + \dots$$
$$b_{12} = \tau_g + (2\,{}_{(0)}\dot{H} - \dot{\varkappa}_n + \varkappa_g\tau_g)x^2 + \dots$$
$$b_{22} = 2\,{}_{(0)}H - \varkappa_n + (2\,{}_{(1)}H - \dot{\tau}_g + 2\varkappa_{g\,(0)}H - 2\varkappa_g\varkappa_n)x^2 + \dots$$

where

$\varkappa_g$ geodesic curvature of C
$\varkappa_n$ normal curvature of C
τ_g geodesic torsion of C
$2H$ $= 2\,{}_{(0)}H + 2\,{}_{(1)}H x^2 + {}_{(2)}H(x^2)^2 + \dots$ mean curvature of middle surface
K $= {}_{(0)}K + {}_{(1)}K x^2 + \frac{1}{2}\,{}_{(2)}K(x^2)^2 + \dots$ Gaussian curvature of middle surface
$(\dot{\ })$ derivative with respect to x^1

Referred to edge-curve coordinates, the three-dimensional equations governing the excess state in the edge zone are called the *Kirchhoff edge-zone equations.*

Asymptotic analysis of these Kirchhoff edge-zone equations and dentification of the excess state with the Kirchhoff edge effect in the ol lowing manner

the Kirchhoff edge effect is represented completely and solely by the subset of in the interior exponentially decreasing asymptotic solutions of the Kirchhoff edge-zone equations,

result in a number of necessary restrictions to be imposed on the edge conditions of the excess state.

The asymptotic analysis consists of reducing the coordinates x^2 and x^3 normal to the edge curve to the measure h of the shell thickness

$$\overset{L}{x^2} = \hat{\varepsilon}\eta; \qquad 0 \leqq \eta < \infty \tag{3.2}$$

$$\overset{L}{x^3} = \hat{\varepsilon}\zeta; \qquad -1 \leqq \zeta \leqq +1 \tag{3.3}$$

and of seeking solutions of the Kirchhoff edge-zone equations in the form of power series in $\hat{\varepsilon}$.

E. g.

$$\hat{w}^i(x^l) = \hat{w}^i_{(k)}(x^1, \eta, \zeta)\,\hat{\varepsilon}^k + \overset{\hat{w}}{R}_m{}^i(x^1, \eta, \zeta, \hat{\varepsilon}). \tag{3.4}$$

If further the expansion functions of the excess stress components $\hat{t}^{ij}$ are replaced successively by appropriate stress functions, the asymptotic equations governing the excess-state solutions of any order in $\hat{\varepsilon}$ turn out to be simple Laplace and biharmonic equations.

$$\nabla^2\Psi_{(k)} = 0 \quad \text{and} \quad \nabla^4\Phi_{(k)} = 0, \quad \text{where} \quad \nabla^2 \equiv \frac{\partial^2}{\partial\eta^2} + \frac{\partial^2}{\partial\zeta^2} \tag{3.5}$$

$\Psi_{(k)}$ and $\Phi_{(k)}$ are stress functions.

The restrictions on the edge conditions for the excess state enable finally the reduction of the prescribed three-dimensional edge conditions to systems of *reduced edge conditions* which are compatible with the character and order of the interior shell equations.

These systems are the following

Dynamic edge conditions

$$\left\{\begin{aligned}
n_B{}^{21} - \varkappa_n m_B{}^{21} + \nu m^{23}_{B,1} &= \hat{\varepsilon}\,{}_{[0]}n_I{}^{21} + \hat{\varepsilon}^3({}_{[2]}n_I{}^{21} - \varkappa_n\,{}_{[2]}m_I{}^{21} + \\
&\quad + \nu\,{}_{[2]}m^{23}_{I,1}) + \hat{\varepsilon}^4\hat{N}^{21}_{(3)} + \dots \\
n_B{}^{22} - \tau_g m_B{}^{21} + \nu\varkappa_g m_B{}^{23} &= \hat{\varepsilon}\,{}_{[0]}n_I{}^{22} + \hat{\varepsilon}^3({}_{[2]}n_I{}^{22} - \tau_g\,{}_{[2]}m_I{}^{21} + \\
&\quad + \nu\varkappa_g\,{}_{[2]}m_I{}^{23}) + \hat{\varepsilon}^4\hat{N}^{22}_{(3)} + \dots \\
n_B{}^{23} + m^{21}_{B,1} + \nu\varkappa_n m_B{}^{23} &= \hat{\varepsilon}\,{}_{(o)}p^2 + \hat{\varepsilon}^3({}_{[2]}n_I{}^{23} + {}_{[2]}m^{21}_{I,1} + \\
&\quad + \nu\varkappa_n\,{}_{[2]}m_I{}^{23}) + \hat{\varepsilon}^4\hat{N}^{23}_{(3)} + \dots \\
m_B{}^{22} &= \hat{\varepsilon}^3\,{}_{[2]}m_I{}^{22} + \hat{\varepsilon}^4\hat{M}^{22}_{(3)} + \dots
\end{aligned}\right.$$

where

n^{2i}, N^{2i}	stress resultants
$m^{2\alpha}$, $M^{2\alpha}$	stress couples
m^{23}	first moment of odd transverse shear stress t^{23} with respect to x^3
index B	prescribed edge loads
index I	interior part of edge-zone stresses
cap $\wedge$	excess or Kirchhoff edge-effect part of edge-zone stresses

Kinematic edge conditions

$$\left\{\begin{aligned}
[w]_B{}^\alpha &= {}_{[0]}[w]_I{}^\alpha + \hat{\varepsilon}^2({}_{[2]}[w]_I{}^\alpha + [\hat{w}]^\alpha_{(2)}) + \hat{\varepsilon}^3[\hat{w}]^\alpha_{(3)} + \dots \qquad \alpha = 1,2 \\
(w)_B{}^\alpha &= \hat{\varepsilon}_{[2]}(w)_I{}^\alpha + \hat{\varepsilon}^2(\hat{w})^\alpha_{(3)} + \dots \qquad \alpha = 1,2
\end{aligned}\right.$$

where

$[w]^i$	average displacements of the normals-to-the-middle surface
$(w)^\alpha$	average rotations of the normals-to-the-middle surface
index B	prescribed edge displacements
index I	interior part of edge-zone displacements
cap $\wedge$	excess or Kirchhoff edge-effect part of edge-zone displacements

The terms of the lower orders in $\hat{\varepsilon}$ on the right-hand sides of the conditions are free of excess solutions. This basically enables the determination, firstly, of the lowest-order interior solutions and secondly, of the lowest-order excess solutions and so on, successively, for the higher-order solutions.

The method adopted for treating the Kirchhoff edge effect is affined to the works (on the same subject) of K. O. FRIEDRICHS and R. F. DRESSLER and of E. L. REISS and S. LOCKE.

4. Asymptotic Analysis of Interior Shell Theory

The interior shell equations and the systems of reduced edge conditions together constitute the *interior shell theory*. The interior shell theory as such is well suited for asymptotic analysis, since the equations and edge conditions contain a small parameter on which the asymptotic analysis may be based directly.

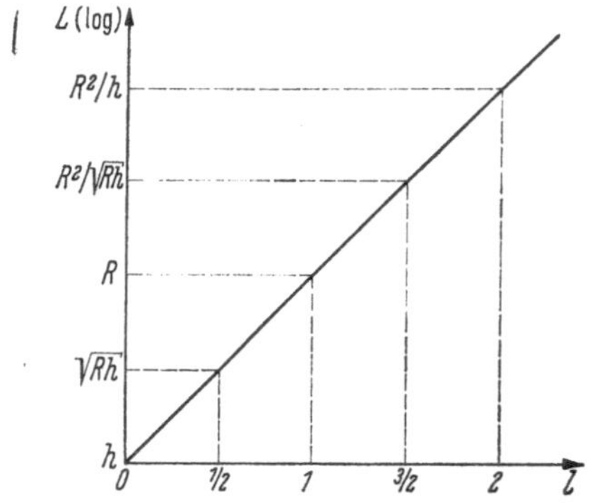

Fig. 3.

The method of asymptotic analysis is summarized on Fig. 3 and below

Geometry scaling

$$\overset{L}{x}{}^{\alpha} = x^{\alpha} \qquad \overset{L}{x}{}^{3} = \hat{\varepsilon}\,\zeta = \varepsilon^{l}\,\zeta \qquad \overset{L}{b}{}_{\beta}{}^{\alpha} = \varepsilon^{r}\,\overset{L_r}{b}{}_{\beta}{}^{\alpha} \qquad \boldsymbol{L}_r = \boldsymbol{R}\,\varepsilon^{1-(r+l)}$$

$$\left\{ \begin{aligned} \frac{\boldsymbol{h}}{\boldsymbol{L}} &= \hat{\varepsilon} = \varepsilon^{l} \\ \frac{\boldsymbol{h}}{\boldsymbol{R}} &= \varepsilon \\ \boldsymbol{L} &= \boldsymbol{R}^{l}\,\boldsymbol{h}^{1-} \end{aligned} \right.$$

where

ζ	normal coordinate; $\lvert\zeta\rvert \leq 1$
l	asymptotic parameter; l (real number) > 0
r	asymptotic parameter; r (real number) $> -l$
index $L(L_r)$	reduced with respect to measure of length $\boldsymbol{L}(\boldsymbol{L}_r)$
$\boldsymbol{R}$	measure of curvature radii
ε	thickness-curvature product

Scaling of displacements, stresses and strains

$$
\begin{aligned}
{}_{[0]}w^\alpha &= \varepsilon^t\left({}_{[t,0]}w^\alpha + {}_{[t+\lambda,0]}w^\alpha \varepsilon^\lambda + {}_{[t+2\lambda,0]}w^\alpha \varepsilon^{2\lambda} + \dots + {}_{[0]}\overset{w}{R}{}^\alpha_{m\lambda}\right)\\
{}_{[0]}w^3 &= \varepsilon^s\left({}_{[s,0]}w^3 + {}_{[s+\lambda,0]}w^3 \varepsilon^\lambda + \dots + {}_{[0]}\overset{w}{R}{}^3_{m\lambda}\right)\\
{}_{[00]}E^{\alpha\beta} &= \varepsilon^\Phi\left({}_{[\Phi,0]}E^{\alpha\beta} + {}_{[\Phi+\lambda,0]}E^{\alpha\beta} \varepsilon^\lambda + \dots + {}_{[0]}\overset{E}{R}{}^{\alpha\beta}_{m\lambda}\right)\\
{}_{(01)}E^{\alpha\beta} &= \varepsilon^\Psi\left({}_{(\Psi,1)}E^{\alpha\beta} + {}_{(\Psi+\lambda,1)}E^{\alpha\beta} \varepsilon^\lambda + \dots + {}_{(1)}\overset{E}{R}{}^{\alpha\beta}_{m\lambda}\right)\\
{}_{[00]}X^{\alpha\beta} &= \varepsilon^\Lambda\left({}_{[\Lambda,0]}X^{\alpha\beta} + {}_{[\Lambda+\lambda,0]}X^{\alpha\beta} \varepsilon^\lambda + \dots + {}_{[0]}\overset{X}{R}{}^{\alpha\beta}_{m\lambda}\right)\\
{}_{(01)}X^{\alpha\beta} &= \varepsilon^\Omega\left({}_{(\Omega,1)}X^{\alpha\beta} + {}_{(\Omega+\lambda,1)}X^{\alpha\beta} \varepsilon^\lambda + \dots + {}_{(1)}\overset{X}{R}{}^{\alpha\beta}_{m\lambda}\right)
\end{aligned}
$$

where

t and s
Φ and Ψ
Λ and Ω
λ
} are asymptotic parameters (positive real numbers)

For the derivatives of displacements, stresses and strains with respect to x^α, asymptotically identical series expansions are assumed.

E. g.

$$
{}_{[0]}w^\alpha_{,\beta} = \varepsilon^t\left({}_{[t,0]}w^\alpha_{,\beta} + {}_{[t+\lambda,0]}w^\alpha_{,\beta}\varepsilon^\lambda + \dots + {}_{[0]}\overset{w}{R}{}^\alpha_{m\lambda,\beta}\right).
$$

With this assumption, the, up to now, arbitrary measure of length $\boldsymbol{L}$ is interpreted as a measure of the derivatives with respect to x^α, or equivalently, of the "*wave lengths*" of the patterns of variation over the middle surface.

$$
\begin{cases}
l = \text{great} \rightarrow \boldsymbol{L} = \text{great: the rate of variation is low}\\
l = \text{small} \rightarrow \boldsymbol{L} = \text{small: the rate of variation is high}
\end{cases}
$$

Hence the parameter l acts as a *contra-variation* index

Besides the thickness parameter $\hat{\varepsilon}$, a second thickness parameter ε, consisting of the product of the measures of shell thickness ($\boldsymbol{h}$) and middle-surface curvatures ($\boldsymbol{R}^{-1}$), is introduced. This second parameter ε is completely determined by the geometry of the shell. Interpreting the measure of length $\boldsymbol{L}$ as a wave length, the firstly introduced parameter $\hat{\varepsilon}$ is determined by the variation of curvatures over the domain of the middle surface and also by the variation of boundary conditions on the upper and the lower face and along the edge of the shell. Hence, $\hat{\varepsilon}$ is more characteristic of what happens to the shell. The connecting link between both parameters $\hat{\varepsilon}$ and ε constitutes the contra-variation index l.

Directly reducing the components of the curvature tensor $b_\beta{}^\alpha$ to the measure $\boldsymbol{R}$ of the curvature radii appears impossible. The insertion of an auxiliary curvature measure $\boldsymbol{L}_r$ proves necessary. The measure $\boldsymbol{L}_r$ allows the curvature coefficients in the resulting asymptotic equations to deviate from the order of magnitude one.

The form of the asymptotic expansions with respect to ε, indicated in the above scalings, is particularly incited by the desire to establish an asymptotic method that is unrestricted by the fact whether the equations in question are linear or not. For that reason, the more familiar form of asymptotic expansion, which is preceded by an exponential function containing the asymptotic parameter in its argument, was out of the question. The form adopted with "real number powers" of the asymptotic parameter is well suited for the analysis of non-linear shell equations also.

To start with, it is natural to assume that a possibly different rate of variation in different directions along the middle surface *does* not influence the ultimate asymptotic results. Therefore the smallest wave length L_1 along the coordinate lines x^2 = constant on the middle surface is assumed to be of the same order of magnitude as the smallest wave length L_2 along the coordinate lines x^1 = constant. Hence

$$L_1 = L_2 = \boldsymbol{L}. \tag{4.1}$$

This type of variation along the middle surface has been called *isotropic variation*.

Analysis of the interior shell equations on the basis of the asymptotic technique described in the foregoing results in series of asymptotic equations which must be satisfied by the consecutive, asymptotic approximations. The resulting equations for the determination of the lowest-order asymptotic approximations correspond to the classical equations of

bending of flat plates and generalized plane stress in flat membranes,
the shallow shell theory and of
inextensional bending and membrane theory of shells.

Their asymptotic features are pictured on Fig. 4. Two limiting cases belonging to the first category are obtained also. The uppermost contains in addition to the plate-bending equations an extra curvature term in the kinematic equations, while the lowermost contains in addition to the plane-stress equations an extra curvature term in the third of the equilibrium equations.

The complete range of wave lengths $L_1 = L_2 = \boldsymbol{L}$ is clearly subdivided in three important sub-regions

$$\left.\begin{array}{ll} l < \frac{1}{2} & : \boldsymbol{L} < \sqrt{\boldsymbol{Rh}} \\ l = 0\left(\frac{1}{2}\right) & : \boldsymbol{L} = 0\left(\sqrt{\boldsymbol{Rh}}\right) \\ l > \frac{1}{2} & : \boldsymbol{L} > \sqrt{\boldsymbol{Rh}} \end{array}\right\} \tag{4.2}$$

The lowest-order asymptotic equations obtained in each of the three sub-regions literally correspond to the well-known classical equations for the respective categories of shell problems listed above. The wedge-shaped

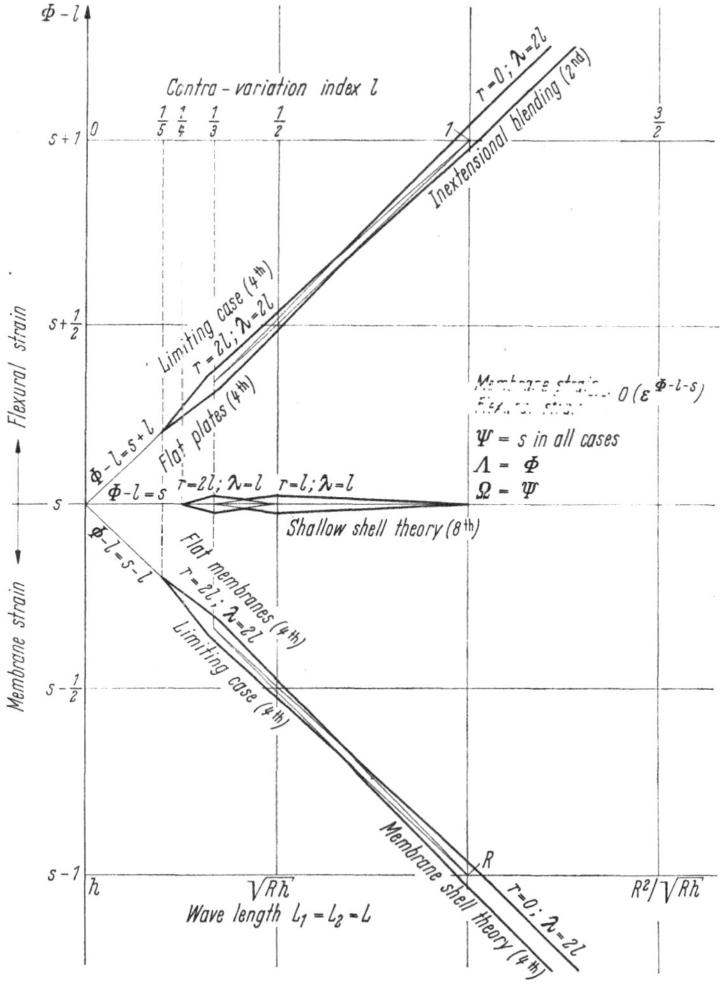

Fig. 4. Asymptotic shell equations; $L_1 = L_2 = L$.

areas represent each a system of asymptotic equations. At the centre of each wedge-shaped area the measure of length L_r is equal to R. Hence the curvature coefficients in the differential equations are of the order of magnitude one. Towards the left-hand vertex of each wedge-shaped area the magnitude of the curvature coefficients gradually decreases from

$O(1)$ to $O(\varepsilon^\lambda)$. Hence the terms in the differential equations containing curvature coefficients shift towards the next higher order in ε. Towards the right-hand vertices exactly the reverse takes place. Curvature terms in the differential equations shift from higher to one order lower in ε. Hence the length of the wedges determines the validity range with respect to $\boldsymbol{L}$ or l of the various systems of interior asymptotic shell equations. The thickness of the wedges is indicative of the relative validity of each system of asymptotic equations.

The order of the various systems of differential equations, when expressed in the displacement component ${}_{[s,0]}w^3$ or ${}_{[t,0]}w^\alpha$, is indicated on Fig. 4 behind each of the group names of lowest-order equations. It is observed that the order of the flat plate and flat membrane systems of asymptotic equations and of the shallow shell theory ($l \leqq 1/2$) agrees with the number of four reduced edge conditions on each edge of the shell. However, the order of the inextensional bending equations and the equations of membrane shell theory ($l > 1/2$) is too low to be able to satisfy all four reduced edge conditions.

Finally it should be observed that the system of shell equations denoted as *shallow shell theory* applies to any shell problem in which the smallest wave length of surface loads and/or edge conditions is of the order of magnitude $\sqrt{\boldsymbol{R}\boldsymbol{h}}$. The applicability of these equations is not restricted to shells which are "shallow" in the geometrical sense only.

The technique adopted of asymptotic analysis by means of the introduction of two asymptotic parameters resembles the technique A. E. GREEN employed in his works of 1962. The fusion of the two parameters by the aid of the contra-variation index l and the introduction of the auxiliary measure $\boldsymbol{L}_r$ for the curvature radii complete this technique. As already explained, the familiar form of asymptotic expansion, which is preceded by an exponential function, is not considered in this work. This is basically divergent from the works of A. L. GOL'DENWEIZER. However, the negative value of the index of variation t GOL'DENWEIZER introduced agrees within an additive constant of the order of magnitude one to the contra-variation index l defined above: $l + t = 0(1)$.

5. Bending Edge Effects

The postulate of isotropic variation of the ultimate asymptotic results, i. e. of displacements, stresses and strains, proves to be too narrow in the cases $l > 1/2$ of slowly varying surface loads and edge conditions, viz. for wave lengths of the order of magnitude $\boldsymbol{R}$ and greater.

In such cases the, let us say, "isotropic" asymptotic equations lose their uniform validity near the edge of the shell. The order of the differential equations is smaller than the number of imposed edge conditions. For the second time an edge effect must be considered, now in relation to the interior shell equations. The edge effect mainly consists of bending strains and stresses fading away exponentially at increasing distance from the edge. For that reason this second edge effect encountered in the asymptotic theory of elastic shells is called the *bending edge effect* of shells.

For the asymptotic analysis of the bending edge effect, the interior shell equations are referred to edge-curve coordinates. The scaling with respect to ε of the geometry of the shell remains the same as indicated above for the isotropic case, except for the coordinate x^2 normal to the edge curve.

$$\overset{L}{x^2} = \varepsilon^b \eta, \quad b \text{ (real number)} \neq 0 \text{ and } b < l \tag{5.1}$$

where η is the coordinate x^2 reduced to the wave length $L_2 \neq \boldsymbol{L}$. The coordinate x^1 along the edge curve remains reduced to the wave length $L_1 = \boldsymbol{L}$. The values of the exponent b, which allow a consistent asymptotic analysis of the interior shell equations, are of the form

$$b = \frac{\pm 2}{n + k + 1} l, \qquad k = 1, 2, 3 \quad \text{and} \quad n = 1, 2, 3 \ldots \tag{5.2}$$

For $n \to \infty$, the isotropic case is retrieved.

The order of magnitude of the wave lengths L_2, which correspond to the above values of the exponent b, is equal to

$$L_2 = 0\left(\boldsymbol{L}\left(\frac{\boldsymbol{R}\,\boldsymbol{h}}{\boldsymbol{L}^2}\right)^{\frac{\pm 1}{k+1}}\right), \qquad k = 1, 2, 3. \tag{5.3}$$

The ε-scaling of the geodesic ($\varkappa_g$) and normal curvatures ($\varkappa_n$ and $2_{(0)}H - \varkappa_n$) and of the geodesic torsion (τ_g) suffers modifications as a result of the asymptotic analysis of the governing equations.

The ε-scaling of the components of displacements, stresses and strains is again similar to the scalings adopted in the isotropic case. It proves necessary only to expand the number of exponents of the prefactors of the asymptotic series.

$$t \to t_\alpha \quad \text{and} \quad \begin{cases} \Phi \to \Phi_{\alpha\beta} \\ \Psi \to \Psi_{\alpha\beta} \end{cases} \quad \text{and} \quad \begin{cases} \Lambda \to \Lambda_{\alpha\beta} \\ \Omega \to \Omega_{\alpha\beta} \end{cases} \tag{5.4}$$

The exponent λ in each case is equal to the modulus of b.

Systematic examination of the interior shell equations on the basis of the asymptotic scheme described above produces several sequences of asymptotic equations for the description of the bending edge effect.

In the same manner as on Fig. 4, the validity ranges with respect to the contra-variation index l of the various systems of bending edge-effect equations are indicated on Fig. 5. The systems of the lowest order in ε are considered only. Each row of wedges I, II, III, and IV on the radii towards the point $(s, 1/2)$ on Fig. 5 represents the equations characterizing

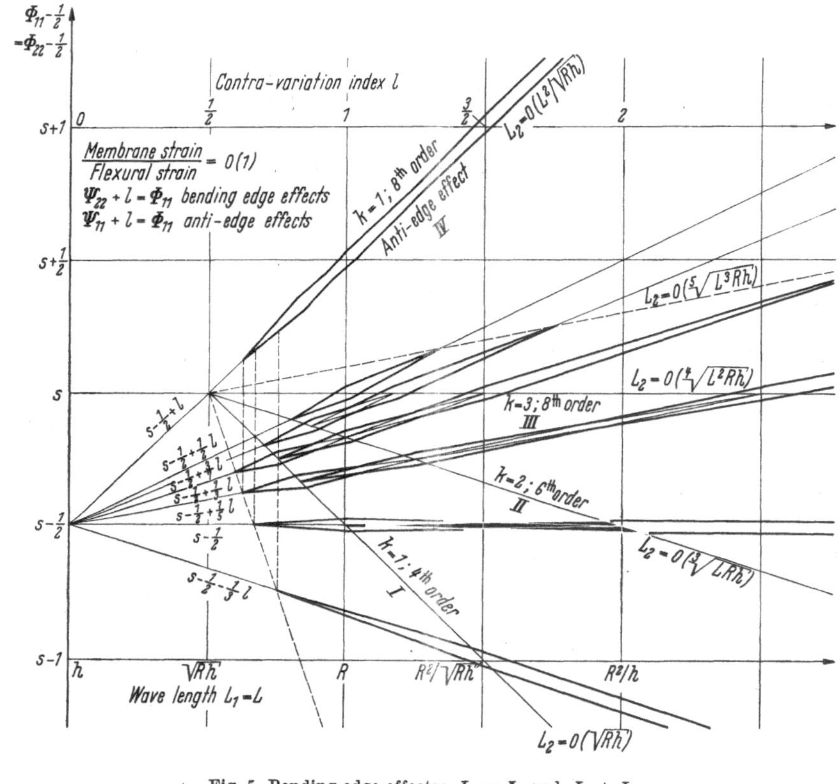

Fig. 5. Bending edge effects; $L_1 = L$ and $L_2 \neq L_1$.

a typical bending edge effect. The $k = 1$ — wedges below on Fig. 5 (I) relate to the edge effect governed by an edge curve which is *not* an asymptotic line of the middle surface. The $k = 2$ — wedges (II) relate to the edge effect of an edge curve which is a simple or a non-geodesic double asymptotic line of the middle surface, while the row of $k = 3$ — wedges (III) relates to geodesic, double asymptotic edge curves. The coinciding $k = 1$ — wedges above on Fig. 5 (IV) finally, represent the systems of equations which describe the edge effect governed by an edge curve coinciding with a geodesic, non-asymptotic line on a middle surface of zero Gaussian curvature. The occurence of this latter edge effect is actual only, if the prescribed edge conditions of the shell are such that inextensional or nearly-inextensional bending of the middle surface is not precluded. With shrinking thickness of the shell this edge effect infinitely expands into the interior of the shell. For this reason, the denomination *anti-edge effect* is adopted.

The bending edge effects all converge in the "isotropic" shallow shell theory for decreasing wave lenghts L_1 along the edge curve. The more than reasonable overlapping of the validity ranges for each type of bending edge effect originates from the higher order approximations.

The lowest-order systems of equations indicated on Fig. 5 almost all coalesce for each type of edge effect. The differences arise in the systems of asymptotic equations to be satisfied by the higher order approximations in ε.

Examples of the lowest-order asymptotic equations governing the various types of bending edge effect are given below

I. *Non-asymptotic edge curve*; $L_2 = O\left(\sqrt{\boldsymbol{Rh}}\right)$

$$\varkappa_n = O\left(\frac{1}{\boldsymbol{R}}\right) \qquad \text{and} \qquad 2\,{}_{(0)}H - \varkappa_n = O\left(\frac{1}{\boldsymbol{R}}\right)$$

$$\tau_g = O\left(\frac{1}{\boldsymbol{R}}\right) \qquad \qquad \varkappa_g = O\left(\frac{1}{\boldsymbol{L}}\right)$$

$$\left[\frac{\partial^4}{\partial \eta^4} + 3(1-\nu^2)\varkappa_n^2\right]_{[s,0]}w^3 = 0$$

II. *Simple asymptotic edge curve*; $L_2 = O\left(\sqrt[3]{\boldsymbol{LRh}}\right)$

$$\varkappa_n = 0 \qquad \text{and} \qquad 2\,{}_{(0)}H - \varkappa_n = O\left(\frac{1}{\boldsymbol{R}}\right)$$

$$\tau_g = O\left(\frac{1}{\boldsymbol{R}}\right) \qquad \qquad \varkappa_g = O\left(\frac{1}{\boldsymbol{L}}\right)$$

$$\left[\frac{\partial^6}{\partial \eta^6} + 3(1-\nu^2)\left\{2\tau_g \frac{\partial}{\partial x^1} - \dot{\tau}_g + 4\varkappa_{g(0)}H - (\dot{\tau}_g - 2\varkappa_{g(0)}H)\frac{\partial}{\partial \eta}\eta\right\} \times \right.$$
$$\left. \times \left\{\frac{\partial}{\partial x^1} 2\tau_g - 2\varkappa_{g(0)}H - (\dot{\tau}_g - 2\varkappa_{g(0)}H)\frac{\partial}{\partial \eta}\eta\right\}\right]_{[s,0]} w^3 = 0$$

if $\dot{\tau}_g = 0$ and $\varkappa_g = 0$: geodesic asymptotic line; or

if $\dot{\tau}_g = 0$ and ${}_{(0)}H = 0$: asymptotic line on minimal surface

$$\left[\frac{\partial^6}{\partial \eta^6} + 3(1-\nu^2)(2\tau_g)^2 \frac{\partial^2}{(\partial x^1)^2}\right]_{[s,0]} w^3 = 0$$

III. Geodesic, double asymptotic edge curve; $L_2 = O\left(\sqrt[4]{L^2 R h}\right)$

$$\varkappa_n = 0 \qquad \text{and} \qquad 2\,{}_{(0)}H - \varkappa_n = O\left(\frac{1}{R}\right)$$

$$\tau_g = 0 \qquad\qquad \varkappa_g = 0$$

$$\left[\frac{\partial^8}{\partial \eta^8} + 3(1-\nu^2)\,(2\,{}_{(0)}H)^2 \frac{\partial^4}{(\partial x^1)^4}\right]_{[s,0]}w^3 = 0$$

and

$$\left[\frac{\partial^8}{\partial \eta^8} + 2(2\,{}_{(0)}H)^2 \frac{\partial^6}{\partial \eta^6} + (2\,{}_{(0)}H)^4 \frac{\partial^4}{\partial \eta^4} + 3(1-\nu^2)\,(2\,{}_{(0)}H)^2 \frac{\partial^4}{(\partial x^1)^4}\right]_{[s,0]}w^3 = 0$$

condition: ${}_{(0)}\dot{H} = 0$

IV. Non-asymptotic edge curve; $L_2 = O\left(\frac{L^2}{\sqrt{R h}}\right)$

$$\varkappa_n = O\left(\frac{1}{R}\right) \qquad \text{and} \qquad 2\,{}_{(0)}H - \varkappa_n = 0$$

$$\tau_g = 0 \qquad\qquad \varkappa_g = 0$$

$$\left[\frac{\partial^8}{(\partial x^1)^8} + 2\varkappa_n^2 \frac{\partial^6}{(\partial x^1)^6} + \varkappa_n^4 \frac{\partial^4}{(\partial x^1)^4} + 3(1-\nu^2)\,\varkappa_n^2 \frac{\partial^4}{\partial \eta^4}\right]_{[s,0]}w^3 = 0$$

condition: $\dot{\varkappa}_n = 0$

The orders four, six and eight of these differential equations, whether complemented with the membrane or inextensional bending equations or not, in any relevant case of slowly varying surface loads and edge conditions, finally, guarantee the satisfaction of all four reduced edge conditions that can be prescribed on each edge of the shell. This completes the asymptotic analysis of the interior shell equations.

The first, fourth order, ordinary differential equation specified above is well-known from the classical theory of bending of circular cylindrical shells subject to rotationally symmetric loads and edge constraints. Also in the case of circular cylindrical shells, A. L. GOL'DENWEIZER derived edge-effect equations on the basis of the classical shell theory, which are identical to the lowest-order, bending edge-effect equations specified above.

6. Epilogue

The foregoing foundation of the linear theory of elastic shells on the three-dimensional, linear theory of elasticity is not yet complete. The first deficiency to be mentioned is the uncertainty that still exists with regard to the real asymptotic character of the series expansions in ε, despite the consistency of the final results. The formal series expansions in ε *have* been called *asymptotic series* and the analyses on the basis of these expansions *have* been called *asymptotic methods*. However, the actual asymptotic properties of the series expansions have nowhere been shown or proven mathematically.

For the present time we must be content with the rather poor fact that adequate mathematical proofs are lacking in practically all such cases.

For the future, however, on the one hand the considerable progress of recent years made in the branch of mathematics that is concerned with singular perturbation problems and the development of advanced methods for the evaluation of concrete estimates for the derivatives of any order of the solutions of elliptic partial differential equations on the other hand, hold out the hopes that within a reasonable number of years the required mathematical proofs may be constructed. The apparent consistency and nearly uniform validity of the formal asymptotic results already obtained constitute confident indications that the series expansions actually possess the asymptotic properties.

The other remaining questions all are related to the above principal theme. A brief review of the remaining questions comprises the following

Non-uniformity of the assumed asymptotic expansions of the Kirchhoff edge-zone equations in the vicinity of corners and singularities on the edge of the shell. Questions concerning uniform asymptotic representations of the Kirchhoff edge effect arise (Asymptotic continuation, Stokes' phenomenon, transition points?);

Non-uniformity of the assumed asymptotic expansions of the interior shell equations at the transitions from one type of edge curve to another type and also at the corners of the shell. Similar questions as before.

Apart from the, for the time being, inevitable deficiences, the formal asymptotic results obtained and methods experienced appear to be very promising for future application to related, unsolved problems. Among the related problems, the following two occupy a dominant place

Derivation of the non-linear elastic shell theory on the basis of the three-dimensional, geometrically non-linear theory of elasticity and

Derivation of the theory of elastic beams and rods on the basis of the interior shell theory.

References

1924 Lichtenstein, L.: Neuere Entwicklung der Theorie partieller Differentialgleichungen zweiter Ordnung vom elliptischen Typus. Enzykl. d. Math. Wiss., Vol. II, 3, III.

1941 Synge, J. L., and W. Z. Chien: The intrinsic theory of elastic shells and plates. Appl. Mech., Theodore von Kármán Anniversary Volume.

1944 Chien, W. Z.: The intrinsic theory of thin shells and plates. Quart. Appl. Math., **I** No. 4, **II** No. 1 and 2.
1949 John, F.: On linear partial differential equations with analytic coefficients. Comm. P. Appl. Math., Vol. 2.
1950 Friedrichs, K. O.: Kirchhoff's Boundary Conditions and the Edge Effect for elastic plates. Proc. Symp. Appl. Math., Vol. III.
1952 Truesdell, C.: The mechanical foundations of elasticity and fluid dynamics. J. Rat. Mech. Anal., 1.
1953 Gol'denweizer, A. L.: Theory of thin elastic shells, New York: Pergamon Press.
1961 Friedrichs, K. O., and R. F. Dressler: A boundary-layer theory for elastic plates. Comm. P. Appl. Math., Vol. 14.
1961 Reiss, E. L., and S. Locke: On the theory of plane stress. Quart. Appl. Math., **19** No. 3.
1962 Green, A. E.: On the linear theory of thin elastic shells. Proc. Roy. Soc., Ser. A, Vol. 266, and Boundary-layer equations in the linear theory of thin elastic shells. Proc. Roy. Soc., Ser. A, Vol. 269.
1965 John, F.: Estimates for the derivatives of the stresses in a thin shell and interior shell equations. Comm. P. Appl. Math., Vol. 18.
1966 Eckhaus, W., and E. M. de Jager: Asymptotic solutions of singular perturbation problems for linear differential equations of elliptic type. Arch. Rat. Mech. Anal., **23**, No. 1.

Discussion

I. N. Vekua: If you use an expansion in an infinite power series in the thickness coordinate, it is impossible to guarantee the convergence of this series.

H. S. Rutten: The local analyticity of the general solutions, which is guaranteed by the properties of the system of elliptic partial differential equations, is expanded in the case of the interior of shells to include the whole thickness range by properly restricting the boundary conditions on the upper and lower face.

E. Reissner: I would like to raise a question concerning use of expansions in powers of the thickness coordinate x^3, which seems to be the basis of your asymptotic procedure. Inasmuch as formal expansions in powers of a thickness coordinate which were used in classical approaches to the subject did not lead to asymptotic results it would be interesting to understand clearly, the essence of your approach, in particular in comparison with other asymptotic considerations such as Professor Gol'denweizer's and that of my own 1963-paper on the subject.

H. S. Rutten: The expansion in powers of x^3 is not a formal expansion but it evolves from the fundamental regularity of the general solution of the three-dimensional equations. This is the same starting point as Professor John adopted for his analysis. He forms estimates of the derivatives of any order of the three-dimensional solution, I explicitly express the higher order derivatives in the lowest order ones with respect to middle surface coordinates.

On the Shell Equations in Complex Form

By

J. L. Sanders, jr.

Harvard University, Cambridge, Mass., USA

Introduction

This paper is concerned with the problem of formulating the general equations of linear thin shell theory in terms of complex combinations of dependent variables in such a way as to arrive at a fourth order system of partial differential equations. Consideration is limited to elastic, isotropic, homogeneous shells with edge loads only.

NOVOZHILOV was the first to formulate such a system of equations for a shell with an arbitrary middle surface subjected to arbitrarily distributed loads. A full account of his theory is given in [1]. A history of the problem and survey of developments up to 1962 was given by NOVOZHILOV in [2]; therefore, no such survey will be attempted here. Further developments of the Novozhilov theory, principally with respect to the "displacement" form of the equations, boundary conditions, and variational principles are contained in a recently published book by CHERNYKH [3]. Related results concerned with extensions of the theory to include effects of anisotropy, inhomogeneity, and thermal strains have been obtained by LIBRESCU and by VISARION and STANESCU in [4, 5], and other papers. In the author's understanding, these extensions lead to a system of equations of a mixed type which contain the operations of partial differentiation and the taking of complex conjugates. In general the "conjugate terms" are not negligible and cannot be eliminated without raising the "order" of the system above four.

The remarkable symmetry in the complete system of shell equations (including stress-stress function relations) discovered by GOL'DENWEIZER and called the static-geometric analogy leads to an almost obvious

Acknowledgement. This work was supported in part by the National Aeronautics and Space Administration under Grant NsG-559, and by the Division of Engineering and Applied Physics, Harvard University.

complex formulation of the equations in the case of vanishing Poission's ratio ($\nu = 0$). In the case $\nu \neq 0$, difficulties arise. NOVOZHILOV, in effect, derived approximate equilibrium and compatibility equations for which the difficulty is avoided. The validity of the approximate equations is not easy to prove in general, although considerable thought and effort has been devoted to establishing their correctness. As far as this author is aware, the validity of the equations has not yet been definitely disproved in any particular case. CHERNYKH derived the same system of equations by a different method. He first arrived at a "mixed" system (of the type noted previously) and then obtained equations (through several transformations) in which the conjugate terms appear multiplied by a small parameter; these terms were dropped at each stage of the transformations. However, it is extremely difficult to justify the dropping of apparently small terms in a system of partial differential equations.

In the present paper the equations of shell theory are obtained in complex form (free of conjugate terms) by a method which avoids the necessity of approximating the equilibrium and compatibility equations or any of the equations except the constitutive relations. This is done at the expense of introducing an auxiliary set of partial differential equations for certain "error terms" in the constitutive relations. The order of magnitude of the error terms does not exceed the order of magnitude of the errors inherent in the constitutive relations due to the fundamental hypotheses of thin shell theory. Solution of this set of shell equations divides into two problems: I. the solution of a determinate fourth order system of equations similar to the Novozhilov equations, but with a few more terms, and II, the solution of a set of equations identical in form to the non-homogeneous equilibrium equations for the error terms previously mentioned.

For general cylinders and for shells of revolution, problem I is reduced (in two ways) to the solution of a single fourth order partial differential equation in a scalar unknown. An alternative form of problem II is derived and a complete solution of problem II without quadratures is obtained for spheres and general cylinders.

Fundamental Equations in Dimensionless Form

Let L be a reference length, the length over which significant changes in the dependent variables occur, sometimes called a wavelength. Let R be a reference radius of curvature, σ a reference stress, and h the (constant) shell thickness. Let the equation of the middle surface of the shell be

$$\tilde{x}^i = \tilde{x}^i(\xi^\alpha)$$

where $\tilde{x}^i$ and $\tilde{\xi}^\alpha$ have dimensions of length, and a change in $\tilde{\xi}^\alpha$ of $O(R)$ corresponds to a distance of $O(R)$. Introduce the following dimensionless variables:

$$\tilde{x}^i = R x^i \qquad \tilde{g}_{\alpha\beta} = g_{\alpha\beta} \text{ (first fundamental form)}$$
$$\tilde{\xi}^\alpha = L \xi^\alpha \qquad R\tilde{b}_{\alpha\beta} = b_{\alpha\beta} \text{ (second fundamental form).}$$

A few of the familiar equations of surfaces in dimensionless form are as follows

$$ds^2 = g_{\alpha\beta}\, d\xi^\alpha\, d\xi^\beta \qquad x^i_{,\alpha} x^i_{,\beta} = \beta\mu g_{\alpha\beta}$$
$$n^i_{,\alpha} = b_\alpha{}^\gamma x^i_{,\gamma} \qquad x^i_{,\alpha\beta} = -\beta\mu\, b_{\alpha\beta} n^i$$

where n^i is the unit normal to the surface, $b_{\alpha\beta}$, the second fundamental form, differs in sign from the usual definition, and a comma denotes covariant differentiation. The dimensionless parameters β and μ are defined as follows

$$\mu = \frac{L^2}{Rh}\sqrt{12(1-\nu^2)} \qquad \beta = \frac{h}{R\sqrt{12(1-\nu^2)}}\,.$$

Dimensionless stress and strain measures, displacements and stress functions (without a tilde) are related to the dimensional variables as follows

$$\tilde{N}_{\alpha\beta} = \sigma h N_{\alpha\beta} \qquad \tilde{E}_{\alpha\beta} = \frac{\sigma}{E} E_{\alpha\beta}$$

$$\tilde{M}_{\alpha\beta} = \frac{\sigma h L^2}{\mu R} M_{\alpha\beta} \qquad \tilde{K}_{\alpha\beta} = \frac{\sigma R\mu}{E L^2} K_{\alpha\beta}$$

$$\tilde{u}_\alpha = \frac{\sigma L}{E} u_\alpha \qquad \tilde{\chi}_\alpha = \frac{\sigma h L^2}{\mu R} \chi_\alpha$$

$$\tilde{w} = \frac{\mu\sigma R}{E} w \qquad \tilde{\psi} = \sigma h L^2 \psi\,.$$

Two special notations which permit the shell equations to be written in a more compact form are introduced next. Define a tensor $B_{\alpha\beta}$ in terms of $b_{\alpha\beta}$ as follows

$$B_{\alpha\beta} = \frac{1}{2}\left(\varepsilon_{\alpha\gamma} b_\beta{}^\gamma + \varepsilon_{\beta\gamma} b_\alpha{}^\gamma\right)$$

where $\varepsilon_{\alpha\beta}$ is the usual permutation tensor. For any second order tensor define a "bar" operations as follows

$$\overline{T}_{\alpha\beta} = \varepsilon_{\alpha\lambda}\varepsilon_{\beta\mu}\, T^{\lambda\mu} \equiv g_{\alpha\beta}\, T_\lambda{}^\lambda - T_{\beta\alpha}\,.$$

Note that $\overline{\overline{T}}_{\alpha\beta} = T_{\alpha\beta}$, $\overline{T}_\alpha{}^\alpha = T_\alpha{}^\alpha$, $\overline{g}_{\alpha\beta} = g_{\alpha\beta}$, $\overline{B}_{\alpha\beta} = -B_{\alpha\beta}$. Hereafter a bar over a second order tensor will always denote this operation.

The following field equations of linear thin shell theory are taken from [6—8] and put in dimensionless form.

Strain-displacement relations

$$E_{\alpha\beta} = \frac{1}{2}(u_{\alpha,\beta} + u_{\beta,\alpha}) + \mu b_{\alpha\beta} w$$

$$K_{\alpha\beta} = \frac{1}{2}(\vartheta_{\alpha,\beta} + \vartheta_{\beta,\alpha}) + \beta B_{\alpha\beta}\omega$$

$$\varphi_\alpha = -w_{,\alpha} + \beta b_\alpha^{\ \gamma} u_\gamma$$

$$\omega = \frac{1}{2}\varepsilon^{\alpha\beta} u_{\beta,\alpha}. \tag{1}$$

Stress-stress function relations

$$\overline{M}_{\alpha\beta} = \frac{1}{2}(\chi_{\alpha,\beta} + \chi_{\beta,\alpha}) + \mu b_{\alpha\beta}\psi$$

$$\overline{N}_{\alpha\beta} = -\frac{1}{2}(\vartheta_{\alpha,\beta} + \vartheta_{\beta,\alpha}) - \beta B_{\alpha\beta}\Omega$$

$$\vartheta_\alpha = -\psi_{,\alpha} + \beta b_\alpha^{\ \gamma}\chi_\gamma$$

$$\Omega = \frac{1}{2}\varepsilon^{\alpha\beta}\chi_{\beta,\alpha}. \tag{2}$$

Equilibrium equations

$$N^{\alpha\beta}_{,\beta} + \beta\, b_\gamma^{\ \alpha} M^{\gamma\beta}_{,\beta} - \frac{1}{2}\beta\varepsilon^{\alpha\beta}(B_{\gamma\delta}M^{\gamma\delta})_{,\beta} = 0$$

$$M^{\alpha\beta}_{,\alpha\beta} - \mu\, b_{\alpha\beta}N^{\alpha\beta} = 0. \tag{3}$$

Compatibility equations

$$\overline{K}^{\alpha\beta}_{,\beta} - \beta\, b_\gamma^{\ \alpha}\overline{E}^{\gamma\beta}_{,\beta} + \frac{1}{2}\beta\,\varepsilon^{\alpha\beta}(B_{\gamma\delta}\overline{E}^{\gamma\delta})_{,\beta} = 0$$

$$\overline{E}^{\alpha\beta}_{,\alpha\beta} + \mu\, b_{\alpha\beta}\overline{K}^{\alpha\beta} = 0. \tag{4}$$

Constitutive relations

$$\overline{E}_{\alpha\beta} = -(1+\nu)\,N_{\alpha\beta} + g_{\alpha\beta}N_\gamma^{\ \gamma}$$

$$M_{\alpha\beta} = -(1-\nu)\,\overline{K}_{\alpha\beta} + g_{\alpha\beta}\overline{K}_\gamma^{\ \gamma}. \tag{5}$$

In the shell theory set down here the tensor measures of stress and strain are symmetric. The similarity in form between pairs of equations is, of course, a manifestation of the static-geometric analogy.

The reference length L appearing in the parameter μ and elsewhere is meant to be chosen in such a way that differenatition does not change the order of magnitude of any of the fundamental variables $N^{\alpha\beta}$, w etc. In some cases (notably for certain solutions of the equations of cylindrical shells) the variables change more rapidly in one direction than in another. In such a case L should be the shorter of the two "wavelengths" so that differentiation will not increase the order of magnitude of a variable. It seems unlikely that the variables $N^{\alpha\beta}$, w etc., can change much less rapidly than the surface variables, $b_{\alpha\beta}$, $B_{\alpha\beta}$ etc.; in any case it will generally be assumed that differentiation of these surface variables does not increase their order of magnitude. The reference stress σ can be chosen such that at least one of the components of $N^{\alpha\beta}$ or $M^{\alpha\beta}$ is about unity (over some region of the middle surface of dimension L, usually) and none of the components are much larger than unity.

Complex Combinations

Let $N'^{\alpha\beta}$, u_α' etc., be some solution to the shell equations and let $N''^{\alpha\beta}$, u''_α etc., be some other unrelated solution. Let

$$N^{\alpha\beta} = N'^{\alpha\beta} + i N''^{\alpha\beta} \tag{6}$$

and similarly for all other variables. The Eqs. (1) to (5), now in terms of complex-valued variables, hold as they stand. Of course, nothing has been gained by this operation. However, if some relation can be introduced between the primed and double-primed solutions, then there is the possibility of some gain. To see how this might be so, consider the case of Poisson's ratio ν equal to zero, and introduce the additional equation

$$N^{\alpha\beta} = i \bar{K}^{\alpha\beta}. \tag{7}$$

From (5) and (7) follows

$$M^{\alpha\beta} = -i \bar{E}^{\alpha\beta} \tag{8}$$

and one can obviously set $w = i\psi$ and $u_\alpha = i\chi_\alpha$. The number of field equations (1 to 5) is cut in half because the equations become identical by pairs, and the order of the system is reduced from eight to four. By separating real and imaginary parts in Eq. (7), one finds

$$N''_{\alpha\beta} = \bar{K}'_{\alpha\beta}; \qquad \bar{K}''_{\alpha\beta} = -N'_{\alpha\beta}$$

so

$$N_{\alpha\beta} = N'_{\alpha\beta} + i\bar{K}'_{\alpha\beta}$$
$$\bar{K}_{\alpha\beta} = \bar{K}'_{\alpha\beta} - iN'_{\alpha\beta}$$

and similarly for other quantities. Thus, the introduction of the relation (7) is equivalent to the introduction of dependent variables which are complex combinations of quantities which are static-geometric analogs of each other. The shell equations were reduced in this way by NOVOZHILOV in [1]. The fact that the additional relation (7) does not result in an insoluble overdetermination of the system of shell equations is obviously due to the static-geometric analogy (which suggests just such a complex combination of the variables). An immediate difficulty is that $\nu = 0$ is not generally a useful approximation.

Suppose that (7) is retained in the case $\nu \neq 0$, then from (5) there follows (instead of (8))

$$M_{\alpha\beta} = -i\bar{E}_{\alpha\beta} + 2\nu\bar{K}_{\alpha\beta}. \tag{9}$$

One can easily verify that the system of Eqs. (1) to (5) plus the relations (7) and (9) is inconsistent. However, consider the VLOSOV approximate equations which are obtained from (1) to (5) by setting $\beta = 0$ (to which should be appended the rule $\varepsilon^{\alpha\beta} T_{\gamma,\alpha\beta} = 0$). By (7) the equation $N^{\alpha\beta}_{,\beta} = 0$ becomes $i\bar{K}^{\alpha\beta}_{,\beta} = 0$ which is true. The equation

$$M^{\alpha\beta}_{,\alpha\beta} - \mu b_{\alpha\beta} N^{\alpha\beta} = 0$$

becomes, by (7) and (9)

$$-i\bar{E}^{\alpha\beta}_{,\alpha\beta} + 2\nu\bar{K}^{\alpha\beta}_{,\alpha\beta} - i\mu b_{\alpha\beta}\bar{K}^{\alpha\beta} = 0$$

which is the same as

$$\bar{E}^{\alpha\beta}_{,\alpha\beta} + \mu b_{\alpha\beta}\bar{K}^{\alpha\beta} = 0$$

since $\bar{K}^{\alpha\beta}_{,\alpha\beta} = 0$.

Thus the compatibility equations are equivalent to the equilibrium equations. The relations $\psi = -iw$ and $\chi_\alpha = -iu_\alpha - 2\nu\, w_{,\alpha}$ between complex stress functions and complex displacements also holds. The complete system of VLOSOV equations can thus be put in complex form with no further approximations and obviously half of the equations and variables could be discarded to reduce the order of the system by half. The reduction to a single fourth order equation in w or ψ is well known.

In the case of a spherical shell ($b_{\alpha\beta} = g_{\alpha\beta}$, $B_{\alpha\beta} = 0$) the introduction of a slight modification to the relation (7) leads to a similar reduction of the exact system of equations. Put

$$N_{\alpha\beta} = i\lambda\bar{K}_{\alpha\beta} \tag{10}$$

where

$$\lambda^2 - 1 - 2i\beta\nu\lambda = 0.$$

Since β is small there is a root $\lambda \approx 1$. The relation (9) is replaced by

$$\bar{E}_{\alpha\beta} = i\lambda M_{\alpha\beta} - 2i\lambda\nu\bar{K}_{\alpha\beta}. \tag{11}$$

For λ a root of the above quadratic equation, one can easily verify that the equilibrium and compatibility equations are equivalent to each other, and moreover the following relations hold

$$w = i\lambda\psi, \qquad \chi_\alpha = -i\lambda u_\alpha - 2\nu w_{,\alpha}.$$

Without much difficulty the system of equations can be reduced to a single fourth order equation similar to one derived by Koiter [9]. Unfortunately, in the case of a general shell, (10) does not lead to a similar result for any value of λ.

Novozhilov [1] in effect derives approximate sets of equilibrium and compatibility equations for which the introduction of (7) does not lead to contradictions. Novozhilov, of course, bases his derivations on his set of shell equations which differ (but in no essential way) from (1) to (4). In his derivation of approximate equations for the complex theory, the equation of moment equilibrium about the normal to the middle surface is not enforced. Such an approximation is known to lead to difficulties with respect to accuracy in some cases, notably in the case of helicoidal shells [10—13]. In any case, some approximations in the equilibrium and compatibility equations are necessary if one wishes to put the shell equations in complex form and retain relation (7). However, in [13] Koiter has shown that the exact equilibrium and compatibility equations cannot, in general, be simplified. That is to say, no term in them is always negligible in all possible cases. Since the Novozhilov equations have been tested in practice and subjected to careful study, there seems to be little doubt that they are sufficiently accurate for most practical cases, at least for general cylinders and shells of revolution. There does, however, seem to be room for reasonable doubt that they are accurate (to within the usual limits of shell theory) in all cases, and in particular for certain problems of the helicoidal shell.

In view of the foregoing it is evident that the relation (7) must be nearly the correct one to introduce if one wishes to put the shell equations in complex form without introducing essential errors, if indeed this is possible. In the present paper a modification of (7) is introduced which leads to the desired result without approximation of the equilibrium or compatibility equations.

Formulation of Complex Equations in the General Case

Since $N_{\alpha\beta} - i\bar{K}^{\alpha\beta}$ is expected to be small (and for other reasons) it is convenient to introduce the "double" complex combinations of variables given below

$$P^{\alpha\beta} = N^{\alpha\beta} + i\bar{K}^{\alpha\beta} \qquad \overset{*}{P}{}^{\alpha\beta} = N^{\alpha\beta} - i\bar{K}^{\alpha\beta}$$

$$Q^{\alpha\beta} = M^{\alpha\beta} - i\bar{E}^{\alpha\beta} \qquad \overset{*}{Q}{}^{\alpha\beta} = M^{\alpha\beta} + i\bar{E}^{\alpha\beta}. \tag{12}$$

(Because of the way in which the stress and strain measures were made dimensionless $P_{\alpha\beta}$ and $Q_{\alpha\beta}$ must be $0(1)$. $P_{\alpha\beta} = 0(1)$ means that at least one of the components of $P_{\alpha\beta}$ has a real or imaginary part which is about unity and no component has a real or imaginary part much greater than unity. In general the real and imaginary parts of $P_{\alpha\beta}$ can differ in order of magnitude. In Eq. (13) below an apparently small term cannot be dropped unless this term is negligible in both the real and imaginary parts of the equation. In the VLOSOV case the real and imaginary parts of $P_{\alpha\beta}$ and $Q_{\alpha\beta}$ are $0(1)$ and the β terms can be dropped).

The (complex) equilibrium and compatibility equations (3) and (4) combine together to give

$$P^{\alpha\beta}_{,\beta} + \beta\, b_\gamma{}^\alpha Q^{\gamma\beta}_{,\beta} - \frac{1}{2}\beta\, \varepsilon^{\alpha\beta}(B_{\gamma\delta}Q^{\gamma\delta})_{,\beta} = 0$$

$$Q^{\alpha\beta}_{,\alpha\beta} - \mu\, b_{\alpha\beta}\, P^{\alpha\beta} = 0 \tag{13}$$

and the same pair of equations in terms of $\overset{*}{P}{}^{\alpha\beta}$ and $\overset{*}{Q}{}^{\alpha\beta}$, hereafter referred to as (13)*. Also define

$$W = \psi - i w \qquad X_\alpha = \chi_\alpha - i u_\alpha$$

$$\Phi_\alpha = \vartheta_\alpha - i\varphi_\alpha \qquad z = \Omega - i\omega \tag{14}$$

along with $W^* = \psi + iw$, etc.

The (complex) stress-stress function relations (2) and strain displacement relations (1) combine in the following form

$$\overline{P}_{\alpha\beta} = -\frac{1}{2}(\Phi_{\alpha,\beta} + \Phi_{\beta,\alpha}) - \beta B_{\alpha\beta} z$$

$$\overline{Q}_{\alpha\beta} = \frac{1}{2}(X_{\alpha,\beta} + X_{\beta,\alpha}) + \mu b_{\alpha\beta} W \tag{15}$$

where

$$\Phi_\alpha = -W_{,\alpha} + \beta b_\alpha{}^\gamma X_\gamma$$

and

$$z = \frac{1}{2}\varepsilon^{\alpha\beta} X_{\beta,\alpha}$$

and the same equations with stars.

In place of the constitutive relations (5) introduce the following modified relations

$$\overset{*}{P}{}^{\alpha\beta} = i\beta\nu R^{\alpha\beta} \tag{16}$$

$$\overset{*}{Q}{}^{\alpha\beta} = -i\nu P^{\alpha\beta} + i\beta\nu S^{\alpha\beta} \tag{17}$$

$$Q^{\alpha\beta} = -i\overline{P}{}^{\alpha\beta} \tag{18}$$

in which $R^{\alpha\beta}$ and $S^{\alpha\beta}$ are, for the present, arbitrary symmetric tensors except for the restriction $R^{\alpha\beta} = 0(1)$ or $0(1/\mu)$, whichever is greater, and the same for $S^{\alpha\beta}$. If, in (16) to (18) one sets $\beta = 0$ and solves for $\bar{E}_{\alpha\beta}$ in terms of $N_{\alpha\beta}$ and $M_{\alpha\beta}$ in terms of $\bar{K}_{\alpha\beta}$ the results are exactly the constitutive relations (5). Equation (16) becomes Eq. (7). With $\beta \neq 0$ the constitutive relations implied by (16) to (18) are (5) with error terms of $0(\beta, \beta/\mu)$. NOVOZHILOV [1] and KOITER [6] have shown that it is permissible to introduce absolute errors of this order in the (dimensionless) constitutive relations without impairing the accuracy of the system of shell equations because the constitutive relations already contain errors of this order from the fundamental assumptions. With $\beta \neq 0$ equation (16) is a modification of the relation (7). As will be shown, the tensors $R_{\alpha\beta}$ and $S_{\alpha\beta}$ can be chosen such that the system of shell equations is not overdetermined in an inconsistent manner by the introduction of one extra tensor relation. If either $\nu = 0$ or $\beta = 0$ the tensors $R_{\alpha\beta}$ and $S_{\alpha\beta}$ drop out of the relations (16) to (18) and the corresponding system of shell equations is consistent, as has already been shown.

Any set of four tensors $P^{\alpha\beta}$, $Q^{\alpha\beta}$, $\overset{*}{P}{}^{\alpha\beta}$ and $\overset{*}{Q}{}^{\alpha\beta}$ which satisfy (13) and (13)* are equivalent to a solution of the shell equations provided $R^{\alpha\beta}$ and $S^{\alpha\beta}$ can be found such that (16) to (18) are satisfied with restrictions on the order of magnitude of $R^{\alpha\beta}$ and $S^{\alpha\beta}$ observed. The problem splits into two parts: problem I, the determination of $P^{\alpha\beta}$; problem II, the determination of $R^{\alpha\beta}$ and $S^{\alpha\beta}$.

Problem I. Use Eq. (18) to eliminate the variable $Q^{\alpha\beta}$ from Eqs. (13). The result is

$$P^{\alpha\beta}_{,\beta} + i\beta b_{\gamma}{}^{\alpha} P^{\gamma\beta}_{,\beta} - i\beta b^{\alpha\beta} P_{,\beta} - \frac{1}{2} i\beta \varepsilon^{\alpha\beta} (B_{\gamma\delta} P^{\gamma\delta})_{,\beta} = 0$$

$$i P^{\alpha\beta}_{,\alpha\beta} - i \nabla^2 P - \mu b_{\alpha\beta} P^{\alpha\beta} = 0 \tag{19}$$

in which $P = P_{\alpha}{}^{\alpha}$ and $\nabla^2 P = g^{\alpha\beta} P_{,\alpha\beta}$. To obtain these equations the following relations have been used.

$$\bar{P}^{\alpha\beta} = g^{\alpha\beta} P - P^{\alpha\beta}, \qquad B_{\alpha\beta} \bar{P}^{\alpha\beta} = \bar{B}_{\alpha\beta} P^{\alpha\beta} = -B_{\alpha\beta} P^{\alpha\beta}.$$

By obvious manipulations (19) reduces to the following (with β^2 terms dropped)

$$P^{\alpha\beta}_{,\beta} - i\beta b^{\alpha\beta} P_{,\beta} - \frac{1}{2} i\beta \varepsilon^{\alpha\beta} (B_{\gamma\delta} P^{\gamma\delta})_{,\beta} = 0$$

$$\nabla^2 P - i\mu b_{\alpha\beta} P^{\alpha\beta} - i\beta (b^{\alpha\beta} P_{,\beta})_{,\alpha} = 0. \tag{20}$$

These equations constitute a determinate fourth order system of equations for the components of the tensor $P^{\alpha\beta}$. If the last term in each of (20) is omitted, then these equations are identical in form (but for a different variable) to the complex shell equations derived by NOVOZHILOV. In certain cases (to be discussed later) these equations have been reduced to a single fourth order equation in a scalar variable.

Problem II. Replace $\overset{*}{P}{}^{\alpha\beta}$ and $\overset{*}{Q}{}^{\alpha\beta}$ in Eqs. (13)* by their expressions in terms of $R^{\alpha\beta}$, $S^{\alpha\beta}$, and $P^{\alpha\beta}$ from (16) and (17). The result is

$$R^{\alpha\beta}_{,\beta} + \beta b_\gamma{}^\alpha S^{\gamma\beta}_{,\beta} - \frac{1}{2}\beta\varepsilon^{\alpha\beta}(B_{\gamma\delta}S^{\gamma\delta})_{,\beta} - b_\gamma{}^\alpha P^{\gamma\beta}_{,\beta} + \frac{1}{2}\varepsilon^{\alpha\beta}(B_{\gamma\delta}P^{\gamma\delta})_{,\beta} = 0$$

$$\beta(S^{\alpha\beta}_{,\alpha\beta} - \mu b_{\alpha\beta}R^{\alpha\beta}) - P^{\alpha\beta}_{,\alpha\beta} = 0\,. \tag{21}$$

These equations can be reduced somewhat by the use of the first of equations (20) with the result

$$R^{\alpha\beta}_{,\beta} + \beta b_\gamma{}^\alpha S^{\gamma\beta}_{,\beta} - \frac{1}{2}\beta\varepsilon^{\alpha\beta}(B_{\gamma\delta}S^{\gamma\delta})_{,\beta} = -\frac{1}{2}\varepsilon^{\alpha\beta}(B_{\gamma\delta}P^{\gamma\delta})_{,\beta}$$

$$S^{\alpha\beta}_{,\alpha\beta} - \mu b_{\alpha\beta}R^{\alpha\beta} = i(b^{\alpha\beta}P_{,\beta})_{,\alpha}\,. \tag{22}$$

These equations have been simplified by the omission of certain β and β^2 terms on the right-hand side, but without claiming that these terms are always negligible. The solution of Eqs. (22) for $R^{\alpha\beta}$ and $S^{\alpha\beta}$ is equivalent to the determination of a particular solution to the non-homogeneous equilibrium equations (regarding $P^{\alpha\beta}$ as known from the solution of (20)). Since the original equations were made dimensionless in such a way that $P^{\alpha\beta}$ is $0(1)$, there should be a particular solution to the Eqs. (22) with $R^{\alpha\beta}$ and $S^{\alpha\beta}$ $0(1, 1/\mu)$.

In a later section problem II is recast in another form and exact solutions without quadratures are obtained in the cases of a sphere and a general cylinder.

Further Results on Problem I

The equations (20) for $P^{\alpha\beta}$ probably cannot be reduced to a single fourth order equation in a scalar variable in the general case. In [1] NOVOZHILOV has reduced the equations for a general cylinder to a single equation in a variable roughly corresponding to the variable P and the equations for a shell of revolution to two coupled second-order equations. Different reductions to a fourth order equation have been made by KOITER [9] for a sphere and by SIMMONDS [14] for a circular cylinder. In the present section equations (20) are reduced to a fourth order equation (in two ways) for the cases of a general cylinder and a shell of revolution (except for a sphere). The sphere is treated as a special case. Before proceeding certain geometrical results are developed.

Geometrical note. Except in the case of a sphere, any second order tensor on a surface can be represented in the following form

$$T^{\alpha\beta} = \lambda^1 g^{\alpha\beta} + \lambda^2 b^{\alpha\beta} + \lambda^3 B^{\alpha\beta} + \lambda^4 \varepsilon^{\alpha\beta} \tag{23}$$

where the λ^i are scalars. The elimination process to be performed is based on the representation of $P^{\alpha\beta}$ in a form similar to this. A modification of (23), possible for a certain class of shells, leads to somewhat simpler results, but is in no way essential to the elimination process. The modified form of (23) for $P^{\alpha\beta}$ is as follows

$$P^{\alpha\beta} = \lambda^1 g^{\alpha\beta} + \lambda^2 \dot{b}^{\alpha\beta} + \lambda^3 \dot{B}^{\alpha\beta} \tag{24}$$

in which $\dot{B}^{\alpha\beta} = \tau B^{\alpha\beta}$ and $\dot{b}^{\alpha\beta} = \tau(b^{\alpha\beta} - \varrho g^{\alpha\beta})$ where $\varrho = \frac{1}{2} b_\alpha{}^\alpha$ is the mean curvature and where τ is determined in such a way that $\dot{b}^{\alpha\beta}_{,\beta} = 0$ and $\dot{B}^{\alpha\beta}_{,\beta} = 0$. Since $B^{\alpha\beta} = \varepsilon^{\alpha\gamma}(b_\gamma{}^\beta - \varrho\delta_\gamma{}^\beta)$ the problem is the same for either tensor. Consider

$$[\tau(b^{\alpha\beta} - \varrho g^{\alpha\beta})]_{,\beta} = \tau_{,\beta}(b^{\alpha\beta} - \varrho g^{\alpha\beta}) + \tau g^{\alpha\beta}\varrho_{,\beta} = 0\,. \tag{25}$$

Use $(b_\alpha{}^\gamma - \varrho\delta_\alpha{}^\gamma)(b^{\alpha\beta} - \varrho g^{\alpha\beta}) = (\varrho^2 - \varkappa)g^{\beta\gamma}$ where $\varkappa$ is the Gaussian curvature to obtain

$$\tau_{,\alpha} + \frac{1}{\varrho^2 - \varkappa}(b_\alpha{}^\gamma - \varrho\delta_\alpha{}^\gamma)\varrho_{,\gamma}\,\tau = 0\,. \tag{26}$$

The condition of integrability of this equation for τ is

$$\varepsilon^{\alpha\beta}\left[\frac{1}{\varrho^2 - \varkappa}(b_\alpha{}^\gamma - \varrho\delta_\alpha{}^\gamma)\varrho_{,\gamma}\right]_{,\beta} = 0$$

or

$$\varepsilon^{\alpha\beta}(b_\alpha{}^\gamma - \varrho\delta_\alpha{}^\gamma)\left(\frac{\varrho_{,\gamma}}{\varrho^2 - \varkappa}\right)_{,\beta} \equiv -B^{\beta\gamma}\left(\frac{\varrho_{,\gamma}}{\varrho^2 - \varkappa}\right)_{,\beta} = 0\,. \tag{27}$$

This condition is satisfied for general cylinders and cones, surfaces of revolution, and any surface of constant mean curvature. For these cases

$$(\log\tau)_{,\alpha} = -\frac{1}{\varrho^2 - \varkappa}(b_\alpha{}^\gamma - \varrho\delta_\alpha{}^\gamma)\varrho_{,\gamma}\,. \tag{28}$$

For general cylinders and shells of revolution in lines-of-curvature coordinates τ, ϱ, $\varkappa$ etc., are functions of one coordinate only and (28) is easily integrated. For cylinders $\tau = 1/\varrho$, for shells of revolution $\tau = 1/r^2\sqrt{\varrho^2 - \varkappa}$ where r is the radius of a parallel circle ($R_2 \sin\vartheta$ in the notation of [1]). A few useful formulas involving $\dot{b}_{\alpha\beta}$ etc., are as follows

$$\zeta \dot{b}_{\alpha\gamma}\dot{b}_{\beta}^{\gamma} = g_{\alpha\beta} \qquad \varepsilon_{\alpha\gamma}\dot{b}_{\beta}^{\gamma} = \dot{B}_{\alpha\beta}$$
$$\zeta \dot{B}_{\alpha\gamma}\dot{B}_{\beta}^{\gamma} = g_{\alpha\beta} \qquad \dot{B}_{\alpha}^{\gamma}\varepsilon_{\gamma\beta} = \dot{b}_{\alpha\beta} \tag{29}$$
$$\zeta \dot{B}_{\alpha\gamma}\dot{b}_{\beta}^{\gamma} = \varepsilon_{\alpha\beta} \qquad \dot{b}_{\alpha}^{\alpha} = 0$$

where

$$\zeta^{-1} = \tau^2(\varrho^2 - \varkappa) \qquad \dot{B}_{\alpha}^{\alpha} = 0.$$

Characteristic Equation in P. Put (see (19))

$$T^{\alpha} = P^{\alpha\beta}_{,\beta} + i\beta b_{\gamma}^{\alpha} P^{\gamma\beta}_{,\beta} - i\beta b^{\alpha\beta}P_{,\beta} - \frac{1}{2}\, i\beta\varepsilon^{\alpha\beta}(B_{\gamma\delta}P^{\gamma\delta})_{,\beta} \tag{30}$$

$$T = iP^{\alpha\beta}_{,\alpha\beta} - i\nabla^2 P - \mu b_{\alpha\beta}P^{\alpha\beta} \tag{31}$$

and define

$$\overline{T}^{\alpha} \equiv T^{\alpha} - i\beta b_{\gamma}^{\alpha} T^{\gamma} = P^{\alpha\beta}_{,\beta} - i\beta b^{\alpha\beta}P_{,\beta} - \frac{1}{2}\, i\beta\varepsilon^{\alpha\beta}(B_{\gamma\delta}P^{\gamma\delta})_{,\beta} \tag{32}$$

in which certain β^2 terms have been omitted. From (31) and (32)

$$T - i\overline{T}^{\alpha}_{,\alpha} = -i\nabla^2 P - \mu b_{\alpha\beta}P^{\alpha\beta} - \beta(b^{\alpha\beta}P_{,\beta})_{,\alpha}. \tag{33}$$

Now put

$$P^{\alpha\beta} = \lambda^1 g^{\alpha\beta} + \lambda^2 \dot{b}^{\alpha\beta} + \lambda^3 \dot{B}^{\alpha\beta} \tag{34}$$

in which

$$\lambda^1 = \frac{1}{2} P$$
$$\lambda^2 = \frac{1}{2}\tau\zeta(b_{\alpha\beta}P^{\alpha\beta} - \varrho P)$$
$$\lambda^3 = \frac{1}{2}\tau\zeta B_{\alpha\beta}P^{\alpha\beta}. \tag{35}$$

From (32) and (34)

$$\overline{T}^{\alpha} = g^{\alpha\beta}\lambda^1_{,\beta} + \dot{b}^{\alpha\beta}\lambda^2_{,\beta} + \dot{B}^{\alpha\beta}\lambda^3_{,\beta} - i\beta b^{\alpha\beta}P_{,\beta} - i\beta\varepsilon^{\alpha\beta}[(\tau\zeta)^{-1}\lambda^3]_{,\beta}. \tag{36}$$

Multiply this equation by $\zeta\dot{b}_{\alpha}^{\gamma}$ to obtain

$$\zeta\dot{b}_{\alpha}^{\gamma}\overline{T}^{\alpha} = \zeta\dot{b}^{\beta\gamma}\lambda^1_{,\beta} + g^{\beta\gamma}\lambda^2_{,\beta} + \varepsilon^{\beta\gamma}\lambda^3_{,\beta} - i\beta\zeta\dot{b}_{\alpha}^{\gamma}b^{\alpha\beta}P_{,\beta} + i\beta\zeta\dot{B}^{\beta\gamma}[(\tau\zeta)^{-1}\lambda^3]_{,\beta}. \tag{37}$$

The first order term in λ^3 can now be eliminated by a differentiation.

$$(\zeta\dot{b}_{\alpha}^{\gamma}\overline{T}^{\alpha})_{,\gamma} = \dot{b}^{\beta\gamma}(\zeta\lambda^1_{,\beta})_{,\gamma} + \nabla^2\lambda^2 - i\beta(\zeta\dot{b}_{\alpha}^{\gamma}b^{\alpha\beta}P_{,\beta})_{,\gamma} + i\beta\dot{B}^{\beta\gamma}\{\zeta[(\tau\zeta)^{-1}\lambda^3]_{,\beta}\}_{,\gamma}. \tag{38}$$

The last term in (38) is equal to the following

$$i\beta \dot{B}^{\beta\gamma}\left\{\tau\zeta\left(\frac{1}{\tau^2\zeta}\lambda^3_{,\beta}\right)_{,\gamma}+\lambda^3\left[\zeta\left(\frac{1}{\tau\zeta}\right)_{,\beta}\right]_{,\gamma}\right\}.$$

The term with $\lambda^3_{,\beta}$ can be eliminated as follows

$$(\zeta\dot{b}_\alpha{}^\gamma\overline{T}^\alpha)_{,\gamma}-i\beta\tau\zeta\left(\frac{1}{\tau^2\zeta}\overline{T}^\alpha\right)_{,\alpha}=\dot{b}^{\beta\gamma}(\zeta\lambda^1_{,\beta})_{,\gamma}+\nabla^2\lambda^2-i\beta(\zeta\dot{b}_\alpha{}^\gamma b^{\alpha\beta}P_{,\beta})_{,\gamma}-$$

$$-i\beta\tau\zeta g^{\alpha\beta}\left(\frac{1}{\tau^2\zeta}\lambda^1_{,\beta}\right)_{,\alpha}-i\beta\tau\zeta\dot{b}^{\alpha\beta}\left(\frac{1}{\tau^2\zeta}\lambda^2_{,\beta}\right)_{,\alpha}+i\beta\dot{B}^{\beta\gamma}\left[\zeta\left(\frac{1}{\tau\zeta}\right)_{,\beta}\right]_{,\gamma}\lambda^3. \quad (39)$$

The last term here, the λ^3 term, cannot in general be eliminated without one more differentiation of this equation, and this would lead finally to a characteristic equation in P of order higher than the fourth. However, for general cylinders and shells of revolution, this term vanishes identically and the elimination of λ^3 is complete except for β^2 terms which are certainly negligible. For later reference note here that a variation of (39) is obtained by multiplying (38) by any $0(\beta)$ scalar and adding to (39). This operation does not reintroduce a λ^3 term except to $0(\beta^2)$.

From (33) and (35) it follows that

$$\lambda^2=-\frac{1}{2}\tau\zeta\varrho P-\frac{i}{2\mu}\tau\zeta\nabla^2P-\frac{\beta}{2\mu}\tau\zeta(b^{\alpha\beta}P_{,\beta})_{,\alpha}+\frac{i}{2\mu}\tau\zeta\overline{T}^\alpha_{,\alpha}-\frac{1}{2\mu}\tau\zeta T. \quad (40)$$

Equation (40) is now used to eliminate λ^2 from (39). The result is the following

$$\nabla^2(\tau\zeta\overline{T}^\alpha_{,\alpha})+i\nabla^2(\tau\zeta T)+2i\mu\dot{b}_\alpha{}^\beta(\zeta\overline{T}^\alpha)_{,\beta}+2\beta\mu\tau\zeta\left(\frac{1}{\tau^2\zeta}\overline{T}^\alpha\right)_{,\alpha}-$$

$$-i\beta\tau\zeta\dot{b}^{\alpha\beta}\left[\frac{1}{\tau^2\zeta}(\tau\zeta\overline{T}^\gamma_{,\gamma})_{,\alpha}\right]_{,\beta}+\beta\tau\zeta\dot{b}^{\alpha\beta}\left[\frac{1}{\tau^2\zeta}(\tau\zeta T)_{,\alpha}\right]_{,\beta}$$

$$=\nabla^2(\tau\zeta\nabla^2P)+i\mu\dot{b}^{\alpha\beta}(\zeta P_{,\alpha})_{,\beta}-i\mu\nabla^2(\tau\zeta\varrho P)+$$

$$+2\beta\mu(\zeta\dot{b}_\alpha{}^\gamma b^{\alpha\beta}P_{,\beta})_{,\gamma}+\beta\mu\tau\zeta g^{\alpha\beta}\left(\frac{1}{\tau^2\zeta}P_{,\alpha}\right)_{,\beta}-\beta\mu\tau\zeta\dot{b}^{\alpha\beta}\times$$

$$\times\left[\frac{1}{\tau^2\zeta}(\tau\zeta\varrho P)_{,\alpha}\right]_{,\beta}-i\beta\nabla^2[\tau\zeta(b^{\alpha\beta}P_{,\alpha})_{,\beta}]-$$

$$-i\beta\tau\zeta\dot{b}^{\alpha\beta}\left[\frac{1}{\tau^2\zeta}(\tau\zeta\nabla^2P)_{,\alpha}\right]_{,\beta}\equiv L(P). \quad (41)$$

The desired characteristic equation is

$$L(P)=0 \quad (42)$$

valid for cylinders and shells of revolution. A tentative simplification of this equation, known to be valid in the case of cylinders from the work of NOVOZHILOV, GOL'DENVEIZER, and SIMMONDS is obtained by dropping the β terms. From now on suppose this has been done; the terms can be restored if further investigation proves it necessary.

The details will not be shown here, but if (38) is multiplied by an $0(\beta)$ scalar and added to (39) this modified equation (39) leads, by the above process, to a characteristic equation which differs from (42) only in the $\beta\mu$ terms. Such variability in the $\beta\mu$ terms (not to be mistaken for arbitrariness) was encountered by SIMMONDS [14] whose paper is enlightening in this regard.

In the case of a cylinder, if one adds $\beta\varrho$ (38) to (39), the characteristic equation becomes

$$\nabla^2\left(\frac{1}{\varrho}\nabla^2 P\right) + 2\beta\mu(b^{\alpha\beta}P_{,\alpha})_{,\beta} - \frac{i\mu}{\varrho}\bar{b}^{\alpha\beta}P_{,\alpha\beta} = 0. \tag{43}$$

This equation is the invariant form of Novozhilov's cylinder equation. It can be obtained by elimination from the Eqs. (20) with the last term in each of these equations missing. The result seems to be peculiar to the cylinder. In general, the effect of the last terms in (20) on the characteristic equation (42) cannot be annulled by any such modification of (39).

Once a solution of the characteristic equation has been found λ^1 and λ^2 are given directly in terms of P by (35) and (40) (with $\overline{T}^\alpha = T = 0$). In order to construct $P^{\alpha\beta}$ using (34) one must find λ^3. This can be done using (36), to the first order by quadratures, and more accurately by an iteration process. Complex displacements X_α and W are then found (if necessary) by integration of the Eqs. (15) and (18); which must be possible since the integrability conditions (20) are satisfied. In practice the various integrations mentioned here might prove to be difficult. An alternative formulation of Problem I, possibly more convenient for application, is developed in the next section.

Alternative Characteristic Equation. The expressions (15) for $P^{\alpha\beta}$ and $Q^{\alpha\beta}$ in terms of X_α and W furnish a complete solution to the "equilibrium" equations (13). If one could find expressions for X_α and W, say in terms of some scalar function φ, such that the relation $Q^{\alpha\beta} = -iP^{\alpha\beta}$ is satisfied, then one would have a general solution to the Eqs. (20). This is possible as proved in the following.

Let A be a simply connected region of the shell middle surface with boundary C. Let $P^{\alpha\beta}$ be an arbitrary symmetric tensor which is continuous in A together with its first four covariant derivatives and such that $P^{\alpha\beta}$ and its first three covariant derivatives vanish on C. Let T^α and T

be defined in terms of $P^{\alpha\beta}$ by (30) and (31). By application of Green's theorem

$$\int\limits_A (T^\alpha X_\alpha' + T W')\, da = -\int\limits_A (\bar{Q}'_{\alpha\beta} + i P'_{\alpha\beta})\, P^{\alpha\beta}\, da . \tag{44}$$

in which $\bar{Q}'_{\alpha\beta}$ and $\bar{P}'_{\alpha\beta}$ are related to X_α' and W' as in equations (15). Let $F(T^\alpha, T)$ be the invariant functional of T^α and T on the left-hand side of (41), and let X_α' and W' be defined in A in terms of a function φ by the equations

$$\int\limits_A F(T^\alpha, T)\, \varphi\, da = -\int\limits_A (T^\alpha X_\alpha' + T W')\, da. \tag{45}$$

One also has

$$\int\limits_A F \varphi\, da = \int\limits_A L(P)\, \varphi\, da = \int\limits_A \bar{L}(\varphi)\, P\, da \tag{46}$$

where $\bar{L}$ is the operator adjoint to L. Now if φ satisfies $\bar{L}(\varphi) = 0$, then $\int\limits_A (\bar{Q}'_{\alpha\beta} + i P'_{\alpha\beta})\, P^{\alpha\beta}\, da = 0$ for arbitrary $P^{\alpha\beta}$ from (44) to (46). It follows that $\bar{Q}'_{\alpha\beta} + i P'_{\alpha\beta} = 0$ in A. Explicit expressions for X_α and W obtained by use of (45) and the simplified F are as follows

$$W = i \tau \zeta \nabla^2 \varphi \tag{47}$$

$$X_\alpha = i W_{,\alpha} + 2 i \mu \zeta (1 - i \beta \varrho)\, \dot{b}_\alpha{}^\beta \varphi_{,\beta} + \frac{4 \beta \mu}{\tau \sqrt{\tau \zeta}} \left(\sqrt{\tau \zeta}\; \varphi\right)_{,\alpha} \tag{48}$$

where

$$\bar{L}(\varphi) = 0 . \tag{49}$$

These lead (from (15)) to an expression for $\bar{P}_{\alpha\beta}$ in terms of φ.

Problem I in the form $Q_{\alpha\beta} = -i \bar{P}_{\alpha\beta}$, regarded as equations for X_α and W, corresponds to what Novozhilov and Chernykh call "the equations in terms of complex displacements". Such a formulation of the problem has been exploited by Chernykh in his book [3].

A Characteristic Equation for the Spherical Shell. In the case of a sphere $b_{\alpha\beta} = g_{\alpha\beta}$, $B_{\alpha\beta} = 0$ and the preceding analysis breaks down. However, in this particularly simple case it is easy to derive a characteristic equation. For a sphere the Eqs. (19) are

$$(1 + i\beta)\, P^{\cdot\beta}_{,\beta} - i \beta g^{\alpha\beta} P_{,\beta} = 0 \tag{50}$$

$$P^{\alpha\beta}_{\cdot\,\beta} - \nabla^2 P + i \mu P = 0 \tag{51}$$

from which it follows that

$$\nabla^2 P - i \mu (1 + i \beta)\, P = 0 . \tag{52}$$

Note that P satisfies a second order equation in this case. The general solution to (50) is

$$P^{\alpha\beta} = \frac{i\beta}{1+i\beta} g^{\alpha\beta} P + U^{\alpha\beta} \tag{53}$$

where $U^{\alpha\beta}_{,\beta} = 0.$ (54)

For a sphere the general solution to (54) is

$$\overline{U}_{\alpha\beta} = \varphi_{,\alpha\beta} + \beta\mu g_{\alpha\beta}\varphi\,. \tag{55}$$

From (53)

$$P = \frac{1+i\beta}{1-i\beta}(\nabla^2\varphi + 2\beta\mu\varphi) \tag{56}$$

and then from (52)

$$[\nabla^2 - i\mu(1+i\beta)](\nabla^2 + 2\beta\mu)\varphi = 0\,. \tag{57}$$

This characteristic equation is identical in form to one derived by KOITER [9]. In terms of φ one has for $\overline{P}_{\alpha\beta}$

$$(1-i\beta)\overline{P}_{\alpha\beta} = (1-i\beta)\varphi_{,\alpha\beta} + (1+i\beta)\beta\mu g_{\alpha\beta}\varphi + i\beta g_{\alpha\beta}\nabla^2\varphi\,. \tag{58}$$

There are obvious possibilities for simplification. The system of equations for a sphere derived earlier in this paper can be reduced in a similar manner, and the present system can be reduced in other ways.

Further Results on Problem II

The complete solution to equations (13)* is given by (15)* for $\overset{*}{P}{}^{\alpha\beta}$ and $\overset{*}{Q}{}^{\alpha\beta}$ in terms of $\overset{*}{W}$ and $\overset{*}{X}_\alpha$. Problem II is equivalent to the determination of $\overset{*}{W}$, $\overset{*}{X}_\alpha$, $R^{\alpha\beta}$ and $S^{\alpha\beta}$ such that (16) and (17) are satisfied with order of magnitude restrictions on $R^{\alpha\beta}$ and $S^{\alpha\beta}$ as previously stated. If these restrictions are dropped (for the present) a solution is easily obtained as follows. Write (17) in the form

$$\overline{\overset{*}{Q}}_{\alpha\beta} = -i\nu\overline{P}_{\alpha\beta} + i\beta\nu\overline{S}_{\alpha\beta} = i\nu\left[\frac{1}{2}(\Phi_{\alpha,\beta} + \Phi_{\beta,\alpha}) + \beta B_{\alpha\beta}z + \beta\overline{S}_{\alpha\beta}\right] \tag{59}$$

and compare to (15)* repeated here

$$\overline{\overset{*}{Q}}_{\alpha\beta} = \frac{1}{2}(\overset{*}{X}_{\alpha,\beta} + \overset{*}{X}_{\beta,\alpha}) + \mu b_{\alpha\beta}\overset{*}{W}.$$

Equation (59) is satisfied if

$$\overset{*}{X}_\alpha = i\nu\Phi_\alpha,\quad \overset{*}{W} = 0 \qquad \text{and} \qquad \overline{S}_{\alpha\beta} = -B_{\alpha\beta}z\,. \tag{60}$$

Now calculate $\overset{*}{\bar{P}}_{\alpha\beta}$ from (15)*, and use (16) to find

$$\bar{R}_{\alpha\beta} = -\frac{1}{2}[(b_\alpha^{\ \gamma}\Phi_\gamma)_{,\beta} + (b_\beta^{\ \gamma}\Phi_\gamma)_{,\alpha}] - \beta B_{\alpha\beta}(\varrho z + \frac{1}{2} i B^{\gamma\delta}\bar{P}_{\gamma\delta}). \quad (61)$$

This determination of $R_{\alpha\beta}$ and $S_{\alpha\beta}$ furnished an exact particular solution to the Eqs. (21). The difficulty with it is that the order-of-magnitude requirements are not met in general. By choice of scaling $P_{\alpha\beta}$ and $Q_{\alpha\beta}$ are $0(1)$, and in the usual case φ is $0(1)$ in equations (47) and (48). From (48)

$$z = \frac{1}{2}\varepsilon^{\alpha\beta}X_{\beta,\alpha} \text{ is } 0(\mu) \qquad \text{and} \qquad S_{\alpha\beta} = B_{\alpha\beta}z \text{ is } 0(\mu)$$

rather than $0(1)$ as required. This is unacceptable, in general, because μ can be large. The general solution for $R_{\alpha\beta}$ and $S_{\alpha\beta}$ is obtained by adding the general solution to the homogeneous equations (22) for $R_{\alpha\beta}$ and $S_{\alpha\beta}$ to the particular solution, and since these equations have the same form as (15) a general solution is known. The result is obviously

$$\bar{S}_{\alpha\beta} = -B_{\alpha\beta}z + \frac{1}{2}(X'_{\alpha,\beta} + X'_{\beta,\alpha}) + \mu b_{\alpha\beta}W' \quad (62)$$

$$\begin{aligned} \bar{R}_{\alpha\beta} = & -\frac{1}{2}[(\Phi_\alpha' + b_\alpha^{\ \gamma}\Phi_\gamma)_{,\beta} + (\Phi_\beta' + b_\beta^{\ \gamma}\Phi_\gamma)_{,\alpha}] - \\ & - \beta B_{\alpha\beta}\left(z' + \varrho z + \frac{1}{2} i B^{\gamma\delta}\bar{P}_{\gamma\delta}\right). \end{aligned} \quad (63)$$

The complete expressions for $\overset{*}{X}_\alpha$ and $\overset{*}{W}$ used to form $\overset{*}{P}_{\alpha\beta}$ and $\overset{*}{Q}_{\alpha\beta}$ are

$$\overset{*}{X}_\alpha = i\nu(\Phi_\alpha + \beta X_\alpha') \quad (64)$$

$$\overset{*}{W} = i\beta\nu W' \quad (65)$$

where X_α' and W' are arbitrary. Problem II will be solved if X_α' and W' can be found to satisfy the order-of-magnitude requirements on $R_{\alpha\beta}$ and $S_{\alpha\beta}$. In the general case the problem in this form does not seem to be an easy one. However, for spheres and cylinders, results for $R_{\alpha\beta}$ and $S_{\alpha\beta}$ without quadratures are readily obtainable.

Spheres. In the case of a sphere $B_{\alpha\beta} = 0$, $b_{\alpha\beta} = g_{\alpha\beta}$ and $X_\alpha' = W' = 0$ leads to the acceptable solution

$$R_{\alpha\beta} = P_{\alpha\beta}, \; S_{\alpha\beta} = 0. \quad (66)$$

A more general solution is

$$R_{\alpha\beta} = (1 + C)P_{\alpha\beta}, \; S_{\alpha\beta} = CQ_{\alpha\beta} \quad (67)$$

and correspondingly

$$\overset{*}{X}_\alpha = i\nu(\Phi_\alpha + \beta C X_\alpha), \; \overset{*}{W} = i\beta\nu C W \tag{68}$$

where C is any $0(1)$ constant. The arbitrariness involved is simply a manifestation of the well known degree of arbitrariness involved in the form of the shell equations.

Circular Cylinders. In this case put

$$X_\alpha{}' = b_\alpha{}^\gamma X_\gamma - X_\alpha, \qquad W' = 0, \text{ and obtain}$$

$$\begin{aligned} X'_{\alpha,\beta} &= b_\alpha{}^\gamma X_{\gamma,\beta} - X_{\alpha,\beta} \\ &= b_\alpha{}^\gamma(\bar{Q}_{\beta\gamma} + \varepsilon_{\beta\gamma} z - \mu b_{\beta\gamma} W) - \bar{Q}_{\alpha\beta} - \varepsilon_{\alpha\beta} z + \mu b_{\alpha\beta} W \end{aligned}$$

$$\frac{1}{2}(X'_{\alpha,\beta} + X'_{\beta,\alpha}) = \frac{1}{2}(b_\alpha{}^\gamma \bar{Q}_{\beta\gamma} + b_\beta{}^\gamma \bar{Q}_{\alpha\gamma}) - \bar{Q}_{\alpha\beta} + B_{\alpha\beta} z.$$

The result for $\bar{S}_{\alpha\beta}$ is

$$\bar{S}_{\alpha\beta} = \frac{1}{2}(b_\alpha{}^\gamma \bar{Q}_{\beta\gamma} + b_\beta{}^\gamma \bar{Q}_{\alpha\gamma}) - \bar{Q}_{\alpha\beta}. \tag{69}$$

Further

$$\Phi_\alpha{}' = \beta b_\alpha{}^\gamma(b_\gamma{}^\delta X_\delta - X_\gamma) = 0$$

$$\begin{aligned} z' &= \frac{1}{2}\varepsilon^{\alpha\beta} b_\beta{}^\gamma X_{\gamma,\alpha} - z = \frac{1}{2}\left(B^{\alpha\gamma} + \frac{1}{2}\varepsilon^{\alpha\gamma}\right) X_{\gamma,\alpha} - z \\ &= \frac{1}{2} B^{\alpha\gamma} \bar{Q}_{\alpha\gamma} - \frac{1}{2} z. \end{aligned}$$

The result for $\bar{R}_{\alpha\beta}$ is

$$\bar{R}_{\alpha\beta} = \frac{1}{2}(b_\alpha{}^\gamma \bar{P}_{\gamma\beta} + b_\beta{}^\gamma \bar{P}_{\gamma\alpha}) - \frac{3}{2} i\beta B_{\alpha\beta} B^{\gamma\delta} \bar{P}_{\gamma\delta} \tag{70}$$

and for $\overset{*}{X}_\alpha$ and $\overset{*}{W}$ there is

$$\overset{*}{X}_\alpha = i\nu(-W_{,\alpha} - \beta X_\alpha + 2\beta b_\alpha{}^\gamma X_\gamma), \qquad \overset{*}{W} = 0. \tag{71}$$

These results can be generalized as in the case of the sphere.

General Cylinders. For cylinders one has the formulas

$$b^\gamma_{\alpha,\beta} = \frac{1}{\varrho} b_\alpha{}^\gamma \varrho_{,\beta} = \frac{1}{\varrho} b_{\alpha\beta} g^{\gamma\delta} \varrho_{,\delta} = \frac{1}{\varrho} b_{\alpha\beta} \varrho^{,\gamma}.$$

The choice

$$\begin{aligned} X_\alpha{}' &= b_\alpha{}^\gamma X_\gamma \\ W' &= 2\varrho W - \frac{1}{\mu\varrho} \varrho_{,\gamma} X^\gamma \end{aligned} \tag{72}$$

leads to the results

$$\overset{*}{X}_\alpha = i\nu(-W_{,\alpha} + 2\beta\, b_\alpha{}^\gamma X^\gamma)$$

$$\overset{*}{W} = i\beta\nu\left(2\varrho W - \frac{1}{\mu\varrho}\,\varrho_{,\gamma} X^\gamma\right)$$

$$\bar{S}_{\alpha\beta} = \frac{1}{2}\,(b_\alpha{}^\gamma \bar{Q}_{\gamma\beta} + b_\beta{}^\gamma \bar{Q}_{\gamma\alpha})$$

$$\bar{R}_{\alpha\beta} = 2\varrho\bar{P}_{\alpha\beta} + \frac{1}{2}\,(b_\alpha{}^\gamma \bar{P}_{\gamma\beta} + b_\beta{}^\gamma \bar{P}_{\gamma\alpha}) - \frac{3}{2}\, i\beta B_{\alpha\beta} B^{\gamma\delta} \bar{P}_{\gamma\delta} + 2\varrho_{,\alpha\beta} W + \\ + 2\varrho_{,\alpha} W_{,\beta} + 2\varrho_{,\beta} W_{,\alpha} + \frac{1}{\varrho}\, b_{\alpha\beta}\varrho^{,\gamma} W_{,\gamma} - \frac{1}{\mu}\left(\frac{1}{\varrho}\,\varrho_{,\gamma} X^\gamma\right)_{,\alpha\beta} - 4\beta b_{\alpha\beta}\varrho_{,\gamma} X^\gamma. \tag{73}$$

The last term in the expression for $\bar{R}_{\alpha\beta}$ is apparently $0(\beta\mu)$ which would violate the order-of-magnitude requirements on $\bar{R}_{\alpha\beta}$ in case $\beta\mu$ is large (assuming such a possibility). A detailed investigation proves that $\beta\varrho_{,\gamma} X^\gamma = 0(1)$ in any case. The analysis is too lengthy to give here, but a brief statement of the facts may be in order. When $\beta\mu$ is large the characteristic equation is approximately (assuming $\varrho_{,y} = 0(1)$)

$$2\beta(\varrho\varphi_{,y})_{,y} - i\varphi_{,xx} = 0 \tag{74}$$

in cartesian coordinates with x in the axial direction. Evidently there are two length scales involved since $\varphi_{,x} \ll \varphi_{,y}$ from (74). To be consistent here the L in μ should be the shorter (circumferential) "wavelength" so that no differentiation increases the order of magnitude of φ. One must use (47) and (48) to calculate X_α, W and from these calculate $\bar{Q}_{\alpha\beta}$. When φ is scaled to make $\bar{Q}_{\alpha\beta} = 0(1)$ it turns out that $\beta\mu\varphi = 0(1)$ and $\beta X_y = 0(1)$. In the unlikely event that $\varrho_{,y}$ is large φ becomes smaller to compensate; in any case $\bar{R}_{\alpha\beta} = 0(1)$.

Concluding Remarks

The linear equations of thin shell theory have been reduced in complex form to a true fourth order system (with no conjugation operation) without approximation of the equilibrium or compatibility equation, but at the expense of introducing an auxiliary system of equations for certain allowable error terms in the constitutive relations. The fourth order system has been reduced to a single fourth order equation in certain cases. Other than this no serious attempt has been made at further simplification of the equations (which might be made in general or in specific cases) by dropping small terms. However, such approximations are usually easier to justify in a single equation than in a system of equations. Extensions of the results to include the case of distributed loads remains to be worked out.

A characteristic equation for the membrane-inextensional bending theories (which are combined in the complex formulation) is obtained by omitting all but the μ terms in $L(P) = 0$ or $\bar{L}(\varphi) = 0$. In this connection the author encountered a paradoxical situation. A characteristic equation for the membrane-inextensional bending (M. I. B.) theory in terms of z (the Weingarten equation) can be obtained for any shape of middle surface. A full account of the theory is to be found in Vekua's book [15]. On the other hand, a characteristic equation (for the M. I. B. theory) in terms of P is difficult if not impossible to obtain except in those cases treated in this paper. One might think that a general characteristic equation for the bending theory in terms of z would be easy to obtain. It turns out that the relation $Q_{\alpha\beta} = -i\bar{P}_{\alpha\beta}$ can be replaced by equivalent equations in terms of rotations Φ_α and z only. In the M. I. B. case $(Q_{\alpha\beta} = 0)\,\Phi_\alpha$ is easily eliminated to obtain the Weingarten equation. The author has been unable to perform the elimination for the bending theory in any case except the circular cylinder, (assuming that higher than fourth derivatives of z are ruled out). An important unanswered question is the following: Can the system (19) be redued to a single fourth order equation in general, and if not, can it be so reduced in cases other than those already found?

References

1. NOVOZHILOV, V. V.: Thin shell theory (translated from second Russian edition) Groningen: Noordhoff 1965.
2. NOVOZHILOV, V. V.: Development of the method of complex transformation in the linear theory of shells during the last fifty years, Proc. of the 4th All-Union Conference on Shells and Plates, October 1962. (NASA Technical Translation F 341).
3. CHERNYKH, K. F.: Linear theory of shells, 1962 (NASA Technical Translation F 441).
4. LIBRESCU, L.: Elastische Mehrschichtenschalen, Rev. de Mécanique Appliquée, 5, No. 5 (1960).
5. VISARION, V., and CR. STANESCU: Quasinvariants of the static-geometric analogy for thin elastic shells, PMM, Vol. 25, 1961.
6. KOITER, W. T.: A consistent first approximation in the general theory of thin elastic shells, Proc. of the IUTAM-Symposium on The Theory of Thin Elastic Shells, Amsterdam: North-Holland 1960.
7. BUDIANSKY, B., and J. L., Jr. SANDERS: On the „Best" first-order linear shell theory, Progress in Applied Mechanics (The Prager Anniversary Volume) Macmillan 1963.
8. SANDERS, J. L., Jr.: An improved first-approximation theory for thin shells, NASA, Report 24, 1959.
9. KOITER, W. T.: A spherical shell under point loads at its poles, Progress in Applied Mechanics, (The Prager Anniversary Volume) Macmillan 1963.

10. COHEN, J. W.: The inadequacy of the classical stress-strain relations for the right helicoidal shell, Proc. of the IUTAM Symposium on The Theory of Thin Elastic Shells, Amsterdam: North-Holland 1960.
11. REISSNER, E.: Note on axially symmetric stress distributions in helicoidal shells, Tollmien Festschrift, Akademie Verlag 1962.
12. O'MATHUNA, D.: Rotationally symmetric deformations in helicoidal shells, J. Math. and Phys. 42, No. 2 (1963).
13. KOITER, W. T.: A Systematic Simplification of the General Equations in the Linear Theory of Thin Shells, Proc. Kon. Nederlandse Akad. von Wetensch., Series B, Vol. 64, 1961.
14. SIMMONDS, J. G.: A set of simple, accurate equations for circular cylindrical elastic shells, Int. J. of Solids and Structures, (Pergamon) Vol. 2 (1966).
15. VEKUA, I. N.: Generalized analytic functions, (English Translation) Pergamon Press 1959. (Distributed in USA by Addison-Wesley).

Discussion

J. G. SIMMONDS: My comment is in the nature of a footnote to Professor SANDERS'presentation. It is not quite spontaneous, because I had a chance to look at Professor SANDERS' paper about a week before this conference. Professor SANDERS has shown that for an arbitrary shell, one may reduce the governing equations, without any loss of accuracy, to three differential equations for the symmetric complex-valued tensor $P^{\alpha\beta}$. He has also shown that, for general cylindrical shells and shells of revolution, it is possible to further reduce these three equations to a single fourth order equation for the scalar invariant $P = P_\alpha^{\ \alpha}$. Without going into any details, I should like to point out that for arbitrary shells of non-zero Gaussian curvature it is possible to reduce SANDERS' three equations for $P^{\alpha\beta}$ to two coupled second order equations for the invariant P and a (complex-valued) "stress function" Ψ. Its reduction is achieved by adding a negligible term to one of SANDERS' equations, and then making use of ideas set forth by Professor VEKUA and by Dr. LIBAI. The detailed deduction is given in the appendix below.

J. L. SANDERS: The reduction by Professor SIMMONDS of the equations for $P^{\alpha\beta}$ to two coupled scalar equations in the general case is, to me, a very interesting and valuable development.

S. LUKASZEWICZ: I would like to mention that I have also occupied myself with the development of simplified shell equations, valid for arbitrary shells. I have compared the equations obtained with the exact equations for spherical and cylindrical shells and found that they are accurate enough. My equations are the following:

$$D(\Delta + 1/R_1^2 + 1/R_2^2)^2 w + \Delta_k \Phi = 2$$

$$\frac{1}{Eh}(\Delta + 1/R_1 R_2)^2 \Phi - \Delta_k w = 0$$

I think, it would be interesting to compare the results obtained by means of Professor SANDERS' simplified equations for an arbitrary shell with those of mine in case of shells of other shapes.

J. G. SIMMONDS: Like Dr. LUKASZEWICZ, I too have investigated the problem of reducing the equations of shell theory to two coupled equations for the normal deflection w and a stress function φ. I found that such a reduction is possible only when both the mean and the Gaussian curvatures of the midsurface are constant.

The only curved surfaces satisfying these conditions are the sphere and the circular cylinder, and this is why Dr. LUKASZEWICZ' method gives accurate results for these two cases.

W. T. KOITER: Following up the comments made by Dr. LUKASZEWICZ and Professor SIMMONDS, I should like first to say that Professor SANDERS' reduction of the shell equations seems of great interest and to hold promise of significant advances. In this connection I may mention perhaps that for circular cylindrical shells and for spherical shells similar results may be obtained by an iterative integration procedure applied to the CODAZZI equations and the equation of equilibrium in the tangential plane. As far as I know, this procedure has not yet been published, although Dr. LUKASZEWICZ's work bears some resemblance to it. As a futher remark I may perhaps add that in the case of cylindrical shells, for which Professor SIMMONDS has expressed a strong preference for the stress function approach, equally simple results can be obtained in the form of displacement equations of equilibrium.

C. R. STEELE: Thank you for an interesting presentation. I would like to be clear on your reference to an approximation in the static-geometric analogy. I believe that the exact equations can be used and that your complex function Q can be obtained from the constitutive relations in terms of P and its conjugate, without the introduction of the auxillary functions R and S. Such a procedure was used by BAKER and CLINE for the axisymmetric problem and by NAGHDI for the general case. Although the conjugate terms remain in the final equations they have negligible influence on the final solution. This may sometimes be proved rigorously by procedures of error estimation for asymptotic solutions.

J. L. SANDERS: An exact reduction of the system of shell equations in complex form to a vector and scalar equation was achieved by CHERNYKH in his book (and somewhat later but independently by NAGHDI). These equations, however, contain the conjugation operation as well as partial differentiations. I agree that functional equations of this type may possibly be amendable to methods of symptotic integration. I do not agree that the conjugate terms are always negligible.

Addendum

J. L. SANDERS: Since the close of the symposium the author has had the opportunity to test the theory developed in the paper on the problem of the helicoidal shell for the case in which stresses are independent of the helix angle. It turns out to be impossible in this case to determine $R^{\alpha\beta}$ and $S^{\alpha\beta}$ such that these quantities are $0(1)$. One must conclude therefore that the theory is not universally applicable.

In this problem of the helicoidal shell the complex formulation would not be of much advantage anyway since the real equations reduce to a second order differential equation for one of the curvature components as shown by COHEN [10]. The Novozhilov equations for this problem are easily reduced to a second order equation for P (T in his notation) which differs essentially from the second order equation for P easily derived from Cohen's equations. A similar observation has previously been made by KOITER [13].

Whether or not the theory in the present paper, when applicable (i.e. when $R^{\alpha\beta}$ and $S^{\alpha\beta}$ of order unity can be found), is ever more accurate than the Novozhilov equations in the same case remains in doubt. However, the present theory does appear to have a "built-in" test for validity.

Appendix

Further Reduction of Equation (20) for Arbitrary Shells of Non-Zero Gaussian Curvature

By

J. G. Simmonds

University of Virginia, Charlottesville, Va., USA

We shall show that for shells of non-zero Gaussian curvature ($\varkappa \neq 0$), the 3 equations (20) for the 3 components of the symmetric tensor $P_{\alpha\beta}$ can be reduced to 2 coupled equations for the invariant $P = P_\alpha{}^\alpha$ and a "stress function" ψ, akin to the membrane stress function of VEKUA [1]. The equations of (semi-)membrane-inextensional bending theory follow directly from the general equations upon setting $L = R$ and neglecting terms of relative order β^2.

The reduction begins with the observation that if we add the negligible term

$$-\beta^2 b_\beta{}^\alpha b_\gamma{}^\beta P^{,\gamma} \tag{A1}$$

to the first line of (20), and multiply the second line of (20) by $-i\beta$, the resulting equations may be written in the form

$$\begin{aligned} T^{\beta\alpha}_{,\beta} + b_\beta{}^\alpha Q^\beta &= 0 \\ Q^\beta_{,\beta} - \beta\mu b_{\alpha\beta} T^{\alpha\beta} &= 0 \end{aligned} \tag{A2}$$

where

$$T^{\beta\alpha} = P^{\beta\alpha} + \frac{1}{2}\, i\beta\varepsilon^{\beta\alpha}(B_{\gamma\delta}P^{\gamma\delta}) \tag{A3}$$

$$Q^\beta = -i\beta(P^{,\beta} - i\beta b_\gamma{}^\beta P^{,\gamma}). \tag{A4}$$

Let the covariant base vectors $x^i_{,\alpha}$ and the unit normal vector n^i be denoted by $\boldsymbol{g}_\alpha$ and $\boldsymbol{n}$, respectively. Two of the fundamental equations of surface theory in dimensionless form, as stated in the body of the

---

Acknowledgement. This research was supported by the National Science Foundation under institutional Grant Gu-1978.

paper, now read

$$\boldsymbol{g}_{\alpha,\beta} = -\beta\mu b_{\alpha\beta}\boldsymbol{n}, \qquad \boldsymbol{n}_{,\beta} = b_\beta{}^\alpha \boldsymbol{g}_\alpha \tag{A5}$$

With the aid of these equations, (A2) may be written in the concise vector form

$$\boldsymbol{T}^\beta_{,\beta} = 0 \tag{A6}$$

where

$$\boldsymbol{T}^\beta = T^{\beta\alpha}\boldsymbol{g}_\alpha + Q^\beta \boldsymbol{n}. \tag{A7}$$

As Gol'denweizer has shown [2], (A6) may be identically satisfied by expressing $\boldsymbol{T}^\beta$ in terms of an arbitrary stress function vector

$$\boldsymbol{F} = f^\alpha \boldsymbol{g}_\alpha + \psi \boldsymbol{n} \tag{A8}$$

as follows:

$$\begin{aligned} \boldsymbol{T}^\beta &= \varepsilon^{\beta\gamma}\boldsymbol{F}_{,\gamma} \\ &= \varepsilon^{\beta\gamma}[(f^\alpha_{,\gamma} + b_\gamma{}^\alpha\psi)\boldsymbol{g}_\alpha + (\psi_{,\gamma} - \beta\mu b_{\alpha\gamma} f^\alpha)\boldsymbol{n}]. \end{aligned} \tag{A9}$$

Comparing (A7) and (A9) we have

$$T^{\beta\alpha} = \varepsilon^{\beta\gamma}(f^\alpha_{,\gamma} + b_\gamma{}^\alpha \psi) \tag{A10}$$

$$Q^\beta = \varepsilon^{\beta\gamma}(\psi_{,\gamma} - \beta\mu b_{\alpha\gamma} f^\alpha). \tag{A11}$$

If $\varkappa \neq 0$, then with the aid of the identity

$$\bar{b}^{\alpha\beta} b_{\beta\gamma} = \varkappa \delta_\gamma{}^\alpha, \tag{A12}$$

we may solve (A11) for f^α in terms of Q^α:

$$f^\alpha = (\beta\mu\varkappa)^{-1}\bar{b}^{\alpha\lambda}(\psi_{,\lambda} + \varepsilon_{\lambda\mu}Q^\mu), \tag{A13}$$

and thus use Q^α in place of f^α in (A10). This possibility was first suggested by Reissner [3], buts its exploitation is due to Libai [4], whose work motivated the method of reduction presented in this Appendix.

Substituting (A13) into (A10), and using (A4) to express Q^β in terms of P, we arrive at the following representation for the unsymmetric tensor $T^{\beta\alpha}$ in terms of the two invariants P and ψ:

$$T^{\beta\alpha} = \varepsilon^{\beta\gamma}\left\{\frac{1}{\beta\mu}\left(\frac{\bar{b}^{\alpha\lambda}\psi_{,\lambda}}{\varkappa}\right)_{,\gamma} + b_\gamma{}^\alpha\psi - \frac{i\varepsilon_{\lambda\mu}}{\mu}\left[\frac{\bar{b}^{\alpha\lambda}}{\varkappa}(P^{,\mu} - i\beta b_\sigma{}^\mu P^{,\sigma})_{,\gamma}\right]\right\}. \tag{A14}$$

Through use of the identity and definition

$$b_{\alpha\lambda}b_\beta{}^\lambda = 2\varrho b_{\alpha\beta} - \varkappa g_{\alpha\beta}, \qquad \bar{b}^{\alpha\beta} = \varepsilon^{\alpha\lambda}\varepsilon^{\beta\mu}b_{\lambda\mu}, \tag{A15}$$

(A 14) may be written

$$T^{\beta\alpha} = \varepsilon^{\beta\gamma}\left\{\frac{1}{\beta\mu}\left(\frac{\bar{b}^{\alpha\lambda}\psi_{,\lambda}}{\varkappa}\right)_{,\gamma} + b_\gamma{}^\alpha\psi - \right.$$
$$\left. - \varepsilon^{\alpha\lambda}\left[\frac{i}{\mu}\cdot\frac{b_\lambda{}^\mu P_{,\mu}}{\varkappa} + \frac{\beta}{\mu}\left(\frac{2\varrho}{\varkappa}\, b_\lambda{}^\mu P_{,\mu} - P_{,\lambda}\right)\right]_{,\gamma}\right\} \equiv$$
$$\equiv \varepsilon^{\beta\gamma}\left\{W_\gamma{}^\alpha(\psi) + S_\gamma{}^\alpha(P)\right\}. \tag{A16}$$

This representation for $T^{\beta\alpha}$ has a particularly simple physical interpretation: The operator $W_\gamma{}^\alpha(\psi)$ represents the pure membrane and pure inextensional bending part of $T^{\beta\alpha}$; the operator $S_\gamma{}^\alpha(P)$ represents that part of $T^{\beta\alpha}$ arising from the coupling of bending and stretching.

From (A 3) we have

$$g_{\beta\alpha}T^{\beta\alpha} = P \tag{A17}$$

$$\varepsilon_{\beta\alpha}T^{\beta\alpha} = i\beta B_{\beta\alpha}P^{\beta\alpha}$$
$$= i\beta B_{\beta\alpha}T^{\beta\alpha}. \tag{A18}$$

When $T^{\beta\alpha}$ is replaced by the right side of (A 16), these two relations become two coupled equations for P and ψ. First, from (A 16) and (A 17) there follows:

$$P = \frac{\varepsilon^{\alpha\beta}}{\beta\mu}\left(\frac{b_\beta{}^\gamma\psi_{,\alpha}}{\varkappa}\right)_{,\gamma} - \frac{i}{\mu}\left(\frac{b^{\alpha\gamma}P_{,\alpha}}{\varkappa}\right)_{,\gamma} + \frac{\beta}{\mu}\left[\nabla^2 P - \left(\frac{2\varrho}{\varkappa}\, b^{\alpha\gamma}P_{,\alpha}\right)_{,\gamma}\right]. \tag{A19}$$

Second, with the relation

$$B_{\beta\alpha}\varepsilon^{\beta\gamma} = b_\alpha{}^\gamma - \varrho\delta_\alpha{}^\gamma \tag{A20}$$

there follows from (A 16) and (A 18)

$$(1 + i\beta\varrho)\left[W_\alpha{}^\alpha(\psi) + S_\alpha{}^\alpha(P)\right] = i\beta b_\alpha{}^\gamma\left[W_\gamma{}^\alpha(\psi) + S_\gamma{}^\alpha(P)\right] \tag{A21}$$

which, when expanded and simplified, can be brought into the form

$$(1 + i\beta\varrho)\left\{W(\psi) - \frac{\varepsilon^{\alpha\lambda}b_\lambda{}^\gamma}{\mu}\left[(i + 2\beta\varrho)\,\frac{P_{,\gamma}}{\varkappa}\right]_{,\alpha}\right\}$$
$$= i\beta\left\{\frac{1}{\beta\mu}\left(\nabla^2\psi - \frac{2\varrho_{,\alpha}}{\varkappa}\,\bar{b}^{\alpha\lambda}\psi_{,\lambda}\right) + (4\varrho^2 - 2\varkappa)\psi + \right.$$
$$\left. + \varepsilon^{\alpha\lambda}b_\lambda{}^\gamma\left[\frac{i}{\mu}\cdot\frac{2\varrho_{,\alpha}}{\varkappa}\,P_{,\gamma} + \frac{\beta}{\mu}\left(\frac{4\varrho\varrho_{,\alpha}}{\varkappa}\,P_{,\gamma} - P_{,\alpha\gamma}\right)\right]\right\} \tag{A22}$$

where

$$W_\alpha{}^\alpha(\psi) \equiv W(\psi) = \frac{\bar{b}^{\alpha\beta}}{\beta\mu}\left(\frac{\psi_{,\alpha}}{\varkappa}\right)_{,\beta} + 2\varrho\psi \tag{A23}$$

is the Weingarten operator.

Semi-Membrane-Inextensional Bending (S. M. I. B.)

A state of S. M. I. B. may be characterized as one for which $L = R$, i.e. $\mu = \beta^{-1}$. Under these conditions, and with the neglect of terms of relative order β^2, (A16), (A19), and (A22) reduce to

$$T^{\beta\alpha} = \varepsilon^{\beta\gamma}\left[\left(\frac{\bar{b}^{\alpha\lambda}\psi_{,\lambda}}{\varkappa}\right)_{,\gamma} + b_\gamma{}^\alpha\psi - i\beta\varepsilon^{\alpha\lambda}\left(\frac{b_\lambda{}^\mu P_{,\mu}}{\varkappa}\right)_{,\gamma}\right] \tag{A24}$$

$$P = \varepsilon^{\alpha\beta}\left(\frac{b_\beta{}^\gamma\psi_{,\alpha}}{\varkappa}\right)_{,\gamma} - i\beta\left(\frac{b^{\alpha\gamma}P_{,\alpha}}{\varkappa}\right)_{,\gamma} \tag{A25}$$

$$(1 + i\beta\varrho)\left[W(\psi) - i\beta\varepsilon^{\alpha\lambda}b_\lambda{}^\gamma\left(\frac{P_{,\gamma}}{\varkappa}\right)_{,\alpha}\right]$$
$$= i\beta\left[\nabla^2\psi - \frac{2\varrho_{,\alpha}}{\varkappa}\bar{b}^{\alpha\lambda}\psi_{,\lambda} + (4\varrho^2 - 2\varkappa)\psi + i2\beta\varepsilon^{\alpha\lambda}b_\lambda{}^\gamma\frac{\varrho_{,\alpha}P_{,\gamma}}{\varkappa}\right]. \tag{A26}$$

Substitution of (A25) into (A26) and the neglect of further terms of relative order β^2, leads to the equation

$$W(\psi) - i\beta V(\psi) = 0 \tag{A27}$$

where

$$V(\psi) = \varepsilon^{\alpha\lambda}\varepsilon^{\sigma\beta}b_\lambda{}^\gamma\left[\frac{1}{\varkappa}\left(\frac{b_\beta{}^\mu\psi_{,\sigma}}{\varkappa}\right)_{,\mu\gamma}\right]_{,\alpha} +$$
$$+ \nabla^2\psi - \frac{2\varrho_{,\alpha}}{\varkappa}\bar{b}^{\alpha\lambda}\psi_{,\lambda} + (4\varrho^2 - 2\varkappa)\psi - \varrho W(\psi). \tag{A28}$$

For shells of infinite extent (A27) represents the extension of Vlasov's semi-membrane theory of cylindrical shells to shells of arbitrary shape. For shells of finite extent, it is permissible to assume a solution of (A27) of the form

$$\psi = \psi_0 + i\beta\psi_1 \tag{A29}$$

where

$$W(\psi_0) = 0 \tag{A30}$$

$$W(\psi_1) = V(\psi_0). \tag{A31}$$

References

1. Vekua, I. N.: Generalized analytic functions (English translation), Pergamon Press (1959) (Distributed in USA by Addison-Wesley Pub. Co.).
2. Gol'denweizer, A. L.: Theory of elastic thin shells (English translation), Pergamon Press 1961, Chap. 2.
3. Reissner, E.: A note on stress functions and compatibility equations in shell theory, Topics in Applied Mechanics, the Schwerin Memorial Volume, Amsterdam: Elsevier 1965 pp. 23—32.
4. Libai, A.: Invariant stress and deformation functions for doubly curved shells, J. Appl. Mech. **34**, 43—48 (1967).

On the Optimum Design of Shells Loaded by Concentrated Forces

By

S. Lukasiewicz

Politechnika Warszawska, Warsaw, Poland

1. Introduction

Concentrated loads cause strong concentrations of stresses in a shell. When the loads are distributed on a small portion of the shell surface, there also appears a concentration of stresses, making the effort of this portion large in comparison to other parts of the structure. Therefore, the designers strengthen the loaded places locally, welding strups, ribs or other strengthening elements to the shell. The present paper deals with the optimum shape of the shell in the vicinity of the loading point ensuring constant strength and minimum weight. The optimum design is pursued without regard to any difficulties or costs that may arise in its manufacture. Nevertheless, it can be the basis of comparison with any proposed structure.

The constant strength can be achieved by designing the thickness of the shell, changing it symmetrically or asymmetrically with respect to the middle surface. In case of a symmetrical change (for ex. by welding strups at both sides of the wall) the shape of the middle surface does not undergo a change. To ensure constant strength it is enough to determine the proper thickness. When the thickness is changed asymmetrically, the shape of the middle surface also changes. The solution of the first case reduces to the solution of the problem for a shell which carries both direct and bending stressses and is usually rather difficult to obtain. In the second case, the bending can be eliminated, so that membrane forces only appear in the shell. By designing the shape of the middle surface, we have more possibilities and at the same time we can achieve a smaller weight of the whole structure.

The solutions obtained with the assumption that both principal stresses are equal and constant at any point of the shell are well known [1, 2]. This assumption determines at once the thickness and the shape of the shell. If, however, we assume only the satisfaction of the condition of constant strength we determine only the thickness and we are still able to choose its shape. Owing to that the shape of the shell can be found from additional conditions, e. g. minimum weight, minimum cost etc.

2. The Shell Loaded by a Concentrated Force and by a Distributed Load

2.1. Let us at first solve the case of a shell loaded by a concentrated force and by a constant external pressure p_0. The unknowns to be found are the following: the internal forces N_φ N_ϑ, the thickness of the shell t and the shape of the shell determined by the function $z(x)$. We have at our disposal the following equations containing the unknown functions:

1. Two differential equations of equilibrium

$$\frac{d}{d\varphi}(x N_\varphi \sin\varphi) = Q x \frac{dx}{d\varphi} \qquad \text{where} \qquad Q = p_r - p_\varphi \operatorname{tg}\varphi \tag{1}$$

$$\frac{N_\varphi}{r_1} + \frac{N_\vartheta}{r_2} = p_r .$$

2. The condition of constant strength. We assume that the material of the shell obeys the Tresca yield criterion (Fig. 1).

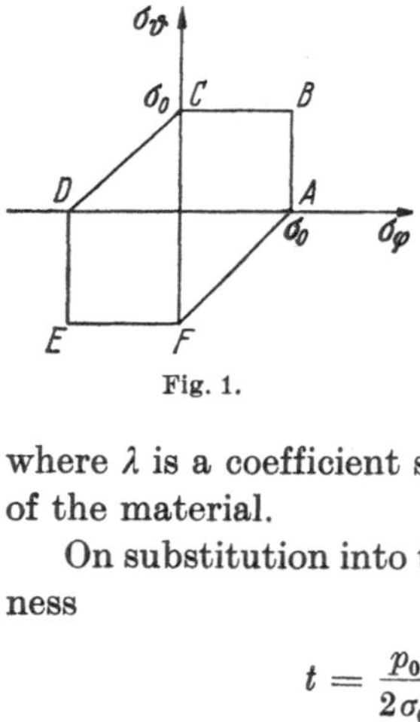

Fig. 1.

3. The condition of minimum weight which depends on the volume of the shell.

The function $z(x)$ can be determined from the condition of minimum volume. Let us assume that $N_\varphi \leq N_\vartheta \leq 0$. Then the stresses are represented by the straight line DE of the hexagon (Fig. 1). We write

$$N_\varphi = -\sigma_0 t; \; N_\vartheta = -\lambda\sigma_0 t$$

where λ is a coefficient such that $0 \leq \lambda \leq 1$ σ_0 is the tensile yield stress of the material.

On substitution into the equations of equilibrium we obtain the thickness

$$t = \frac{p_0}{2\sigma_0} \frac{\varrho^2 + x^2}{x} \frac{\sqrt{1 + z'^2}}{z'}, \qquad \varrho^2 = \frac{P}{\pi p_0}. \tag{2}$$

The volume of the shell is

$$V = \frac{\pi p_0}{\sigma_0} \int\limits_0^{x_1} (\varrho^2 + x^2)\left(\frac{1}{z'} + z'\right) dx. \tag{3}$$

The above functional has a minimum when the following Euler equation is satisfied

$$\frac{d}{dx}\left[(\varrho^2 + x^2)\left(-\frac{1}{z'^2} + 1\right)\right] = 0 \tag{4}$$

which yields

$$z' = \pm\sqrt{\frac{\varrho^2 + x^2}{\varrho^2 + B + x^2}} \tag{5}$$

where B is an arbitrary constant. The Eq. (6) can be integrated. We obtain (for $B < 0$).

$$z = -\varrho^2 E\left[gd\left(\operatorname{arcsin} h\,\frac{x_1}{\sqrt{\varrho^2 + B}}\right), \left(\frac{\varrho^2 + B}{\varrho^2}\right)^{1/2}\right] + c$$

$$gd(x) = \int\limits_0^x \frac{dt}{\cosh t} \tag{6}$$

where $E(x)$ is an elliptic function. For $x = 0$ the angle of the slope depends on the constant B and is $z' = \pm(\varrho^2/\varrho^2 + B)^{1/2}$. This angle has a real value when $\varrho^2 + B > 0$. The thickness takes the following form

$$t = \frac{p_0}{2\sigma_0}\,\frac{1}{x}\,(2\varrho^2 + B + 2x^2)^{1/2}(\varrho^2 + x^2)^{1/2}. \tag{7}$$

The above solution is valid only when $0 \leq \lambda \leq 1$. Therefore the value of λ should be evaluated. It can be obtained from the second equilibrium equation. On transformation we have

$$\lambda = \frac{2x^2}{x^2 + \varrho^2} - \frac{x\,z''}{(1 + z'^2)z'}.$$

Substitution of z'' and z' from (5) yields

$$\lambda = \frac{x^2}{\varrho^2 + x^2} \cdot \frac{4\varrho^2 + B + 4x^2}{2\varrho^2 + B + 2x^2}. \tag{8}$$

λ satisfies the condition $0 \leq \lambda \leq 1$ when $0 \leq x^4 \leq \varrho^2\left(\varrho^2 + \frac{B}{2}\right)$. The constants B and C make it possible to determine the shape of the shell in accordance with the boundary conditions. Some examples have been given in Fig. 2.

Let us assume now that height of the shell is not determined and its edge lies on the line $x = x_1$. The condition of minimum volume gives

$$(\varrho^2 + x^2)\left(-\frac{1}{z'^2} + 1\right)_{x=x_1} = 0; \qquad z' = 1.$$

In conclusion we observe that the minimum volume is obtained for a conical shape with the slope angle $\alpha = 45°$. The thickness should change as

$$t = \frac{p_0}{2\sigma_0} \frac{\varrho^2 + x^2}{x}. \tag{9}$$

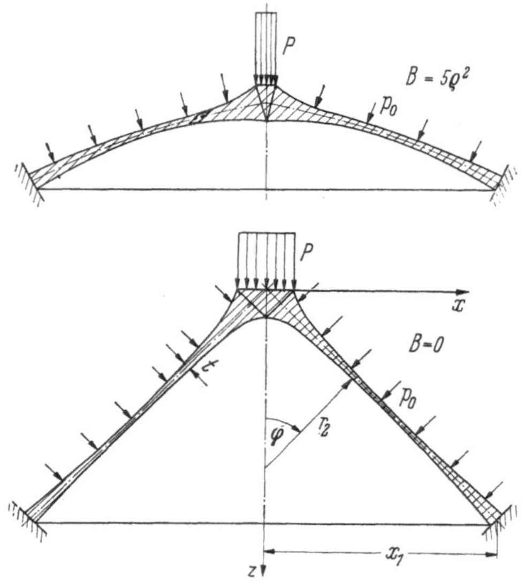

Fig. 2.

When the shell is subjected only to one concentrated force we have

$$V = \frac{P}{\sigma_0} \int_0^{x_1} \left(\frac{1}{z'} + z'\right) dx. \tag{10}$$

The above functional is a minimum when $z' = \text{const}$. The absolute minimum of volume is for $z' = 1$, i. e. for a conical shell with thickness $t = P/2\sigma_0 x$.

2.2. Now we consider the second example- the case of a load uniformly distributed in the horisontal plane. The components of the load are $p_r = -p_0 \cos^2\varphi$; $p_\varphi = p_0 \cos\varphi \sin\varphi$ and $Q_r = -p_0$. The first equilibrium equation does not change and therefore the thickness of the shell and the volume are expressed by means of the same formula as before. The only change is in the equation for λ. After some manipulation we obtain

$$\lambda = x^2/(\varrho^2 + x^2). \tag{11}$$

We see that for arbitrary x the condition $0 \leq \lambda \leq 1$ is fulfilled. The shape and the thickness of the shell are as before.

2.3. Now we consider the case when the concentrated force P acts in the opposite direction of the distributed load. We assume that the force N_φ is a compressive force and N_ϑ is a tensile force. Such a state of stresses is presented by the straight line CD in Fig. 1. These forces can be defined in the following way: $N_\varphi = -\lambda \sigma_0 t$; $N_\vartheta = (1-\lambda)\sigma_0 t$; $0 \leq \lambda \leq 1$. Then we obtain the thickness and the parameter λ by solving the equations of equilibrium. We have after some manipulations

$$t = \frac{p_0}{2\sigma_0}\left[\left(1 + \frac{r_2}{r_1}\right)\left(\frac{\varrho^2 - x^2}{x^2}\right) + 2\right]\frac{(1+z'^2)^{1/2}}{z'}. \tag{12}$$

The volume of the shell is

$$V = \frac{\pi p_0}{\sigma_0}\int\limits_0^{x_1} \frac{1+z'^2}{z'}\left[\frac{x z''(\varrho^2 - x^2)}{z'(1+z'^2)} + \varrho^2 + x^2\right] dx. \tag{13}$$

Its minimum is when the Euler equation is satisfied

$$\frac{1 - z'^2}{z'^2}(\varrho^2 + x^2) + \frac{\varrho^2 - 3x^2}{z'^2} = B.$$

Then

$$z' = \sqrt{\frac{2\varrho^2 - 2x^2}{\varrho^2 + B + x^2}}. \tag{14}$$

The shape of the shell can be obtained from (14) in the form of elliptic integral. The thickness of the shell takes the form

$$t = \frac{p_0}{2\sqrt{2}\,\sigma_0}\,\frac{1}{x}\,\frac{3\varrho^4 + B\varrho^2 - x^4}{(3\varrho^2 + B - x^2)^{1/2}\,(\varrho^2 - x^2)^{1/2}}. \tag{15}$$

The constants B, C enable us to obtain the shape of the shell required by the boundary conditions. When the height of the shell is not determined in advance, we obtain the minimum weight for $B = 0$. The parameter λ is determined by means of the second equilibrium equation (1). We obtain after some transformations

$$\lambda = \frac{(\varrho^2 - x^2)(3\varrho^2 + B - x^2)}{3\varrho^4 + \varrho^2 B - x^4}$$

for $x = 0$; $\lambda = 1$, for $x = \varrho$; $\lambda = 0$.

The above solution is valid only if $0 \leq x \leq \varrho$; for $x = \varrho$, $\lambda = 0$, i. e. when the force $N\varphi = 0$ and when the shell is in equilibrium under the action of a concentrated force and a distributed load. The shape of the shell is given in Fig. 3. We observe that the thickness of the shell grows to infinity when $x \to \varrho$. The reason for this phenomenon is that the radii of curvature $r_1 \to 0$ and $r_2 \to \infty$ at this point. But the surface of this part of cross-section of the shell is limited. We can imagine that the shell is strengthened there by a ring of certain cross-section, which carries the tension in a parallel circle.

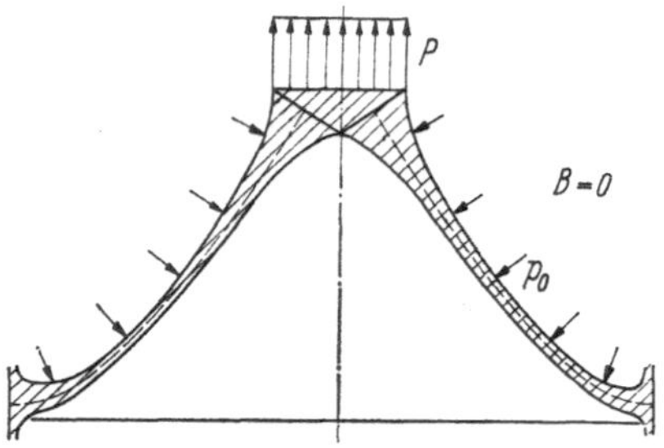

Fig. 3.

2.4. In case of a uniformly distributed load in the horizontal plane we find the shape of the shell in the similar way.

$$z' = \pm \sqrt{\frac{2\varrho^2 - 2x^2}{\varrho^2 + B - x^2}}. \tag{16}$$

The thickness is

$$t = \frac{p_0}{2\sqrt{2}\,\sigma_0} \frac{1}{x} \left[\frac{2(\varrho^2 - x^2) + (\varrho^2 + x^2)(\varrho^2 + B - x^2)}{3\varrho^2 + B - 3x^2} - \frac{Bx^2}{\varrho^2 + B - x^2} \right] \frac{1}{(\varrho^2 - x^2)^{1/2}}.$$

The absolute minimum occurs when $B = 0$ which gives $z' = \sqrt{2} = \text{const.}$ It follows that the conical shell is also the structure of minimum weight. The slope angle of the shell is $\alpha = 54° 44'$ and the thickness

$$t = \frac{p_0}{2\sqrt{6}\,\sigma_0} \left(\frac{3\varrho^2}{x} + x \right) \quad \text{(see Fig. 4).}$$

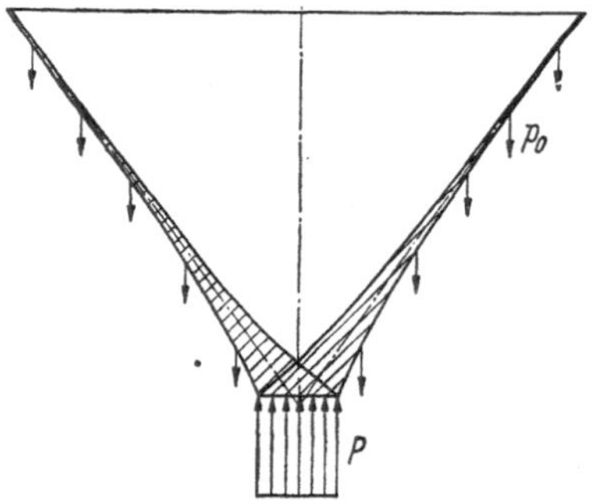

Fig. 4.

The shallow shell can be obtained if we assume large B.

Neglecting x in the denominator we have an approximate expression

$$z' = \frac{\sqrt{2}(\varrho^2 - x^2)^{1/2}}{\left(\frac{\varrho^2}{2} + B\right)^{1/2}},$$

$$z = \frac{\sqrt{2}}{\left(\frac{\varrho^2}{2} + B\right)^{1/2}} \left[\frac{x}{2}(\varrho^2 - x^2)^{1/2} + \frac{\varrho^2}{2} \arcsin \frac{x}{\varrho} \right] + C.$$

3. The Shell Carrying its Own Weight

3.1. Let us now consider a shell loaded by a concentrated normal force and by its own weight. If γ is the specific weight of the material, the components of load per unit area of the surface are

$$p_\varphi = \gamma t \sin\varphi; \qquad p_r = -\gamma t \cos\varphi. \qquad \text{Then } Q = -\frac{\gamma t}{\cos\varphi}.$$

We assume that both internal forces are compressive and $N_\varphi \leq N_\vartheta \leq \leq 0$. Then this state of stresses is expressed by the line DE in the hexagon — Fig. 1. Then $N_\varphi = -t\sigma_0$, $N_\vartheta = -\lambda t \sigma_0$ where $0 \leq \lambda \leq 1$. On substitution into first equilibrium condition (1) we have

$$\frac{d}{dx}\left(xt\frac{z'}{\sqrt{1+z'^2}}\right) = \frac{\gamma}{\sigma_0} xt\sqrt{1+z'^2}. \tag{17}$$

This equation can be integrated:

$$t = C\frac{\sqrt{1+z'^2}}{xz'} \exp\frac{\gamma}{\sigma_0}\int\left(\frac{1}{z'}+z'\right)dx. \tag{18}$$

C is here an arbitrary constant which can be found comparing formulae (3), and (18). For $\gamma = 0$ we should obtain the result of the shell loaded only by the concentrated force P. The volume of the shell is

$$V = \frac{P}{\gamma}\int_0^{x_1}\left[\left(\frac{1}{z'}+z'\right)\exp\frac{\gamma}{\sigma_0}\int\left(\frac{1}{z'}+z'\right)dx\right]dx.$$

The above functional can be integrated:

$$V = \frac{P}{\gamma}\left[\exp\frac{\gamma}{\sigma_0}\int_0^{x_1}\left(\frac{1}{z'}+z'\right)dx - 1\right]. \tag{19}$$

If we assume that $z' > 0$ when $0 \leq x \leq x_1$, then the integral in the above expression is positive. The functional V has its minimum when this integral is a minimum. That gives $z = \text{const}$ i. e. a conical shell. The absolute minimum is when $z' = 1$. The thickness of the shell for $z' = 1$ is

$$t = \frac{P}{\sqrt{2}\pi\sigma_0}\cdot\frac{e^{\frac{2\gamma}{\sigma_0}x}}{x}.$$

The parameter λ is determined by the following equation:

$$0 \leq \lambda = \frac{\gamma}{\sigma_0}\frac{x}{z'} - \frac{xz''}{z'(1+z'^2)} \leq 1. \tag{20}$$

Therefore $z' = 1$

$$x \leq \frac{\sigma_0}{\gamma}$$

The shape of the shell is similar to that given in Fig. 1. When $P = 0$ and the shell is loaded only by its own weight we also obtain the minimum weight for a conical shape. In order to compare the obtained results with [2] we transform expression (18) observing equation (20). Then

$$\frac{\gamma}{\sigma_0}\frac{1}{z'} = \frac{\lambda}{x} + \frac{z''}{z'(1+z'^2)} = \frac{\sqrt{1+z'^2}}{xz'}\frac{d}{dx}\left(\frac{z'x}{\sqrt{1+z'^2}}\right) + \frac{\lambda-1}{x}.$$

Introducing it into Eq. (18) we have

$$t = t_0 \exp \int \left(\frac{\gamma}{\sigma_0} z' - \frac{\lambda - 1}{x}\right) dx$$

Assuming $\lambda = 1$ we find the known solution $t = t_0 e^{\frac{\gamma}{\sigma_0} z}$, we see that only in this case the thickness of the shell is limited at the top. The shell is not a conical one.

3.2. Now we consider the case when the force P acts in the opposite direction to the weight. Assuming that $N_\varphi = \lambda\sigma_0 t > 0$ and $N_\vartheta = -(1-\lambda)\,\sigma_0 t < 0$ we find on substitution into Eq. (1) the thickness

$$t = C\frac{\sqrt{1+z'^2}}{\lambda x z'} \exp\left[-\frac{\gamma}{\sigma_0}\int\left(\frac{1}{z'} + z'\right)\frac{dx}{\lambda}\right]. \tag{21}$$

The volume of the shell is on integration

$$V = 2\pi C\left\{1 - \exp\left[-\frac{\gamma}{\sigma_0}\int_0^{x_1}\left(\frac{1}{z'} + z'\right)\frac{dx}{\lambda}\right]\right\}. \tag{22}$$

Assuming that $z' > 0$ when $0 \leq x \leq x_1$ we have the minimum of the functional V when the integral in (22) has its minimum.

Parameter λ can be evaluated from the equation

$$\frac{1}{\lambda} = \frac{\left(z' - \frac{\gamma x}{\sigma_0}\right)(1+z'^2)}{z'(1+z'^2) + xz''}. \tag{23}$$

Introducing it into Eq. (22) we find the following differential equation

$$z'^2\left[z'^2(1+B) - 2(1+B)\frac{\gamma}{\sigma_0}x - 2 + B\left(\frac{\gamma}{\sigma_0}x\right)^2\right] = 0$$

where B is an arbitrary constant. This equation has one solution $z'^2 = 0$ which corresponds to infinitely thick flat plate. Then there remains the algebraical equation of second order with respect to z'. Solving and integrating we find the shape for which the weight has an absolute minimum. From the variational conditions for the edge $x = x_1$ we find that $B = 0$. The solution takes the form

$$z' = \frac{\gamma x}{\sigma_0} \pm \sqrt{\frac{\gamma x}{\sigma_0} + 2}\,. \tag{24}$$

Since $z' > 0$ we take the sign $+$.

By introducing the non-dimensional coordinates $u = \frac{\gamma x}{\sqrt{2}\sigma_0}$ we have

$$z = \frac{\sigma_0}{\gamma}\left[u^2 + u\sqrt{u^2+1} + \ln\left(u + \sqrt{u^2+1}\right)\right] \tag{25}$$

and the thickness

$$t = \frac{P}{2\sigma_0 \pi x} f_2(u) \exp\left[f_1(u)\right]$$

where

$$f_1(u) = 2\left(u^2 + u\sqrt{u^2+1}\right) + \ln\left(u^2 + u\sqrt{u^2+1} + 1\right)$$

$$f_2(u) = \frac{(4u^2+3)\sqrt{u^2+1} + 4u^3 + 5u}{\sqrt{2}\,(u^2+1)\,(4u^2 + 4u\sqrt{u^2+1} + 3)^{1/2}} \tag{26}$$

The shape of the shell is given in Fig. 5. Introducing $z(x)$ from Eq. (25) into Eq. (23) we obtain the coefficient λ equal to 1 for $x = 0$ and 0,4 for $x = \infty$. It means that the condition $0 \leq \lambda \leq 1$ is always satisfied.

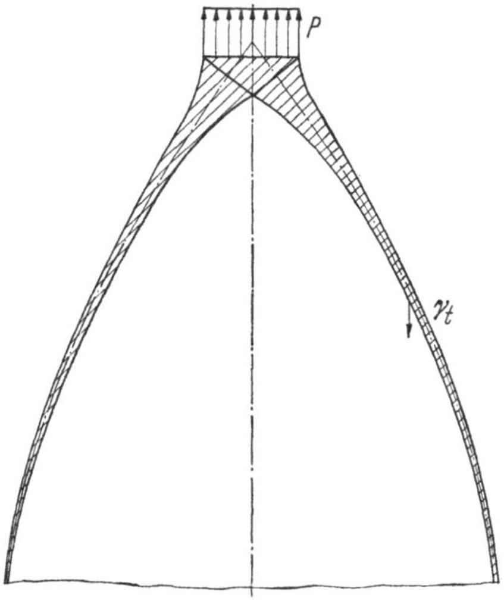

Fig. 5.

The results presented in this paper have been obtained by means of the membrane theory of shells. The equations are exact enough only when the thickness of the shell does not change considerably and rapidly. Therefore the results for the region near the top may not be quite accurate. However, the theory is able to give an idea of the optimum shape of the shell. Direct application of the obtained results may lead to certain difficulties in practice. At large distances from the top the thickness of the shell determined by the given formulae is very small and the manufacture of such a shell would be impossible for technological reasons. In these areas the shell can buckle. This has not been considered during the designing of the shell.

References

1. Tölke, F.: Über Rotationsschalen gleicher Festigkeit für konstanten Innen- oder Außendruck, Z. angew. Math. Mech. **19**, 338–343 (1939).
2. Margareus, G.: Die Kuppel gleicher Festigkeit, Bauing. **20**, 232–234 (1939).
3. Ziegler, H.: Kuppeln gleicher Festigkeit, Ing. Arch. **26**, 378–382 (1958).

Discussion

V. Křupka: The author solves his problem on the assumption that the membrane theory holds. However, local bending stresses are of considerable importance in reality. Sometimes also the mode of application of the force to the shell must be taken into consideration. Very often the shell is not directly subjected to a concentrated load but via an elastic or rigid die. These problems must be worth while solving. A certain idealization is contained in the fact that local stability is totally neglected. Both these effects can essentially affect a real optimum design of a construction.

Boussinesq's theoretical solution of the problem of an absolutely rigid die acting on semi infinite elastic solid has found its application in many engineering fields. Even in the theory of shells very often we deal with problems of the die pressed against the cylindrical surface of the shell.

A general theoretical solution would be sure to lead to a complicated solution of integral equations, as well as it did in Boussinesq's solution. I don't think any solution of this kind has been yet published. All papers deal with shells acted directly upon by forces and moments [3, 4].

That is why we are going to explain a very effectively solution based on the half differential method which solves the mentioned problem using the digital computer.

Let us show the fundamentals of the method in investigating the saddle die in contact with the cylindrical shell. Instead of a die in continuous contact with the shell we assume the forces to be transferred by means of concentrated supports (Fig. 1) located on a circumferential circle. We will neglect the influence of the shear and consider the radial displacement only. Then in each support we obtain a statically indeterminate force X, which is resultant of the distributed load acting over an area of dimensions Δ and $2b$ in circumferential and axial direction respectively (Fig. 1). The distributed load is assumed to be uniform over all the mentioned area.

We may put down the continuity condition for contact of the die with shell using the matrix notation

$$\boldsymbol{A}_t \boldsymbol{X} - c \boldsymbol{A}_d = 0 \tag{1}$$

where

$\boldsymbol{A}$ square matrix of displacement of the shell loaded by unit value forces
$\boldsymbol{A}_d$ column matrix of displacement of the die as a rigid body
$\boldsymbol{X}$ column matrix of statically indeterminate forces $x_1, x_2 \ldots x_n$.
c displacement of the die in the symetry axis.

Let us assume the resultant vertical force Q to act on the die and let us solve the symetrical problem. The matrix A_t becomes

$$A_t = \left| \begin{array}{lllll} 0{,}5\,\delta_0; & \delta_1; & \delta_2; & \ldots; & \delta_n \\ \delta_1; & \delta_0 + \delta_2; & \delta_1 + \delta_3; & \ldots; & \delta_{n-1} + \delta_{n-1} \\ \delta_2; & \delta_1 + \delta_3; & \delta_0 + \delta_4; & \ldots; & \delta_{n-2} + \delta_{n-2} \\ \vdots & \vdots & \vdots & \vdots & \vdots \\ \delta_n; & \delta_{n-1} + \delta_{n+1}; & \delta_{n-2} + \delta_{n-2}; & \ldots; & \delta_0 + \delta_{2n} \end{array} \right| \tag{2}$$

where

δ_0 the radial displacement of the shell in the point of action of the unit force
δ_1 the radial displacement in the point distant Δ from the point of action of the unit force
δ_n the radial displacement in the point distant $n\Delta$ from the point of action of the unit force.

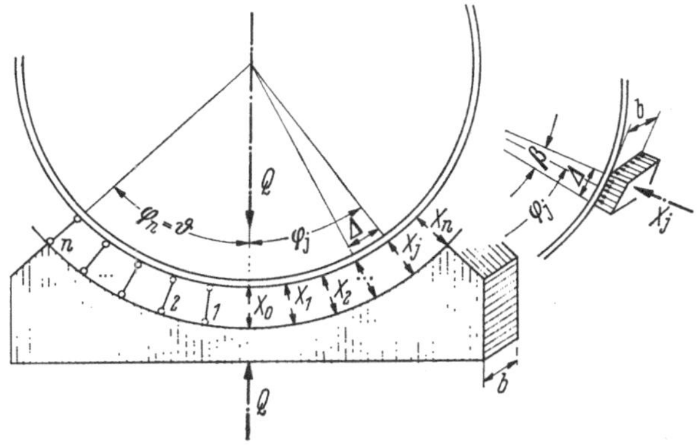

Fig. 1.

For long cylindrical shells the above mentioned relations were derived in [2, p. 236] on the basis of the semibending theory. Thus for the radial displacement we obtain:

$$\delta = f_w \frac{1}{E r} \left(\frac{r}{t} \right)^{5/2} \tag{3}$$

where

$$f_w = \frac{6}{\pi k} \sum_{2}^{m} \frac{\cos m\beta \cos m\varphi}{m(m^2-1)^2 \beta} \left(1 - e^{-k_m k} \cos k_m k\right)$$

$$k_m = \frac{m(m^2-1)^{1/2}}{48^{1/4}} \tag{4}$$

$$k = \frac{b}{r} \sqrt{\frac{t}{r}}.$$

Here

$2b$ denotes the width of the die
β denotes circumferential angle corresponding to one Δ
m equels to 2,3 ... m.

The column matrix $\boldsymbol{A}_d$ represents the displacement of the absolutely rigid die in the direction of the force Q. For small displacement it becomes:

$$\boldsymbol{A}_d = \begin{vmatrix} 0{,}5 \\ \cos\varphi_1 \\ \cos\varphi_2 \\ \vdots \\ \cos\varphi_n \end{vmatrix} \tag{5}$$

From Eq. (1) we get then

$$\boldsymbol{X} = c\boldsymbol{A}_t^{-1}\boldsymbol{A}_d \tag{6}$$

Statically indeterminate forces X have been obtained as the multiplies of the displacement c

$$X = c\overline{X} \tag{7}$$

We can get this displacement from the following equilibrium equation:

$$Q = c(\overline{X}_0 + 2\overline{X}_1 \cos\varphi_1 + \cdots + 2\overline{X}_n \cos\varphi_n). \tag{8}$$

From this it follows that

$$c = \frac{Q}{\overline{X}_0 + 2\overline{X}_1 \cos\varphi_1 + \cdots + 2X_n \cos\varphi_n}. \tag{9}$$

Thus substituting c from Eq. (9) in the Eq. (6) we get

$$\boldsymbol{X} = \frac{Q}{\overline{X}_0 + 2\overline{X}_1 \cos\varphi_1 + \cdots + 2X_n \cos\varphi_n} \boldsymbol{A}_t^{-1}\boldsymbol{A}_d \tag{10}$$

from which individual X can be evaluated.

Having obtained the values of X forces we can find the stresses in the shell. According to [2] the membrane axial stress σ_x and circumferential bending stress σ_s along the circumference due to the unit force applied in $\varphi = 0$ will be:

$$\sigma_x = f_X \frac{1}{t^2}\left(\frac{t}{r}\right)^{1/2} \tag{11}$$

$$\sigma_S = f_S \frac{1}{t^2}\left(\frac{t}{r}\right)^{1/2} \tag{12}$$

where

$$f_x = \frac{\sqrt{3}}{\pi k} \sum_2^m \frac{\sin m\beta \sin m\varphi}{m(m^2-1)\beta} e^{-k_m k} \sin k_m k \tag{13}$$

$$f_S = \frac{S}{\pi k} \sum_2^m \frac{\sin m\beta \sin m\varphi}{m(m^2-1)\beta} (1 - e^{-k_m k} \cos k_m k). \tag{14}$$

In an arbitrary point on the circumference we then obtain the corresponding coefficient F as a sum of effects of all the statically indeterminate forces

$$F = X_0 f_j + X_1(f_{j-1} + f_{j+1}) + X_2(f_{j-2} + f_{j+2}) + \ldots + X_n(f_{j-n} + f_{j+n}) \tag{15}$$

where F denotes either relation for radial displacement w or stresses (11) and (12)
j denotes the number of differences Δ measured from the beginning.

In this way for an arbitrary point on the circumference of the shell we obtain the expressions for radial displacement w, stress σ_x and σ_s in the following final form

$$w = F_w \frac{Q}{E' r} \left(\frac{r}{t}\right)^{5/2} \tag{16}$$

$$\sigma_x = F_x \frac{Q}{t^2} \left(\frac{t}{r}\right)^{1/2} \tag{17}$$

$$\sigma_S = F_S \frac{Q}{t^2} \left(\frac{t}{r}\right)^{1/3}. \tag{18}$$

The complete algorithm has been programmed for digital computor MINSK 22. Let us give here some results. In Fig. 2 resultant bending stress is shown for shell of rigidity $k = 0{,}001$ and the die with the $\vartheta = 5°$. The difference has been chosen 1°. In Fig. 2 on the left, the distribution of the stresses is shown under the assumed action of uniformly distributed load with the same resultant Q and the same angle $\vartheta = 5°$ as the case was with rigid die. On the right we can see the corresponding results for the same shell acted upon by the die. The stress distribution is different in principle.

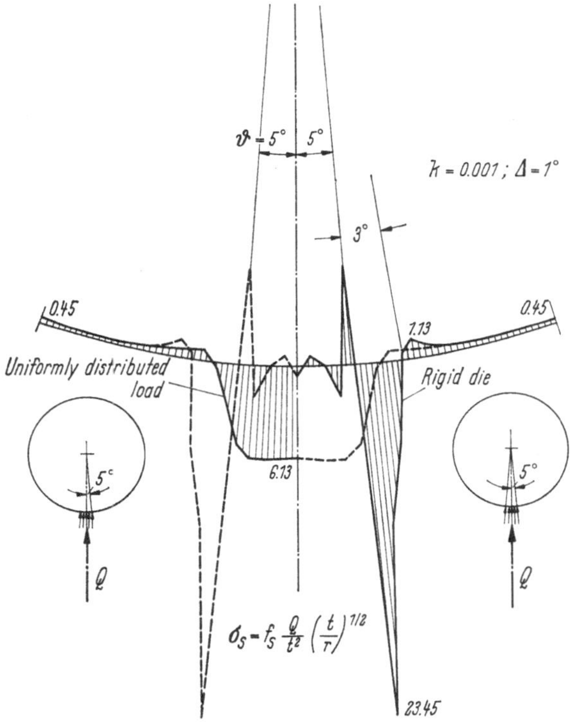

Fig. 2.

Maximum stresses due to the action of the uniformly distributed load appear approximately in the middle of the loaded area. In contradiction to that the stresses due to the action of the die have their maximum at the edge of it. The results are similar to those of Boussinesq. But to a great extent they depend on the proper rigidity of the shell k. This can be well seen in Fig. 3, where the stress distribution and reactions between the shell and the saddle support are drawn for different rigidity of the shell ($k = 0{,}001$ and $k = 0{,}01$) and $\vartheta = 45°$. They have been calculated for $\varDelta = 5°$. It is evident that for a higher rigidity k and smaller angle ϑ the wall of the shell tends to rais above the die in its middle zone.

$$X = K_x Q \qquad w = F_w \frac{Q}{E't}\left(\frac{r}{t}\right)^{5/2} \qquad \sigma_x = F_x \frac{Q}{t^2}\left(\frac{t}{r}\right)^{1/2} \qquad \sigma_s = F_s \frac{Q}{t^2}\left(\frac{t}{r}\right)^{1/2}$$

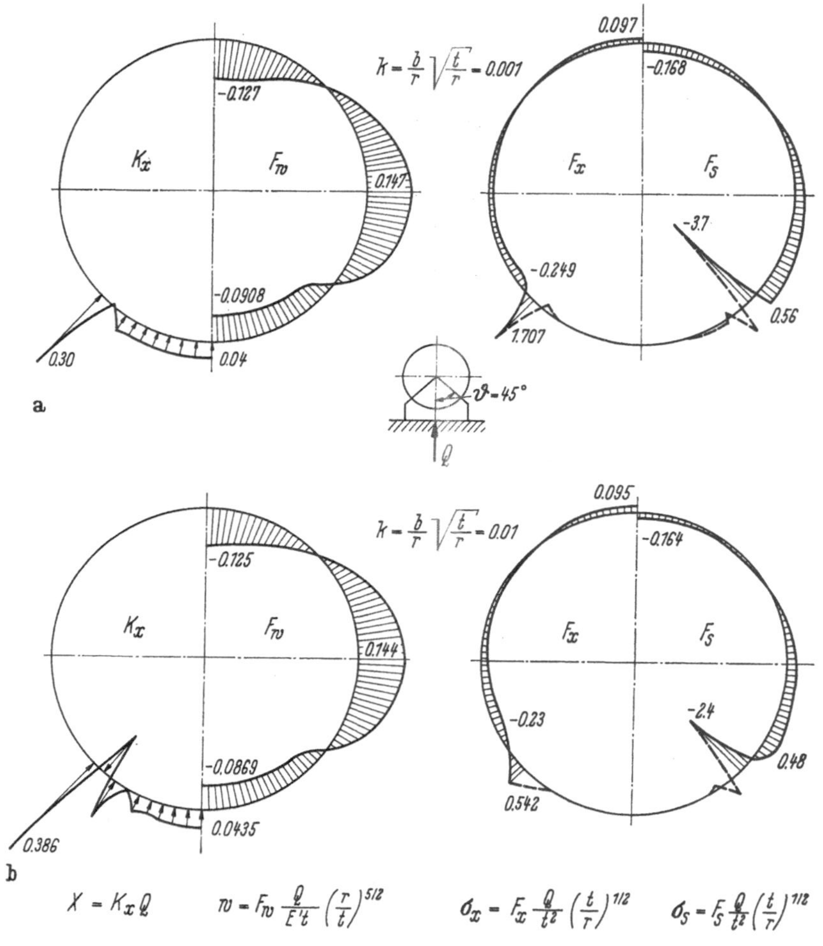

Fig. 3a and b.

The method that has been just explained enables the solution even of that problem. The programmes needed including the curves for different saddle rigidity have been already prepared.

To prevent arising of stress peaks at the edge of the die either the rigidity or the shape of it may be changed. Also this fact can be taken into account in this method. Then the Eq. (1) shall be changed; the displacement matrix of the saddle support shall be added to that of the shall.

References

1. TIMOSHENKO, S., and J. N. GOODIER: Theory of Elasticiy, New York: MacGraw-Hill 1951, p. 371.
2. KŘUPKA, V.: Statical analysis of Cylindrical Thinwalled Metal Vessels and Tubes, SNTL Praha 1957 (in czech.).
3. BIJLAARD, P. P.: Stresses from Local Loadings . . ., Trans. ASME Vol. 77, 1955, pp. 805–816.
4. DAREVSKIJ, V. M.: Contact Problems in the Theory of Shells, The VIth Symposium on Theory of Shells in Baku, 1966 (in russian), pp. 927–934.

Finite Deformation of Membranes and Shells under Localized Loading

By

W. Nachbar

University of California, San Diego, Cal., USA

1. Introduction

This paper is concerned mainly with localized loading on prestressed elastic membranes or sheets. By localized loading will be meant that the characteristic length of the surface area over which the lateral load on the membrane is acting is small in comparison with other characteristic lengths (apart from sheet thickness) of the problem. A limiting case of localized loading is the concentrated load, and one of the purposes of this study of localized loading is to examine the meaning of a concentrated load solution by looking at the limit of a solution for localized loading. In this section, some of the relevant literature on this subject is reviewed, and the attempt is made to put into proper context the results that are obtained in subsequent sections.

In their well-known paper on nonlinear elastic membrane theory [1], Bromberg and Stoker refer to the problem of an initially unstrained and flat elastic sheet, held fixed at its boundaries and stressed by an applied normal pressure, as the problem of Föppl-Hencky. The remark ([1], footnote, p. 250) that "problems for the case in which the boundary displacements . . . are not zero (i. e., the case of the initially stretched sheet) appear not to have been treated". Because such problems involve nonlinear differential equations, it is expected that solutions in general will be generated only by numerical methods or by techniques of successive approximation. The present paper will deal with a particular problem of an initially flat sheet under normally applied loads, the solution for which, even for the case of initial stretching, is expressible in a closed form involving only elementary functions.

Acknowledgement. The author is grateful to Mr. N. M. Bhatia and Mr. D. L. Mingori for their assistance in preparation of the figures and checking the algebra.

This work was performed at Stanford University with the sponsorship of the U. S. Army Research Office under Grant DA-ARO(D) 31-124-G299.

The FÖPPL-HENCKY problem for the annular membrane, fixed at the outer boundary and uniformly loaded only along the inner boundary, was treated by SCHWERIN [2]. SCHWERIN showed that, due to the absence of distributed lateral pressure on the membrane surface, the governing equations for the rotatianally symmetric state reduced to an especially simple differential equation. This equation is integrable by quadratures. A solution satisfying the given boundary conditions was then constructed by determining values for the two constants of integration. The latter were found from the roots of two transcendental equations which, in SCHWERIN's paper, were solved graphically. SCHWERIN arrived at the interesting conclusion that for $\nu > 1/3$, where ν is Poisson's ratio, real solutions of this problem are not possible under the supposition of axial symmetry. SCHWERIN suggested that this phenomenon was due to the appearance of radial wrinkles in the membrane.

In the analytical portion of their recent paper, JAHSMAN et al. [3] have followed substantially the steps in the SCHWERIN analysis in extending this analysis to the case where a fixed distributed tension, rather than a zero displacement, is imposed along the outer boundary of the membrane. JAHSMAN observed that the integration constants need not be real in order to generate real solutions; SCHWERIN assumed that they were real. JAHSMAN remarks that Schwerin's solution of the Föppl-Hencky problem can be extended to all physically meaningful values of Poisson's ratio by allowing one of the integration constants to take on purely imaginary values.

These analyses are incomplete in several respects. BOTH [2] and [3] deal with formal solutions to boundary value problems and do not investigate the question of existence of the solution for the complete, meaningful range of the various physical parameters. Furthermore, the equations with which they formulate the problem contain inherently the assumption that the angle β, which measures rotation of the tangent to the midsurface, obeys the inequality $\beta^2 \ll 1$. The general consistency of their solution with this "moderate β" assumption is not investigated. In addition, the solution in [3] is obtained only for a boundary condition prescribed at the inner radius of the membrane which assumes the existence of a limit as this radius tends toward zero: Consequently, the behavior of the membrane stresses in the vicinity of a finite inner boundary cannot then be determined from these results.

The previous work cited above is extended in several ways in the present paper. The equations for finite, rotationally symmetric deformations of an initially flat, annular membrane are deduced as a special case of the equations developed by REISSNER [4] for thin shells of revolution. These equations, as well as those used in [1—3], involve inherently the assumption of linear, isotropic elasticity of a stiff material for which the strains must be small compared to unity. However, Reissner's equations are valid for arbitrary magnitudes of β. It is shown at the conclusion of section 2 below (see Eq. (2.32)) that for general β an equation is ultimately deduced which is already relatively simple. With the introduction of the moderate β assumption, this equation simplifies still further (see Eq. (3.6)) to the form of the equation found by SCHWERIN.

In sec. 3, general properties of solutions to the Schwerin equation are investigated. It is shown that solutions which incorporate the boundary conditions of a fixed and non-negative in-plane displacement at the outer edge are so related to solutions for the boundary condition of fixed, tensile edge-force distribution at the outer edge that it is necessary to investigate in detail only the latter problem. In sec. 4, an investigation of the existence and uniqueness of the solution is made for the case of initial distributed tension of magnitude H_0 at the outer edge together with zero in-plane displacement at the inner boundary. The latter condition corresponds physically to constraining the inner boundary by attaching to it a rigid circular disk and then applying normal load to the disk. The resultant normal load is a force P directed along the axis of symmetry. For given geometry and elastic constants, the solution is found to be characterized by a single load parameter F_0 which is proportional to $H_0 P^{-2/3}$. Existence of a unique solution is found for all values of $F_o \geqq F_{oc}$, where F_{oc} is a positive number to be calculated, for all ratios of inner to outer radius between 0 and 1, and for all values of ν between 0 and 1/2.

Stability considerations limit the validity of these equilibrium solutions. Buckling of stretched membranes is called "wrinkling", a term adopted by STEIN and HEDGEPETH [5], and the criterion adopted for wrinkling is that in a membrane it is not possible to have negative (compressive) principal stresses. Hence, the vanishing of the minimum principal stress is taken to be the necessary and sufficient condition for incipient wrinkling. This condition is the basis for a theory of stress fields in the wrinkled region of flat stretched membranes that has been developed by CHEREPANOV [6] and, independently, by STEIN and HEDGEPETH [5]. In addition, an approximate theory for strains and displacements in a wrinkled region is also developed in [5]. The wrinkling condition stated above has not been demonstrated by a formal stability analysis. However, it may be noted that, for rotationally symmetric problems in an initially flat sheet, it has been shown [1] that the second variation of the potential energy is proportional to the radial component of stress, σ_r. Thus, for σ_r negative, the second variation is also negative, indicating that the equilibrium solution is not stable.

The radial stress component σ_r is shown below (sec. 3, subparagraph 3) to be always positive for initial tension in the membrane. Hence, the condition for incipient wrinkling to be used here is the vanishing of the circumferential stress component σ_θ. The same criterion was used also in [3]. The wrinkling stability criterion is determined here by a load parameter value F_{ow}, which possesses the property that at $F = F_{ow}$, $\sigma_\theta = 0$ at one boundary, while for $F > F_{ow}$, σ_θ is positive (tensile) everywhere. It is shown also that in all cases $F_{ow} > F_{oc}$.

Some interesting properties of the rigid circular disk solution are found in section 5 below for the case in which ε, the ratio of the inner to outer radius, is assumed small, and where it is assumed that incipient yielding is reached when σ_r at the inner boundary reaches the yield value σ_Y. Strains are still required to be small everywhere. Then there is found an optimum value $H_0 = H_{\text{opt}}$ for the initial tension. Only at this value of tension is the membrane able to carry all central loads P less than the ultimate value $P = P_{\text{ult}}$ without reaching the point either of incipient wrinkling or of incipient yielding. If $H \lesseqgtr H_{\text{opt}}$, the least upper bound of the values of P which do not produce either incipient yielding or incipient wrinkling will be less than P_{ult}. It is found that H_{opt} is generally small compared to the tension which would first produce yielding in the flat membrane.

In addition, for ε small, the restriction of the rigid circular disk solution to $F_o \geqq F_{ow}$, and to small strains only, is found to imply that the rotations always remain moderate. Moderate rotation is originally introduced as an independent assumption. However, if ε is close to unity, (a "thin ring" membrane), this implication is found not to hold, and the full nonlinear equations would have to be used to determine the solution for sufficiently large central load. This result demonstrates again the close relation between the assumptions of small strains and of moderate rotations, as has been found in other problems, but it also demonstrates that the two assumptions are not necessarily dependent in all problems.

A recent paper by Clark and Narayanaswamy [7] has treated the Schwerin problem of the flat annular membrane without initial tension and under a transverse load along the inner boundary. Equations valid for arbitrary β are used in their analysis, and numerical solutions are obtained using step by step integration of the nonlinear differential equation on a large digital computer. It is found that β at the loaded boundary can be of the order of 90° while all strain components are still small compared to unity.

A transverse load applied to rigid disk attached to the annular membrane represents one method of applying a localized load. The localized load could also be considered to be applied as a uniform transverse pressure over a circular area at the center of a complete membrane. There is no transverse load in the membrane between this circular area and the outer boundary. A formal series solution to the equations for moderate β was recently given for this problem by COSTELLO and STIPPES [8].

Application of a localized load as a uniform transverse pressure is difficult to reproduce experimentally, however. The experiments on indentation of stretched MYLAR membranes with circular boundaries that are reported in [3] use a rigid indenter with a hemispherical head which is pressed into the center of a prestressed sheet. The solution to such a problem can be imagined as a combination of the following two solutions: the solution to the problem of a prestressed membrane wrapped around a rigid sphere; the solution of an annular membrane problem, as is considered in the present paper, but with boundary conditions at the inner boundary such as to assure equilibrium and compatibility with the stresses and displacements of the first solution. This composite solution was carried out, using the general method described in the present paper, by BHATIA and NACHBAR [9].

A detailed numerical analysis of stresses produced by indentation, and a comparison of the prediction of theory with experiment, showed that stresses in the membrane under the indenter reached the yield surface with quite small values of transverse load on the indenter. The indenter problem described above has also been analyzed by BHATIA and NACHBAR [10] under the assumption that the membrane material is elastic-perectly-plastic (ELPP) and obeys the Tresca yield condition with the associated glow law. It is found that the membrane can sustain very much greater transverse loads with a contained plastic regime around the indenter. Agreement between the predictions of this ELPP solution for stresses and deflections and the data from the experiments [3] is good. Even the burst or penetration load is predicted well on the basis of an inelastic instability.

There are almost no analytic solutions to finite indentation problems for curved membranes and shells. The equations that describe such problems are not difficult to write down from existing literature, and so the solutions for specific cases would be obtainable at least numerically. An exception is the recent note by ROSETTOS [11] in which asymptotic methods are used to analyze the finite indentation of a pressurized spherical membrane.

2. Equations for Finite, Axisymmetrical Displacements of a Flat Sheet

Equations are now developed for the deformation of an initiallyflat membrane which is free from surface traction and is loaded uniformly along its edges. In the unstrained state, the membrane is in the form of an annulus of inner radius b and outer radius a. The geometry and nomenclature are indicated in Fig. 2.1. The radial coordinate is r; displacements normal and parallel to the initial plane are w and u, respectively; circumferential and radial stress components are σ_θ and σ_r, respectively; circumferential and radial strain components are ε_θ and ε_r, respectively. The angle of rotation of the tangent to the midsurface from the initial plane is β. The membrane has uniform thickness h and elastic moduli E, ν. Nondimensional stress variables s_r and s_θ being defined as

$$s_r = \frac{\sigma_r}{E} \quad \text{and} \quad s_\theta = \frac{\sigma_\theta}{E}, \tag{2.1}$$

the governing equations are [4] those of equilibrium,

$$\frac{d}{dr}(r s_r \cos\beta) = s_\theta \tag{2.2}$$

$$\frac{d}{dr}(r s_r \sin\beta) = 0 \tag{2.3}$$

compatibility,

$$\frac{d}{dr}(r\varepsilon_\theta) - \varepsilon_r + (1 + \varepsilon_r)(1 - \cos\beta) = 0 \tag{2.4}$$

elasticity ($0 \leqq \nu \leqq 1/2$),

$$\varepsilon_r = s_r - \nu s_\theta \tag{2.5}$$

$$\varepsilon_\theta = s_\theta - \nu s_r \tag{2.6}$$

and strain-displacement,

$$\varepsilon_\theta = \frac{u}{r} \tag{2.7}$$

$$\frac{dw}{dr} = \sin\beta. \tag{2.8}$$

In Eqs. (2.2), (2.3), (2.4) and (2.8), the small strain assumption

$$|\varepsilon_\theta| \ll 1, \qquad |\varepsilon_r| \ll 1 \tag{2.9}$$

has been consistently applied. The validity of this assymption as it has been applied to the derivative terms in Eqs. (2.2) and (2.3) is discussed in Appendix A.

The outer edge of the membrane can be assigned through a rigid body displacement to remain always in the initial plane. The resultant P of the normal components of the distributed loading along the deformed inner edge must be in equilibrium there with the normal component of the membrane stress resultants (see Fig. 2.1(c)). We will consider $P > 0$, with $P = 0$ as limiting case. These two boundary conditions are written as

$$w(a) = 0 \tag{2.10}$$

$$P = 2\pi b E h (s_r \sin\beta)_{r=b}. \tag{2.11}$$

We write b rather than $b + u(b)$ in Eq. (2.11) to be consistent with (2.9).

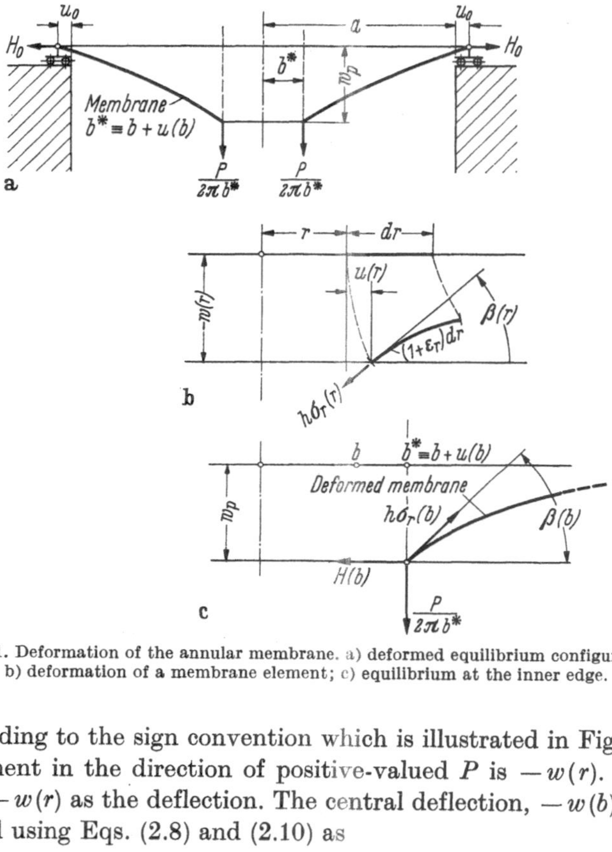

Fig. 2.1. Deformation of the annular membrane. a) deformed equilibrium configuration; b) deformation of a membrane element; c) equilibrium at the inner edge.

According to the sign convention which is illustrated in Fig. 2.1, the displacement in the direction of positive-valued P is $-w(r)$. We shall refer to $-w(r)$ as the deflection. The central deflection, $-w(b) \equiv w_p$, is expressed using Eqs. (2.8) and (2.10) as

$$w_p = \int_b^a \sin\beta \, dr \tag{2.12}$$

and the deflection at any point is given by

$$-w(r) = \int_r^a \sin\beta \, dr. \tag{2.13}$$

Two nondimensional parameters ε and ϱ, and a new independent variable y, are now introduced:

$$\varepsilon \equiv b/a \qquad 0 < \varepsilon < 1 \tag{2.14}$$

$$\varrho \equiv \frac{P}{2\pi E h b} \qquad \varrho > 0 \tag{2.15}$$

$$y \equiv (r/a)^2 \qquad \varepsilon^2 \leqq y \leqq 1\,. \tag{2.16}$$

Henceforth, functional dependence on y rather than r will be understood unless explicitly noted to the contrary.

With substitution from Eq. (2.16), the differential operator in Eqs. (2.2), (2.3) and (2.4) transforms as

$$\frac{d}{dr}(rz) = 2y^{1/2}\frac{d}{dy}(y^{1/2}z) \tag{2.17}$$

where z is any function. The integral of Eq. (2.3) which satisfies the boundary condition, Eq. (2.11), is then expressed as

$$y^{1/2}s_r \sin\beta = \text{constant} = \varepsilon\varrho \qquad \varepsilon^2 \leqq y \leqq 1\,. \tag{2.18}$$

It follows from Eqs. (2.5) and (2.6) that solutions to Eqs. (2.2), (2.3) and (2.4) can be a priori consistent with the small strain assumption (2.9) only if the conditions

$$|s_r| \ll 1 \quad \text{and} \quad |s_\theta| \ll 1 \tag{2.19}$$

are satisfied. Now Eq. (2.18) at $y = \varepsilon^2$ gives the condition $\varrho \leqq |s_r(\varepsilon^2)|$. In view of (2.19), therefore, consistent solutions will be expected from the small-strain formulation of this problem only if

$$\varrho \ll 1\,. \tag{2.20}$$

It is also seen from Eq. (2.18) that $\sin\beta$ can have no zero in $\varepsilon^2 \leqq y \leqq 1$.

Equations (2.2), (2.17) and (2.18) give

$$s_\theta = 2y^{1/2}\frac{d}{dy}(y^{1/2}s_r\cos\beta) = 2\varepsilon\varrho y^{1/2}\frac{d}{dy}(\operatorname{ctn}\beta) \tag{2.21}$$

and Eq. (2.18) can be written as

$$s_r = \varepsilon\varrho y^{1/2}\csc\beta \tag{2.22}$$

If the functions $f(y)$ and $\chi(y)$ are defined as

$$f = y^{1/2}\operatorname{ctn}\beta \tag{2.23}$$

$$\chi = \sec\beta = 1 \tag{2.24}$$

then Eq. (2.21) becomes

$$s_\theta = \varepsilon \varrho \left[2 \frac{df}{dy} - \frac{f}{y} \right] \tag{2.25}$$

and Eq. (2.22) becomes

$$s_r = \varepsilon \varrho y^{1/2} \csc \beta = \varepsilon \varrho \frac{f}{y} (1 + \chi). \tag{2.26}$$

Evident relations between f and χ are derivable from Eqs. (2.23) and (2.24), viz.

$$f^2 = \frac{y}{\chi(2+\chi)} \quad \text{and} \quad 1 - \cos\beta = \frac{\chi}{1+\chi}. \tag{2.27a, b}$$

Thus, the equations of equilibrium are satisfied by all differentiable functions f and χ which satisfy Eq. (2.27a).

A differential equation relating f and χ is now obtained from the compatibility equation, Eq. (2.4). As a preliminary step, multiply both sides of Eq. (2.2) by $\cos\beta$, and both sides of Eq. (2.3) by $\sin\beta$, and add these two equations. The sum can be expressed as

$$\frac{d}{dr}(r s_r) = s_\theta \cos\beta. \tag{2.28}$$

If Eq. (2.4), together with Eqs. (2.5) and (2.6), is written as

$$\frac{d}{dr}[r(s_\theta + s_r) - (1+\nu) r s_r] - (s_r - \nu s_\theta) + (1 + s_r - \nu s_\theta)(1 - \cos\beta) = 0$$

then with the use of Eq. (2.28), this equation can also be expressed as

$$r \frac{d}{dr}(s_\theta + s_r) + (1+\nu) s_\theta (1 - \cos\beta) + (1 + s_r - \nu s_\theta)(1 - \cos\beta) = 0.$$

Upon collection of terms and use of Eq. (2.17), this equation takes the form

$$2y \frac{d}{dy}(s_\theta + s_r) + (1 + s_\theta + s_r)(1 - \cos\beta) = 0. \tag{2.29}$$

Now Eqs. (2.25) and (2.26) give

$$s_\theta + s_r = \varepsilon \varrho \left(2 \frac{df}{dy} + \frac{f\chi}{y} \right). \tag{2.30}$$

This relation and Eq. (2.24) are substituted into Eq. (2.29), and then Eq. (2.29) is rearranged into the following form:

$$4 \varepsilon \varrho y \frac{d^2 f}{dy^2} + \frac{\chi}{1+\chi} = - \left[2 \varepsilon \varrho y \frac{d}{dy}\left(\frac{f\chi}{y}\right) + (s_\theta + s_r) \frac{\chi}{1+\chi} \right]. \tag{2.31}$$

Consider now the terms within the square bracket on the right-hand side of Eq. (2.31). Expansion of the derivative gives

$$2\varepsilon\varrho y \frac{d}{dy}\left(\frac{f\chi}{y}\right) = -2\varepsilon\varrho\frac{f\chi}{y} + 2\varepsilon\varrho\chi\frac{df}{dy} + 2\varepsilon\varrho f\frac{d\chi}{dy}. \tag{2.32}$$

When the logarithmic derivative of both sides of Eq. (2.27a) is taken, there is obtained

$$\frac{2}{f}\frac{df}{dy} = \frac{1}{y} - \frac{2(1+\chi)}{\chi(2+\chi)}\frac{d\chi}{dy},$$

and so the derivative of χ is found to the expressible as

$$2f\frac{d\chi}{dy} = \frac{\chi}{1+\chi}\cdot(2+\chi)\cdot\left(\frac{f}{y} - 2\frac{df}{dy}\right), \tag{2.33}$$

and, with the use of Eq. (2.25), as

$$2\varepsilon\varrho f\frac{d\chi}{dy} = -\frac{\chi(2+\chi)}{1+\chi}s_\theta.$$

The sum of terms within the square bracket of Eq. (2.31) can the be expressed as

$$[\cdot] = \frac{\chi}{1+\chi}\left[-2\varepsilon\varrho\frac{f(1+\chi)}{y} + 2\varepsilon\varrho(1+\chi)\frac{df}{dy} + s_r - (1+\chi)s_\theta\right],$$

and it is readily seen, by using Eqs. (2.25) and (2.26), that this sum is identically zero.

The differential equation which, together with Eq. (2.27a), determines the solution to this problem is therefore from Eq. (2.31) as

$$\frac{d^2f}{dy^2} + (4\varepsilon\varrho y)^{-1}\frac{\chi}{1+\chi} = 0 \qquad \varepsilon^2 < y < 1 \tag{2.34}$$

Equations (2.27a) and (2.34) are valid for arbitrarily large displacements (e. g., $0 \leqq |\chi| < \infty$) so long as the strains are small.

Two boundary conditions are generally required to specify a unique integral $f(y)$ of Eq. (2.34). These will take the form of edge conditions involving u, β and H. H is the horizontal component of the stress resultant,

$$\frac{H}{Eh} = \frac{\sigma_r h}{Eh}\cos\beta \equiv s_r\cos\beta = \varepsilon\varrho\frac{f}{y}. \tag{2.35}$$

The displacement u is expressed by means of Eqs. (2.7), (2.6), and (2.25) and (2.26) as follows:

$$\frac{u}{a} = y^{1/2}\,\varepsilon_\theta = y^{1/2}\,(s_\theta - \nu s_r)$$
$$= \varepsilon\varrho\left[2y^{1/2}\frac{df}{dy} - y^{1/2} f(1+\nu+\nu\chi)\right]. \tag{2.36}$$

Since Eqs. (2.18) and (2.26) give

$$\sin\beta = \varepsilon\varrho\,(y^{1/2}s_r)^{-1} = y^{1/2}\,|f(1+\chi)|^{-1}, \tag{2.37}$$

then it follows from Eq. (2.12) and (2.13) that the deflection is given by:

$$-\frac{w(y)}{a} = \int_y^1 \frac{\sin\beta}{2\eta^{1/2}}\,d\eta = \frac{1}{2}\int_y^1 \frac{d\eta}{f(1+\chi)}; \tag{2.38}$$

$$w_p = \frac{1}{2}\int_{\varepsilon^2}^1 \frac{d\eta}{f(1+\chi)}. \tag{2.39}$$

3. Problem for Moderate β

If for $\varrho > 0$ the function $F(y)$ is defined as

$$F = 2^{2/3}\,(\varepsilon\varrho)^{1/3} f \tag{3.1}$$

then Eqs. (2.32) and (2.30) take the following form

$$\frac{d^2F}{dy^2} + [(4\varepsilon\varrho)^{2/3}y]^{-1}\frac{\chi}{1+\chi} = 0 \qquad \varepsilon^2 < y < 1 \tag{3.2}$$

$$F = (4\varepsilon\varrho)^{1/3}y^{1/2}\chi^{-1/2}\,(2+\chi)^{-1/2}. \tag{3.3}$$

We now impose the additional assumption of moderate but finite β by assuming that

$$\chi(y) \ll 1 \qquad \varepsilon^2 \leqq y \leqq 1. \tag{3.4}$$

At the conclusion of Section 5 it will be shown that, for a particular class of problems, the solutions to the simplified equations obtained through imposition of (3.4) are consistent with the requirement of (3.4). With this assumption, Eqs. (3.3) and (3.2) become

$$F^2 = (4\varepsilon\varrho)^{2/3}\,y\,(2\chi)^{-1} \tag{3.5}$$

$$\frac{d^2F}{dy^2} + \frac{1}{2F^2} = 0 \qquad \varepsilon^2 < y < 1. \tag{3.6}$$

Similarly, Eqs. (2.24), (2.25), (2.26), (2.33), (2.35), and (2.23) take the following forms under assumption (3.4)

$$\frac{u}{a} = \left(\frac{1}{2}\varepsilon\varrho\right)^{2/3} y^{1/2}\left[2\frac{dF}{dy} - (1+\nu)\frac{F}{y}\right], \tag{3.7}$$

$$s_\theta = \left(\frac{1}{2}\varepsilon\varrho\right)^{2/3}\left(2\frac{dF}{dy} - \frac{F}{y}\right), \tag{3.8}$$

$$s_r = \left(\frac{1}{2}\varepsilon\varrho\right)^{2/3}\frac{F}{y}, \tag{3.9}$$

$$\frac{H}{Eh} = \left(\frac{1}{2}\varepsilon\varrho\right)^{2/3}\frac{F}{y}, \tag{3.10}$$

$$\frac{w}{a} = -\left(\frac{1}{2}\varepsilon\varrho\right)^{1/3}\int_y^1 \frac{dy}{F}, \tag{3.11}$$

$$\frac{w_p}{a} = -\frac{w(\varepsilon^2)}{a} = \left(\frac{1}{2}\varepsilon\varrho\right)^{1/3}\int_{\varepsilon^2}^1 \frac{dy}{F}, \tag{3.12}$$

$$\beta = (2\chi)^{1/2} = (4\varepsilon\varrho)^{1/3} y^{1/2} F^{-1}. \tag{3.13}$$

Equation (3.6) requires that d^2F/dy^2 exist everywhere in $\varepsilon^2 < y < 1$. Hence, if $F(y)$ is a solution of Eq. (3.6), then $F(y)$, $\frac{dF}{dy}$ and $\frac{d^2F}{dy^2}$ are continuous in $\varepsilon^2 < y < 1$, and $F(y)$ can have no zero in $\varepsilon^2 < y < 1$. Physical conditions and consistency with previously introduced assumptions imply the following further restrictions on $F(y)$ in order that $E(y)$ represent the solution to a membrane problem.

1. The inequality (3.4), which is implied in Eq. (3.6), can be expressed by use of Eq. (3.5) as

$$\chi(y) = \left(\frac{1}{2}\varepsilon\varrho\right)^{2/3} 2y[F(y)]^{-2} \ll 1 \qquad \varepsilon^2 \leqq y \leqq 1. \tag{3.14}$$

It then follows from (3.14) that $|F(y)|$ is bounded away from zero on $\varepsilon^2 \leqq y \leqq 1$.

2. Since initial tension is applied either by positive edge forces or edge displacement, then Eq. (3.10) shows that $F(y)$ must be positive in the neighborhood of $y = 1$ at least for sufficiently small $\varrho > 0$. A physically acceptable $F(y)$ must be continuous in ϱ and thus positive for all y and ϱ.

3. The reduced radial stress s_r is positive everywhere, since $F(y)$ is positive. Moreover, by (2.18) s_r must have an upper bound on $\varepsilon^2 \leqq y \leqq 1$,

and it follows then that for every physically acceptable $F(y)$ there must exist two positive numbers M_1 and M_2, $M_2 > M_1$, such that

$$0 < M_1 \leqq F(y) \leqq M_2 < \infty \qquad \varepsilon^2 \leqq y \leqq y. \tag{3.15}$$

4. Application of (3.15) to Eq. (3.6) shows that dF/dy must be monotone strictly decreasing, and therefore dF/dy can have at most one zero on $\varepsilon^2 < y < 1$. This conclusion implies that if $dF/dy \geqq 0$ at $y = 1$, then $dF/dy > 0$ for $\varepsilon^2 < y < 1$, while if $dF/dy < 0$ at $y = 1$, then $dF/dy < 0$ in some left neighborhood of $y = 1$.

We observe at this point that the foregoing conditions do not relate to the stability of the rotationally symmetric equilibrium determined by $F(y)$. The wrinkling stability criterion stated in the introduction reduces now to the statement

$$s_\theta \geqq 0 \qquad \varepsilon^2 \leqq y \leqq 1. \tag{3.16}$$

Since the sum of the principal stresses can be expressed by Eqs. (3.8) and (3.9) as

$$s_\theta + s_r = 2\left(\frac{1}{2}\varepsilon\varrho\right)^{2/3}\frac{dF}{dy}, \tag{3.17}$$

it then follows that the equilibrium is unstable unless $dF/dy \geqq 0$ for $\varepsilon^2 \leqq y \leqq 1$. This condition does not necessarily imply (3.16), however, since s_r is positive.

The solutions which correspond to stable equilibrium according to criterion (3.16) are to be called stable solutions. Stable solutions are included in the class of allowable solutions which is now defined.

Definition: An allowable solution $F(y)$ to Eq. (3.6) is positive and bounded (c. f. (3.15)) for all y on $\varepsilon^2 \leqq y \leqq 1$, and at the right-hand limit obeys the condition

$$dF/dy \geqq 0 \qquad \text{at } y = 1. \tag{3.18}$$

Henceforth it will be understood that $F(y)$ refers to an allowable solutions unless stated to the contrary. From the preceding discussion, it follows that $F(y)$ is monotone strictly increasing on $\varepsilon^2 \leqq y \leqq 1$.

It follows from Eq. (3.13) that for an allowable solution and $\varrho > 0$, $\beta(y) > 0$ on $\varepsilon^2 \leqq y \leqq 1$. Furthermore, with differentiation of both sides of Eq. (3.13) and with the use of Eq. (3.8) we may write

$$\frac{d\beta}{dy} = (4\varepsilon\varrho)^{1/3}\frac{y^{1/2}}{2F^2}\left(\frac{F}{y} - 2\frac{dF}{dy}\right) = -\left(\frac{1}{2}\varepsilon\varrho\right)^{-1/3}\frac{y^{1/2}}{F^2}s_\theta. \tag{3.19}$$

Equation (3.19) shows that $\beta(y)$ is monotone strictly decreasing for a stable solution.

Monotone behavior in y cannot be deduced in general for $s_\theta(y)$ and $s_r(y)$. However, with the use of Eqs. (3.6) and (3.17) it can be shown that $(s_\theta + s_r)$ is monotone strictly decreasing.

Equation (3.6) may be solved by quadratures. Noting that dF/dy is positive in $\varepsilon^2 \leqq y < 1$ (paragraph 4 above), we write

$$\frac{dF}{dy}\frac{d^2F}{dy^2} + \frac{1}{2F^2}\frac{dF}{dy} = \frac{d}{dy}\left[\left(\frac{dF}{dy}\right)^2\right] - \frac{d}{dy}\left(\frac{1}{F}\right) = 0.$$

Hence

$$\left(\frac{dF}{dy}\right)^2 - \frac{1}{F} = C_0$$

where C_0 is a real constant, and so

$$\frac{dF}{dy} = (F^{-1} + C_0)^{1/2} \qquad \varepsilon^2 < y < 1. \tag{3.20}$$

The quantity within the brackets on the right-hand side of Eq. (3.20) is always positive if $C_0 \geqq 0$. If $C_0 < 0$, on the other hand, then inequality (3.18) implies a positive lower bound for $F(1)$:

$$\frac{1}{F(1)} + C_0 \geqq 0. \tag{3.21}$$

Using Eqs. (3.8) and (3.20), we see that the wrinkling stability condition (3.16) can be written as

$$\frac{1}{F} + C_0 \geqq \frac{F^2}{4y^2} \qquad \varepsilon^2 \leqq y \leqq 1 \tag{3.22}$$

which in comparison to (3.21) again shows that stable solutions are necessarily allowable solutions.

With the use of Eqs. (3.6) and (3.20), additional relations are found for $w(y)$ and w_p which avoid the necessity of performing the integration indicated in Eqs. (3.11) and (3.12). We may write

$$\begin{aligned}\int_y^1 \frac{dy}{F} &= -2\int_y^1 F\frac{d^2F}{dy^2}\,dy\\ &= -\left(2F\frac{dF}{dy}\right)\Big|_y^1 + 2\int_y^1\left(\frac{dF}{dy}\right)^2 dy\\ &= -\left(2F\frac{dF}{dy}\right)\Big|_y^1 + 2\int_y^1\frac{dy}{F} + 2C_0(1-y)\end{aligned}$$

and therefore

$$\int_y^1 \frac{dy}{F} = 2\left\{F_0\left(\frac{1}{F_0} + C_0\right)^{1/2} - F(y)\left[\frac{1}{F(y)} + C_0\right]^{1/2} - C_0(1-y)\right\}. \quad (3.23)$$

By the use of Eq. (3.23) we are able to express $w(y)$ and w_p directly in terms of the values of $F(y)$.

An equation for determining $F(y)$ is obtained by integration of Eq. (3.20):

$$y = \int_{F(C_1)}^{F(y)} \left(\frac{1}{v} + C_0\right)^{-1/2} dv + C_1. \quad (3.24)$$

The constants C_0 and C_1 are to be determined from additional boundary conditions. Up to the present point in the discussion, boundary conditions other than Eqs. (2.9) and (2.10) have not been specified (apart from the condition that prestresses must produce tension in the initially flat membrane). Hence the foregoing results regarding solutions of Eq. (3.6), and in particular regarding allowable solutions, are independent of boundary conditions.

The boundary conditions with which we will be principally concerned for the remainder of this paper are

$$u(\varepsilon^2) = 0 \quad (3.25\text{a})$$

$$F(1) = F_0 \quad (3.25\text{b})$$

where

$$F_0 \equiv \left(\frac{1}{2}\varepsilon\varrho\right)^{-2/3} \frac{H_0}{Eh} \equiv \frac{H_0}{Eh}\left(\frac{4\pi E h a}{P}\right)^{2/3}. \quad (3.26)$$

When an edge force distribution $H(1) = H_0$, $H_0 > 0$, is applied to the flat membrane and is held constant as the normal load resultant P is changed, then F_0 varies as $P^{-2/3}$. As is described in the introduction above, the condition $u(\varepsilon^2) = 0$ is equivalent to applying the normal load to a rigid central disk of radius $r = b$ (see Figs. 1a, c). By use of Eqs. (3.7) and (3.20), Eq. (3.25a) is expressed as

$$\varepsilon^2 (F^{-1} + C_0)^{1/2} = \frac{1}{2}(1+\nu)F; \qquad y = \varepsilon^2. \quad (3.27)$$

A solution satisfying Eqs. (3.24) and (3.25) can exist for some triad of values of C_0, ε^2 and ν only if Eq. (3.27) has a positive root. A positive root of Eq. (3.27) is called $g(\ )$, and condition (3.25a) is equivalent to

$$F(\varepsilon^2] = g(C_0, \varepsilon^2, \nu). \quad (3.28)$$

Hence, Eq. (3.24) becomes

$$1 - y = \int_{F(y)}^{F(1)} \left(\frac{1}{v} + C_0\right)^{-1/2} dv, \tag{3.29}$$

and the equation

$$1 - \varepsilon^2 = \int_{g(C_0,\varepsilon^2,\nu)}^{F(1)} \left(\frac{1}{v} + C_0\right)^{-1/2} dv \tag{3.30}$$

serves to determine the single remaining integration constant C_0. It is evident that Eq. (3.30) determines a functional relation between C_0, $F(1)$, ε^2 and ν. The following section will be devoted to establishing the existence of a unique $F(y)$ satisfying Eqs. (3.25) for given F_0, ε^2 and ν.

The boundary condition which replaces Eq. (3.25b) for a fixed displacement of the outer edge can be stated as

$$u(1) = u_0, \qquad u_0 > 0 \tag{3.31}$$

or, using Eqs. (3.7) and (3.20), as

$$(F^{-1} + C_0)^{1/2} - \frac{1}{2}(1+\nu)\,F = \left(\frac{1}{2}\varepsilon\varrho\right)^{-2/3} \frac{u_0}{2a} \equiv \mu; \qquad y = 1. \tag{3.32}$$

A positive root of Eq. (3.32), if one exists, is a function of C_0, ν and μ, and it will be called $G_1(\ )$. Thus Eq. (3.25b) may be replaced by the condition

$$F(1) = G_1(C_0, \mu, \nu); \qquad \mu > 0. \tag{3.33}$$

Since boundary condition (3.25a) must also be satisfied, the functional relation involving C_0, $F(1)$, ε^2 and ν is again determined by Eq. (3.30) just as it was for the fixed edge force condition. Suppose that for the fixed-edge-force problem we have determined this relation, and suppose that it can be represented explicitly through a function $G_2(\ )$:

$$C_0 = G_2\big(F(1), \varepsilon^2, \nu\big). \tag{3.34}$$

Then by the simultaneous solution of Eqs. (3.33) and (3.34) we are able to determine the functions $J_1(\ )$ and $J_2(\ )$ such that

$$F(1) = J_1(\mu, \varepsilon^2, \nu), \tag{3.35}$$

$$C_0 = J_2(\mu, \varepsilon^2, \nu). \tag{3.36}$$

The solution $F(y)$ under boundary conditions (3.25a) and (3.31) will therefore depend parametrically, through Eqs. (3.25) and (3.36), upon μ, ε^2 and ν only. Hence, for the fixed-edge-displacement problem, μ is a measure of the central load P, as was F_0 in the fixed-edge-force problem, and μ varies also as $P^{-2/3}$. Finally the problem of determining $F(y)$ for fixed μ, ε^2, ν can be regarded as a particular fixed-edge-force problem in which

$$F_0 = J_1(\mu, \varepsilon^2, \nu) = \text{constant}. \tag{3.37}$$

4. Existence of Allowable Solutions for Fixed Edge Forces

The proof of the following theorem is easily obtained using the results of sec. 3.

Theorem 4.1. Given ε^2 in $0 < \varepsilon^2 < 1$, ν in $0 \leqq \nu \leqq 1/2$ and F_0 in $0 < F_0 < \infty$, if there exist real numbers C_0 and g which satisfy the three conditions

$$1 - \varepsilon^2 = \int_g^{F_0} \left(\frac{1}{v} + C_0\right)^{-1/2} dv, \tag{4.1}$$

$$\varepsilon^2 \left(\frac{1}{g} + C_0\right)^{1/2} = \frac{1}{2}(1 + \nu)g, \tag{4.2}$$

$$g > 0 \tag{4.3}$$

then $F(y)$ defined by Eq. (3.29) is an allowable solution to Eq. (3.6) and satisfies the boundary conditions of Eqs. (3.25).

In this section we will be concerned with establishing the existence of C_0 and g. The values for ε^2 and ν play no role in the work of this section, and therefore these parameters are considered as having fixed values. They will not be identified in the functional relations. Thus the positive root of Eq. (4.2) is called $g(C_0)$, and the function to be established by Eq. (4.1) is called $C_0(F_0)$.

Let Eq. (4.2) be written as

$$g^3 - AC_0 g - A = 0 \tag{4.4}$$

where A is defined as

$$A \equiv \left(\frac{2\varepsilon^2}{1 + \nu}\right)^2. \tag{4.5}$$

It follows from Descartes rule of signs [5] that there is only one positive root g of Eq. (4.4) for any real C_0. Further, it is readily shown by using Eq. (4.4) that the derivative of the root g with respect to C_0 is always positive, viz:

$$\frac{dg}{dC_0} = \frac{Ag}{3g^2 - AC_0} = \frac{Ag^2}{2g^3 + A}. \tag{4.6}$$

Thus, $g(C_0)$ is a continuous, monotone, strictly increasing function of C_0 for $-\infty < C_0 < \infty$, and is follows from Eqs. (4.4) to (4.6) that (for a fixed value of A):

$$g(C_0) \sim (-C_0)^{-1} \quad \text{as} \quad C_0 \to -\infty \tag{4.7a}$$

$$g(C_0) = A^{1/3} \quad\quad C_0 = 0 \tag{4.7b}$$

$$g(C_0) \sim A^{1/2} C_0^{1/2} \left[1 + \frac{1}{2} A^{-1/2} C_0^{-3/2}\right] \quad \text{as} \quad C_0 \to \infty. \tag{4.7c}$$

Explicit expressions for $g(C_0)$ are obtained from the well-known formulas for the roots of a cubic equation, as follows:

If $C_0 \leqq 0$, let $\bar{C}_0 = -C_0$, $g(C_0) \equiv \bar{g}(\bar{C}_0)$:

$$\bar{g}(\bar{C}_0) = \left(\frac{A}{2}\right)^{1/3} \left\{\left[1 + \sqrt{1 + \left(\frac{A}{2}\right)\left(\frac{2\bar{C}_0}{3}\right)^3}\,\right]^{1/3} - \left[\sqrt{1 + \left(\frac{A}{2}\right)\left(\frac{2\bar{C}_0}{3}\right)^3} - 1\right]^{1/3}\right\}. \tag{4.8a}$$

If $0 \leqq C_0 \leqq \frac{3}{2}\left(\frac{A}{2}\right)^{-1/3}$:

$$g(C_0) = \left(\frac{A}{2}\right)^{1/3} \left\{\left[1 + \sqrt{1 - \left(\frac{A}{2}\right)\left(\frac{2C_0}{3}\right)^3}\,\right]^{1/3} + \left[1 - \sqrt{1 - \left(\frac{A}{2}\right)\left(\frac{2C_0}{3}\right)^3}\,\right]^{1/3}\right\}. \tag{4.8b}$$

If $C_0 \geqq \frac{3}{2}\left(\frac{A}{2}\right)^{-1/3}$:

$$g(C_0) = 2\sqrt{\left(\frac{A}{2}\right)\left(\frac{2C_0}{3}\right)} \cos\left\{\frac{1}{3} \cos^{-1}\left[\left(\frac{A}{2}\right)^{-1/2}\left(\frac{2C_0}{3}\right)^{-3/2}\right]\right\}. \tag{4.8c}$$

The principal value, $\cos^{-1}(0) = 1$, is implied in Eq. (4.8c), so that when $C_0 = \frac{3}{2}\left(\frac{A}{2}\right)^{-1/3}$, $g = 2^{2/3} X^{1/3}$.

Let $I(F_0, C_0)$ be defined as

$$I(F_0, C_0) = \int_{g(C_0)}^{F_0} \left(\frac{1}{v} + C_0\right)^{-1/2} dv. \tag{4.9}$$

Then Eq. (4.1) is written as

$$I(F_0, C_0) = 1 - \varepsilon^2. \tag{4.10}$$

Attention may evidently be restricted to those values of F_0 and C_0 for which $I(F_0, C_0)$ is real. Since Eq. (4.4) can be written as

$$C_0 g = X^{-1} g^3 - 1 \tag{4.11}$$

then

$$C_0 g > -1 \tag{4.12}$$

for all C_0 in $-\infty < C_0 < \infty$. It follows then that the condition (cf. (3.21))

$$F_0^{-1} + C_0 > 0 \tag{4.13}$$

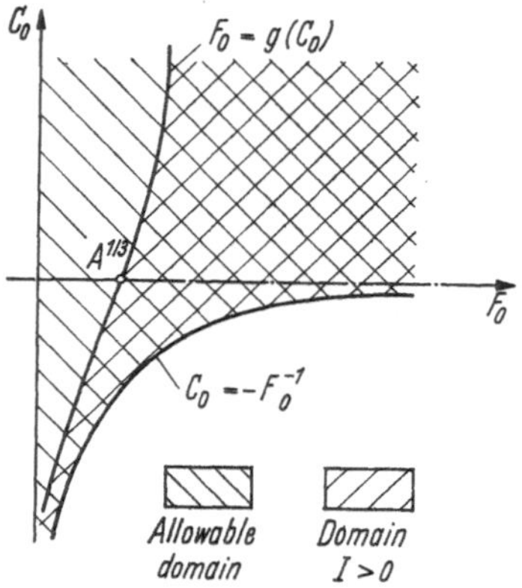

Fig. 4.1. Domain of definition of $I(F_0, C_0)$.

is sufficient to insure that $I(F_0, C_0)$ is a real-valued, proper integral on the points of the half plane $F_0 > 0$ which lie above the curve $C_0 = -F_0^{-1}$ (Fig. 4.1). These points will be called allowable points.

If there exists a root F_0, C_0 of Eq. (4.10), then $I(F_0, C_0)$ is positive. For $I > 0$ it is necessary and sufficient that $F_0 > g(C_0)$, and, therefore, the points on which I is positive are the allowable points lying to the right of the curve $F_0 = g(C_0)$ in the F_0, C_0 plane, as is shown in Fig. 4.1. For any fixed C_0, there exists a range of values of F_0 on which $I > 0$. If $C_0 \geqq 0$, this range is just $F_0 > g(C_0)$, but if $C_0 < 0$, the range for F_0 is finite:

$$g(C_0) < F_0 \leqq -C_0^{-1}. \tag{4.14}$$

If the magnitude of the range of F_0 for $C_0 < 0$ is called R_0, then

$$R_0 = (-C_0)^{-1}(1 + C_0 g).$$

Inequality (4.12) implies $R_0 > 0$ for finite C_0, and (4.7a) implies $R_0 \to 0$ as $C_0 \to -\infty$.

In the interior of the allowable domain $I(F_0, C_0)$ is continuously differentiable, and its differential is

$$dI = \frac{\partial I}{\partial F_0} dF_0 + \frac{\partial I}{\partial C_0} dC_0 = \left(\frac{1}{F_0} + C_0\right)^{-1/2} dF_0 -$$
$$- \left[\frac{1}{2}\int\limits_g^{F_0} \left(\frac{1}{v} + C_0\right)^{-3/2} dv + \left(\frac{1}{g} + C_0\right)^{-1/2} \frac{dg}{dC_0}\right] dC_0. \tag{4.15}$$

It follows from inequality (4.13) that $\delta I/\delta F_0$ is positive. Furthermore, if I is positive, then $\delta I/\delta C_0$ is negative. It may be concluded, therefore, that in the allowable domain I is monotone strictly increasing with F_0 for fixed C_0. In the smaller domain in which I is positive, I is also monotone strictly decreasing with C_0 for fixed F_0.

It will now be shown that for any fixed value $C_0 = C_{01}$, $C_{01} \geqq 0$, Eq. (4.10) has a unique root $F_0 = F_{01}$. It follows from Eq. (4.9) that

$$\lim_{F_0 \to \infty} I(F_0, C_{01}) = \infty$$

and also that $I(g(C_{01})C_{01}) = 0$. Since $I(F_0, C_{01})$ is a continuous, monotone strictly increasing function of F_0, then the existence of F_{01} immediately follows. By the implicit function theorem, the points C_{01}, F_{01} for $C_{01} \geqq 0$ may be shown to form a continuous curve lying entirely within the $I > 0$ domain. This curve may be described by a continuous, monotone strictly increasing function $C_0(F_0)$, because the derivative along this curve, as obtained from setting $dI = 0$ in Eq. (4.15), is

$$\frac{dC_0}{dF_0} = \left\{\left(\frac{1}{F_0} + C_0\right)^{1/2} \left[\frac{1}{2}\int_g^{F_0} \left(\frac{1}{v} + C_0\right)^{-3/2} dv + \left(\frac{1}{g} + C_0\right)^{-1/2} \frac{dg}{dC_0}\right]\right\}^{-1} \tag{4.16}$$

and is finite and positive for all $I > 0$. The range of F_0 for $0 \leqq C_0(F_0) < \infty$ is evidently contained in $g(\mathrm{T}_0) < F_0 < \infty$.

For $C_0 > 0$, the integrand in $I(F_0, C_0)$ may be reduced to a standard form by the substitution $\eta^2 = C_0 v$. Equation (4.10) becomes, after the integration is performed,

$$\begin{aligned} 1 - \varepsilon^2 &= (C_0)^{-3/2}\left[\eta\sqrt{1+\eta^2} - \ln\left(\eta + \sqrt{1+\eta^2}\right)\right]_{(C_0 g)^{1/2}}^{(C_0 F_0)^{1/2}} \\ &= (C_0)^{-3/2}\left\{[C_0 F_0(1 + C_0 F_0)]^{1/2} - [C_0 g(1 + C_0 g)]^{1/2} - \right. \\ &\quad \left. - \ln\left[\frac{(C_0F_0)^{1/2} + (1 + C_0F_0)^{1/2}}{(C_0 g)^{1/2} + (1 + C_0 g)^{1/2}}\right]\right\}. \end{aligned} \tag{4.17}$$

For the particular case $C_0 = 0$, we will call $F_0 = F_0^*$, the root of Eq. (4.10), and we obtain directly from Eq. (4.10)

$$F_0^* = \left[\frac{3}{2}(1 - \varepsilon^2) + A^{1/2}\right]^{2/3} = \left(\frac{3}{2}\right)^{2/3}\left\{1 + \varepsilon^2\left[\frac{4}{3(1+\nu)} - 1\right]\right\}^{2/3}. \tag{4.18}$$

For $\varepsilon^2 \ll 1$, $F_0^* \cong (3/2)^{2/3} = 1.310371$. It follows from Eq. [4.18) that F_0^* lies between the limits

$$\left(\frac{4}{3}\right)^{2/3} < F_0^* < 2^{2/3} \tag{4.19}$$

where the lower limit corresponds to $\varepsilon = 1$, $\nu = 1/2$ and the upper limit to $\varepsilon = 1$, $\nu = 0$.

The existence of a root of Eq. (4.10) for $C_0 < 0$ (Fig. 4.1) will now be investigated. It can be concluded from the monotone properties of $I(F_0, C_0)$ that, given any $C_0 = C_{01} < 0$, there is at most one root $F_0 = F_{01}$. Since the point $(F_0^*, 0)$ lies within the domain in which $I > 0$, let the curve $C_0(F_0)$, which was determined for $C_0 \geqq 0$, be extended continuously for $C_0 < 0$ as the integral of a differential equation Eq. (4.16). The root of Eq. (4.10) for $C_0 < 0$ must lie on this extended curve. Three possible situations could occur:

A. The extended curve $C_0(F_0)$ for $C_0 < 0$ intersects the curve $C_0 = -F_0^{-1}$ at some $F_0 = F_{0c}$, $0 < F_{0c} < F_0^*$, and lies within the domain $I > 0$ for $F_0 > F_{0c}$. Furthermore, $C_0(F_0)$ does not exist for $F_0 < F_{0c}$.

B. $C_0(F_0)$ intersects $C_0 = F_0^{-1}$ at $F_0 = F_{0c}$ and can be extended for $F_0 < F_{0c}$.

C. $C_0(F_0)$ does not intersect the curve $C_0 = -F_0^{-1}$ for any $F_0 > 0$. Thus the curve lies inside the "well" of the crosshatched domain in Fig. 4.1.

We will show that (A) above is correct and that (B) and (C) are not.

For $C_0 < 0$ the integrand in $I(F_0, C_0)$ may be reduced to a standard form by the substitutions $\bar{C}_0 = -C_0$, $\bar{g}(\bar{C}_0) = g(C_0)$, $\eta^2 = \bar{C}_0 v$. Equation (4.10) is then written as

$$\begin{aligned} 1 - \varepsilon^2 &= 2(\bar{C}_0)^{-3/2}\left[-\frac{\eta}{2}(1-\eta^2)^{1/2} + \frac{1}{2}\sin^{-1}\eta\right]_{(\bar{C}_0 g)^{1/2}}^{(\bar{C}_0 F_0)^{1/2}} \\ &= (\bar{C}_0)^{-3/2}\left\{[\bar{C}_0\bar{g}(1-\bar{C}_0\bar{g})]^{1/2} - [\bar{C}_0 F_0(1-\bar{C}_0 F_0)]^{1/2} + \right. \\ &\quad \left. + \sin^{-1}(\bar{C}_0 F_0)^{1/2} - \sin^{-1}(\bar{C}_0\bar{g})^{1/2}\right\}. \end{aligned} \tag{4.20}$$

It follows from (4.12) that, for $\bar{C}_0 > 0$,

$$0 < \bar{C}_0\bar{g} < 1. \tag{4.21}$$

Hence, the right-hand side of Eq. (4.20) is continuous within the allowable region for $\bar{C}_0 > 0$ and (as an improver integral) on the boundary curve $\bar{C}_0 = F_0^{-1}$ as well. On this curve, the right-hand side of Eq. (4.20) will be called $I_n(\bar{C}_0)$:

$$\begin{aligned} I(-C_0^{-1}, C_0) \equiv I_n(\bar{C}_0) &= (\bar{C}_0)^{-3/2}\left\{[\bar{C}_0\bar{g}(1-\bar{C}_0\bar{g})]^{1/2} + \right. \\ &\quad \left. + \frac{\pi}{2} - \sin^{-1}(\bar{C}_0\bar{g})^{1/2}\right\}. \end{aligned} \tag{4.22}$$

The derivative of $I_n(\overline{C}_0)$ can be expressed as

$$\frac{dI_n}{d\overline{C}_0} = -\left[\frac{3}{2}\,\overline{C}_0^{\,-1} I_n + \overline{C}_0^{\,-3/2}\left(\frac{\overline{C}_0\overline{g}}{1-\overline{C}_0\overline{g}}\right)^{1/2}\frac{d(\overline{C}_0\overline{g})}{d\overline{C}_0}\right]. \tag{4.23}$$

Inequality (4.21) shows that $I_n(\overline{C}_0)$ is positive for all $\overline{C}_0 > 0$. Moreover it is seen from Eq. (4.22) that

$$\lim_{\overline{C}_0\to 0} I_n(\overline{C}_0) = +\infty, \tag{4.24a}$$

$$\lim_{\overline{C}_0\to +\infty} I_n(\overline{C}_0) = 0. \tag{4.24b}$$

The last follows using also (4.7a). In addition, it can be seen from Eq. (4.6) that $d(\overline{C}_0\overline{g})/d\overline{C}_0$ is positive. Hence, applying these results to the right-hand side of Eq. (4.23) shows that $dI_n/d\overline{C}_0$ is negative, and thus $I_n(\overline{C}_0)$ is a monotone, strictly decreasing function for all $\overline{C}_0 > 0$. In view of this and of (4.24), the equation

$$I_n(\overline{C}_0) = 1 - \varepsilon^2 \tag{4.25}$$

has been shown to have a unique root $\overline{C}_0 = \overline{C}_{0c}$, $0 < C_{0c} < \infty$. Furthermore, because $I(F_0, -\overline{C}_0) \leqq I_n(\overline{C}_0)$ for F_0 obeying (4.14), then Eq. (4.10) can have no root for $C_0 < -\overline{C}_{0c}$. Thus possibility (A) is confirmed, and possibilities (B) and (*C*) are refuted. It is noted, however, that (4.24b) implies $\overline{C}_{0c} \to \infty$ as $\varepsilon \to 1$.

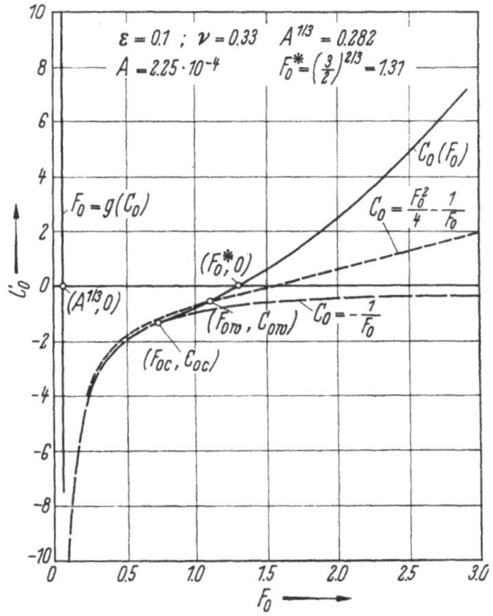

Fig. 4.2. Behavior of $C_0(F_0)$ for representative values of parameters.

The intersection of $C_0(F_0)$ with $-F_0^{-1}$ is tangential, as is shown in Fig. 4.2. To demonstrate this result in general, we observe from Eq. (4.16) that for $F_0 > F_{0c}$ and $C_0 < 0$:

$$\lim_{F_0\to F_{0c}}\left(\frac{dC_0}{dF_0}\right) = \lim_{F_0\to F_{0c}}\left\{\frac{1}{2}\left[\frac{1}{F_0} + C_0(F_0)\right]^{1/2}\int_g^{F_0}\left[\frac{1}{v} + C_0(F_0)\right]^{-3/2} dv\right\}^{-1} \tag{4.26}$$

and that

$$\int_{g}^{F_0}\left[\frac{1}{v}+C_0(F_0)\right]^{-3/2}dv = 2v^2\left[\frac{1}{v}+C_0(F_0)\right]^{-1/2}\Big|_{g}^{F_0} - $$
$$- 4\int_{g}^{F_0}\left[\frac{1}{v}+C_0(F_0)\right]^{-1/2} v\,dv\,.$$

Only the first term on the right-hand side above will give a non-zero limit in Eq. (4.26) and so

$$\lim_{F_0\to F_{0c}}\left(\frac{dC_0}{dF_0}\right) = F_{0c}^{-2}\,. \tag{4.27}$$

Thus the slope of $C_0(F_0)$ is equal to the slope of the curve $C_0 = -F_0^{-1}$ at $F_0 = F_{0c}$.

There does not appear to be any physical discontinuity associated with the values C_{0c}, F_{0c}. For example, with the use of Eq. (4.27), it can be shown that $dw_p/dF_0 < 0$ at $F_0 = F_{0c}$. Formal expressions for solutions to Eq. (3.6) obeying the boundary conditions of Eqs. (3.25) can in fact, be written for $F_0 < F_{0c}$. These solutions, which evidently cannot be allowable solutions, are of the form

$$y - \varepsilon^2 = \int_{g}^{F(y)}\left(\frac{1}{v}+C_0\right)^{-1/2}dv \qquad \varepsilon^2 \leqq y \leqq y_1 < 1\,, \tag{4.28a}$$

$$y - y_1 = \int_{F(y)}^{F(y_1)}\left(\frac{1}{v}+C_0\right)^{-1/2}dv \qquad y_1 \leqq y \leqq 1\,. \tag{4.28b}$$

The solutions $F(y)$ defined by Eqs. (4.28) have a maximum value at $y = y_1$, at which $dF/dy = 0$.

Two transcendental equations involving the unknowns y_1 and C_0 can be obtained from Eqs. (4.28a, b). These two equations would replace Eq. (4.1). However, solutions of these equations must violate the wrinkling stability criterion (3.16), and for this reason the existence of solutions of the boundary-value problem for $F_0 < F_{0c}$ has not been investigated.

5. Properties of Solutions for Fixed Edge Forces

The following theorem summarizes the main results of Section 4

Theorem 5.1. Given ε in $0 < \varepsilon < 1$, ν in $0 \leqq \nu \leqq 1/2$, Eqs. (4.1) to (4.3) are satisfied uniquely on an interval $0 < F_{0c} \leqq F_0 < \infty$ by a function $C_0 = C_0(F_0)$ which is continuously differentiable and monotone strictly increasing with respect to F_0, and F_{0c} is determined uniquely by Eq. (4.1) and the equation $C_0(F_{0c}) = -F_{0c}^{-1}$.

Theorem (5.1) with theorem (4.1) establishes a unique function $F(y)$, and properties of this function, together with the functions in Eqs. (3.7) to (3.13), will be examined in this section. The first investigation concerns the limiting behavior of $F(y)$ for large F_0 (ε and ν fixed) and relates the stresses and normal displacement obtained from $F(y)$ to the results of plane, linear elasticity and also to a linear theory of the normal displacement of membranes. First, the following corollary will be stated; the proof is readily derived from Eq. (4.10).

Theorem 5.1, Corollary. $C_0(F_0)$ increases without bound as F_0 tends towards infinity.

In view of this and (4.7c), it can be concluded that $(g\,C_0)^{-1}$ tends towards zero as F_0 tends towards infinity. Thus Eq. (3.29) can be approximated by expanding the integrand in powers of $(v\,C_0)^{-1}$ and integrating term-by-term.

$$1 - y = C_0^{-1/2}\Big[F_0 - F(y) - \frac{1}{2}\,C_0^{-1}\ln\frac{F_0}{F(y)} - \\ - \frac{3}{8}\,C_0^{-2}(F_0^{-1} - F(y)^{-1}) - \ldots\Big]. \tag{5.1}$$

If C_0 is so large that

$$(A\,C_0^3)^{1/2} \gg 1 \tag{5.2a}$$

then (4.7c) shows that

$$g(C_0) \sim A^{1/2}C_0^{1/2}. \tag{5.2b}$$

Since $F_0 > F_{0c} \cong 0(1)$, then (5.2a, b) imply

$$C_0^2F_0g \sim C_0F_0(A\,C_0^3)^{1/2} \gg 1. \tag{5.2c}$$

Thus (5.2a) implies (5.2c) which in turn implies that terms beyond the logarithmic term can be neglected in the expansion of Eq. (5.1). This is done in Eq. (5.3a) below. With $F(\varepsilon^2) = g(C_0)$, Eq. (5.3a) then yields Eq. (5.3b) as an approximation to Eq. (4.10).

$$\frac{C_0^{1/2}}{F_0}(1 - y) = 1 - \frac{F(y)}{F_0} - \frac{1}{2F_0C_0}\ln\frac{F_0}{F(y)}, \tag{5.3a}$$

$$\left[1 + \frac{1-\nu}{1+\nu}\varepsilon^2\right]\frac{C_0^{1/2}}{F_0} = 1 + \frac{1}{2F_0C_0}\ln\left(A^{1/2}\frac{C_0^{1/2}}{F_0}\right). \tag{5.3b}$$

If F_0 is chosen so large that the logarithmic terms in Eqs. (5.3a) and (5.3b) may be neglected, then the following approximation for $F(y)$ is obtained from these equations

$$\frac{F(y)}{F_0} = \frac{y + \left(\frac{1-\nu}{1+\nu}\right)\varepsilon^2}{1 + \left(\frac{1-\nu}{1+\nu}\right)\varepsilon^2}. \tag{5.4}$$

With the use of Eqs. (3.26) and (5.4), the expressions for s_r and s_θ from Eqs. (3.8) and (3.9) become

$$s_r = \left(\frac{H_0}{Eh}\right)\frac{F(y)}{F_0 y} = \frac{H_0}{Eh}\left[\frac{1 + \frac{1-\nu}{1+\nu}\frac{b^2}{r^2}}{1 + \frac{1-\nu}{1+\nu}\frac{b^2}{a^2}}\right], \tag{5.5}$$

$$s_\theta = \frac{H_0}{Eh}\left[2\frac{d}{dy}\left(\frac{F}{F_0}\right) - \frac{1}{y}\frac{F}{F_0}\right] = \frac{H_0}{Eh}\left[\frac{1 - \frac{1-\nu}{1+\nu}\frac{b^2}{r^2}}{1 + \frac{1-\nu}{1+\nu}\frac{b^2}{a^2}}\right]. \tag{5.6}$$

These expressions for the stresses are identical to those derived for the stretching of a sheet in plane elasticity [6] (where $P = 0$).

The complete equation for the normal displacement, $w(y)$, is obtained by combining Eqs. (3.11) and (3.23):

$$\frac{w(y)}{a} = -2\left(\frac{H_0/Eh}{F_0}\right)^{1/2}\left\{F_0\left(\frac{1}{F_0} + C_0\right)^{1/2} - \right.$$
$$\left. - F(y)\left[\frac{1}{F(y)} + C_0\right]^{1/2} - C_0(1-y)\right\}. \tag{5.7}$$

Expansion of the radicals in Eq. (5.7) and neglect of higher-order terms under the assumption that $C_0^2 F_0 g \gg 1$ leads to the approximate expression

$$\frac{w(y)}{a} \doteq -2\left(\frac{H_0}{Eh}\right)^{1/2}(C_0 F_0)^{1/2}\left[1 - \frac{F(y)}{F_0} - \frac{(1-y)C_0^{1/2}}{F_0}\right].$$

This becomes, upon substitution from Eq. (5.3a),

$$\frac{w(y)}{a} = -\left(\frac{H_0/Eh}{F_0 C_0}\right)^{1/2}\ln\frac{F_0}{F(y)}. \tag{5.8}$$

Note that retention of the logarithmic term in Eqs. (5.3) is required to obtain Eq. (5.8). Thus Eq. (5.8) is a higher-order approximation than are Eqs. (5.4), (5.5) and (5.6). It will not reduce the order of approximation, however, to use Eq. (5.4) and Eq. (5.3a), the latter without the logarithmic term, in writing Eq. (5.8) as

$$\frac{w(y)}{a} = \frac{P}{4\pi a H_0}\left(1 + \frac{1-\nu}{1+\nu}\varepsilon^2\right)\ln\frac{y + \frac{1-\nu}{1+\nu}\varepsilon^2}{1 + \frac{1-\nu}{1+\nu}\varepsilon^2} \qquad \varepsilon^2 \leqq y \leqq 1. \tag{5.9}$$

The identity used in Eq. (5.9),

$$\frac{P}{4\pi a H_0} \equiv (H_0/E h)^{1/2} F_0^{-3/2} \tag{5.10}$$

is derivable from Eq. (3.26).

In the linear theory which is used mainly for the study of the small vibrations of stretched membranes [7], the equation determining the deflected shape $w(r)$ of a statically-loaded, initially-flat membrane is, in cylindrical coordinates,

$$\nabla^2 w = \frac{1}{r}\frac{d}{dr}\left(r\frac{dw}{dr}\right) = -\frac{p(r)}{H_0}.$$

Here $p(r)$ is the normal pressure. The deflection of the membrane under only a concentrated central load is the point logarithmic potential:

$$\frac{w(r)}{a} = \frac{P}{2\pi a H_0}\ln\frac{r}{a} \qquad 0 < r < a. \tag{5.11}$$

For fixed P and $\varepsilon \to 0$, the expressions on the right-hand sides of Eqs. (5.9) and (5.11) approach equality. However, it is false to conclude from this that Eq. (5.11) is the limiting case obtained from the nonlinear theory. Equation (5.9) has been derived assuming that ε is fixed and positive and that $P \to 0$. The limit of Eq.(5.9), assuming that P is fixed and positive and that $\varepsilon \to 0$, is in violation of assumptions made in the expansion leading to Eq. (5.9). The true limit for $w(b)$ under these conditions will be derived below from the non-linear theory (cf. Eq. (5.17)).

The next topic considered is the behavior of maximum values. We shall first state several theorems, the proofs for which are given in the Appendix.

Theorem 5.2. If $s_\theta(1) \geqq 0$, then $s_\theta(y) > 0$ for $\varepsilon^2 < y < 1$; also, $s_\theta(\varepsilon^2) \geqq 0$, the equality holding only if $\nu = 0$.

Consequently, the wrinkling stability criterion (3.16) is satisfied everywhere if it is satisfied at $y = 1$. All solutions for which $F_0 \geqq F_{0w}$, where F_{0w} is the largest root of

$$\frac{1}{F_0} + C_0(F_0) = \frac{F_0^2}{4}, \tag{5.12}$$

(cf. Eq. (3.22)) will be stable solutions satisfying (3.16). Conversely, if F_{0w} is the unique root of Eq. (5.12), then all solutions for which $F_0 < < F_{0w}$ are unstable, and F_{0w} is the stability limit. The numerical solution of Eq. (5.12) shows that there is a unique root F_{0w} for each ε and ν

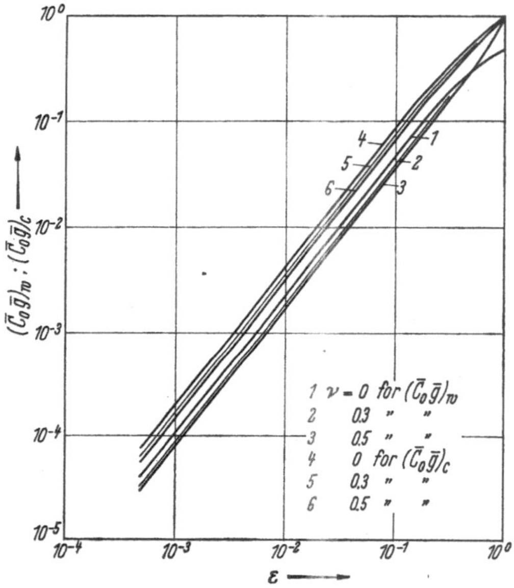

Fig. 5.1. F_{0w} vs. ε for several values of ν.

(see Fig. 5.1 and also the comments in the Appendix, part III). (Since the left-hand side of Eq. (5.12) is a monotone, strictly increasing function of F_0, it follows that $F_{0w} > F_{0c}$. It can also be seen by applying Eq. (5.3b) that for every $A > 0$, F_{0w} is finite.)

Theorem 5.3. For every stable solution in $\varepsilon^2 \leqq y \leqq 1$:

(a) $s_r(y)$ is monotone nonincreasing

(b) $s_\theta(y) \leqq s_r(y)$

(c) σ_r at $r = b$ is the largest stress magnitude.

The conclusions of Theorem (5.3) do not necessarily apply to all boundary conditions on Eq. (3.6), however, but only to those for which $s_\theta(\varepsilon^2) < s_r(\varepsilon^2)$, as for Eq. (3.25a).

We consider next the behavior of $s_r(\varepsilon^2)$ and $w(\varepsilon^2) \equiv -w_p$ as ε tends towards zero. For any fixed $C_0 \geqq C_{0c}$, and for ε so small that $A^{1/3}|C_0| \ll 1$, the positive root of Eq. (4.4) can be represented as

$$g = A^{1/3}\left[1 + \frac{1}{3} A^{1/3} C_0 + 0(A C_0^3)\right]. \tag{5.13}$$

Then

$$F(\varepsilon^2) = g \sim A^{1/3} \equiv \left(\frac{2}{1+\nu}\right)^{2/3} \varepsilon^{4/3} \tag{5.14}$$

and from Eq. (3.9)

$$s_r(\varepsilon^2) \doteq [(1+\nu)^{-1}\varrho]^{2/3} \equiv \left[\frac{P}{2\pi(1+\nu)Ehb}\right]^{2/3} \tag{5.15}$$

and from Eq. (5.7)

$$\frac{w_p}{a} \doteq 2\left(\frac{H_0/Eh}{F_0}\right)^{1/2}\left[F_0\left(\frac{1}{F_0} + C_0\right)^{1/2} - C_0\right]. \tag{5.16}$$

The conclusions to be drawn from Eqs. (5.15) and (5.16) are that, for a fixed $P > 0$ and a vanishing ε, the nonlinear theory predicts that the maximum stress becomes infinite as $b^{-2/3}$ but that the central deflection w_p remains finite in the limit. It is remarkable that these conclusions are exactly the reverse of the predictions of the linear theory, Eqs. (5.5) and (5.8), in which, under the same circumstances, the stresses remain finite and the deflection becomes logarithmically infinite. To prove these conclusions, however, it is necessary to show that $|C_{0w}|$ has at least a finite upper bound as ε tends to zero. The existence of both a limiting value of C_{0w} and a limit function $C_0(F_0)$ is assumed in the analysis of Jahsman, Field and Holmes [3]. Verification of the existence of this limit follows from analysis of the present paper. Since for sufficiently small ε, Eqs. (4.2), (4.6) and (5.13) give

$$\left(\frac{1}{g} + C_0\right)^{-1/2} \frac{dg}{dC_0} = \frac{A^{1/2}}{g} \frac{A g^2}{2g^3 + A} \doteq \frac{1}{3} A^{5/6},$$

then the limit function $C_0(F_0)$ is the integral of

$$\frac{dC_0}{dF_0} = \left[\frac{1}{2}\left(\frac{1}{F_0} + C_0\right)^{1/2} \int_0^{F_0} \left(\frac{1}{v} + C_0\right)^{-3/2} dv\right]^{-1} \tag{5.17}$$

where Eq. (4.16) has been approximated.

The dependence of $C_0(F_0)$ upon ε is so slight for very small ε that, for lesser values of ε, the change in the curve drawn in Fig. 4.2 for $\varepsilon = 0.1$ would be hardly noticeable on this scale. The computed dependence of F_{0w} upon ε for several values of ν is shown in Fig. 5.1[1].

The final topic considered in this section is the behavior of a subclass of allowable solutions which obey additional conditions imposed by considerations of stability and by the necessity for consistency of a solution with the basic assumptions made in the development of the equations. The three conditions to be satisfied are those of small strains stability and moderate rotations. The small strain condition (2.18) can, in view of Theorem 5.3, be replaced by the requirement $\sigma_r(\varepsilon^2) < \sigma_Y/E$, where σ_Y is the elastic limit or yield stress in uniaxial tension for the material. It is assumed then that $\sigma_Y/E \ll 1$. The stability condition, in view of Theorem 5.2, is just $F_0 \geqq F_{0w}$. Since Eqs. (3.19) and (3.13) establish the monotonicity of $X(y)$, then the condition (3.14) can be replaced by the requirement $X(\varepsilon^2) \ll 1$. Therefore, the three necessary conditions can be expressed in analytical form as $\boldsymbol{A}$, $\boldsymbol{B}$ and $\boldsymbol{C}$:

A. Small Strains and Yield Limit Condition (Tresca)

$$s_r(\varepsilon^2) = 2^{-2/3}\varepsilon^{-4/3}\varrho^{2/3}g < \frac{\sigma_Y}{E} \ll 1$$

B. Wrinkling Stability

$$F_0 = \left(\frac{H_0}{Eh}\right)2^{2/3}\varepsilon^{-2/3}\varrho^{-2/3} \leqq F_{0w}$$

C. Moderate Rotations

$$X(\varepsilon^2) = 2^{4/3}\varepsilon^{8/3}\varrho^{2/3}g^{-2} \ll 1.$$

The statements in the following theorem are proven in the Appendix.

---

[1] The limit value of C_{0c} for vanishing ε can be readily obtained analytically. Since (4.21) holds for all $\overline{C}_0 > 0$, and since

$$\sin^{-1}(\overline{C}_0\bar{g})^{1/2} = \frac{1}{2}\sin^{-1}\{2[\overline{C}_0\bar{g}(1-\overline{C}_0\bar{g})]^{1/2}\}$$

then using the expansion of $\sin^{-1}$ in Eq. (4.22), we find the equation for $\overline{C}_0 = \overline{C}_{0c}$ to be

$$I_n(\overline{C}_0) \cong \frac{\pi}{2}(\overline{C}_0)^{-3/2}\{1 + 0[[\overline{C}_0\bar{g}(1-\overline{C}_0\bar{g})]^{1/2}]\} \cong 1$$

Therefore, as $\varepsilon \to 0$, $\overline{C}_{0c} \to (\pi/2)^{2/3}$ and $F_{0c} \to (2/\pi)^{2/3}$.

Theorem 5.4. Assume, unless stated otherwise, a material for which $\sigma_Y/E \ll 1$, a value of H_0 satisfying

$$0 < \frac{H_0}{Eh} < \frac{1}{2}\frac{\sigma_Y}{E}[1 + \nu + (1 - \nu)\varepsilon^2], \tag{5.18}$$

and $0 < \varepsilon < 1$, $0 \leqq \nu \leqq 1/2$.

1. For every set of values (σ_Y/E), (H_0/Eh), ε and ν, there corresponds a positive number ϱ_m such that the solution to the problem defined by Eq. (3.6) and boundary conditions (3.25), also satisfies $\boldsymbol{A}, \boldsymbol{B}$ and $\boldsymbol{C}$ for all ϱ on the interval $0 \leqq \varrho \leqq \varrho_m$.

2. If $\varepsilon^{4/3} \ll 1$, then the following statements, i), ii) and iiii) hold.

i) $\boldsymbol{A} + \boldsymbol{B} \Rightarrow \boldsymbol{C}$, i.e., satisfaction of the small strain and the wrinkle stability condition implies moderate rotations.

ii) If H_0 equals H_{opt}, defined as

$$\frac{H_{\text{opt}}}{Eh} = \left(\frac{1+\nu}{2}\right)^{2/3} \varepsilon^{2/3} \frac{\sigma_Y}{E} F_{0w}, \tag{5.19}$$

then the membrane can carry all central loads $P < P_{\text{ult}}$, where P_{ult} is defined as

$$\frac{P_{\text{ult}}}{2\pi Ehb} \equiv \varrho_{\text{ult}} = (1 + \nu)\left(\frac{\sigma_Y}{E}\right)^{3/2}, \tag{5.20}$$

without violating $\boldsymbol{A}$ or $\boldsymbol{B}$. Both $\boldsymbol{A}$ and $\boldsymbol{B}$ are violated if $P > P_{\text{ult}}$. The central deflection under $H_0 = H_{\text{opt}}$ and $P = P_{\text{ult}}$ is

$$w_p = 2\left(\frac{H_{\text{opt}}}{Eh}\right)^{1/2}\left(\frac{1}{4}F_{0w}^{3/2} + F_{0w}^{-3/2}\right). \tag{5.21}$$

As is seen from Fig. 5.1, the limiting value of F_{0w} for $\varepsilon \to 0$ is 1.06676.

iii) P_{ult} is the least upper bound for values of P which the membrane can support without violating $\boldsymbol{A}$ or $\boldsymbol{B}$.

(a) If $H_0 < H_{\text{opt}}$, then the maximum value of P, which occurs at $F_0 = F_{0u}$, is less than P_{ult}; and

(b) If $H_0 > H_{\text{opt}}$, then $\boldsymbol{A}$ is violated for $F_0 = F_{0w}$. The least value of F_0 for which both $\boldsymbol{A}$ and $\boldsymbol{B}$ can simultaneously be satisfied is greater than F_{0w}, and consequently P for this F_0 is less than P_{ult}.

3. If ε is sufficiently close to 1, then $\boldsymbol{A} + \boldsymbol{B} \Rightarrow \boldsymbol{C}$; i. e., condition $\boldsymbol{C}$ will be violated for values of P at which $\boldsymbol{A}$ and $\boldsymbol{B}$ are satisfied. To solve the problem completely for this geometry requires integration of Eqs. (2.30) and (2.32).

Appendix

Proofs of Theorems in Section 5

I. Proof of Theorem 5.3. With the use of Eqs. (3.6), (3.8), (3.19) and (3.17), it can be shown that for $\varepsilon^2 < y < 1$

$$\frac{ds_r}{dy} = \frac{s_\theta - s_r}{2y}, \tag{A1}$$

$$\frac{ds_\theta}{dy} = -\left(\frac{1}{2}\varepsilon\varrho\right)^{2/3}\frac{1}{F^2} - \frac{ds_r}{dy}. \tag{A2}$$

Equations (2.5), (2.7) and (3.25) imply

$$s_\theta(\varepsilon^2) = \nu s_r(\varepsilon^2). \tag{A3}$$

Therefore, $s_\theta(\varepsilon^2) < s_r(\varepsilon^2)$, and if $s_\theta(y) \leqq s_r(y)$ for all y, Eq. (A1) shows that $s_r(y)$ would be monotone nonincreasing. Suppose the contrary. There would then exist a $y = y_1$, $\varepsilon^2 < y_1 < 1$, such that $s_\theta(y_1) = s_r(y_1)$, and in some right neighborhood of y_1, $s_\theta(y) > s_r(y)$. However, at $y = y_1$, Eqs. (A1) and (A2) show that $ds_r/dy = 0$ and $ds_\theta/dy < 0$. Hence, $s_\theta(y) < s_r(y)$ immediately to the right of $y = y_1$. Thus we arrive at a contradiction and prove parts (a) and (b) of the theorem. Part (c) follows because $ds_r/dy < 0$ at $y = \varepsilon^2$, and because $s_\theta \geqq 0$.

II. Proof of Theorem 5.2. This theorem would be obvious if s_θ were always monotone strictly decreasing with respect to y for fixed F_0. That this is not the general case can be readily seen by examining ds_θ/dy at $y = \varepsilon^2$. Equations (3.9), (3.28), (4.4), (4.5), (A1) and (A2) may be combined to give

$$\left.\frac{ds_\theta}{dy}\right|_{\varepsilon^2} = \left(\frac{1}{2}\varepsilon\varrho\right)^{2/3} g^{-2}\left[\frac{2(1-\nu)}{(1+\nu)^2}(1 + C_0 g) - 1\right]. \tag{A4}$$

The right-hand side above will always be positive if $\nu < 0.236$. The theorem can be proven in the following way, however, for $0 < \varepsilon < 1$ and $0 \leqq \nu \leqq 1/2$. It is convenient in this development to denote explicitly the dependence of the functions on both y and F_0.

The analysis of sections 3 and 4 shows that $s_\theta = s_\theta(y, F_0)$ is a continuous functions of both y and F_0 in a domain $[D | F_{0c} < F_0 < \infty, \varepsilon^2 \leqq y \leqq 1]$. Hence, for each F_0 in D, $s_\theta(y, F_0)$ takes on a minimum value over y; call this minimum value $s(F_0)$, and let $y = \bar{y}(F_0)$ be such that $s_\theta(\bar{y}, F_0) = s(F_0)$.

Assume now that $s(F_0) \leqq 0$, and that $\bar{y}$ is on the open interval $(\varepsilon^2, 1)$. Then since s_θ has continuous second partial derivatives with respect to y, it must follow that

$$\text{at} \quad y = \bar{y}: \; s_\theta \leqq 0, \qquad \frac{\partial s_\theta}{\partial y} = 0, \qquad \frac{\partial^2 s_\theta}{\partial y^2} > 0. \tag{A5}$$

One differentiation of both sides of Eq. (3.8) gives

$$\frac{\partial s_\theta}{\partial y} = \left(\frac{1}{2}\varepsilon\varrho\right)^{2/3}\left(\frac{1}{y}\frac{dF}{dy} - \frac{1}{F^2}\right) - \frac{s_\theta}{y} \tag{A6}$$

and a second differentiation gives

$$\frac{\partial^2 s_\theta}{\partial y^2} = \left(\frac{1}{2}\varepsilon\varrho\right)^{2/3}\left(\frac{2}{F^3}\frac{dF}{dy} - \frac{1}{y^2}\frac{dF}{dy} - \frac{1}{2yF^2}\right) + \frac{s_\theta}{y^2} - \frac{1}{y}\frac{\partial s_\theta}{\partial y}.$$

When substitutions are made into the above equation from Eqs. (3.8) and (A6), and terms are combined, the following equation is obtained:

$$\frac{\partial^2 s_\theta}{\partial y^2} = \frac{s_\theta}{F^3} - \frac{2}{y}\frac{\partial s_\theta}{\partial y} - \frac{1}{2}\left(\frac{1}{2}\varepsilon\varrho\right)^{2/3} y^{-1}F^{-2}. \tag{A7}$$

At $y = \bar{y}$, Eq. (A7) shows that $\partial^2 s_\theta/\partial y^2 < 0$, thus contradicting the last requirement of (A 5).

Consequently, if $s(F_0) \leqq 0$, then $\bar{y}$ must be one of the end points. If $\nu > 0$, then Eq. (A3) shows that $\bar{y}$ can only be 1, and the theorem follows. If $\nu = 0$, then $s_\theta(\varepsilon^2, F_0) = 0$, but, from Eq. (A4), $\frac{\partial s_\theta}{\partial y}(\varepsilon^2, F_0) > 0$.

We have not been able to show analytically that $s_\theta(1, F_0)$ is monotone increasing with F_0, or even to show the weaker condition that $\frac{\partial s_\theta}{\partial F_0}(1, F_{0w}) > 0$. The latter condition, together with Eq. (5.12), would characterize F_{0w} in case there were multiple roots of Eq. (5.12). Numerical solutions have shown, however, that F_{0w} is the unique root of Eq. (5.12), and since $s_\theta(1, F_{0c}) < 0$, it follows that all solutions for which $F_0 < F_{0w}$ are unstable.

III. Proof of Theorem 5.4. 1) With the use of Eq. (3.26) in the form

$$\varrho^{2/3} = 2^{2/3}\varepsilon^{-2/3}\left(\frac{H_0}{Eh}\right)F_0^{-1} \tag{A8}$$

we can write

$$\chi(\varepsilon^2) = 4\varepsilon^2\left(\frac{H_0}{Eh}\right)[F_0^{-1}g(C_0)^{-2}]. \tag{A9}$$

It follows from Theorem 5.1, Corollary, and from Eq. (4.7c), that a lower limit for F_0 can always be chosen so that $\boldsymbol{C}$ is satisfied for all larger values of F_0. This lower bound can evidently also satisfy $\boldsymbol{B}$.

If F_0 is chosen sufficiently large, $s_r(\varepsilon^2)$ will approach the value given by Eq. (5.5),

$$s_r(\varepsilon^2) \to \frac{2H_0}{Eh}(1+\nu) + (1-\nu)\varepsilon^2]^{-1},$$

as a limit. Since (H_0/Eh) obeys condition (5.18), then a lower bound for F_0, or equivalently an upper value ϱ_m for ϱ, can be found so that $\boldsymbol{A}$ is also satisfied.

2i) The least upper bound on ϱ is given from $\boldsymbol{A}$ as

$$\text{l. u. b. } [\varrho^{2/3}] = \frac{\sigma_Y}{E} 2^{2/3}\varepsilon^{4/3} g^{-1}. \tag{A10}$$

If the expression on the right is substituted in place of $\varrho^{2/3}$ into the expression for $\chi(\varepsilon^2)$ in $\boldsymbol{C}$, it is found thereby that when $\boldsymbol{A}$ is satisfied, then $\chi(\varepsilon^2)$ is bounded as

$$\chi(\varepsilon^2) = 2^{4/3}\varepsilon^{8/3}\varrho^{2/3}g^{-2} < 4\frac{\varepsilon^4}{g^3}\frac{\sigma_Y}{E}. \tag{A11}$$

The minimum value of g occurs at $C_0 = C_{0w}$, and for $\varepsilon^{4/3} \ll 1$, C_{0w} is of order unity. Therefore, if $\varepsilon^{4/3} \ll 1$, then $A^{1/3}|C_{0w}| \ll 1$, and so the first term of the expansion on the right-hand side of Eq. (5.13),

$$g^3 \doteq A = \frac{4\varepsilon^4}{(1+\nu)^2} \tag{A12}$$

can be used in the right-hand side of inequality (A11). Therefore, $\chi(\varepsilon^2)$ is bounded as

$$\chi(\varepsilon^2) < (1+\nu)^2\frac{\sigma_Y}{E} \tag{A13}$$

and so $\boldsymbol{A}$ and $\boldsymbol{B}$ imply $\boldsymbol{C}$.

2ii) When the small ε expression for $g(C_{0w})$ from Eq. (A12) is substituted into the right-hand side of Eq. (A10), and both sides are raised to the (3/2) power, it is found that

$$\text{l. u. b. } (\varrho) = \varrho_{\text{ult}} \tag{A14}$$

where ϱ_{ult} is defined by Eq. (5.20). Hence $\boldsymbol{A}$ is satisfied for all ϱ which obey the condition $\varrho < \varrho_{\text{ult}}$. Conversely, $\boldsymbol{A}$ is violated if $\varrho > \varrho_{\text{ult}}$.

Since Eq. (3.26) can be expressed as

$$\frac{H_0}{Eh} = 2^{-2/3}\varepsilon^{2/3}\varrho^{2/3}F_0$$

then, using Eqs. (5.19) and (5.20), we can show that

$$\frac{H_0}{H_{\text{opt}}} = \left(\frac{\varrho}{\varrho_{\text{ult}}}\right)^{2/3}\frac{F_0}{F_{0w}}. \tag{A 15}$$

If $H_0 = H_{\text{opt}}$, then $\boldsymbol{B}$ will be satisfied for all $\varrho < \varrho_{\text{ult}}$. Conversely, we have seen that $\boldsymbol{A}$ is violated, and as can be seen from Eq. (A 15) $\boldsymbol{B}$ is also violated, if $H_0 = H_{\text{opt}}$, $\varrho > \varrho_{\text{ult}}$.

The maximum value of the central deflection, given by Eq. (5.21), is obtained from Eqs. (5.7) and (5.12) by neglecting small terms and setting $F_0 = F_{0w}$.

2iii) If $H_0 < H_{\text{opt}}$, then the maximum value of ϱ at $F_0 = F_{0w}$ is, from Eq. (A 15),

$$\varrho = \left(\frac{H_0}{H_{\text{opt}}}\right)^{3/2}\varrho_{\text{ult}} < \varrho_{\text{ult}}. \tag{A 16}$$

Therefore, the load-carrying capacity of the membrane without wrinkling is reduced below its maximum. If $H_0 < H_{\text{opt}}$, on the other hand, then at $F_0 = F_{0w}$, $\varrho > \varrho_{\text{ult}}$. This violates $\boldsymbol{A}$, and F_0 must be increased to satisfy A. The relation

$$\frac{s_r(\varepsilon)^2}{s_Y} = \left(\frac{\varrho}{\varrho_{\text{ult}}}\right)^{2/3}\frac{g(C_0)}{g(C_{0w})} < 1 \tag{A 17}$$

follows from A and from the definition of ϱ_{ult}, Eqs. (5.20), (A 10) and (A 14). We know from Part (1) of this theorem, that (A 17) will be satisfied for $\varrho \leqq \varrho_m$. Since $g(C_0) > g(C_{0w})$ if $F_0 > F_{0w}$, it follows from (A 17) that

$$\varrho_m < \left[\frac{g(C_{0w})}{g(_{0m})}\right]^{3/2}\varrho_{\text{ult}}.$$

(3) Since we can express

$$g^2 = 4^{2/3}\varepsilon^{8/3}(1+\nu)^{-4/3}\left[1 + C_0 g\right]^{2/3} \tag{A 18}$$

from Eq. (4.4) then $\chi(\varepsilon^2)$ at the wrinkling stability limit can be written as

$$\chi(\varepsilon^2) = 2^{-2/3}(1+\nu)^{4/3}\varrho^{2/3}\left[1 + C_0 g\right]_w^{-2/3}. \tag{A 19}$$

As $\varepsilon \to 1$, $(C_0 g)_w \to (C_0 g)_c \to -1$ (see Fig. A. 1). Hence, if ε is sufficiently close to 1, $\chi(\varepsilon^2)$ will violate C for all ϱ greater than any preassigned, small positive number.

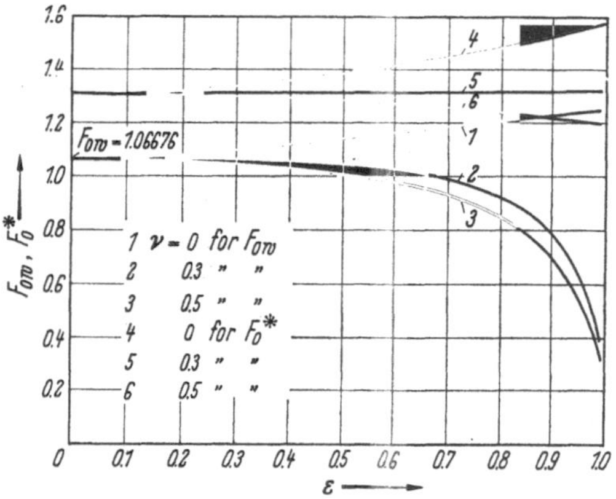

Fig. A. 1. $(\overline{C}_0\overline{g})_w$ and $(\overline{C}_0\overline{g})_c$ vs. ε for several values of ν.

References

1. Bromberg, E., and J. J. Stoker: Non-linear theory of curved elastic sheets, Q. Appl. Math., vol. III, 246–265 (1945).
2. Schwerin, E.: Über Spannungen und Formänderungen kreisringförmiger Membranen, Z. Tech. Phys., 12, 651–659 (1929).
3. Jahsman, W. E., F. A. Field, and A. M. C. Holmes: Finite Deformations in a prestressed, centrally loaded, circular elastic membrane, Proc. of the Fourth U. S. National Congress of Applied Mechanics, A. S. M. E., 1962 (in press).
4. Reissner, E.: Rotationally symmetric problems of thin shells of revolution, Proc. of the Third U. S. National Congress of Applied Mechanics, A. S. M. E., 1958 pp. 51–69.
5. Stein, M., and J. M. Hedgepeth: Analysis of partly wrinkled membranes, N. A. S. A. Technical Note D-813, Washington D. C., July 1961.
6. Cherepanov, G. P.: On the buckling of a membrane containing holes, PMM, Vol. 27, No. 2, 1963 pp. 275–286.
7. Clark, R. A., and O. S. Narayanaswamy: Nonlinear membrane problems for elastic shells of revolution, Proc. Symposium on the Theory of Shells to Honor Lloyd Hamilton Donnell, ed. by D. Muster, Huston, Texas, 1967 pp. 81 bis 110.
8. Gostello, G. A., and M. C. Stippes: Large symmetric deflections of circular membranes, Development in Mechanics, V. 2, part 2, Proc. of the Eight Midwestern Mechanics Conference, ed. by S. Ostrach and R. H. Scanlon, Pergamon Press 1965 pp. 77–83.
9. Bhatia, N. M., and W. Nachbar: Finite indentation of an elastic membrane by a spherical indenter, Int. J. of Nonlinear Mech., 1968 (in press).
10. Bhatia, N. M., and W. Nachbar: Finite indentation of elastic perfectly plastic membranes by a spherical indenter.

11. ROSETTOS, J. N.: An asymptotic analysis for large deflections of pressurized shallow spherical membrane shells. AIAA J., vol. 4, No. 6, 1966 pp. 1121—1123.
12. WILLERS, FR. A.: Practical Analysis, translated by R. T. Beyer, New York: Dover 1948 p. 238.
13. TIMOSHENKO, S., and J. N. GOODIER: Theory of Elasticity, 2nd ed., New York: McGraw-Hill, 1951, sec. 26.
14. MORSE, P. M.: Vibration and Sound, 2nd ed., New York: McGraw-Hill, 1948, sec. 17.

Discussion

F. P. J. RIMROTT: I would like to ask a question in connection with the experimental verification of the deflections. Did you determine the yield surface independently, or did you compute the requisite yield surface values from your experimental results?

W. NACHBAR: The yield stress for the mylar membrane material actually used in the indentation experiments of JAHSMAN, FIELD and HOLMES was measured by uniaxial tension tests on strips of mylar cut from the main sheet. Values of yeld stress measured on several strips in different directions on the sheet averaged 12.000 psi, with no significant differences found due to direction. However, it was difficult in these tests to hold the thin material tightly in the grips of the testing machine, and so this value for the yield stress was not relied upon exclusively.

The photographs that enabled measurement of the radius "b" and the tangent angle β_b at the point of tengency, and also the separate measurement of the indenter load at the time the photograph is taken, allow calculation from statics of the value of σ_r at the point of tangency. These values as calculated ranged from 11.000 to nearly 13.000 psi. Since σ_r is somewhat less than the uniaxial tension yield stress when the elastic-plastic boundary is in the free region, the value $\sigma_y = 13.000$ psi was chosen for the yield stress.

It should be noted, however, that because the spatial gradients of the stress components are large in the vicinity of the indenter for small indenters, an error in determining the yield stress will have relatively little effect on the calculations of the elastic-plastic boundary and of the deflection profile. This conclusion was verified by calculation of several deflection profiles for different values of yield stress.

R. W. LEONARD: It might be noted that the theory of partly wrinkled membranes of STEIN and HEDGEPETH, which appears to be identical to that of CHEREPANOV, has been verified quantitatively in tests on thin Mylar specimens at NASA Langley. The tests included a thin, pressurized cylinder in bending and the stretched, circular membrane in torsion shown by Professor NACHBAR in his last slide. The tests showed excellent agreement with the theory in comparisons of measured rotations versus applied moment for rotations well into the nonlinear range.

R. A. CLARK: At the Houston shell theory symposium in April, 1966, I presented a paper with NARAYANASWAMY which discussed the annular membrane. We presumed that one could have a transverse edge load on the inner circle of the annulus. Such a load causes a 90 degree rotation (which is certainly a large rotation) and leads to a much more interesting mathematical boundary value problem. We were able to get analytic bounds on the solution of that problem and by similar techniques one can prove the existence and uniqueness of a mathematical solution. It is of interest that the problem doesn't fit into most of the standard theories for non-linear boundary value problems.

Post-Buckling Behavior of Cylinders in Torsion

By

B. Budiansky

Harvard University, Cambridge, Mass., USA

Abstract. The initial post-buckling behavior and the consequent imperfection sensitivity of thin-walled cylinders subjected to torsion are studied on the basis of the Kármán-Donnell equations. A perturbation analysis consistent with the general theory of the post-buckling behavior of structures leads to eigth-order systems of complex, ordinary differential equations that are solved numerically for several sets of boundary conditions.

Introduction

Many studies concerned with the post-buckling behavior of shells have had as their aim the determination of the magnitudes of external loads associated with buckles of finite depth. The discovery, usually by approximate energy methods, of post-buckling loads smaller than the classical, initial buckling loads has justifiably been regarded as evidence that the classical loads may constitute unconservative estimates of buckling strength. Occasionally, calculations by similar approximate techniques have also been made for shells having assumed imperfections. The torsion of circular cylinders, perfect and imperfect, has been studied in this way by Loo [1] and Nash [2].

A different approach, pioneered by Koiter [3, 4], seeks to determine the *initial* post-buckling behavior; an asymptotically exact calculation then requires only the solution of linear problems to ascertain whether the applied load increases or decreases immediately after buckling. A drop in load after buckling of a perfect shell implies that the shell is *imperfection-sensitive* in the sense that a small initial geometrical imperfection in the shell would cause it to undergo snap-buckling at a load

---

Acknowledgement. The writer is grateful to Mr. James C. Frauenthal for checking the cylinder analysis, writing the Fortran IV program, and calculating the numerical results.

This work was supported in part by the National Aeronautics and Space Administration under Grant NsG-559, and by the Division of Engineering and Applied Physics, Harvard University.

that is lower than the classical buckling load. Several shell problems have recently been treated in this way [5—9]; this series of studies is continued herein with a somewhat new twist, since torsion has not been considered previously. The shell theory used as a basis for the analysis is that attributed in [10] to DONNELL, MUSHTARI and VLASOV; as applied to a circular cylinder, the governing differential equations, when written in terms of a normal displacement and an Airy stress function, are usually called the KÁRMÁN-DONNELL equations.

In order to make this paper self-contained, and to provide a convenient compendium of formulas for future use, the general theory of the post-buckling behavior of structures [3] will be quickly redeveloped for DONNELL-MUSHTARI-VLASOV shell theory. With a slight change of viewpoint (see [10]) the results will also be applicable to shallow-shell theory.

Post-Buckling Theory

The equilibrium relations of the non-linear shell theory of DONNELL-MUSHTARI-VLASOV follow from the variational statment

$$\iint_A (M^{\alpha\beta}\delta K_{\alpha\beta} + N^{\alpha\beta}\delta E_{\alpha\beta})\, dA = EVW \tag{1}$$

where $M^{\alpha\beta}$ and $N^{\alpha\beta}$ are symmetrical stress-couples and stress-resultants, respectively; $K_{\alpha\beta}$ and $E_{\alpha\beta}$ are bending and membrane strains defined as

$$K_{\alpha\beta} = -W_{,\alpha\beta} \tag{2}$$

$$E_{\alpha\beta} = \frac{1}{2}[U_{\alpha,\beta} + U_{\beta,\alpha}] + b_{\alpha\beta}W + \frac{1}{2}W_{,\alpha}W_{,\beta} \tag{3}$$

in terms of the tangential displacements U_α and the normal displacement W; and $b_{\alpha\beta}$ is the curvature tensor of the undeformed shell[1]. Commas denote covariant differentiation with respect to general surface coordinates ξ^1, ξ^2. The right-hand side of (1) represents the external virtual work of prescribed surface and edge loads, assumed constant-directional in this paper. The assertion (1) is supposed to hold for all δU_α and δW that do not violate boundary conditions; the calculus of variations then leads to the equilibrium equations

$$M^{\alpha\beta}_{,\alpha\beta} - b_{\alpha\beta}N^{\alpha\beta} + N^{\alpha\beta}W_{,\alpha\beta} - p = 0 \tag{4}$$

$$N^{\alpha\beta}_{,\alpha} = 0 \tag{5}$$

---

[1] Here $b_{\alpha\beta}$ is defined (unconventionally) as $\frac{\partial \vec{X}}{\partial \xi^\alpha} \cdot \frac{\partial \vec{N}}{\partial \xi^\beta}$ in terms of the position vector $\vec{X}(\xi^1, \xi^2)$ and the unit (outward) surface normal $\vec{N}(\xi^1, \xi^2)$.

where p is external (inward) pressure, and tangential surface loads are assumed to be absent.

By writing the stress-resultants in terms of an Airy stress function F as

$$N^{\alpha\beta} = \varepsilon^{\alpha\omega}\varepsilon^{\beta\gamma}F_{,\omega\gamma} \tag{6}$$

where $\varepsilon^{\alpha\omega}$ is the alternating tensor, Eqs. (5) are satisfied identically[1]. Then, with the use of the usual constitutive relations

$$\begin{aligned} M^{\alpha\beta} &= \frac{Et^3}{12(1-\nu^2)}\left[(1-\nu)K^{\alpha\beta} + \nu K_\gamma{}^\gamma g^{\alpha\beta}\right] \\ N^{\alpha\beta} &= \frac{Et}{1-\nu^2}\left[(1-\nu)E^{\alpha\beta} + \nu E_\gamma{}^\gamma g^{\alpha\beta}\right] \end{aligned} \tag{7}$$

where E is Young's modulus, ν is Poisson's ratio, t is the shell thickness, and $g^{\alpha\beta}$ is the metric tensor, Eqs. (4), (2), and (3) provide the equilibrium and compatibility equations

$$D\nabla^4 W + \varepsilon^{\alpha\omega}\varepsilon^{\beta\gamma}b_{\alpha\beta}F_{,\omega\gamma} + p = \varepsilon^{\alpha\omega}\varepsilon^{\beta\gamma}F_{,\omega\gamma}W_{,\alpha\beta} \tag{8}$$

$$\left(\frac{1}{Et}\right)\nabla^4 F - \varepsilon^{\alpha\omega}\varepsilon^{\beta\gamma}b_{\alpha\beta}W_{,\omega\gamma} = -\frac{1}{2}\varepsilon^{\alpha\omega}\varepsilon^{\beta\gamma}W_{,\alpha\beta}W_{,\omega\gamma} \tag{9}$$

where

$$D = \frac{Et^3}{12(1-\nu^2)}.$$

In the case of cylindrical shells, Eqs. (8) and (9) become the familiar Karman-Donnell equations.

It will be assumed that the prescribed loading is such that a membrane state governed by *linear* theory is always available as a solution (at least approximately). Thus, if λ denotes a scalar measure of the magnitude of the external loading, it is stipulated that

$$\begin{aligned} N^{\alpha\beta} &= \lambda \overset{0}{N}{}^{\alpha\beta} = \lambda\varepsilon^{\alpha\omega}\varepsilon^{\beta\gamma}\overset{0}{F}_{,\omega\gamma} \\ M^{\alpha\beta} &= 0 \\ W_{,\alpha} &= 0 \end{aligned} \tag{10}$$

satisfies the variational equation (1). Hence, this variational equation can be written as

$$\iint_A \left[M^{\alpha\beta}\delta K_{\alpha\beta} + N^{\alpha\beta}\delta e_{\alpha\beta} + N^{\alpha\beta}W_{,\alpha}\delta W_{,\beta}\right]dA = \lambda\iint_A \overset{0}{N}{}^{\alpha\beta}\delta e_{\alpha\beta}\,dA \tag{11}$$

---

[1] This is strictly correct only if the indices denoting successive covariant differentiation may be interchanged, but such interchange is legitimate in shells of zero Gaussian curvature, and in other shells introduces errors no larger than those already inherent in Donnell-Mushtari-Vlasov theory. (See [11] for discussion of this point). This remark applies also to the derivation of Eq. (9).

where

$$e_{\alpha\beta} = \frac{1}{2}\,(U_{\alpha,\beta} + U_{\beta,\alpha}) + b_{\alpha\beta} W \tag{12}$$

is the linear part of $E_{\alpha\beta}$. Now suppose that a bifurcation off the equilibrium state (10) can occur at the critical load λ_c, and write the expansion

$$\begin{bmatrix} U_\alpha \\ W \\ N^{\alpha\beta} \\ F \\ M^{\alpha\beta} \\ K_{\alpha\beta} \\ E_{\alpha\beta} \\ e_{\alpha\beta} \end{bmatrix} = \lambda \begin{bmatrix} \overset{0}{U}_\alpha \\ \overset{0}{W} \\ \overset{0}{N}{}^{\alpha\beta} \\ \overset{0}{F} \\ 0 \\ 0 \\ \overset{0}{E}_{\alpha\beta} \\ \overset{0}{e}_{\alpha\beta} \end{bmatrix} + \varepsilon \begin{bmatrix} \overset{(1)}{U}_\alpha \\ \overset{(1)}{W} \\ \overset{(1)}{N}{}^{\alpha\beta} \\ \overset{(1)}{F} \\ \overset{(1)}{M}{}^{\alpha\beta} \\ \overset{(1)}{K}_{\alpha\beta} \\ \overset{(1)}{E}_{\alpha\beta} \\ \overset{(1)}{e}_{\alpha\beta} \end{bmatrix} + \varepsilon^2 \begin{bmatrix} \overset{(2)}{U}_\alpha \\ \overset{(2)}{W} \\ \overset{(2)}{N}{}^{\alpha\beta} \\ \overset{(2)}{F} \\ \overset{(2)}{M}{}^{\alpha\beta} \\ \overset{(2)}{K}_{\alpha\beta} \\ \overset{(2)}{E}_{\alpha\beta} \\ \overset{(2)}{e}_{\alpha\beta} \end{bmatrix} + \cdots \tag{13}$$

where $\lambda \to \lambda_c$ as $\varepsilon \to 0$; the second column on the right is the classical buckling mode, assumed unique, and normalized in magnitude in some definite way; and the succeeding members of the expansion are all orthogonal to this buckling mode in some predetermined fashion[1]. Thus, in a well-defined manner, ε can be considered to represent the contribution of the classical buckling mode to the buckled state.

The elements *within* each of the columns of (13) are constrained by the *linear* relations (2), (6), (7) and (12). The non-linear strain displacement equation (3) couples the columns, since

$$\overset{0}{E}_{\alpha\beta} = \overset{0}{e}_{\alpha\beta}$$
$$\overset{(1)}{E}_{\alpha\beta} = \overset{(1)}{e}_{\alpha\beta} \tag{14}$$

but

$$\overset{(2)}{E}_{\alpha\beta} = \overset{(2)}{e}_{\alpha\beta} + \frac{1}{2}\,\overset{(1)}{W}_{,\alpha}\overset{(1)}{W}_{,\beta}$$

[1] The only restriction that has to be placed on the choice of orthogonality condition is that it must not be possible for the buckling mode to be orthogonal to itself!

and so on. Finally, the variational statement (11), reduced by the use of (10), gives

$$\iint_A [\overset{(1)}{M}{}^{\alpha\beta}\delta K_{\alpha\beta} + \overset{(1)}{N}{}^{\alpha\beta}\delta e_{\alpha\beta} + \lambda \overset{0}{N}{}^{\alpha\beta}\overset{(1)}{W}_{,\alpha}\delta W_{,\beta}]dA +$$
$$+ \varepsilon \iint_A [\overset{(2)}{M}{}^{\alpha\beta}\delta K_{\alpha\beta} + \overset{(2)}{N}{}^{\alpha\beta}\delta e_{\alpha\beta} + (\lambda \overset{0}{N}{}^{\alpha\beta}\overset{(2)}{W}_{,\alpha} + \overset{(1)}{N}{}^{\alpha\beta}\overset{(1)}{W}_{,\alpha})\delta W_{,\beta}]dA +$$
$$+ \varepsilon^2 \iint_A [\overset{(3)}{M}{}^{\alpha\beta}\delta K_{\alpha\beta} + \overset{(3)}{N}{}^{\alpha\beta}\delta e_{\alpha\beta} + (\lambda \overset{0}{N}{}^{\alpha\beta}\overset{(3)}{W}_{,\alpha} + \overset{(1)}{N}{}^{\alpha\beta}\overset{(2)}{W}_{,\alpha} +$$
$$+ \overset{(2)}{N}{}^{\alpha\beta}\overset{(1)}{W}_{,\alpha})\delta W_{,\beta}]dA + \cdots = 0. \tag{15}$$

A variational equation for the buckling mode and load, obtained from (15) by letting $\varepsilon \to 0$, is

$$\iint_A [\overset{(1)}{M}{}^{\alpha\beta}\delta K_{\alpha\beta} + \overset{(1)}{N}{}^{\alpha\beta}\delta e_{\alpha\beta} + \lambda_c \overset{0}{N}{}^{\alpha\beta}\overset{(1)}{W}_{,\alpha}\delta W_{,\beta}]dA = 0 \tag{16}$$

and a consequence of this relation is the "energy" equation

$$\iint_A [\overset{(1)}{M}{}^{\alpha\beta}\overset{(1)}{K}_{\alpha\beta} + \overset{(1)}{N}{}^{\alpha\beta}\overset{(1)}{e}_{\alpha\beta}]dA = -\lambda_c \iint_A \overset{0}{N}{}^{\alpha\beta}\overset{(1)}{W}_{,\alpha}\overset{(1)}{W}_{,\beta}dA. \tag{17}$$

The aforementioned orthogonality condition will be chosen as

$$\iint_A [\overset{(1)}{M}{}^{\alpha\beta}\overset{(j)}{K}_{\alpha\beta} + \overset{(1)}{N}{}^{\alpha\beta}\overset{(j)}{e}_{\alpha\beta}]dA = -\lambda_c \iint_A \overset{0}{N}{}^{\alpha\beta}\overset{(1)}{W}_{,\alpha}\overset{(j)}{W}_{,\beta}dA = 0 \tag{18}$$

for $j > 1$; by (16), it also holds for $j = 0$.

Note that the constitutive equations (7) imply the symmetry relations

$$\overset{(i)}{M}{}^{\alpha\beta}\overset{(j)}{K}_{\alpha\beta} = \overset{(j)}{M}{}^{\alpha\beta}\overset{(i)}{K}_{\alpha\beta}$$
$$\overset{(i)}{N}{}^{\alpha\beta}\overset{(j)}{E}_{\alpha\beta} = \overset{(j)}{N}{}^{\alpha\beta}\overset{(i)}{E}_{\alpha\beta} \tag{19}$$

for all i and j. With the use of (17), (18), and (19) the choice $\delta U_\alpha = \overset{(1)}{U}_\alpha$ and $\delta W = \overset{(1)}{W}$ in (15) then gives

$$(\lambda - \lambda_c)\iint_A \overset{0}{N}{}^{\alpha\beta}\overset{(1)}{W}_{,\alpha}\overset{(1)}{W}_{,\beta}dA + \frac{3\varepsilon}{2}\iint_A \overset{(1)}{N}{}^{\alpha\beta}\overset{(1)}{W}_{,\alpha}\overset{(1)}{W}_{,\beta}dA +$$
$$+ \varepsilon^2 \iint_A [2\overset{(1)}{N}{}^{\alpha\beta}\overset{(1)}{W}_{,\alpha}\overset{(2)}{W}_{,\beta} + \overset{(2)}{N}{}^{\alpha\beta}\overset{(1)}{W}_{,\alpha}\overset{(1)}{W}_{,\beta}]dA + \cdots = 0. \tag{20}$$

Hence

$$\frac{\lambda}{\lambda_c} = 1 + a\varepsilon + b\varepsilon^2 + \cdots \tag{21}$$

where

$$a = -\frac{3}{2\lambda_c} \frac{\iint\limits_A \overset{(1)}{N}{}^{\alpha\beta} \overset{(1)}{W}_{,\alpha} \overset{(1)}{W}_{,\beta}\, dA}{\iint\limits_A \overset{0}{N}{}^{\alpha\beta} \overset{(1)}{W}_{,\alpha} \overset{(1)}{W}_{,\beta}\, dA} \tag{22}$$

$$b = -\frac{1}{\lambda_c} \frac{\iint\limits_A [2\overset{(1)}{N}{}^{\alpha\beta} \overset{(1)}{W}_{,\alpha} \overset{(2)}{W}_{,\beta} + \overset{(2)}{N}{}^{\alpha\beta} \overset{(1)}{W}_{,\alpha} \overset{(1)}{W}_{,\beta}]\, dA}{\iint\limits_A \overset{0}{N}{}^{\alpha\beta} \overset{(1)}{W}_{,\alpha} \overset{(1)}{W}_{,\beta}\, dA}. \tag{23}$$

A variational equation governing the elements of the ε^2 column in (13) can easily be written, but will not be needed. Instead, it is easier to work directly with the perturbation equations for $\overset{(i)}{W}$ and $\overset{(i)}{F}$ $(i = 1, 2, \ldots)$ that can be derived directly from (8) and (9) as

$$\left.\begin{aligned} &D\nabla^4 \overset{(1)}{W} + \varepsilon^{\alpha\omega}\varepsilon^{\beta\gamma} b_{\alpha\beta} \overset{(1)}{F}_{,\omega\gamma} - \lambda_c \overset{0}{N}{}^{\alpha\beta} \overset{(1)}{W}_{,\alpha\beta} = 0 \\ &\left(\frac{1}{Et}\right) \nabla^4 \overset{(1)}{F} - \varepsilon^{\alpha\omega}\varepsilon^{\beta\gamma} b_{\alpha\beta} \overset{(1)}{W}_{,\omega\gamma} = 0 \end{aligned}\right\} \tag{24}$$

$$\left.\begin{aligned} &D\nabla^4 \overset{(2)}{W} + \varepsilon^{\alpha\omega}\varepsilon^{\beta\gamma} b_{\alpha\beta} \overset{(2)}{F}_{,\omega\gamma} - \lambda_c \overset{0}{N}{}^{\alpha\beta} \overset{(2)}{W}_{,\alpha\beta} \\ &\qquad = \varepsilon^{\alpha\omega}\varepsilon^{\beta\gamma} \overset{(1)}{F}_{,\omega\gamma} \overset{(1)}{W}_{,\alpha\beta} + a\lambda_c \overset{0}{N}{}^{\alpha\beta} \overset{(1)}{W}_{,\alpha\beta} \\ &\left(\frac{1}{Et}\right) \nabla^4 \overset{(2)}{F} - \varepsilon^{\alpha\omega}\varepsilon^{\beta\gamma} b_{\alpha\beta} \overset{(2)}{W}_{,\omega\gamma} = -\frac{1}{2} \varepsilon^{\alpha\omega}\varepsilon^{\beta\gamma} \overset{(1)}{W}_{,\alpha\beta} \overset{(1)}{W}_{,\omega\gamma} \end{aligned}\right\} \tag{25}$$

and so on.

The coefficient a depends only on $\overset{(1)}{W}$, $\overset{(1)}{F}$, and vanishes whenever post-buckling behavior is independent of the sign of the buckling mode. When $a = 0$, the initial post-buckling behavior depends on b, the evaluation of which requires the determination of $\overset{(2)}{W}$ and $\overset{(2)}{F}$.

If $a = 0$ and $b < 0$, the shell is imperfection-sensitive. This is demonstrable from a repetition of the analysis, with (3) replaced by

$$E_{\alpha\beta} = e_{\alpha\beta} + \frac{1}{2} W_{,\alpha} W_{,\beta} + \frac{1}{2} (W_{,\alpha} \tilde{W}_{,\beta} + \tilde{W}_{,\alpha} W_{,\beta}) \tag{26}$$

where $\tilde{W}$ is an initial normal displacement. If $\tilde{W}$ is chosen as $\tilde{\varepsilon}\overset{(1)}{W}$, a singular perturbation expansion of the form

$$\begin{bmatrix} W \\ F \end{bmatrix} = \lambda \begin{bmatrix} \overset{0}{W} \\ \overset{0}{F} \end{bmatrix} + \varepsilon \begin{bmatrix} \overset{(1)}{W} \\ \overset{(1)}{F} \end{bmatrix} + \varepsilon^2 \begin{bmatrix} \overset{(2)}{W} \\ \overset{(2)}{F} \end{bmatrix} + \cdots$$
$$+ \tilde{\varepsilon}\varepsilon \begin{bmatrix} \overset{(1,1)}{W} \\ \overset{(1,1)}{F} \end{bmatrix} + \tilde{\varepsilon}\varepsilon^2 \begin{bmatrix} \overset{(2,1)}{W} \\ \overset{(2,1)}{F} \end{bmatrix} + \cdots$$
$$+ \tilde{\varepsilon}^2\varepsilon \begin{bmatrix} \overset{(1,2)}{W} \\ \overset{(1,2)}{F} \end{bmatrix} + \cdots \tag{27}$$

becomes appropriate, where $\lim\limits_{\varepsilon\to 0}[\lim\limits_{\tilde{\varepsilon}\to 0} \lambda] = \lambda_c$ but $\lim\limits_{\varepsilon\to 0} \lambda = 0$ for $\tilde{\varepsilon} \neq 0$. It is then found that

$$(1 - \lambda/\lambda_c)\,\varepsilon + a\varepsilon^2 + b\varepsilon^3 + \cdots = (\lambda/\lambda_c)\,\tilde{\varepsilon} + \cdots. \tag{28}$$

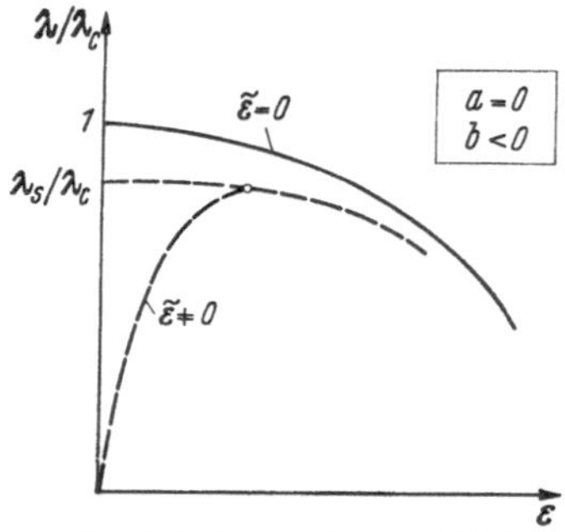

Fig. 1. Influence of small initial imperfection.

Sketches of λ/λ_c vs. ε, given by (21) and (28) with $a = 0$ and $b < 0$, are shown in Fig. 1 for $\tilde{\varepsilon} = 0$ and $\tilde{\varepsilon} \neq 0$. Let λ_s denote the maximum value of λ in the case $\tilde{\varepsilon} \neq 0$; then, as KOITER first showed in [3],

$$\left(1 - \frac{\lambda_s}{\lambda_c}\right)^{3/2} = \frac{3\sqrt{3}}{2}\,|\tilde{\varepsilon}|\,\sqrt{-b}\left(\frac{\lambda_s}{\lambda_c}\right) \tag{29}$$

so that $\left(1 - \frac{\lambda_s}{\lambda_c}\right) = 0(\tilde{\varepsilon})^{2/3}$, and small initial imperfections can induce large reductions in buckling strength.

It may be desirable to calculate the post-buckling variation of λ with the generalized displacement defined by

$$\Delta = \iint\limits_A \overset{0}{N}{}^{\alpha\beta} e_{\alpha\beta}\, dA. \tag{30}$$

(The significance of Δ is that $(\lambda\Delta)$ represents the decrease in potential energy of the applied loads.) Now let

$$\Delta_0 = \iint\limits_A \overset{0}{N}{}^{\alpha\beta} \overset{0}{e}_{\alpha\beta}\, dA \tag{31}$$

and note that letting $\delta e_{\alpha\beta} = \overset{0}{e}_{\alpha\beta}$, $\delta W = 0$ in (15) shows that

$$\iint_A \overset{(j)}{N}{}^{\alpha\beta} \overset{0}{e}_{\alpha\beta} \, dA = 0 \tag{32}$$

for all $j \neq 0$. Substitution of (13) into (30), and the use of (14) and (19) leads to

$$\Delta = \lambda \Delta_0 - \frac{\varepsilon^2}{2} \iint_A \overset{0}{N}{}^{\alpha\beta} \overset{(1)}{W}_{,\alpha} \overset{(1)}{W}_{,\beta} \, dA + \cdots. \tag{33}$$

If $a = 0$, $\varepsilon^2 \approx (\lambda - \lambda_c)/(\lambda_c b)$, and therefore the initial post-buckling stiffness $d\lambda/d\Delta$ at $\lambda = \lambda_c$ is

$$\left(\frac{d\lambda}{d\Delta}\right)_{\lambda=\lambda_c} = \left[\Delta_0 - \frac{1}{2b\lambda_c} \iint_A \overset{0}{N}{}^{\alpha\beta} \overset{(1)}{W}_{,\alpha} \overset{(1)}{W}_{,\beta} \, dA\right]^{-1}.$$

Since $d\lambda/d\Delta = \Delta_0^{-1}$ *before* buckling, the ratio K of the initial postbuckling stiffness to the prebuckling stiffness is

$$K = \left[1 - \frac{\iint_A \overset{0}{N}{}^{\alpha\beta} \overset{(1)}{W}_{,\alpha} \overset{(1)}{W}_{,\beta} \, dA}{2b\lambda_c \iint_A \overset{0}{N}{}^{\alpha\beta} \overset{0}{e}_{\alpha\beta} \, dA}\right]^{-1}. \tag{34}$$

All of the results obtained are applicable in shallow-shell theory, wherein the middle surface of the shell lies a distance $z(\xi^1, \xi^2)$ above a plane, the curvature tensor $b_{\alpha\beta}$ is taken as $-z_{,\alpha\beta}$, and the metric tensor is taken as that of the (ξ^1, ξ^2) system in the reference plane rather than in the surface.

Cylinder Analysis

Differential Equations. For circular cylinders under torsion (Fig. 2) the KÁRMÁN-DONNELL equations are

$$\left.\begin{aligned} D\nabla^4 W + \left(\frac{1}{R}\right) F_{,xx} &= S(F, W) \\ \nabla^4 F - \frac{Et}{R} W_{,xx} &= -\frac{Et}{2} S(W, W) \end{aligned}\right\} \tag{35}$$

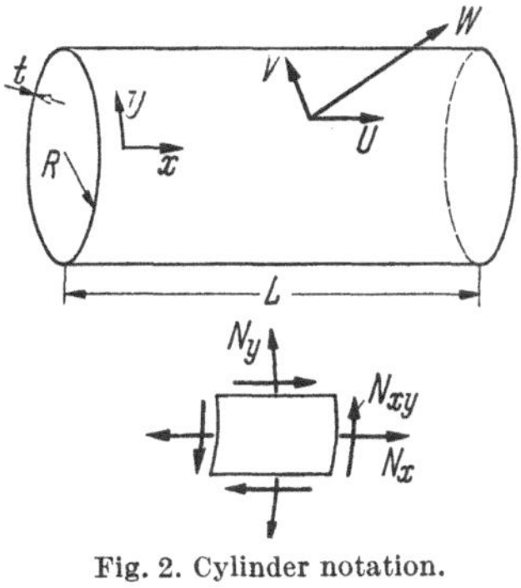

Fig. 2. Cylinder notation.

where $S(P, Q) = P_{,xx} Q_{,yy} + P_{,yy} Q_{,xx} - 2P_{,xy} Q_{,xy}$. The average shear stress $\tau \equiv \frac{1}{t} (N_{xy})av.$ will be chosen to play the role of the generalized load λ; hence, with

$$\overset{0}{F} = -txy \tag{36}$$

the appropriate perturbation expansion is

$$\begin{bmatrix} W \\ \\ F \end{bmatrix} = \tau \begin{bmatrix} 0 \\ \overset{0}{F} \end{bmatrix} + \varepsilon \begin{bmatrix} \overset{(1)}{W} \\ \overset{(1)}{F} \end{bmatrix} + \varepsilon^2 \begin{bmatrix} \overset{(2)}{W} \\ \overset{(2)}{F} \end{bmatrix} + \cdots. \tag{37}$$

The buckling equations (24) become

$$\left.\begin{aligned} D\nabla^4\overset{(1)}{W} + \frac{1}{R}\overset{(1)}{F}_{,xx} - 2\tau_c\overset{(1)}{W}_{,xy} &= 0 \\ \nabla^4\overset{(1)}{F} - \frac{Et}{R}\overset{(1)}{W}_{,xx} &= 0 \end{aligned}\right\} \tag{38}$$

and, with the anticipation that $a = 0$, Eqs. (25) are

$$\left.\begin{aligned} D\nabla^4\overset{(2)}{W} + \frac{1}{R}\overset{(2)}{F}_{,xx} - 2\tau_c\overset{(2)}{W}_{,xy} &= S(\overset{(1)}{F}, \overset{(1)}{W}) \\ \nabla^4\overset{(2)}{F} - \frac{Et}{R}\overset{(2)}{W}_{,xx} &= -\frac{Et}{2} S(\overset{(1)}{W}, \overset{(1)}{W}). \end{aligned}\right\} \tag{39}$$

Let $\xi = x/L$, $\eta = y/L$, and write

$$\left.\begin{aligned} \overset{(1)}{W} &= t\, Re[w_1(\xi) e^{i\alpha\eta}] \\ \overset{(1)}{F} &= \left(\frac{Et^2L^2}{R}\right) Re[f_1(\xi) e^{i\alpha\eta}] \end{aligned}\right\} \tag{40}$$

wherein α is a circumferential wave number making $\alpha R/L$ an integer. It can now be verified that the coefficient a, given by Eq. (21), does, in fact, vanish. Substitution of (40) into (38) gives

$$\left.\begin{aligned} & w_1'''' - 2\alpha^2 w_1'' - 2i\alpha\Lambda_c w_1' + \alpha^4 w_1 + 12Z^2 f_1'' = 0 \\ & f_1'''' - 2\alpha^2 f_1'' + \alpha^4 f_1 - w_1'' = 0 \end{aligned}\right\} \tag{41}$$

where Z is Batdorf's parameter [12]

$$Z = \frac{L^2}{Rt}\sqrt{1-\nu^2} \tag{42}$$

and

$$\Lambda_c = \frac{\tau_c L^2 t}{D}. \tag{43}$$

The functional forms of $S(\overset{(1)}{F}, \overset{(1)}{W})$ and $S(\overset{(1)}{W}, \overset{(1)}{W})$ indicate that the solution of (39) can be written

$$\left.\begin{aligned} \overset{(2)}{W} &= \frac{t}{Z}\sqrt{1-\nu^2}\,\{w_{20}(\xi) + Re[w_{22}(\xi)\,e^{2i\alpha\eta}]\} \\ \overset{(2)}{F} &= \frac{Et^2L^2}{RZ}\sqrt{1-\nu^2}\,\{f_{20}(\xi) + Re[f_{22})(\xi)\,e^{2i\alpha\eta}]\} \end{aligned}\right\} \tag{44}$$

and then, with a bar to denote complex conjugate,

$$w_{20}'''' + 12Z^2 f_{20}'' = -6\alpha^2 Z^2\, Re[(f_1 w_1)''] \tag{45}$$

$$f_{20}'''' - w_{20}'' = \frac{\alpha^2}{4}(w_1\overline{w}_1)'' \tag{46}$$

$$\left.\begin{aligned} & w_{22}'''' - 8\alpha^2 w_{22}'' - 4i\Lambda_c\,\alpha w_{22}' + 16\alpha^4 w_2 + 12Z^2 f_{22}'' \\ &\quad = -6\alpha^2 Z^2 (f_1 w_1'' + f_1'' w_1 - 2f_1' w_1') \\ & f_{22}'''' - 8\alpha^2 f_{22}'' + 16\alpha^4 f_2 - w_2'' = \frac{\alpha^2}{2}[w_1 w_1'' - (w_1')^2]. \end{aligned}\right\} \tag{47}$$

Note that a contribution to $\{\overset{(2)}{W}, \overset{(2)}{F}\}$ proportional to $\{\overset{(1)}{W}, \overset{(1)}{F}\}$ is ruled out by the orthogonality condition

$$\int_0^L dx \int_0^{2\pi R} dy\,[\overset{(1)}{W}_{,x}\overset{(2)}{W}_{,y} + \overset{(1)}{W}_{,y}\overset{(2)}{W}_{,x}] = 0.$$

From the condition that the circumferential displacement V be single valued, it follows that

$$\int_0^{2\pi R} V_{,y}\,dy = \int_0^{2\pi R}\left[\frac{1}{Et}(N_y - \nu N_x) - \frac{W}{R} - \frac{1}{2}(W_{,y})^2\right]dy = 0$$

whence, in the absence of a net end thrust,

$$f_{20}'' = w_{20} + \frac{\alpha^2}{4}(w_1\overline{w}_1). \tag{48}$$

This is consistent with (46), and now f_{20}'' can be eliminated from (45), which becomes

$$w_{20}'''' + 12Z^2 W_{20} = -3\alpha^2[w_1\overline{w}_1 + 2\,Re(f_1 w_1)'']. \tag{49}$$

Note that w_{20}, f_{20} are real, whereas W_1, f_1, w_{22}, f_{22} are complex.

Boundary Conditions. Calculations will be made for the following three sets of boundary conditions at $x = 0, L$ (see Fig. 2):

$$\text{I.}\quad W = W_{,xx} = 0 \qquad \text{(simple support)}$$

$$N_x = V_{,y} = 0; \quad \frac{1}{2\pi R}\oint_0^{2\pi R} N_{xy}\,dy = t\tau$$

$$\text{II.}\quad W = W_{,x} = 0 \qquad \text{(clamped)}$$

$$N_x = V_{,y} = 0; \quad \frac{1}{2\pi R}\oint_0^{2\pi R} N_{xy}\,dy = t\tau$$

$$\text{III.}\quad W = W_{,x} = 0 \qquad \text{(clamped)}$$

$$U_{,y} = V_{,y} = 0; \quad \frac{1}{2\pi R}\oint_0^{2\pi R} N_{xy}\,dy = t\tau; \quad \oint_0^{2\pi R} N_x\,dy = 0.$$

These sets of conditions are designated S3, C3, and C1 by YAMAKI and KODAMA [13] in their recent study of torsional buckling.

With these boundary conditions, it is easily shown that there is no loss in generality in assuming that the real parts of all functions of ξ are symmetrical, and their imaginary parts antisymmetrical, in the interval $0 \leq \xi \leq 1$.

Buckling Problem. Donnell's early approximate solutions for torsional buckling of cylinders [14] were somewhat improved upon by BATDORF, STEIN, and SCHILDCROUT [12]; but the most accurate available solutions to the eigenvalue problem (41) for a variety of boundary conditions are the essentially exact ones presented by YAMAKI and KODAMA. In the present work, however, a unified numerical approach was used in both the buckling and post-buckling problems, and the buckling results of YAMAKI and KODAMA served as a check on the accuracy of the procedure.

A standard finite-difference scheme was used to approximate Eqs. (41). The consequent difference equations, together with appropriate versions of the boundary conditions I, II, and III in terms of w_1 and f_1, provided eigenvalue problems in matrix form that were easily solved with the help of complex arithmetic programmed by means of Fortran IV on an IBM 7094 computer. (Details of the finite-difference analysis are given in the Appendix). The eigenvalues Λ_c were minimized with respect to the circumferential wave number α on the basis of the simplifying assumption that α could vary continuously; the results thus found for Λ_c and the minimizing values of α are tabulated in Tables 1, 2, and 3 for the three

Table 1. *Case I:* $W = W_{,xx} = N_x = V_{,y} = 0$

Z	Λ_c	α	$\frac{b}{1-\nu^2}$	K
0	52.67	2.51	.2139	.671
1	53.18	2.57	.1508	.582
3	56.51	2.86	−.0832	−2.00
10	74.29	3.97	−.2364	3.42
30	124.5	5.90	−.1583	−7.51
100	271.6	9.29	−.0756	− .913
300	602.9	13.39	−.0342	− .407
1,000	1482.	19.75	−.0122	− .175
10,000	8389.	37.4	−.0015	− .031

Table 2. *Case II:* $W = W_{,x} = N_x = V_{,y} = 0$

Z	Λ_c	α	$\frac{b}{1-\nu^2}$	K
0	88.58	3.79	.2513	.673
1	88.76	3.80	.2405	.663
3	89.88	3.87	.1694	.576
10	99.58	4.45	−.0682	− .938
30	141.9	6.13	−.1385	−6.23
100	286.4	9.44	−.0833	−1.212
300	617.6	13.78	−.0346	− .394
1,000	1502.	19.94	−.0124	− .179
10,000	8422.	37.5	−.0015	− .052

Table 3. *Case III:* $W = W_{,x} = U_{,y} = V_{,y} = 0$

$$\left(\oint N_x\, dy = 0\right)$$

Z	Λ_c	α	$\frac{b}{1-\nu^2}$	K
0	88.58	3.79	.256	.677
1	88.78	3.79	.247	.669
3	90.01	3.88	.192	.606
10	100.8	4.49	−.0042	− .030
30	146.6	6.35	−.0961	−1.249
100	298.5	9.86	−.0742	− .797
300	648.4	14.69	−.0350	− .342
1,000	1593.	21.45	−.0135	− .176
10,000	9094.	41.9	−.0017	− .052

sets of boundary conditions considered. Comparison of these results with those of YAMAKI and KODAMA (who actually tabulate $k_s = \Lambda_c/\pi^2$ and $\beta = \alpha/\pi$) revealed discrepancies of only small fractions of one per cent[1].

Plots of Λ_c vs. Z are shown in Fig. 3. YAMAKI and KODAMA discuss the limitations on these results associated with the occurrence of small numbers of circumferential waves, and they also make inferences concerning the relative effects of various types of boundary constraint.

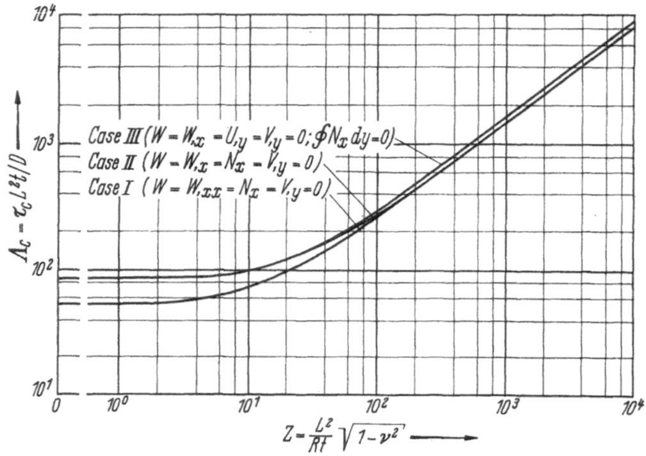

Fig. 3. Torsional buckling coefficient Λ_c.

Post-buckling Problem ($\overset{(2)}{W}$, $\overset{(2)}{F}$). The differential equations (47) and (49) for w_{22}, f_{22}, and w_{20} were also discretized and, with Λ_c, α, and the non-homogeneous right-hand sides available from the solution of the buckling problem, were solved numerically. (Again, details are given in the Appendix.) In all cases, the eigenfunctions w_1, f_1 used in the non-homogeneous terms were normalized to make $|w_1|_{\max} = 1$.

Evaluation of Post-buckling Coefficient b. Invoking the initial stipulation that $\alpha R/L$ be an integer, and exploiting the symmetry properties of the buckling and post-buckling modes, permits the formula (23) for b

---

[1] In case III, it is necessary to assume a value for Poisson's ratio. In this paper, $\nu = \frac{1}{3}$, whereas Yamaki and Kodama used $\nu = .3$; but they also made a study of the influence of ν which indicates that this small difference is quite negligible.

to be transformed to

$$\frac{b}{1-\nu^2} = \frac{3\alpha}{\Lambda_c}\left\{\int_0^{1/2} Re[-4\bar{f}_1 w_1' w_{20}' - 2 f_1 w_1' \overline{w}_{22}'' + 4 f_1'' w_1 \overline{w}_{22} - 4 f_1' w_1' \overline{w}_{22} - 4\bar{f}_1' w_1 w_{20}' + 2 f_1' w_1 \overline{w}_{22}' - 4 f_{22}(\overline{w}_1')^2 + 2 f_{20}'' w_1 \overline{w}_1 - f_{22}''(\overline{w}_1)^2 - 4 f_{22}' \overline{w}_1' \overline{w}_1]\, d\xi\right\}\left\{\int_0^{1/2} \mathrm{Im}\,(w_1 \overline{w}_1')\, d\xi\right\}^{-1} \tag{50}$$

in terms of the buckling solutions w_1, f_1 and the post-buckling solutions w_{20}, w_{22}, and f_{22}. The function f_{20}'' in (50) is given by (49) in terms of w_{20} and w_1. There is, of course, an approximation inherent in using these solutions, as well as the corresponding values of Λ_c and α, in the formula (50), since the requirement that $\alpha R/L$ be an integer was relaxed in their calculation.

The integrals in (50) were evaluated numerically; the results for $\frac{b}{1-\nu^2}$ are given in Tables 1, 2, 3 and are plotted against Z in Fig. 4.

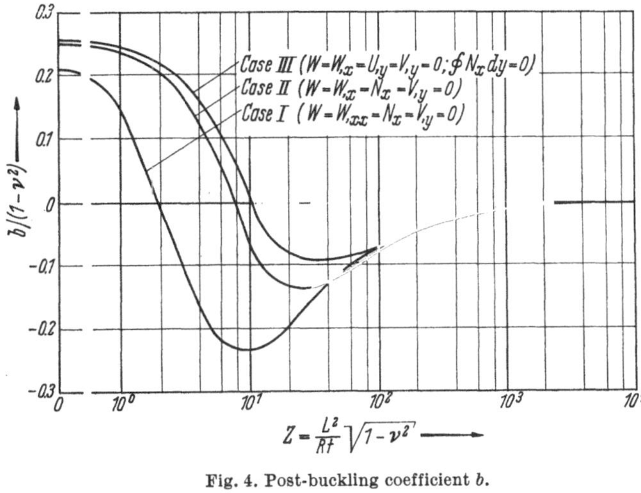

Fig. 4. Post-buckling coefficient b.

Evaluation of Post-buckling Coefficient K. Let γ be the *apparent* shear strain defined by $\gamma L = V(L) - V(0)$. Then, in the present problem, the post-buckling stiffness ratio K has the interpretation

$$K = \frac{1}{G}\left(\frac{d\tau}{d\gamma}\right) = \frac{d(\tau/\tau_c)}{d(\gamma/\gamma_c)} \tag{51}$$

where $G = \frac{E}{2(1+\nu)}$ is the shear modulus, and $\tau_c = G\gamma_c$. Eq. (39) can be manipulated into the form

$$K = \left[1 + \frac{6\alpha(1-\nu)}{\Lambda_c b} \int_0^{1/2} \mathrm{Im}\,(w_1 \overline{w}_1')\, d\xi\right]^{-1} \tag{52}$$

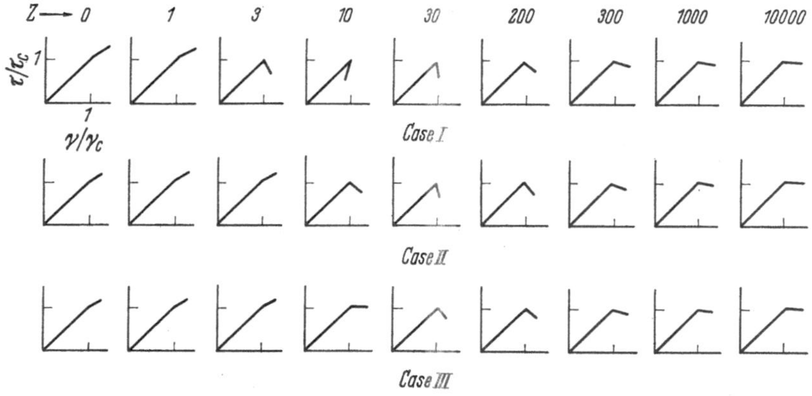

Fig. 5. Initial post-buckling relations between τ/τ_c and γ/γ_c.

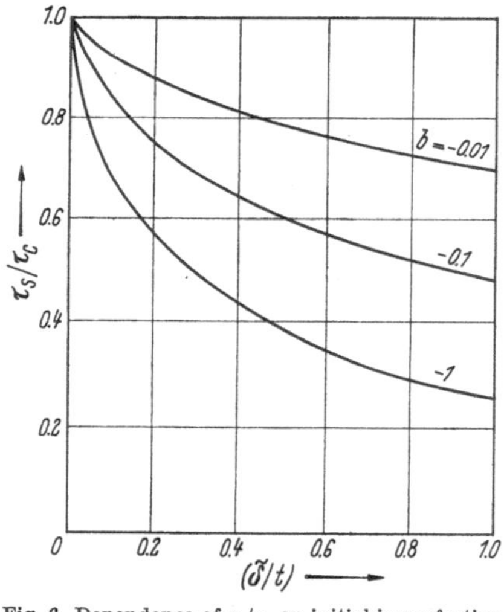

Fig. 6. Dependence of τ_s/τ_c on initial imperfection and on b.

and, for the choice $\nu = \frac{1}{3}$, provides the results in Tables 1, 2, 3. Fig. (5) shows a series of plots of τ/τ_c vs. γ/γ_c in which the initial post-buckling slopes K are shown to scale.

Discussion of Results

It is found by interpolation that the post-buckling coefficient b is negative, and hence that imperfection-sensitivity exists, in the following ranges of Z:

$$\begin{aligned} &\text{Case I:} && Z > 2.0\\ &\text{Case II:} && Z > 8.0\\ &\text{Case III:} && Z > 9.7. \end{aligned} \tag{53}$$

In all three cases, the magnitude of b tends to zero as Z becomes very large. The magnitudes of the actual buckling stresses τ_s of the cylinders in these ranges of Z cannot, of course, be predicted, since they depend on the imperfections as well as on b. However, a rough indication of the extent to which buckling strengths might be degraded is afforded by Eq. (29). Because of the way the buckling mode $\overset{(1)}{W}$ was normalized, the imperfection parameter $\bar{\varepsilon}$ in this equation can be identified with $\bar{\delta}/t$, where $\bar{\delta}$ is the maximum amplitude of that part of the initial deflection that is in the shape of this mode. The curves in Fig. (6), based on Eq. (29), show

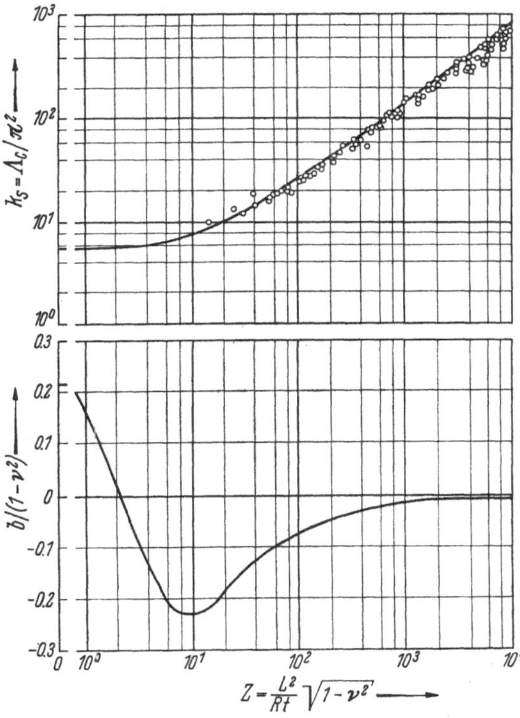

Fig. 7. Comparsion of theory and experiment.

how $\tau_s/\tau_c\,(=\lambda_s/\lambda_c)$ varies with $\bar{\delta}/\tau$ for $b = -.01, -.1$, and -1. It may be recalled (see Figure 4) that for $Z > 100$, $(-b)$ remains less than .1, and only in case I does it ever exceed .2, near $Z = 10$. Accordingly, it might be deduced from Fig. 6 that, with reasonable fabrication care to keep $\bar{\delta}/t$ less than, say, .5, τ_s should rarely be less than about 70% of τ_c.

This is more or less consistent with the experimental data collected from various sources by Batdorf, Stein, and Schildcrout in [12] and reproduced in Fig. 7 on a plot of $k_s = \Lambda_c/\pi^2$ vs. Z for case I, together with the curve for b. Unfortunately, there is little data in the range of maximum imperfection-sensitivity predicted for $Z \sim 10$. (Admittedly, thin cylinder proportions get awkward at such values of Z). Furthermore, one might expect to see a general improvement between classical theory and experiment as Z increases to large values, but this does not seem to occur. However, this kind of expectation may not be justified, since, as is evident from Eq. (29) and Fig. 6, the effect of a beneficial decrease in the magnitude of $(-b)$ could easily be nullified by a simultaneous small increase in $\bar{\delta}/t$.

The graphs in Fig. 5 have an interesting consequence. In Case I there is evidently a small range of Z around $Z = 10$, in which the apparent shear strain γ, as well as the average shear stress τ, decreases after buckling. This does not occur in cases II and III. Whenever $b < 0$, snap buckling can be expected under a monotonically increasing torque; but if $bK < 0$ snapping would also occur under an imposed monotonically increasing relative rotation of the cylinder ends. It might be amusing to try to verify this last prediction experimentally; but it should be noted that it applies only to perfect cylinders, and sufficiently large imperfections might prevent the occurrence of this kind of snapping.

Appendix

Numerical Analysis

Buckling Problem. The differential equations (41) were approximated, in the interval $0 \leq \xi \leq \frac{1}{2}$ by the N finite-difference matrix equations

$$\bar{A}(\alpha) z_{n-1} + B(\alpha) z_n + A(\alpha) z_{n+1} = 0 \qquad (n = 1, 2, \ldots N) \qquad \text{(A1)}$$

where

$$z = \begin{bmatrix} w_1 \\ w_1'' \\ f_1 \\ f_1'' \end{bmatrix} \qquad \text{(A2)}$$

$$A(\alpha) = \begin{bmatrix} \Delta^{-2} & 0 & 0 & 0 \\ -i\alpha\Lambda_c\Delta^{-1} & \Delta^{-2} & 0 & 0 \\ 0 & 0 & \Delta^{-2} & 0 \\ 0 & 0 & 0 & \Delta^{-2} \end{bmatrix} \qquad \text{(A3)}$$

$$B(\alpha) = \begin{bmatrix} -2\Delta^{-2} & -1 & 0 & 0 \\ \alpha^4 & -2(\alpha^2 - \Delta^{-2}) & 0 & 12Z^2 \\ 0 & 0 & -2\Delta^{-2} & -1 \\ 0 & -1 & \alpha^4 & -2(\alpha^2 - \Delta^{-2}) \end{bmatrix} \tag{A4}$$

$$\Delta = \frac{1}{2N} \tag{A5}$$

and the bar denotes complex conjugate. In Potter's method [15] (essentially Gaussian elimination) the equations

$$z_n = -P_n z_{n+1} \qquad (n = 0, 1, 2, \ldots N) \tag{A6}$$

are written, and then (A1) provides the recurrence matrix relation

$$P_n = [B - \bar{A} P_{n-1}]^{-1} A \qquad (n = 1, 2, \ldots N). \tag{A7}$$

In Case I, the boundary conditions imply $w_1 = w_1'' = f_1 = f_1'' = 0$ at $\xi = 0$, whence $z_0 = 0$. With the initial condition $P_0 = 0$, (A7) is used to get $P_1, P_2, \ldots P_N$.

Without loss of generality, the symmetry conditions

$$\text{Im}\,(z) = Re(z') = 0 \quad \text{at} \quad \xi = \frac{1}{2}$$

are now imposed. These conditions imply

$$\begin{aligned} z_N - \bar{z}_N &= 0 \\ z_{N+1} + \bar{z}_{N+1} - z_{N-1} - \bar{z}_{N-1} &= 0 \end{aligned} \tag{A8}$$

from which it is deduced that

$$[P_N^{-1} + \bar{P}_N^{-1} - P_{N-1} - \bar{P}_{N-1}]\, z_N = 0. \tag{A9}$$

For a given α, the eigenvalue Λ_c must therefore satisfy the 4×4 determinantal equation

$$|P_N^{-1} + \bar{P}_{N-1} - P_{N-1} - \bar{P}_{N-1}| = 0. \tag{A10}$$

Once the eigenvalue Λ_c (minimized with respect to α) was found, z_N was calculated, to within an arbitrary factor, from (A9), and a "backward" sweep based on (A6) provided all the other z's making up the eigenmode. Then z was normalized to make $|w_1|_{\max} = 1$.

For the other two cases, the boundary conditions on w_1, f_1 are easily shown to be:

$$\text{Case II}: w_1 = w_1' = f_1 = f_1'' = 0$$

$$\text{Case III}: w_1 = w_1' = f_1'' + \nu\alpha^2 f_1 = f_1''' - (2 + \nu)\,\alpha^2 f_1' = 0.$$

The only new things needed to handle these cases are appropriate initial values of the matrix P_0. In both cases the procedure used was to introduce a fictitious station at $n = -1$, write the boundary conditions in the form

$$-Gz_{-1} + Hz_0 + Gz_1 = 0 \tag{A11}$$

and also write the difference equation (A1) at $n = 0$. Then by elimination of z_{-1} it follows that

$$P_0 = (H + GA^{-1}B)^{-1}\,(G + GA^{-1}A). \tag{A12}$$

In Case II,

$$G = \begin{bmatrix} 0 & 0 & 0 & 0 \\ 1 & 0 & 0 & 0 \\ 0 & 0 & 0 & 0 \\ 0 & 0 & 0 & 0 \end{bmatrix} \qquad H = \begin{bmatrix} 1 & 0 & 0 & 0 \\ 0 & 0 & 0 & 0 \\ 0 & 0 & 1 & 0 \\ 0 & 0 & 0 & 1 \end{bmatrix} \tag{A13}$$

and in Case III,

$$G = \begin{bmatrix} 0 & 0 & 0 & 0 \\ -1 & 0 & 0 & 0 \\ 0 & 0 & -(2+\nu)\alpha^2 & 0 \\ 0 & 0 & 0 & 1 \end{bmatrix} \qquad H = \begin{bmatrix} 1 & 0 & 0 & 0 \\ 0 & 0 & 0 & 0 \\ 0 & 0 & \nu\alpha^2 & 1 \\ 0 & 0 & 0 & 0 \end{bmatrix}. \tag{A14}$$

Post-buckling Problem. The differential equations (47) are approximated by

$$A(2\alpha)\,\zeta_{n-1} + B(2\alpha)\zeta_n + A(2\alpha)\zeta_{n+1} = g_n \tag{A15}$$

with

$$\zeta = \begin{bmatrix} w_2 \\ w_2'' \\ f_2 \\ f_2'' \end{bmatrix} \qquad g = \begin{bmatrix} 0 \\ -6\alpha^2 Z^2 (f_1 w_1'' + f_1'' w_1 - 2 f_1' w_1') \\ 0 \\ \frac{\alpha^2}{2}\,[w_1 w_1'' - (w_1')^2] \end{bmatrix} \tag{A16}$$

Potters' method now says

$$\zeta_n = -Q_n \zeta_{n+1} + \chi_n \tag{A17}$$

and (A15) gives

$$Q_n = [B(2\alpha) - \bar{A}(2\alpha) Q_{n-1}]^{-1} A(2\alpha) \tag{A18}$$

$$\chi_n = [B(2\alpha) - \bar{A}(2\alpha) Q_{n-1}]^{-1} [g_n - \bar{A}(2\alpha)_{n-1} \chi_{n-1}]. \tag{A19}$$

The initial values Q_0 are the same as those for P_0 in the buckling problem, except that α is replaced by 2α wherever it appears; $\chi_0 = 0$ in all cases. The symmetry conditions (A8) now give

$$\zeta_N = [Q_{N-1} + \bar{Q}_{N-1} - Q_N{}^{-1} - \bar{Q}_N{}^{-1}]^{-1} [\chi_{N-1} + \bar{\chi}_{N-1} - P_N{}^{-1} \chi_N - \bar{P}_N{}^{-1} \bar{\chi}_N] \tag{A20}$$

and then all the ζ's are found by means of a backward sweep based on (A17).

An obvious 2×2 matrix analogue of these procedures was used to calculate w_{20} from Eq. (49). The number of intervals used in the calculations varied from $N = 50$ to $N = 90$.

References

1. Loo, T. T.: Effects of large deflections and imperfections on the elastic buckling of cylinders under torsion and axial compression, Proc. Second U. S. Nat. Cong. Appl. Mech. 1954 pp. 345–357.
2. Nash, W. A.: Buckling of initially imperfect cylindrical shells subject to torsion, J. Appl. Mech. **24**, 125–130 (1957).
3. Koiter, W. T.: On the stability of elastic equilibrium, (in Dutch), Thesis, Delft, Amsterdam 1945.
4. Koiter, W. T.: Buckling and post-buckling behavior of a cylindrical panel under axial compression, Report S. 476, Nat. Luchvaartlab., Amsterdam 1956.
5. Hutchinson, J. W.: Initial post-buckling behavior of toroidal shell segments, Int. J. Solids Structures, **3**, 97–115 (1967).
6. Hutchinson, J. W., and J. C. Amazigo: Imperfection-sensitivity of eccentrically stiffened cylindrical shells, AIAA J., **5**, No. 3, 392–401 (1967).
7. Hutchinson, J. W.: Imperfection-sensitivity of externally pressurized spherical shells, J. Appl. Mech. March 1967, 49–55.
8. Budiansky, B., and J. C. Amazigo: Initial post-buckling behavior of cylindrical shells under external pressure, Report SM-15, Harvard University, August 1967. (To be pub. in J. Math. Phys. Sept. 1968).
9. Hutchinson, J. W.: Buckling and initial post-buckling behavior of oval cylindrical shells under axial compression. J. Appl. Mech. March 1968, 66–72.
10. Sanders, J. L.: Non-linear theories for thin shells, Quart. App. Math., **21**, 21–36 (1963).
11. Koiter, W. T.: On the non-linear theory of thin elastic shells, Proc. Kon. Ned. Ak. Wet., Amsterdam, B **69**, 1961, pp. 1–54.

12. BATDORF, S. B., M. STEIN, and M. SCHILDCROUT: Critical stress of thin-walled cylinders in torsion, NACA TN No. 1344, 1947.
13. YAMAKI, N., and S. KODAMA: Buckling of circular cylindrical shells under torsion, Report 2, Rep. Inst. High Sp. Mech., Japan, vol. 18, 1966/1967, pp. 121–142.
14. DONNELL, L. H.: Stability of thin-walled tubes under torsion, NACA Report No. 479, 1933.
15. POTTERS, M. L.: A matrix method for the solution of a second order difference equation in two variables, Report MR 19, Mathematisch Centrum, Amsterdam 1955.

Discussion

N. J. HOFF: I would like to make a few remarks from the standpoint of the user of these equations, that is from the standpoint of the man who wants to solve them rather than derive them. From this standpoint I wish to thank Professor BUDIANSKY for having presented us with a very concise and clear derivation of a set of mosti convenient equations.

It might also be of interest to note that one of my doctoral students, NARASIMHAN, has calculated values of the ratio ϱ of the maximal stress an imperfect circular cylindrical shell can support to the critical stress of the classical small-displacement theory. The loading was uniformly distributed axial compression, the equations defining the problem were the KÁRMÁN-DONNELL equations, the boundary conditions the SSl simple support conditions $[w = w_{,xx} = (\sigma_x)_{ad} = (\tau_{xy})_{ad} = 0$ when $x = 0, L]$, and the initial deviations from the exact cylindrical shape were taken in the form of the buckling shape calculated from the small-displacement theory for the SSl conditions. The solution was obtained with the aid of a digital computer by integrating finite difference equations as had been done by FISCHER in Germany and STEIN in the USA.

The results show that the values of ϱ drop less rapidly with increasing amplitude of the inaccuracy than in the classical case designated as SS3 $[w = w_{,xx} = (\sigma_x)_{ad} = v = 0$ when $x = 0, L]$. For the values of the deviation amplitude expected in practice the ϱ values for SS1 do not differ significantly from those obtained for SS3. For the variation of ϱ with the radius-to-thickness ratio I would like to refer to my remarks to Professor SINGER's paper.

W. T. KOITER: I should like to join Professor HOFF in complimenting Professor BUDIANSKY on his simple and clear presentation of the theory of initial post-buckling behaviour, as applied to shallow buckling modes. My first question is whether it has been verified that the displacements are single-valued. The need for such a verification, not mentioned in the lecture, probably due to lack of time, is a minor disadvantage of the equations in terms of a stress function.

My second question refers to a study now being carried out by one of my students, who examines the post-buckling behaviour of a cylindrical shell stiffened by longitudinal stringers, employing Fourier-expansion in the solution of the equation. My question is whether Professor BUDIANSKY has found the numerical solution by finite differences to be a more effective approach than a solution in terms of Fourier-series?

B. BUDIANSKY: Single-valuedness of the displacements *was* indeed enforced, as discussed in the text of the paper, and, as you suspected, was not mentioned in the oral presentation to save time. To answer your last question, I am afraid the Fourier series *is* less suitable; in fact I first started to do this problem by the Fourier series approach and I just lost patience since the calculations just got too heavy. They can be pushed through but the finite difference scheme is much more attractive to me now.

W. F. THIELEMANN: I understood from your lecture that the sensitivity to initial imperfections becomes less strong for long cylinders. I have found experimentally that the sensitivity in long cylinders is actually stronger than for cylinders of middle length. Is there any explanation for this?

B. BUDIANSKY: A possible explanation (discussed in the text of the paper) is simply that the buckling strength degradation depends not only on b but also on the ratio of the magnitude of the imperfection to the thickness. Now, it may well be that as Z increased (perhaps as the result of thickness decreasing) the ratio of mperfection of thickness also went up and so the beneficial effect of smaller b was ounterbalanced by the effect of increasing imperfection.

The Influence of Stiffener Geometry and Spacing on the Buckling of Axially Compressed Cylindrical and Conical Shells

By

J. Singer

Technion — Israel Institute of Technology, Haifa, Israel

Abstract. An experimental and theoretical study of the buckling of closely stiffened cylindrical and conical shells under axial compression has been undertaken to determine the influence of the stiffener geometry and spacing on the applicability of linear theory. Tests on integrally ring-stiffened cylinders, in which the spacing, cross-sectional area and eccentricity of the stiffeners is varied are described. The bounds of general instability are first determined by an elementary analysis of sub-shells and panels between stiffeners, in conjunction with "smeared" stiffener theory. The interaction between stiffeners and shell is then investigated with a linear discrete-stiffener theory. The experimental results are correlated with theory and approximate design criteria are developed. Experimental results and conclusions of other investigators are also discussed. The results of a test program of integrally ring-stiffened conical shells are briefly discussed and correlated with the results obtained for cylindrical shells.

The structural efficiency of closely stiffened cylindrical shells is then studied in view of the observed bounds of applicability of linear theory.

1. Introduction

Closely stiffened shells are usually analysed by linear theory, with fairly good agreement between experiment and theory. A closer look at the experimental verification of linear theory shows that the agreement is good for closely and heavily stiffened shells whereas it is poor for moderate or sparse stiffening.

Acknowledgement. The author would like to thank Messrs. T. Weller, J. Frum, S. Nachmani, S. Regenstreif, A. Klausner, A. Greenwald and G. Shiner for their assistance during the course of the tests, to Miss A. Adler and Mrs. M. Herbst-Rosiano for assistance with computations and the staff of the Technion Computing Center for their valuable help.

This work was sponsored in part by the Air Force Office of Scientific Research OAR under Contract AF 61 (052)-905 through the European Office of Aerospace Research, United States Air Force. Based on Technion (TAE Report No. 68).

This is especially noticeable in the case of the worst "offender" of classical buckling theory — the cylindrical shell under axial compression. The very large discrepancies between experimental and theoretical buckling loads observed in unstiffened cylinders motivated extensive study of the problem (see for example [1 or 2]) as well as a major effort to stabilize the shell by internal pressure, stiffening or sandwich construction (see for example [3]). The designers also developed empirical "knock down factors", that were revised as the number of tests increased and summarized in various empirical formulae (see [4—6]). The axially compressed cylinder is therefore very suitable for study of the validity of linear theory in stiffened shells. A study of conical shells under axial compression may lend additional support to the results obtained in cylinders.

For many years the usual approach to the stability analysis of a stiffened shell was to replace it by an equivalent orthotropic shell (see for example [7], [8 or 9]). Such an approach, however, does not permit taking into account the eccentricity of stiffeners. As stiffeners became heavier, the importance of these eccentricity effects, observed already earlier [10, 11], was realized and their influence studied.

Our group at the Technion has in recent years studied the behavior of stiffened cylindrical and conical shells with a linear theory that includes eccentricity effects (see [12—14]). For cylindrical shells, the main merit of this theory, in which the stiffeners are "smeared", or "distributed" over the entire shell, lies in its simplicity that has led to its adoption also by other investigators, in particular at NASA (for example [15, 16]) and at Lockheed (for example [17]). For conical shells this theory, [18, 19], though less simple is manageable and even permitted some optimization studies [13, 20].

These studies and their conclusions are, however, meaningful only if linear theory predicts the buckling load adequately and if the discreteness of the stiffeners has no noticeable effect. The second problem is the easier one, since it can be attacked by a linear discrete-stiffener-theory. Previous investigators, [21—23], have shown that for ring stiffened cylindrical shells the discreteness effect is of importance only when the number of rings is very small, but eccentricity was not taken into account by them and stringers were not considered. Hence a more detailed discrete-stiffener analysis seems warranted.

The first problem — the adequacy of linear theory is more formidable and can be conclusively settled only by tests. The investigators differ in their opinions. In 1962 VAN DER NEUT [22] considered, on the basis of a logical expected reduction in imperfection sensitivity, "that linear theory is adequate for the investigation of general instability of stiffened shells". Recently HOFF [3] has pointed out that "it is perhaps premature to state that the small displacement theory is rigorously applicable to reinforced shells" and HUTCHINSON and AMAZIGO [24] have presented imperfection sensitivity studies to "indicate to what extent the classical buckling results can be considered reliable".

GERARD and his group developed a linear theory for orthotropic cylindrical shells [9, 25] and then embarked on an extensive test program [25, 26] to show that linear orthotropic theory is adequate. Already in 1962 they pointed out the remarkable agreement of the buckling load under axial compression for Pugliese's integrally machined ring-stiffened cylinder, tested in 1959 [27], with linear orthotropic theory. Their later tests on carefully manufactured, and one can add beautiful, ring-and stringer-stiffened cylindrical shells [27] supported their contention of the adequacy of linear theory for closely stiffened shells. From one aspect, however, these otherwise excellent tests are inadequate — they were too closely stiffened. As the width of the stiffeners equalled the distance between them, they represented really thick cylinders with longitudinal or circumferential slots rather than stiffened thin cylinders. One cannot, therefore, rely on these tests to settle the problem of applicability of linear theory for reinforced cylindrical shells. A similar difficulty arises with another series of excellent tests in which GARKISH [28] investigated the pronounced eccentricity effect that appears in longitudinally stiffened cylinders under axial compression. Here the stiffeners are back to back "L" sections, and the width of unstiffened skin is only 1/10 of the total width of the stringer.

Another series of tests on internal ring-stiffened cylindrical shells [29], though primarily concerned with buckling under hydrostatic pressure loading, includes some tests with predominant axial loading that support the adequacy of linear theory. As these tests, however, are for shells of rather low (R/h) values, about 250, and the wall thickness of the relevant specimens exhibit variations of up to $+36\%$ and -27% of the weighted average, they are not included in the discussion.

Three series of careful tests of stringer-stiffened cylinders present important evidence for evaluation of the applicability of linear theory. Card's tests with heavily integral-stiffened shells [30], the tests by PETERSON, DOW, CARD and JONES on more moderately stiffened shells, [31, 16], and Katz's tests on large scale moderately stiffened cylinders [32]. Card's tests are by now the "classical" evidence of validity of linear theory for heavy stiffening and the importance of the eccentricity effect. Katz's tests are equally important when they are judiciously correlated with linear theory, as they give some indication on the stringer area required for linear theory to become valid. Some recent tests on ringstiffened cylinders reported by ALMROTH [33] and a stringer-stiffened cylindrical panel tested by LEN'KO [34] are additional evidence for the discussion.

It may be pointed out, that in stiffened shells the boundary conditions may be even more important than in isotropic shells, since two additional effects have to be considered. The eccentricity of the applied axial load (or end moment effects) may be important here [35, 32, 36], and the in-plane boundary conditions have different effects on internal and external stiffeners [37].

If one now concurs with van der Neut's opinion and feels "that linear theory is adequate for closely spaced stiffeners", and turns to the experimental evidence for support, one cannot discern the influence of the various geometrical parameters on the applicability of linear theory. A primary aim of the present investigation is to bring forth the predominant parameters. Answers are needed to the questions what is "closely stiffened" and how "heavy" have stiffeners to be.

The problem is not only one of validity of a theory. Since the main stability contribution of stiffeners in cylindrical and conical shells under axial compression, be they rings or stringers, is the raising of the buckling load to the classical one, the problem is one of structural efficiency. The question of how heavily does it pay to stiffen has also to be considered.

2. Theoretical Considerations

The discussion in the present paper is limited to ring or stringer-stiffened shells. Other forms of stabilization, such as internal pressure, sandwich construction, corrugated skin, 45° waffle-stiffening and orthotropic materials are not considered, though they may be of equal practical importance and hence justify serious study. The stiffeners are considered to be rigidly joined to the shell and the conclusions apply therefore best to integrally stiffened shells. Furthermore, stiffeners are here considered to be of the same material as the shell though the theory can easily accommodate different materials (see for example [14 or 16]) and different materials may sometimes be more efficient, for example [38].

2.1. Ring-Stiffened Cylindrical Shells

A ring-stiffened cylindrical shell under axial compression may fail in two forms of instability, local buckling of the sub-shell between the rings or general instability of the stiffened shell as a whole. In both forms of instability axisymmetric or asymmetric modes may occur, depending on the geometry of the shell. Furthermore, there may be a noticeable restraining effect of the rings on the local buckling and there may be interaction between the two forms of instability that may lower the general instability load.

An elementary linear analysis of the buckling of an axially compressed ring-stiffened cylindrical shell considers the sub-shell separately as a simply supported isotropic shell and then examines the general instability of a shell reinforced by the "distributed" or "smeared" stiffeness of the rings.

Linear theory predicts that short simply supported sub-shells with a Batdorf parameter $Z' = (1 - \nu^2)^{1/2}\,(a^2/Rh) < 2.85$ will buckle into an axisymmetric pattern with one axial half-wave, see [39] or [40], for which

$$P_{cr} = P_{cl}\,[1 + (12Z'^2/\pi^4)]/0.702\,Z' \tag{1}$$

where P_{cl} is the classical buckling load

$$P_{cl} = [3(1 - \nu^2)]^{-1/2}\,2\pi h^2 E\,. \tag{2}$$

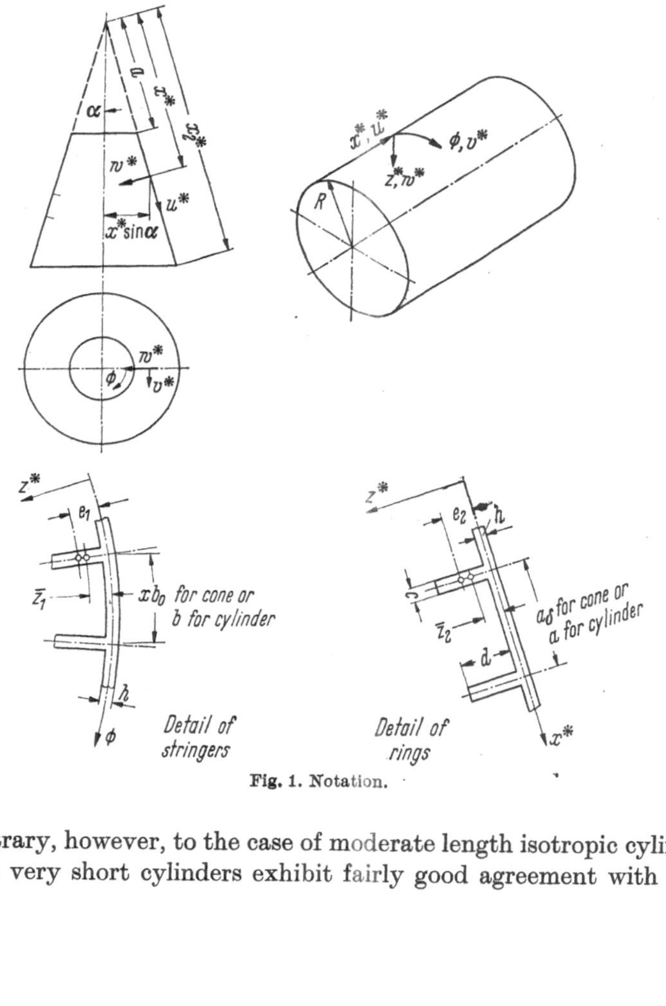

Fig. 1. Notation.

Contrary, however, to the case of moderate length isotropic cylinders, tests on very short cylinders exhibit fairly good agreement with linear theory, see for example Fig. 4 of [6], Figs. 6 to 8 of [41] or Fig. 8.7 of [42]. The axisymmetric buckling pattern of the short shells is probably the main reason for their "linear behavior", for the axisymmetric mode has a stable post-buckling behavior that makes it insensitive to imperfections, see [8]. The more pronounced influence of the boundary conditions, in particular rotational restraint, is the second reason.

Since for a clamped cylindrical shell no closed form solution is available even in linear theory, only an approximate estimate of the sub-shell geometry that ensures an axisymmetric buckling mode can be made. With aid of the second approximation of the Galerkin solution of Donnell's equations by BATDORF, SCHILDCROUT and STEIN [43] it is found that when $Z < 11.6$ the axisymmetric buckling pattern predominates. Though the rings, however heavy, will never represent fully clamped conditions, an indication of an upper bound for possible "linear" behavior of the sub-shells is thus obtained.

The elementary analysis, therefore, immediately yields a conservative criterion to ensure local "linear" behavior. By taking "classical" simple supports as the weakest practical boundary conditions of the sub-shells one obtains for safe ring spacing

$$\frac{a}{h} < [2.85(1-\nu^2)^{-1/2}\,(R/h)]^{1/2}. \tag{3}$$

Since "linear" behavior extends beyond the simple support axisymmetric range, Eq. (3) is rather conservative.

The rings actually provide elastic rotational restraints that stiffen the sub-shells also within the framework of linear theory. The rings can provide two types of restraint: resistance to twist if the sub-shell buckles non-axisymmetrically and the cross-sections of ring resist rotation one relative to the other, or resistance to "rolling over" of ring that subjects the ring to out of plane bending [44]. The rings usually offer more restraint to twist than to "rolling over" which promotes the tendency towards axisymmetric buckling.

These rotational restraints are usually neglected in buckling analysis of cylindrical shells, though they were included in Reynold's careful study of ring stiffened shells under hydrostatic pressure [45], where appreciable restraint was observed. Here, the effect of the rotational restraints is analysed by a simple one-term Rayleigh-Ritz approach, that is an extension of earlier studies on the effect of axial restraint, [46, 47]. The details are given in Appendix A of [48]. The restraints are expressed as non-dimensional spring coefficients. For torsional resistance the spring coefficient is

$$(k_T/ER^4) = GJ/ER^4 = J/2(1+\nu)R^4 \tag{4}$$

where $k_T = GJ$ is the torsional moment per twist per unit length (hence the dimensions of k_T are FL^2) and J is the torsional constant of the ring cross section. The effect of the torsional restraint appears in the final expression of the buckling load

$$(P_{cr}/Eh^2) = 2\pi \left(\frac{1}{12(1-\nu^2)(R/h)} \frac{[n^2\beta^2 + t^2]^2}{n^2\beta^2} + \right. \\ \left. + (R/h) \frac{n^2\beta^2}{[n^2\beta^2 + t^2]^2} + (k_T/ER^4)\, 2t^2 (R/h)^2 (R/L) \right). \quad (5)$$

For resistance to "rolling over" in the axisymmetric buckling mode the spring coefficient is, as in [44]

$$k_R = (EI_R/R^2) \quad (6)$$

where I_R is the moment of inertia of the cross-section for the centroidal axis in the plane of the ring. The buckling load is given by

$$(P_{cr}/Eh^2) = 2\pi \left(\frac{n^2\beta^2}{12(1-\nu^2)(R/h)} + \frac{(R/h)}{n^2\beta^2} + (k_R/ER^2)\, 4 (R/h)^2 (R/L) \right). \quad (7)$$

As the torsional restraint, when effective, is usually one order of magnitude larger than the restraint to "rolling over", the latter is neglected in the asymmetric buckling mode.

In the case of unrestrained sub-shells general instability can occur within the framework of linear theory, only when the sub-shells are in the short "axisymmetric mode" range, $Z' < 2.85$ where P_{cr} is given by Eq. (1). Otherwise the unstiffened sub-shell will always buckle locally at a lower load than the whole stiffened shell, as the critical load, according to Eq. (2) does not depend on the length of the shell. The rotational restraint determines therefore an upper bound to the ring spacing by the requirement that

$$P_{\text{general instability}} < P_{\text{local restrained}} \,. \quad (8)$$

In most of the specimens tested in the present program the influence of the rotational restraint was found to be very small. As an example of a noticeable influence of the rotational restraint three of the ring-stiffened shells of [33] are considered. In Table 1 the geometry of the shells and sub-shells are given. For simple supports (zero rotational restraint) an asymmetric local buckling pattern should appear in column (1). The heavy rings, however, offer considerable torsional resistance and hence in calculations of P_{cr} from Eq. (4) $t \to 0$ or, in other words, the sub-shell is forced to buckle axisymmetrically, column (2).

The resistance to "rolling over", that was not taken into account in the asymmetrical mode, has to be considered now, yielding column (4). It is seen that $P_{\text{rest. axisym.}} > P_{Gs}$, the "smeared" general instability load, and that even $P_{\text{axisym.}}$, column (2), not considering the resistance to „rolling over", is not less than P_{Gs}, column (6). This example shows that by forcing the subshells to buckle axisymmetrically when $Z' > 2.85$, the torsional resistance of the rings increases the safe ring spacing. Hence heavier rings, especially with high torsional stiffness, may be advantageous. Note also in Table 1 that the test results are not far from the predicted buckling loads. The imperfection sensitivity study of [33], which concludes that very short cylinders are insensitive to initial imperfections, lends further support to the expected "linear" behavior of the subshells.

A linear theory analysis for general instability of stiffened cylindrical shells under axial compression is given in [14]. It is shown there that with inside rings non-axisymmetric buckling will occur and the positive eccentricity will lower the buckling load below that for centrally placed rings. With outside rings, however, the increase in buckling load that would result from the negative eccentricity if the shell were to buckle in a chess-board pattern is not realized, since the shell now buckles in the ring-shape pattern which is unaffected by eccentricity and yields a lower buckling load.

For a ring-stiffened shell with outside rings, that buckles in the axisymmetric mode and with many waves

Table 1. *Effect of restraints on local buckling in ring-stiffened shells* [33]

$R = 8''$ $E = 10.5 \times 10^6$ psi $k_T/ER^4 = 4.20 \times 10^{-9}$

$L/R = 2.03$ $\nu = 0.33$ $k_R/ER^2 = 1.63 \times 10^{-8}$

Shell Almroth [33]	R/h	A_2/ah	e_2/h	a/h	Z'	Buckling Load Lbs. (1) $P_{\text{classical}}$ (Asym.) Lbs.	(2) $P_{\text{axisym.}}$ Lbs.	(3) $P_{\text{restrained}}/P_{\text{axisym.}}$	(4) $P_{\text{rest. axisym.}}$ Lbs.	(5) P_{exp} Lbs.	(6) P_{Gs} Lbs.
A 1	395	.878	3.0	57	7.764	16150	17370	1.107	19230	15300	17060
A 2	384	.900	2.93	56	7.710	16960	18350	1.100	20180	14830	17950
A 3	415	.834	3.16	60	8.189	14570	15340	1.124	17240	15030	15400

in the axial direction, the general instability load can be computed from a simple formula [14]

$$P_{Gs} = [3(1 - \nu^2)]^{-1/2}\, 2\pi h^3 E [1 + (A_2/ah)]^{1/2} = P_{cl}[1 + (A_2/ah)]^{1/2}. \quad (9)$$

In passing, it may be pointed out that the reduction in buckling for internal rings may be offset if the rings have high torsional stiffness, as then less energy will be absorbed in the axisymmetric buckling mode, that is unaffected by eccentricity, than in the non-axisymmetric mode that

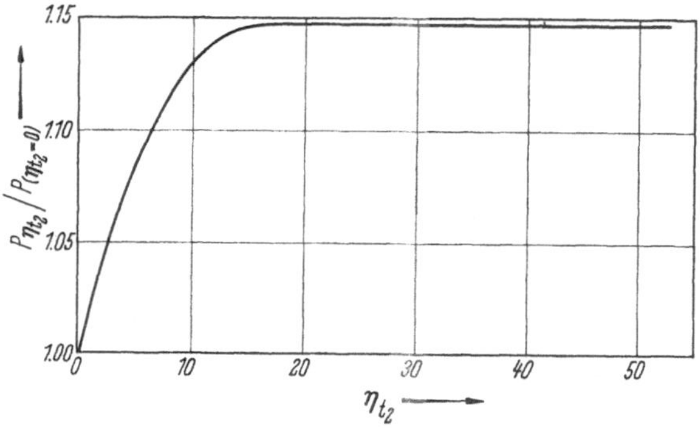

Fig. 2. The influence of torsional rigidity on critical axial load of internal-ring-stiffened cylinders.

involves torsion of the rings. Some computations for a typical shell ($L/R = 0.5$, $R/h = 500$, $A_2/ah = 0.5$, $e_2/h = 5$ and $\nu = 0.3$) show (see Fig. 2) that when $\eta_{t_2} > 15$ the shell buckles axisymmetrically with an accompanying increase of 14.7% in the general instability load, or — in other words — complete recovery of the reduction due to internal placing of the rings. For the typical shell considered, $\eta_{t_2} = 15$ can be obtained in practice (for a ring spacing of $a/h = 40$) by rings of tubular cross-section with a diameter of about 7h and a wall thickness of h.

In order to investigate the effect of discreteness of the rings. the buckling of ring-stiffened cylindrical shells is analysed by a linear "discrete" theory. Instead of being "smeared", the rings are now considered as linear discontinuities represented by the Dirac delta function, but otherwise the analysis is similar to that of [14]. It may be noted that the delta function representation of rings has been employed by other investigators, [49, 21, 23], but without consideration of the eccentricity of the rings. The details of the method used here, that is based on the formulation of [50], are given in [51], where also extensive parametric studies are discussed.

It should be pointed out that the Dirac delta function representation is satisfactory only as long as the width of the stiffeners is not comparable to the distance between them. Hence it could not be applied, for example, to the very closely stiffened shells of [26], and its reliability becomes doubtful in any shell with very wide stiffeners.

Though for buckling under hydrostatic pressure appreciable load reductions were found in [51] for discrete rings, the discreteness effect was always found to be very small for ring-stiffened cylinders under axial compression. A similar conclusion was reached in [23]. For completeness, however, the "discrete" buckling load, in addition to the "smeared" one, is computed for some of the test cylinders.

Hence, if the ring-spacing and the rotational restraint due to rings ensure axisymmetric local buckling and if the rings are placed on the outside or have high torsional stiffness to compensate for internal placing, an initially stable axisymmetric general instability should dominate and tests should agree well with predicted buckling loads.

2.2. Stringer-Stiffened Cylindrical Shells

A stringer-stiffened cylindrical shell under axial compression may again fail in two forms of instability, local buckling of the panel between the stringers or general instability of the stiffened shell as a whole. Axisymmetric buckling modes will occur, in both forms of instability, only for short shells, and hence has to be considered only in the case of stringer-stiffened shells reinforced also by strong rings. The present discussion is therefore limited to asymmetric modes, that include for general instability the $n = 1$ or "longitudinal" buckling modes mentioned in [26, 52]. There may be an appreciable restraining effect of the stringers on the local buckling and there may be interaction between local and general instability.

The elementary analysis again separates the consideration of buckling of the panels between the stringers and the study of the general instability of the "smeared-stringer" shell.

The buckling and initial post buckling behavior of cylindrical panels has been studied by Koiter [53] for stringers that exert no rotational restraint on the panel. A wide panel will buckle in the same mode and the same critical stress as the corresponding complete unstiffened cylindrical shell. A narrow panel, when the angle between the equally spaced stringers $\varphi_0 < \pi/m$ where

$$m = (1/2)\,[12(1 - \nu^2)]^{1/4} \left(\frac{R}{h}\right)^{1/2}$$

will have a higher critical stress, as is well known.

The stiffening of the panel due to "narrowness" may be written, [53] or [48],

$$\frac{(\sigma_{cr})_{\text{narrow panel}}}{(\sigma_{cr})_{\text{complete cylinder}}} = (1/\theta^2)\,(1 + \theta^4) \tag{10}$$

where

$$\theta = \frac{m\varphi_0}{\pi} = \frac{[12(1-\nu^2)]^{1/4}}{2\pi}\, b\,(Rh)^{-1/2}.$$

The "narrowness" of the panel between stringers of a longitudinally stiffened cylindrical shell is somewhat analogous to the "shortness" of the sub-shell in a ring-stiffened shell.

The "narrowness" of the panel has, however, an even more important influence on the buckling load carried, than the stiffening predicted by linear theory as given by Eq. (10). A wide panel, $\varphi_0 > \pi/m$, buckles as an unstiffened cylinder and hence the usual large discrepancies between experimental and linear theory buckling loads will appear. By investigating the initial post-buckling behavior of narrow panels and the effect of initial imperfections, KOITER [53] showed that θ is a suitable parameter for estimation of the expected "linear" buckling behavior of the panel. For perfect panels, the change from stable "plate type" behavior to unstable "cylindrical shell type" behavior occurs at $\theta = 0.64$ for prescribed load. Since the post-buckling tangent changes its direction rapidly after the transition, stiffened cylinders with $\theta < 0.64$ are advisable for predominance of general instability.

The limiting value of $\theta = 0.64$ is, however, a conservative value since Koiter's analysis assumes zero torsional constraint. The finite torsional stiffness of the stringers will raise the limiting θ, as already pointed out by KOITER, and will also stiffen the panel within the bounds of linear theory. An approximate estimate of the "linear" stiffening due to the rotational restraint of the stringers can be obtained from an analysis of the buckling of elastically restrained plates [54] in conjunction with an analysis of a clamped long curved panel [55]. A more precise analysis, that is an extension of the "discrete stiffener" theory of [51], is now being carried out at the Technion.

The general instability of the "smeared-stringer" stiffened cylindrical shell under axial compression is discussed in detail in [14]. For classical simple supports and for "classical" clamped shells. For other boundary conditions the method of [37] can be applied. Slightly less accurate methods for calculation of the general instability are also given in [56, 57].

The effect of the discreteness of the stringers is again investigated by the linear "discrete" theory of [51] extended to stringer-stiffened shells and calculations are in progress. Preliminary results indicate that for thin shells of practical dimensions the discreteness effect is negligible. This is not surprising on account of the large number of stringers required to prevent local buckling.

3. Experimental Investigation of Stiffened Cylindrical Shells

Since the adequacy of linear theory can be reliably ascertained only by test, one has to turn to experiments if one aims at definition of some bounds of applicability of the theory. The experimental program is only briefly described here and details are given in [48, 58].

3.1. Test Apparatus and Procedure

The load frame and the test set-up is shown in Fig. 3. The load is applied by two hydraulic jacks, controlled from an Amsler universal testing machine to a beam that moves a central load-transfer shaft with a thrust bearing on which the lower supporting disc fits. The upper

Fig. 3. Test set-up for stiffened cylindrical shells under axial compression.

supporting disc reacts against a load cell that records the actual load applied to the test specimen. Motion along the vertical axis is preserved by a guide pin and mating sleeve fixed to the upper and lower supporting discs (except in two tests in which the sleeve was removed).

Many strain gages (24—48) were distributed over each specimen. The strain gages served to assist in the detection of incipient buckling and to check the symmetry of load. Strain measurements were recorded on a B & F multi-channel strain plotter.

The specimens are not clamped to the supporting discs. They are just put on the lower disc, which has a low central location platform with a clearance of about h, and the similar top disc is just put on top of the specimen. To prevent end moment effects, see [32, 35, 36], the end rings of the specimens have ridges of width h that represent a continuation of the shell, see Fig. 4. The boundary conditions are therefore not far from classical simple supports, probably somewhere between SS3 and SS4 (in the notation of [59]).

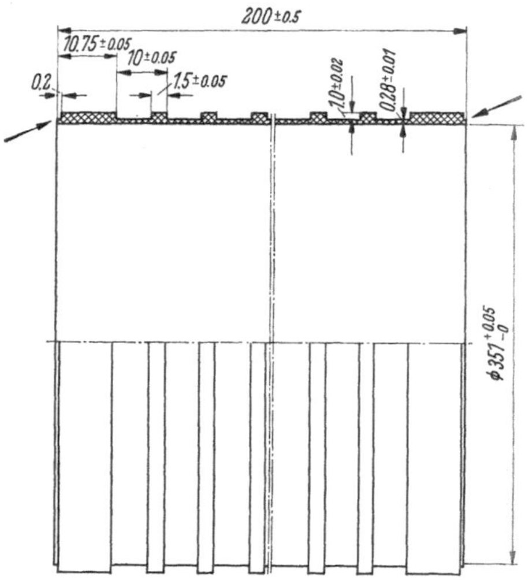

Fig. 4. Typical specimen (MZ-2).

The dimensions of the shells are carefully measured before each test. The thickness is measured at about

Table 2. *Ring-stiffened cylindrical shells — $R = 175.6$ mm*

Shell No.	Mean Thickness h [2] [mm]	Thickness Deviation [± mm]	a [mm]	a/h	R/h	Number of Rings	e_2/h	A_2/ah	I_{22}/ah^3	Z'	P'_{cl} unstiffened [kg]
MZ — 1 [1]	.283	− (.01) + (.02)	10	35.3	620	17	− 2.28	.535	.568	1.92	6090
2	.280	− (.01) + (.01)	10	35.7	627	17	− 2.32	.546	.604	1.94	5960
3	.285	− (.01) + (.02)	15	52.6	616	11	− 2.31	.361	.393	4.29	6180
4	.284	− (.01) + (.02)	15	52.8	618	11	− 2.30	.359	.386	4.30	6130
5	.274	− (.01) + (.02)	7	25.6	641	25	− 1.56	.373	.148	.97	5710
6	.253	− (.02) + (.02)	7	27.7	694	25	− 1.68	.405	.187	1.05	4870
7	.268	− (.02) + (.02)	10	37.3	655	17	− 2.06	.373	.301	2.03	5460
8	.258	− (.02) + (.02)	10	38.8	681	17	− 2.04	.370	.293	2.11	5060
9	.281	− (.01) + (.02)	7	24.9	625	25	− 1.29	.271	.057	.95	6000
10	.281	− (.01) + (.01)	7	24.9	625	25	− 1.29	.271	.057	.95	6000
11	.271	− (.01) + (.01)	10	36.9	648	17	− 1.34	.202	.048	2.00	5580
12	.270	− (.01) + (.01)	10	37.0	650	17	− 1.33	.198	.045	2.01	5540
13	.246	− (.01) + (.01)	7	28.5	714	25	− 1.77	.435	.235	1.08	4600
14	.236	− (.01) + (.01)	7	29.7	744	25	− 1.12	.211	.027	1.13	4230
15	.249	− (.01) + (.01)	8	32.1	705	22	− 1.01	.165	0.14	1.40	4710
16	.295	− (.01) + (.01)	8	27.1	595	22	− 0.88	.123	.006	1.18	6620
17	.255	− (.01) + (.02)	8	31.4	689	21	− 1.38	.264	.068	1.36	4940
18 [3]	.245	− (.02) + (.025)	8	32.7	717	21	− 1.59	.328	.130	1.42	4560

[1] In the test of shell MZ — 1 a dynamic load was introduced inadvertently and hence it is not considered a valid test point.

[2] The last figure in h is only approximate.

300 points for each shell and the stiffener dimensions were checked with a special gage for 4 shells at about 200 locations. Out-of-roundness is measured at 5 vertical stations prior to each test after the shell is in position, and correlation between initial out-of-roundness and strain gage readings is studied (see [48]).

3.2. Test Specimens

18 integrally ring-stiffened cylindrical shells were tested in the present test program. The dimensions of the shells as defined in Fig. 1 are given in Table 2. The specimens were machined from AISI 4130 steel alloy drawn tubes with a 1/4″ wall thickness. The mechanical properties of the material were measured on many specimens. The average value of Young's modulus found is $E = 2.0 \times 10^4$ kg/mm² (or 28.5×10^6 psi), the usual value of 4130 steel, and the yield stress $\sigma_{yp} > 50$ kg/mm² (or 71 000 psi) is considerably above the buckling stresses. Poisson's ratio is taken as $\nu = 0.3$.

In the interest of precision, the machining process is divided into stages. In the final stages the shell is mounted on a special "cooled" aluminium mandrel. When liquid air is poured into the reservoir of the

dimensions of specimens and Results

$L = 200$ mm $\quad L/R = 1.14 \quad \nu = 0.3 \quad E = 2 \times 10^4$ kg/mm²

P_{Gs} Approx. [kg]	P_{Gs} [14] [kg]	n	P_{discrete}	$P_{\text{Loc.}}$ s. s. [kg]	$P_{\text{Loc.}}$ corrected for "Springs" [kg]	P_{exp} [kg]	t_{exp}	t_{iers}	P_{exp} Southwell [kg]	$\frac{P_{\text{exp}}}{P_{Gs}}$	$\frac{P_{\text{exp}}}{P_{\text{Local}}}$
7540	7580	18	—	6560	6640	5060	9	2	—	.670	.762
7410	7430	18	—	6400	6490	6260	9	2	6570	.844	.965
7210	7230	18	—	6700	6760	5900	9	—	6470	.818	.872
7150	7170	18	—	6660	6720	5940	9	3	6560	.830	.884
6690	6690	19	6680	9340	9390	6070	13	2	6390	.877	
5770	5770	19	5720	7480	7530	4510	12	2	4790	.781	
6400	6410	18	—	5780	5830	4670	12	2	—	.729	.801
5920	5920	19	—	5290	5320	4420	12	2	5240	.746	.831
6770	6780	18	—	10030	10060	5780	12–13	2	—	.854	
6770	6780	18	—	10030	10060	6260	13	3	6980	.925	
6120	6130	18	—	5930	5950	5150	13	3	5920	.841	.866
6070	6080	18	—	5870	5890	4400	13	3	4710	.724	.747
5510	5510	20	5450	6930	6960	5370	13	2	5730	.952	
4660	4660	19	4650	6180	6200	3140	12	2	3430	.673	
5090	5090	18	5070	5960	5980	3850	14	2–3	—	.757	
7010	7030	17	7020	9360	9380	5950	14	3	6380	.848	
5560	5560	18	—	6350	6380	4650	16	3	4810	.837	
5260	5260	19	—	6710	6740	3630	15	2	3960	.691	

[3] Shell MZ 18 had more pronounced non-uniformities in thickness than the other test specimes

mandrel its diameter contracts 0.4 mm, enabling the shell to slide onto the mandrel. After returning to room temperature the shell sits well on the mandrel and permits accurate machining. After completion the shell is removed from the mandrel by another liquid air "cooling" and a second shell is immediately mounted. This technique, combined with extreme care in the machining and continuous measurements, has resulted in precise specimens in which the deviation of thickness (the most sensitive dimension) of the shell does not exceed 5% of the average in the worst case and is usually within 2.5%.

3.3. Ring Stiffened Cylinders

Table 2 presents the important geometric parameters of the 18 ring-stiffened cylindrical shells tested, as well as the experimental and calculated buckling loads. The dimensions of the specimens have been chosen to yield the largest feasible (R/h) ratio and an average of 660 was achieved. The ring spacing is chosen small enough for the sub-shell to be in the axisymmetric range, or slightly above it to ensure local "ring-buckling" behavior. Except in shells Nos. MZ 3 and 4, the local shell geometry parameter $Z' < 2.85$. In Shells MZ 3 and 4, however, the torsional stiffness of the rings is sufficiently large to force the sub-shell to buckle axisymmetrically in spite of $Z' > 2.85$. This is the same behavior as observed in the examples of Table 1, and indicates that Eq. (3) is a very conservative criterion and may be exceeded, provided the rings possess adequate torsional stiffness. The linear theory buckling load for very short shell exceeds the "classical moderate-length" buckling loads appreciable, see Eq. (1). Hence for shells Nos. MZ 5, 6, 9, 10, 13, 14, 15, 16, 17 and 18 local buckling is remote. In shells Nos. MZ 1, 2, 3, 4, 7, 8, 11 and 12 on the other hand, local buckling may be possible since, even after taking into account the resistance to "rolling over" (which is small in the test specimens — of the order of 1%) the local buckling load is slightly below the general instability load. In general, in the shells which were likely to buckle locally, the local buckling load was close enough to the general instability to make detection of local buckling behavior in the tests hardly feasible.

The general instability load is computed for some shells also by the "discrete theory" of [51], although the difference between "discrete" and "smeared" theory amounts here only to about 1% or less. For shells beyond the "cut off point", for which local buckling dominates, the "discrete" general instability loads are not given.

At first sight, one may wonder why the ring spacing was not kept very low in all specimens to ensure predominance of general instability.

As the aim of the present study was, however, to find the bounds of "linearity", the nominal design dimensions for some shells were intentionally chosen to be in the "doubtful" region. In Table 2, the experimental buckling loads for shells Nos. MZ 1, 2, 7, 8, 11 and 12 are also correlated with predicted local buckling loads. In the discussion, however, the correlation for all shells is for the general instability load that is of primary interest.

Before turning to the buckling loads carried by the specimens, one may discuss their observed buckling behavior. Fig. 5 shows a typical shell before and after failure. For all the shells, buckling occurred suddenly, and was rather violent in some. Visually, only the large displacement diamond patterns were detected. In shells MZ 1—14 they changed rapidly into typical plastic deformation patterns with sharp yield hinges. In some tests "travelling" of the diamond patterns could be seen momentarily, but in general the pattern appeared practically simultaneously around the whole circumference. The large inertia of the loading system is probably the main cause of the violence of the buckling process. In Shells MZ 15—18, a distance tube was incorporated to arrest the displacements much earlier, when the axial shortening was approximately 3 times the linear prebuckling one. Most of the plastic deformation was thereby eliminated (see [48]).

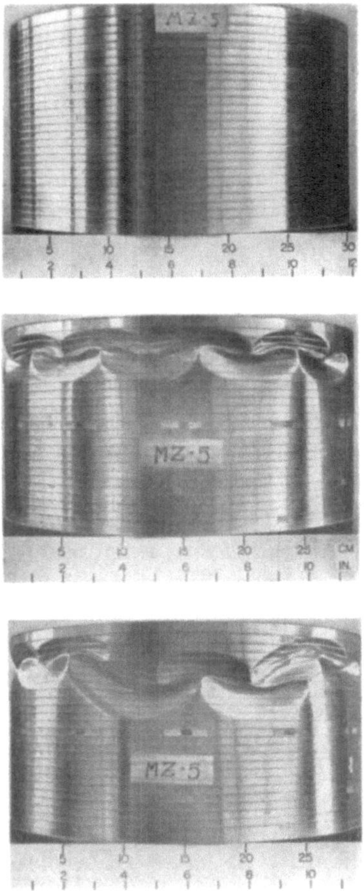

Fig. 5. Specimen MZ 5.

The buckling pattern is fairly uniform in most specimens, except some helical "climbing" of the diamonds in some shells, somewhat reminiscent of buckling patterns observed in pressurized unstiffened cylinders, see [8 or 41]. No axisymmetrical buckling patterns were observed even for shells that supported 95 percent of the linear theory buckling load P_{Gs} (in some of the ring-stiffened conical shells, discussed in Section 4, ring buckling was observed, but only for very heavy rings). It may, however, be possible that an axisymmetric pattern appears before the deformations become large, similar to the very shallow ripples indicating incipient buckling that were observed in pressurized cylinders [41]. Such an initial, briefly occuring axisymmetric pattern would fit the predictions of orthotropic theory [8 or 26].

The strain gages that "covered" the specimens proved to be excellent indicators of incipient buckling. In spite of the suddenness of the actual buckling, most of the gages showed signs of near-buckling. Hence buckling could sometimes be predicted from the gages during the test to within 5% of the load. From the strain gage readings Southwell plots could readily be made for most specimens. The computations follow the method proposed in [60] and the mean of the "perfect shell" buckling loads are given in Table 2. The results support the claims of HORTON and his associates [61] with regard to the applicability of Southwell's method.

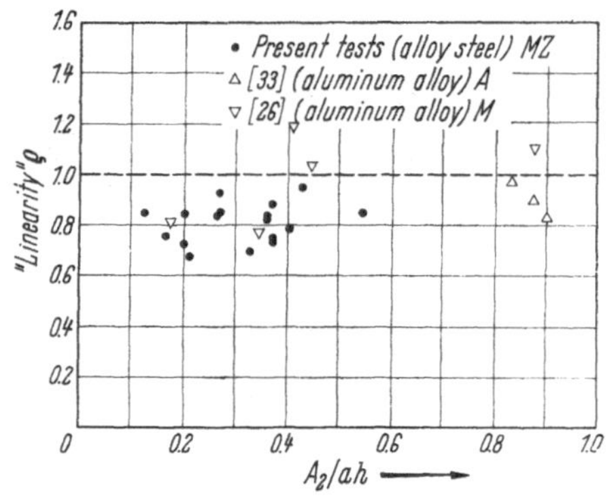

Fig. 6. "Linearity" of ring-stiffened cylindrical shells as function of ring area.

An additional interesting preliminary result of the extensive use of strain gages is the spread over the shell of the indication of incipient buckling. One does not notice isolated local indications of near-buckling, but the gages become "lively" at many locations simultaneously. In general, all, or nearly all, the strain gages along a circumference deviated in one direction (see [48] for details) indicating axisymmetric deformation. With some stretching of the imagination one could "see" in these widespread indications of incipient buckling the initial axisymmetric pattern that is missing, or some confirmation that initial buckling has a complete periodic pattern as most theories assume and which the usual diamond pattern contradicts! Obviously, much more substantial evidence is needed before one could make a definite claim, but it is an interesting thought. In one attempt to provide more evidence, the "Southwell loads" obtained from strain gages at various locations were compared and in most of the shells studied the "Southwell loads" at locations far away from the final buckling pattern location were similar to those near it.

The primary purpose of the present test series was to confirm the validity of linear theory for analysis of stiffened cylindrical shells and to obtain bounds on the stiffener parameters for upholding this validity. The ratio $\varrho = (P_{exp}/P_{Gs})$, also referred to as "linearity", is therefore the primary criterion. For the present tests ϱ is given in Table 2.

Two of the stiffener geometry parameters, the ring area ratio (A_2/ah) and the ring spacing (a/h) should predominate. The eccentricity (e_2/h), as well as (I_{22}/ah^3) do not affect the buckling of a shell with external rings,

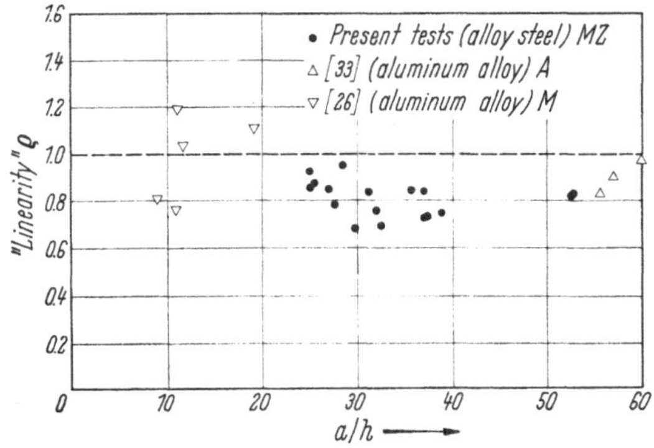

Fig. 7. "Linearity" of ring-stiffened cylindrical shells as function of ring spacing.

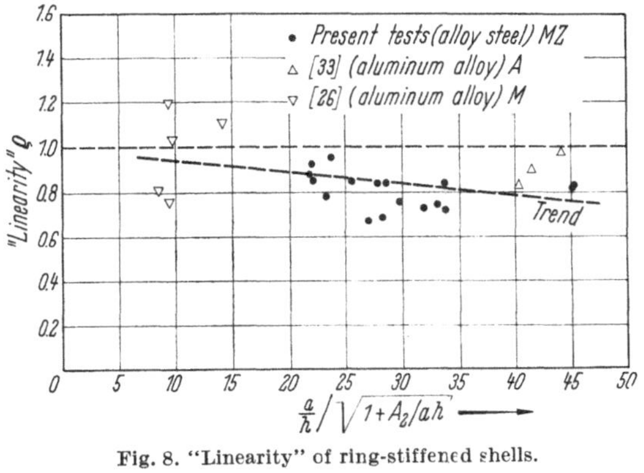

Fig. 8. "Linearity" of ring-stiffened shells.

at least in theory. Hence ϱ is plotted against (A_2/ah) in Fig. 6 and against (a/h) in Fig. 7. The ϱ obtained from three ring-stiffened shells of [33] and from some of the shells of [26] are also plotted for comparison in the figures. A slight drop in ϱ for ring area ratios below 0.2 appears in Fig. 6 and there seems to be no additional gain for $(A_2/ah) > 0.4$. There is a tendency towards a lower ϱ with increasing (a/h) in Fig. 7, offset only by the three shells of [33] which, however, have much heavier rings that restrain the sub-shell appreciably, as has already been discussed. One of the attempts to arrive at a combined parameter is shown in Fig. 8. The values of ϱ obtained in the present tests and in the experiments of other investigators on integral ring-stiffened cylindrical shells show that linear theory can be considered valid even for relatively weak rings, provided the ring spacing is not large enough to promote premature local buckling. The applicability of linear theory for ring stiffened shells under axial compression appears to be similar to that under external pressure or torsion. However, if one carries out structural efficiency studies (see Section 5) one finds that only light rings are advantageous.

3.4. Stringer-Stiffened Cylinders

A series of tests on stringer-stiffened cylinders of similar dimensions to the ring-stiffened cylinders tested has been initiated and will be reported separately [58]. The discussion will hence be limited to an evaluation of the tests of other investigators [16, 30, 31, 32, 34].

The measure of "linearity" ϱ is again plotted versus the two primary stiffener geometry parameters, area ratio (A_1/bh) and spacing (b/h) in Figs. 9 and 10 (see also Table 5 of [48]). In the case of stringer-stiffened shells, however, the eccentricity (e_1/h) and (I_{11}/bh^3) are very important.

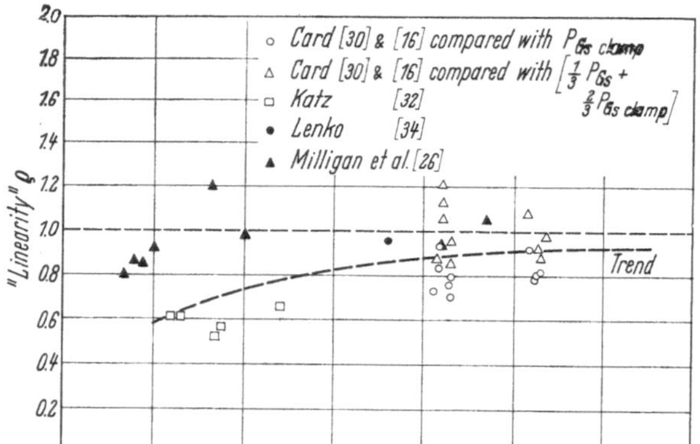

Fig. 9. "Linearity" of stringer-stiffened cylindrical shells as function of (A_1/bh).

Furthermore the boundary conditions, in particular clamping, are here more important (see for example (14, 16 or 56]). Hence the test results of [30, 31], where the boundary conditions approached clamped ends, are compared instead of with P_{Gs}, with $P_{Gs\,\mathrm{clamp}}$ and an arbitrarily chosen partial clamping $[(1/3)P_{Gs} + (2/3)P_{Gs\,\mathrm{clamp}}]$. The comparison with $P_{Gs\,\mathrm{clamp}}$ is conservative since clamping is not complete.

The results show that linear theory is also applicable to integral stringer-stiffened cylindrical shells, except for very weak stringers and wide stringer-spacing. In Fig. 10, the decrease in ϱ with increase in (b/h) is pronounced. Heavy, stringers show higher "linearity". Preliminary

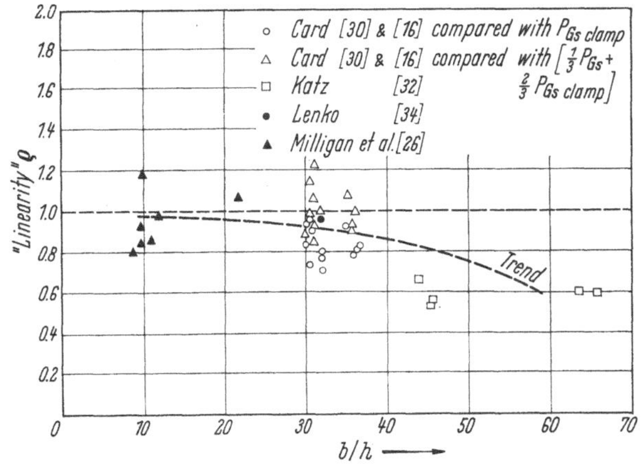

Fig. 10. "Linearity" of stringer-stiffened cylindrical shells as function of (b/h).

structural efficiency studies (see Section 5) point, however, towards weaker stringers. Further study of the relation between (A_1/bh) and (b/h), as well as the influence of (e_1/h), is therefore needed.

3.5. Further Remarks on Stiffened Cylindrical Shells

Recent imperfection studies [24] maintain that the reliability of linear theory for stiffened shells is doubtful for certain types of stiffening and shell geometries due to increased imperfection sensitivity. For ring-stiffened cylindrical shells, and in particular for shells with outside rings, Hutchinson and Amazigo [24] expect that the "shell may buckle at axial loads which are well below the classical buckling loads". The results of the present tests and those of other investigators discussed above do not support this prediction. As for stringer-stiffened cylinders tested in [16, 30, 31, 32] these can be classified as light to medium stiffening in the definition of Fig. 3 of [24]. The geometries of the shells tested (their Z) fall in the range, in Fig. 3 of [24], where imperfection sensitivity effects for outside stiffening should not be severe. This may be the reason for the high ϱ values found in the evaluation of these tests. In Garkish's recent tests [28], cylinders with outside stringers buckled at loads that were appreciably below the prediction of linear theory. However, as the Z of the specimens puts them in a range in Fig. 3 of [24] where the imperfection sensitivity effects are hardly influenced by eccentricity, these low results do not lend support to the contention of [24], and are due to other causes. Therefore one can also conclude for stringer-stiffened cylindrical shells, although with less certainty, that tests to date do not verify the fears of [24].

The "knock-down factor" as ϱ is sometimes called, is much larger for integrally stiffened shells than for unstiffened ones. If one compares some of the "knock-down factors" commonly used with the ϱ found in tests of integrally stiffened shells, one finds the commonly used factors very conservative (see Table 4 of [48]).

4. Ring-Stiffened Conical Shells

Before the present test program on cylindrical shells was initiated, a similar test program on ring-stiffened conical shells was carried out at the Technion, which continued parallel to the present program. The details are reported in [20, 62], and here only the main results are given to support the conclusion arrived at for cylindrical shells.

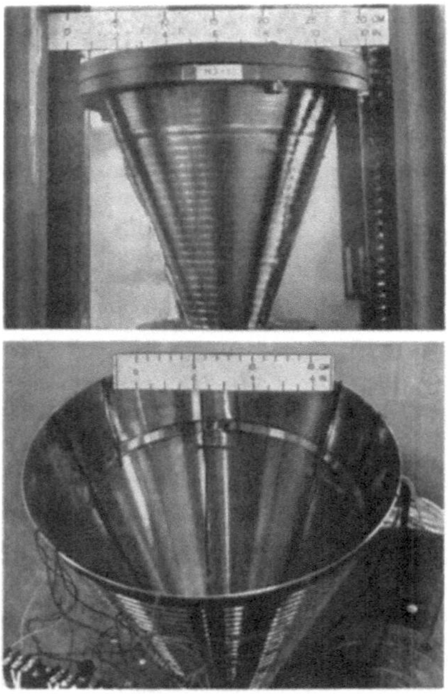

Fig. 11. Symmetrical buckling patterns in heavily stiffened conical shells.

The test arrangement for ring-stiffened conical shells is similar to that for cylindrical shells. The boundary conditions are different, however, since the cones are clamped in fixtures as explained in [20, 62]. The test specimens were machined from 17-7 PH steel blanks that are first shear-spun to form thick cones. The accuracy of the machined shells was of the same order as that of the cylindrical shells, though slightly less — the thickness variations were up to $\pm 5\%$. Fig. 11 shows one of the ring-stiffened conical shells tested. The rings were much heavier in some of the conical shells, for example $(A_2/a_0 h) = 3$ in the shell shown in the figure, and the heavier rings promoted axisymmetric buckling as can clearly be seen.

The test results on conical shells are compared in [20, 62] with an approximate formula for general instability of ring-stiffened conical shells, which is a combination of Eq. (9) for ring-stiffened cylinders and an approximate formula for unstiffened conical shells proposed by SEIDE [63].

$$P_{Gs\,\mathrm{cone}} = 2\pi h^2 E[3(1-\nu^2)]^{-1/2}(1 + A_2/a_0 h)^{1/2}\cos^2\alpha \qquad (11)$$

where α is the cone angle.

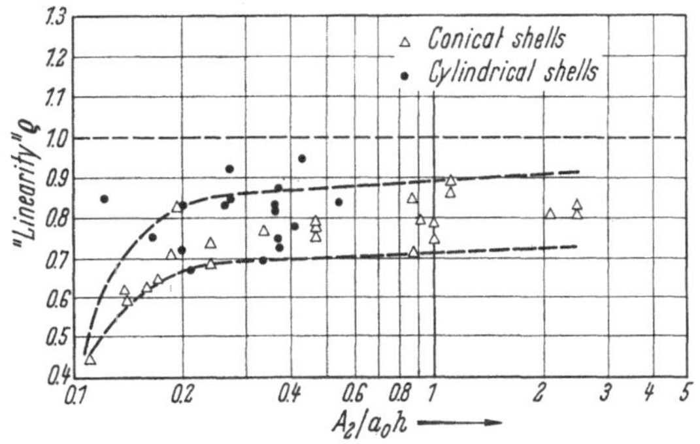

Fig. 12. Effect of ring area on "Linearity" – Comparison of conical and cylindrical shells.

The validity of Eq. (11) has been checked with the theory of [18] for some typical shells.

In Fig. 12 the variation of ϱ versus (A_1/a_0h) is shown for the shells tested. It can be seen that above $(A_2/a_0h) = 0.2$ good "linearity" is obtained and that little is gained by increasing the ring area ratio beyond 1.0. Structural efficiency again advocates lighter rings. The results of the present tests on cylindrical shells are superimposed on the results for conical shells in the figure, and, although of slightly better "linearity", the cylinders fit the trend of the conical shells well.

5. Structural Efficiency of Stiffened Cylindrical Shells

The experimental study reported here and the work of other investigators that has been discussed show that cylindrical shells with closely spaced stiffeners buckle at axial loads not far from those predicted by linear theory. The next step is to compare closely-stiffened shells with equal weight unstiffened shells.

First, one has to establish a convenient standard of comparison. Since the buckling loads of unstiffened axially compressed cylindrical shells are much below the predictions of linear theory and no reliable theoretical estimate is available, one has to rely on empirical formulae that show the primary dependence of the buckling coefficient on (R/h). A very simple formula has been proposed by PFLÜGER [5] for $R/h > 200$,

$$(P_B/P_{cl}) = 1/[1 + (R/100h)]^{1/2} \tag{12}$$

that, in addition to its simplicity, has the additional merit — for the purpose of comparison — of being unconservative for most test data. In Fig. 26 of [48] Pflüger's formula, Eq. (12), is superimposed on test results obtained by 14 investigators and found to be an upper bound for practically all the shells tested. Hence P_B from Eq. (12) is a suitable buckling load for the "equivalent" unstiffened cylinders with which the stiffened cylinders are compared.

For ring-stiffened cylindrical shells with outside rings, axisymmetric buckling predominates and the simple formula for the general instability load, Eq. (9), makes the comparison with the "equivalent" unstiffened shell very easy. The thickness of the equivalent shell (of identical weight) is

$$\bar{h} = [1 + (A_2/ah)]h \tag{13}$$

and, if Pflüger's empirical formula Eq. (12) is employed, the buckling load of the equivalent shell is given by

$$(\bar{P}_B/P_{cl}) = (\bar{h}/h)^2 [1 + (R/100\bar{h})]^{-1/2}. \tag{14}$$

Hence the efficiency of externally ring-stiffened cylindrical shells is given by

$$\eta = (P_{Gs}/\bar{P}_B) = \frac{\varrho[\Delta_R + (R/100h)]^{1/2}}{\Delta_R^{2}} \tag{15}$$

where ϱ is the fraction of the "linear" load achieved and

$$\Delta_R = 1 + (A_2/ah). \tag{16}$$

With Eq. (15) design curves can readily be drawn that give η versus (R/h) for various values of ϱ and (A_2/ah). In Fig. 13 a typical set of such curves is presented. It is immediately seen that even when only 60% "linearity" is achieved, weak ring-stiffening is very efficient for thin shells (large R/h); or, in other words, thin shells with many closely spaced rings (to prevent local buckling) and external rings of small cross-sectional area carry axial compression very efficiently.

For internal rings, asymmetrical buckling occurs and Eq. (9) is no longer valid, unless the rings have very high torsional stiffness. The critical load parameter λ has therefore to be computed from Eq. (6) of

[14], and the efficiency for internally ring-stiffened cylindrical shells becomes

$$\eta = \frac{\varrho\lambda}{8[3(1-\nu^2)]^{1/2}} \cdot \frac{[\Delta_R + (R/100h)]^{1/2}}{(R/h)\,\Delta_R^{2.5}} \tag{17}$$

where Δ_R is again given by Eq. (16). Design curves can then be drawn for internal rings. Internal rings will, naturally, be less efficient stiffeners than external ones, see [14].

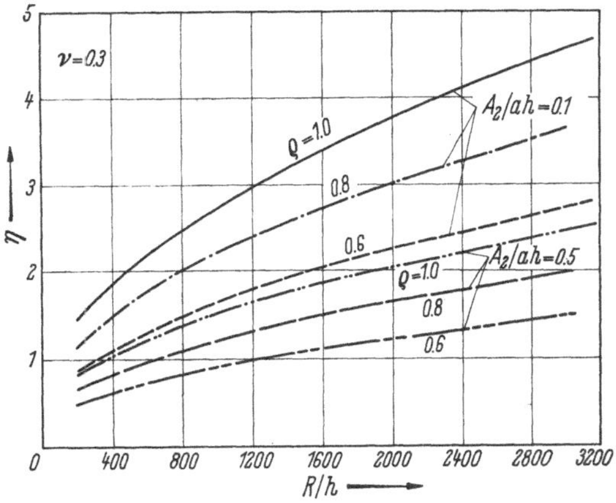

Fig. 13. Structural efficiency of ring-stiffened cylindrical shells (outside rings).

For stringer-stiffened cylindrical shells a similar expression can be obtained for the structural efficiency,

$$\eta = \frac{\varrho\lambda}{8[3(1-\nu^2)]^{1/2}} \cdot \frac{[\Delta_s + (R/100h)]^{1/2}}{(R/h)\,\Delta_s^{2.5}} \tag{18}$$

where

$$\Delta_s = 1 + (A_1/bh)\,. \tag{19}$$

Some typical design curves for external stringers are presented in Fig. 14. Again the stiffening is seen to be more efficient for lighter stringers and a thinner shell, except for (R/h) below 500 where the efficiency may rise again slightly as (R/h) decreases.

For optimization of the stiffened shell, be it stringer-or ring-stiffened, one has to balance the likely "linearity" obtained for various stiffener areas with the η for the respective stiffener area and shell (R/h). From the present tests and the results of other investigators, the range of $0.2 < A_2/ah < 0.5$ appears most promising for rings and $0.3 < A_1/bh < 0.8$ for stringers, provided, obviously, that the stiffener spacing is small enough to eliminate local buckling. Figure 15 shows this for the ring-stiffened shells of the present tests. Further study is needed for better definition of these ranges.

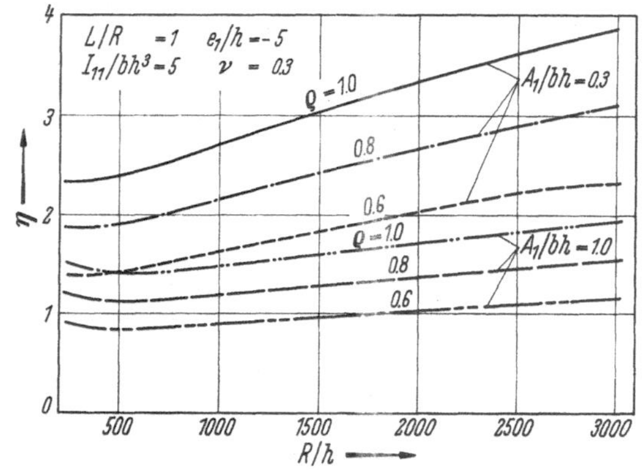

Fig. 14. Structural efficiency of stringer-stiffened cylindrical shells (outside stringers).

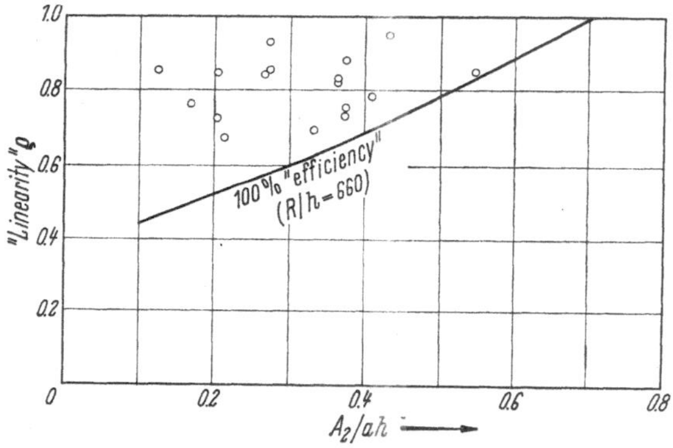

Fig. 15. "Linearity" of ring-stiffened cylindrical shells tested showing their structural efficiency.

6. Conclusions

The present tests and the results of other investigators discussed lead to the conclusion that linear theory is applicable to integrally ring-and stringer-stiffened cylindrical shells under axial compression, provided the stiffeners are closely spaced. Applicability of linear theory means here that buckling loads can be estimated with the same reliability as say for unstiffened cylindrical shells under external pressure. Small design factors for imperfections of up to 30% are not excluded, but the customary large "knock down factors" are absent even for weak integral stiffening. Hence stiffener eccentricity effects and optimization studies based on linear theory may be relied upon.

Structural efficiency studies, which include the reduced "linearity" observed in tests, show that even ring-stiffening is advantageous under axial compression. The dominant geometrical parameters that determine the "linearity" are the stiffener spacing and stiffener cross-sectional area. Ring spacing that ensures axisymmetric local buckling, or stringer spacing with Koiter's $\theta < 0.64$, will give the required predominance of general instability; and stiffener cross-sectional areas that yield $0.2 < (A_2/ah) < 0{,}5$ for rings, or $0.3 < A/_1bh < 0.8$ for stringers, appear most promising from the structural efficiency point of view.

The strain gage recordings in the tests suggest that the initial buckling pattern covers the whole shell and differs basically from the final visually observed pattern.

References

1. Fung, Y. C., and E. E. Sechler: Instability of thin elastic shells, Structural Mechanics, Proc. of the First Symposium on Naval Structural Mechanics, Stanford University, August 1958, ed. by Goodier, T. N., and N. J. Hoff Pergamon Press 1960 pp. 115—168.
2. Hoff, N. J.: The perplexing behavior of thin circular cylindrical shells in axial compression, Second Theodore von Karman Memorial Lecture, Proc. 8th Israel Annual Conference on Aviation and Astronautics, Israel J. Techn. 4 No. 1, 1—28 (February 1966).
3. Hoff, N. J.: Thin shells in aerospace structures, 4th AIAA von Karman Lecture, Astronautics and Aeronautics, **2**, 26—45 (February 1967).
4. Kanemitsu, S., and N. M. Nojima: Axial compression tests of thin circular cylinders, M. S. Thesis, California Institute of Technology, 1939.
5. Pflüger, A.: Zur praktischen Berechnung der axial gedrückten Kreiszylinderschale, Der Stahlbau, **32**, No. 6, 161—165 (June 1963).
6. Weingarten, V. I., E. J. Morgan and P. Seide: Elastic stability of thin walled cylindrical and conical shells under axial compression, AIAA Journal, **3**, No. 3, 500—505 (March 1965).
7. Bodner, S. R.: General instability of a ring-stiffened cyindrical shell under hydrostatic pressure, J. of App. Mech. **24**, No. 2, 269—277 (June 1957).
8. Thielemann, W. F.: New developments in the nonlinear theories of the buckling of cylindrical shells, Aeronautics and Astronautics, Proc. of the Durand Centennial Conference 1959 Oxford: Pergamon Press 1960 pp. 76—119.
9. Becker, H., and G. Gerard: Elastic stability of orthotropic shells, J. of the Aerospace Sciences, **29**, No. 5, 505—512 (May 1962).
10. Flügge, W.: Die Stabilität der Kreiszylinderschale, Ing.-Arch. **3**, 463—506 (1932).
11. van der Neut, A.: The general instability of stiffened cylindrical shells under axial compression, Report S-314, National Luchtvaartlaboratorium, Amsterdam, Report and Transactions, Vol. 13, S. 57—84, 1947.
12. Baruch, M., and J. Singer: Effect of eccentricity of stiffeners on the general instability of stiffened cylindrical shells under hydrostatic pressure, J. of Mech. Eng. Sci. **5**, No. 1, 23—27 (March 1963).

13. SINGER, J., and M. BARUCH: Recent studies on optimization for elastic stability of cylindrical and conical shells, presented at the Royal Aeronautical Society Centenary Congress, London, September 12—16, 1966. Aerospace Proceedings 1966, pp. 751—782.
14. SINGER, J., M. BARUCH and O. HARARI: On the stability of eccentrically stiffened cylindrical shells under axial compression. Int. J. of Solids and Structures, **3**, No. 4, 445—470 (July 1967); also TAE Report No. 44, Technion Research and Development Foundation, Haifa, Israel, December 1965.
15. BLOCK, D. L., M. F. CARD and M. M. MIKULAS: Buckling of eccentrically stiffened orthotropic cylinders, NASA TN D-2960, August, 1965.
16. CARD, M. F., and R. M. JONES: Experimental and theoretical results for buckling of eccentrically stiffened cylinders, NASA TND-3639, October, 1966.
17. BURNS, A. B.: Structural optimization of axially compressed cylinders, considering ring-stringer eccentricity effects, J. of Spacecraft and Rockets, **3**, No. 8, 1263—1268 (August 1966).
18. BARUCH, M., and J. SINGER: General instability of stiffened circular conical shells under hydrostatic pressure, The Aeronautical Quarterly, Vol. 26, Part 2, May 1965 p. 187; also TAE Report 28, Technion Research and Development Foundation, Haifa, Israel, June 1963.
19. BARUCH, M., J. SINGER and O. HARARI: General instability of conical shells with non-uniformly spaced stiffeners under hydrostatic pressure, Proc. of the 7th Israel Annual Conference on Aviation and Astronautics, Israel J. of Techn. **3**, No. 1, 62—71 (February 1965).
20. SINGER, J., A. BERKOVITS, T. WELLER, O. ISHAI, M. BARUCH and O. HARARI: Experimental and theoretical studies on buckling of conical and cylindrical shells under combined loading, TAE Report 48, Technion Research and Development Foundation, Haifa, Israel, June 1966.
21. MOE, J.: Stability of ring-reinforced cylindrical shells under lateral Pressure, Publications, International Association for Bridge and Structural Engineering, Vol. 18, 1958, pp. 113—136.
22. VAN DER NEUT, A.: General instability of orthogonally stiffened cylindrical shells, Collected Papers on Instability of Shell Structures 1962, NASA TND-1510, December 1962, pp. 309—319.
23. BLOCK, D. L.: Influence of ring stiffeners on instability of orthotropic cylinders in axial compression, NASA TND-2482, October 1964.
24. HUTCHINSON, J. W., and J. C. AMAZIGO: Imperfection-sensitivity of eccentrically stiffened cylindrical shells, AIAA Journal, **5**, No. 3, 392—401 (March 1967).
25. GERARD, G.: Elastic and plastic stability of orthotropic cylinders Collected Papers on Instability of Shell Structures 1962 NASA TND-1510, December 1962, pp. 277—295.
26. MILLIGAN, R., G. GERARD, C. LAKSHMIKANTHAM and H. BECKER: General instability of orthotropic stiffened cylinders under axial compression AIAA Journal, **4**, No. 11, 1906—1913 (November 1966); also Report AFFDL-TR-65-161, Air Force Flight Dynamics Laboratory, USAF, Wright Patterson Air Force Base, Ohio, July 1965.
27. PUGLIESE, P. J.: Tank compression test-model DM-18, Douglas Aircraft Co., Report SM-27650, February 1959.
28. GARKISCH, H. D.: Experimentelle Untersuchung des Beulverhaltens von Kreiszylinderschalen mit exzentrischen Längsversteifungen (to be published as a DFL Forschungsbericht 1967).

29. MIDGLEY, W. R., and A. E. JOHNSON: Experimental buckling of internal integral ring-stiffened cylinders, Experimental Mechanics, 7, No. 4, 145—153 (April 1967).
30. CARD, M. F.: Preliminary results of compression tests on cylinders with eccentric longitudinal stiffeners, NASA TMX-104, September, 1964.
31. PETERSON, J. P., and M. B. DOW: Compression tests on circular cylinders stiffened longitudinally by closely spaced Z-section stringers, NASA MEMO 2-12-59L, 1959.
32. KATZ, L.: Compression tests on integrally stiffened cylinders, NASA TMX-55315, August 1965.
33. ALMROTH, B. O.: Influence of imperfections and edge restraint on the buckling of axially compressed cylinders, presented at the AIAA/ASME 7th Structures and Materials Conference, Cocoa Beach, Florida, April 18—20, 1966.
34. LEN'KO, O. N.: The stability of orthotropic cylindrical shells, Raschet Prostranstvennykh Konstruktsii, Issue IV, pp. 499—524, Moscow 1958, Translation NASA TT F-9826, July 1963.
35. DE LUZIO, A., C. E. STUHLMAN and B. O. ALMROTH: Influence of stiffener eccentricity and end moment on stability of cylinders in compression, AIAA Journal, 4, No. 5, 872—877 (May 1966).
36. BLOCK, D. L.: Influence of prebuckling deformations, ring stiffeners and load eccentricity on the buckling of stiffened cylinders, presented at the AIAA/ASME 8th Structures, Structural Dynamics and Materials Conference, Palm Springs, California, March 29—31, 1967.
37. SOONG, T. C.: Influence of boundary constraints on the buckling of eccentrically stiffened orthotropic cylinders, presented at the 7th International Symposium on Space Technology and Science, Tokyo, May, 1967.
38. BURNS, A. B.: Optimization cylinders with contrasting materials and various ring/stringer configurations, J. of Spacecraft and Rockets, 4, No. 3, 375—385 (March 1967).
39. BATDORF, S. R.: A simplified method of elastic stability analysis for thin cylindrical shells, NACA Report 874, 1947.
40. TIMOSHENKO, S. P., and G. M. GERE: Theory of elastic stability, second ed., New York: McGraw-Hill 1961 pp. 465—467.
41. HARRIS, L. A., H. S. SUER, W. T. SKENE and R. J. BENJAMIN: The stability of thin-walled unstiffened circular cylinders under axial compression, including the effects of internal pressure, J. Aeron. Sci. 24, No. 8, 587—596 (August 1957).
42. GERARD, G.: Introduction to structural stability theory, New York: McGraw-Hill 1962 p. 145.
43. BATDORF, S. B., M. SCHILDCROUT and M. STEIN: Critical stress in thin-walled cylinders in axial compression, NACA Report 887, 1947.
44. FLÜGGE, W.: Stresses in Shells, Berlin/Heidelberg/New York: Springer 1960, p. 428 and p. 480.
45. REYNOLDS, T. E.: Elastic lobar buckling of ring-supported cylindrical shells under hydrostatic pressure, David Taylor Model Basin Report 1614, September 1962.
46. SINGER, J.: The effect of axial constraint on the instability of thin circular cylindrical shells under external pressure, J. Appl. Mech. 27, No. 4, 737—739 (December 1960).
47. SINGER, J.: The effect of axial constraint on the instability of thin cylindrical shells under uniform axial compression, Int. J. Mech. Sci., 4, No. 2, 253—258 (1962).

48. SINGER, J.: The influence of stiffener geometry and spacing on the buckling of axially compressed cylindrical and conical shells, TAE Report 68, Technion Research and Development Foundation, Haifa, Israel, October 1967.
49. STEIN, M., S. L. SANDERS and H. CRATE: Critical stress of ring-stiffened cylinders NACA Report 989, 1951.
50. BARUCH, M.: Equilibrium and stability equations for discretely stiffened shells, Israel J. Techn., **3**, No. 2, 138—146 (June 1965).
51. SINGER, J., and R. HAFTKA: Buckling of discretely ring-stiffened cylindrical shells, TAE Report 67, Technion Research and Development Foundation, Haifa, Israel, August 1967.
52. APPEL, H.: Buckling Modes for Orthotropic Circular Cylinders under Axial Compression for various Combinations of Stiffness Parameters, Deutsche Forschungsanstalt für Luft- und Raumfahrt, Report DLR FN 65—47.
53. KOITER, W. T.: Buckling and post-buckling behavior of a cylindrical panel under axial compression, Report S, 476, National Luchtvaartlaboratorium, Amsterdam, Reports and Transactions, Vol. 20, 1956.
54. LUNDQUIST, E. E., and E. Z. STOWELL: Compressive stress for flat rectangular plates supported along all edges and elastically restrained against rotation along the unloaded edges, NACA Report 733, 1942.
55. LEGGETT, D. M.: The buckling of a long curved panel under axial compression, Reports and Memorandum No. 1899, Ministry of Aviation, ARC 1942.
56. HEDGEPETH, J. M., and D. B. HALL: Stability of stiffened cylinders, AIAA Journal, **3**, No. 12, 2275—2286 (December 1965).
57. GEIER, B.: Beullasten versteifter Kreiszylinderschalen, Jahrbuch WGLR 1965, Vieweg, 1966 pp. 440—447.
58. SINGER, J. and T. WELLER: Experimental studies on buckling of integrally stiffened cylindrical shells under axial compression, to be published as a TAE Report, Technion Research and Development Foundation, Haifa, Israel.
59. HOFF, N. J., and T. C. SOONG: Buckling of circular cylindrical shells in axial compression, Int. J. of Mech. Sci. **7**, No. 7, 489—520 (July 1965).
60. GALLETLY, G. D., and T. E. REYNOLDS: A simple extension of Southwell's method for determining the elastic general instability pressure for ring-stiffened cylinders subject to external hydrostatic pressure, Proc. of the Society for Experimental Stress Analysis, Vol. 13, No. 2, 1956 pp. 141—152.
61. HORTON, W. H., F. L. CUNDARI and R. W. JOHNSON: The analysis of experimental data obtained from stability studies on elastic column and plate structures, Proc. of the 9th Israel Annual Conference on Aviation and Astronautics, Israel J. Techn., **5**, Nos. 1—2, 104—113 (February 1967).
62. WELLER, T., and J. SINGER: Further experimental studies on buckling of ring-stiffened conical shells under axial compression, TAE Report 70, Technion Research and Development Foundation, Haifa, Israel, November 1967.
63. SEIDE, P.: Axisymmetrical buckling of circular cones under axial compression, J. Appl. Mech., **23**, No. 4, 625—628 (December 1956).

Discussion

N. J. HOFF: One explanation of the fact that the values of the ratios ϱ of the experimental buckling stress to the critical stress of the classical small-displacement theory obtained with stiffened shells are much closer to unity than those measured in tests on unstiffened shells, is the low value of the effective radius-to-thickness ratio $(R/h)_{\text{eff}}$. This should be calculated from the effective wall thickness to take into account the average bending rigidity of the wall-and-stiffener combination. The effective value of R/h has been in the neighborhood of 100 with stiffened cylindrical shells and in this region ϱ is much higher than at $R/h = 800$ or $R/h = 2000$ which are values characterizing many unstiffened test specimens.

The reason for this dependence of ϱ on R/h is not difficult to understand. In a first approximation the permissible inaccuracy, or workshop tolerance, for the manufactured articles is proportional to the maximal dimension of the article, which is the diameter $2R$ in the case of the cylindrical shell. This means that δ/h, the deviation amplitude divided by the wall thickness, is proportional to R/h. But ϱ decreases rapidly with increasing δ/h according to the KOITER theory [In the case of stiffened shells δ/h should be replaced by $(\delta/h)_{\text{eff}}$]. When this assumption was introduced in KOITER's formula, good agreement was obtained with experimental results as shown in the two papers listed as references (2) and (3) in Professor SINGER's paper.

J. SINGER: The reduction in effective value of (R/h) is only a partial explanation for the high values of ϱ obtained in tests of integrally stiffened shells. Comparison of ϱ_{eff} calculated for the suggested effective (R/h) with the ϱ obtained in the present tests and those of Refs. [26], [30] and [32] shows that the test values are in most cases about twice ϱ_{eff}. This is in particular true for closely spaced stiffeners of relatively small cross-sectional area, which show high "linearity" with only a small reduction in effective (R/h). Details of the comparison are given in Ref. [48] of the paper.

J. W. HUTCHINSON: I would like to attach a note of caution concerning Professor SINGER's remarks on the adequacy of the linear (classical) buckling analysis as applied to stiffened cylinders. Professor SINGER's specimens were machined with considerable precision and his tests were carried out under laboratory conditions; even so, two specimens buckled at loads just below 70% of the classical buckling load and about half of the specimens buckled at around 80% of the classical. The catastrophic character of the buckling as evident from the collapsed shells is indicative of a highly unstable postbuckling behaviour. This strongly suggests that a more imperfect specimen (more typical, perhaps, of an actual structure in service) might buckle at a load somewhat lower than those recorded in Professor SINGER's test series.

J. SINGER: In our tests we tried to be careful in two ways. On one hand we aimed at good specimens and this helped us to approach "linear" loads. On the other hand we were also careful not to introduce "beneficial" errors in the boundary condition, that compensate in many cases for the less perfect structure in practice.

Also, it may be noted that most of the specimens buckled above 80%, and practically all of those that buckled below were either weakly stiffened or had slightly large ring spacing. We included "weak" specimens, since we were not trying to design efficient stiffened shells but aimed at the weakest stiffening that would still promote buckling of the shell near the linear theory load.

Hence I want to defend linear theory. I feel that the tests show that it is applicable also to closely integrally stiffened cylindrical and conical shells under axial compression, with the same reliability, as for unstiffened cylindrical shells under external pressure or torsion.

On the Postbuckling Equilibrium and Stability of Thinwalled Circular Cylinders under Axial Compression

By

W. F. Thielemann and **M. E. Esslinger**

Deutsche Forschungsanstalt für Luft- und Raumfahrt e. V.,
Braunschweig, Germany.

1. Introduction

The design of thinwalled compressed circular cylinders is a difficult task, since the buckling loads of actual cylinders scatter over a wide region.

It is known that the scatter of the buckling loads is caused by initial imperfections and that the sensibility of a thinwalled cylinder to such imperfections is closely related to its postbuckling behavior. Consequently for investigations of the sensibility to initial imperfections, the postbuckling behavior of the cylinder has to be studied. In some cases the load carrying capacity of a buckled cylinder is of practical interest too.

The first theoretical investigations on the postbuckling behavior of thinwalled circular cylinders have been performed by VON KÁRMÁN and TSIEN [1] in 1941. These investigators calculated the postbuckling states of equilibrium of an axially compressed cylinder and found postbuckling loads far below the buckling load. The calculations were based on a number of simplifications; the most important one amongst these was, that the boundary conditions were disregarded.

In the following twentyfive years many investigators tried to improve the calculations of VON KÁRMÁN and TSIEN in order to find the smallest load which the axially compressed cylinder carries in the postbuckling region. It was intended to define this postbuckling load as lower limit of the scatter region of the buckling loads. This search for the smallest postbuckling load came to an end, when HOFF, MADSON and MAYERS [2] found that in their calculations the minimum postbuckling load tended towards zero with improved functions for the solution of the governing differential equations. Then systematic experimental postbuckling investigations [3] on cylinders with different lengths revealed, that the postbuckling behavior depends strongly on the length of the cylinder.

Disregarding the boundary conditions in the previous mentioned postbuckling calculations means disregarding the length of the treated cylinder, and thus resulted in the postbuckling load of the infinitely long cylinder, which are zero.

KOITER [4] outlined another method for the determination of the sensibility of the cylinder to initial imperfections. He studied the postbuckling behavior in the immediate neighbourhood of the buckling load: The more the postbuckling load decreases with increasing depth of the buckles, the more sensible is the cylinder to initial imperfections.

In this paper a new proposal is presented to find design loads for actual thinwalled cylinders. In the first part of the paper the experimental postbuckling behavior is discussed. Thereby some fundamental regularities are found and on this basis the theoretical problem is formulated.

In the second part of the paper this problem is treated mathematically. A procedure for the determination of postbuckling states and their stability is described. The application of the procedure is demonstrated using simple examples and the obtained numerical results are compared with test results.

Finally an outlook is given to the calculation of design curves for thinwalled cylinders under axisymmetric loads.

2. Notations

$A_{11}, A_{12}, A_{22}, A_{33}$	[cm/kg]	stiffness coefficients for membrane stresses defined in Appendix, section 1.2
B_{jm}	[kg/cm³]	combination of stiffness coefficients and wave numbers, defined in Appendix, section 1.4
$D_{11}, D_{12}, D_{22}, D_{33}$	[kg cm]	stiffness coefficients for bending and twisting stresses, defined in Appendix, section 1.3
E	[kg/cm²]	Young's modulus
M_x, M_y, M_{xy}	[cmkg/cm]	bending and twisting moments, respectively, per unit length
N_x, N_y, N_{xy}	[kg/cm]	membrane forces per unit length
N_{x0}	[kg/cm]	axial membrane force per unit length at the edges of the cylinder, positive for tension
R_x	[kg/cm]	radial shear force per unit length at the edges of the cylinder
$T_{11}, T_{12}, T_{22}, T_{33}$	[cm/kg]	reciprocal stiffness coefficients for bending and twisting stresses, defined in Appendix, section 1.3
$V_{\tilde{j}\tilde{m}}$	[1/kgcm³]	combination of stiffness coefficients and, wave numbers, defined in Appendix, section 1.4
a_{00}	[cm]	radial displacement, constant over the entire surface of the cylinder, Eq. (9)

a_{jm}	[cm]	free coefficients of the series, Eq. (13), for the radial displacement w
$b_{\tilde{j}\tilde{m}}$	[kgcm]	free coefficients of the series, Eq. (12), for the stress function Φ
$f_j(y)$, $f_m(x)$		functions of the series, Eq. (13), for the radial displacement w
$g_{\tilde{j}}(y)$, $g_{\tilde{m}}(x)$		functions of the series, Eq. (12), for the stress function Φ
j		multiplication factor for the number of waves in circumferential direction
l	[cm]	length of the cylinder
m		half wave number in longitudinal direction
n		wave number in circumferential direction
p	[kg/cm²]	radial pressure (positive for internal pressure)
r	[cm]	radius of the cylinder
t	[cm]	wall thickness of the cylinder
u, v, w	[cm]	axial, circumferential and radial displacements (w positive inward)
$u_0 = \Delta l$	[cm]	shortening of the entire cylinder
x, y	[cm]	axial and circumferential coordinates (x beginning in the middle of the cylinder)
Φ	[cmkg]	stress function, defined in Appendix, section 1.1
Φ_b	[cmkg]	periodic part of the stress function Φ, defined by Eq. (14)
Π		total potential energy, defined by Eq. (1)
Π_i		potential energy of internal forces, defined by Eq. (2)
Π_e		potential energy of external loads, defined by Eq. (3)
Π_{res}		resulting energy, defined by Eq. (2)
$\Pi_{\text{compl.}}$		complementary energy, defined by Eq. (2)
ε_x, ε_y, γ_{xy}		axial and circumferential normal and shear strains, defined in Appendix, section 1.2

Indices

j appended below to the right	multiplication factor for the number of the buckles in circumferential direction
m appended below to the right	half wave number in longitudinal direction

Indices of j and m

˜ above the letter	concerning the stress function
e appended below to the right	end value

3. Discussion of the Postbuckling Problem

3.1. Preliminary Remarks

Throughout this paper the investigations are restricted to cylinders, which buckle with one tier of waves, extending over the whole length of the cylinder. Examples of postbuckling patterns of such cylinders are shown in Fig. 1.

Fig. 1. Postbuckling patterns of longitudinally stiffened cylinders under axial load, $l = 130$ mm, $r = 100$ mm.

One-tier buckling and postbuckling patterns occur at short cylinders for all types of stiffening and at long cylinders, if the combination of the stiffness parameters satisfies certain conditions. These conditions have been specified by GERARD [5] and APPEL [6] for the buckling of cylinders with simply supported edges.

3.2. Experiments on Longitudinally Stiffened Cylinders

The characteristic buckling and postbuckling behavior of thinwalled, circular, axially compressed cylinders, buckling with one-tier patterns, will be discussed in this section on the basis of test results, obtained for short longitudinally stiffened cylinders.

GARKISCH [7] investigated the buckling and postbuckling behavior of longitudinally eccentrically stiffened Mylar cylinders. The test specimens consisted of a thin skin with a large number of stringers, glued to the skin inside or outside of the cylinder (see Fig. 2). The edges of the cylinders were cast into rigid endplates. These endplates were parallely guided in a shortening controlled, stiff test facility.

Fig. 2 shows the characteristic buckling and postbuckling behavior of an outside and of an inside stiffened cylinder.

The experimental buckling and postbuckling loads are non-dimensionalized with respect to the buckling loads of the ideal cylinders, $P_{cr} = 703$ kg for the outside stiffened and $P_{cr} = 495$ kg for the inside stiffened cylinder. The buckling loads of the ideal cylinders and the corresponding wave

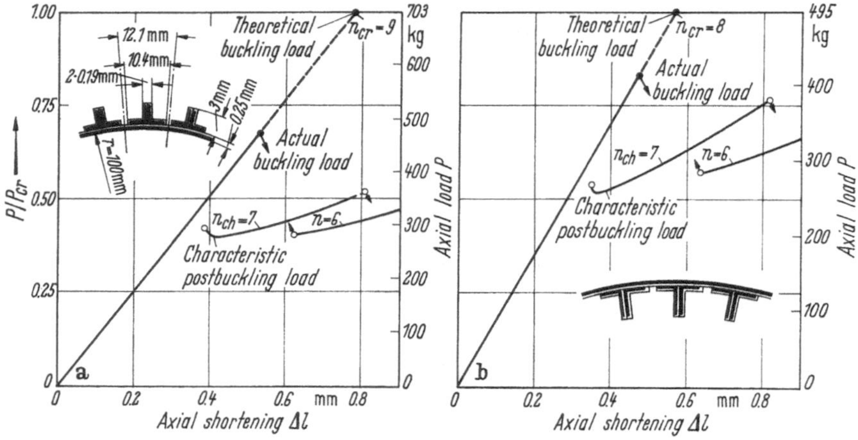

Fig. 2. Experimental postbuckling curves of longitudinally outside and inside stiffened axially compressed Mylar cylinders, $l = 200$ mm.

numbers n_{cr} have been calculated by GEIER [8] for cylinders with rigidly clamped edges under the assumption of a pure state of membrane stresses in the prebuckling region. The stiffness coefficients of the shell were obtained from measurements, as far as possible.

After buckling, both cylinders jumped into a periodic one-tier postbuckling pattern with $n = 7$ waves in circumferential direction. If the shortening was further increased the buckled cylinders attained a secondary buckling load and snapped through into another one-tier pattern with a circumferential wave number reduced by one. If the shortening was reduced, the cylinders reassumed the unbuckled state.

It is important to note, that for both cylinders no postbuckling patterns with more than $n = 7$ waves in circumferential direction could be found in the tests. Obviously the states of equilibrium connected with these patterns are unstable.

The postbuckling pattern with the highest circumferential wave number n for which stable states of equilibrium exist, will be called "characteristic postbuckling pattern" in the following, and the connected postbuckling curve will be called "characteristic postbuckling curve". The minimum load of the characteristic postbuckling curve will be denoted "characteristic postbuckling load".

It is remarkable that for both cylinders the circumferential wave number of the characteristic postbuckling pattern is in close vicinity of the circumferential wave number of the theoretical buckling pattern.

The outside stiffened cylinder is rather sensible to initial imperfections, since the experimental buckling load is only 68% of the buckling load predicted by the linear theory. The inside stiffened cylinder is obviously less sensible to initial imperfections, since its experimental buckling load amounts to 83% of the theoretical buckling load.

From comparison of Fig. 2a and 2b it can be deduced that the higher theoretical buckling load of the outside stiffened cylinder is compensated by its higher sensibility to initial imperfections, so that the experimental buckling loads as well as the characteristic postbuckling loads are almost the same for both types of stiffening.

It is interesting to note that a relation between the non-dimensionalized characteristic postbuckling load and the sensibility to initial imperfections can be established. For the outside stiffened cylinder, which is rather sensible to initial imperfections, the characteristic postbuckling load is 41% of the theoretical buckling load, whereas for the

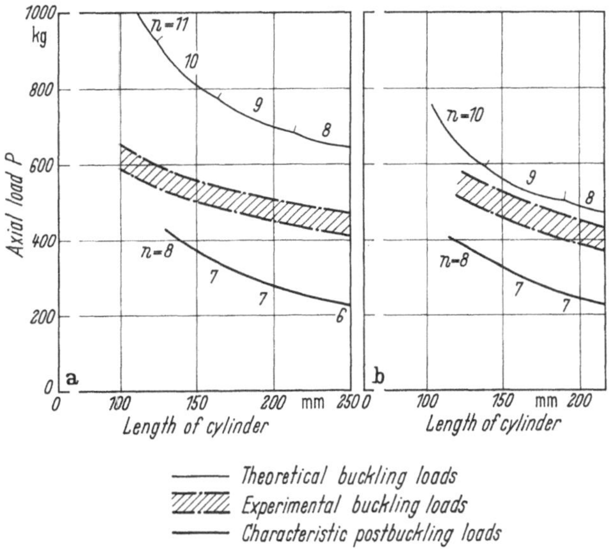

Fig. 3. Experimental and theoretical buckling and postbuckling loads of longitudinally stiffened axially compressed Mylar cylinders of different length.

inside stiffened cylinder, which is less sensible to initial imperfections, the characteristic postbuckling load amounts to 53% of the theoretical buckling load. These experimental results verify the well known fact

[1, 4], that postbuckling loads, which are small relative to the theoretical buckling load, indicate high sensibility of the cylinder to initial imperfections.

Fig. 3 shows the influence of the length of the cylinder on the buckling and postbuckling behavior of longitudinally excentrically stiffened cylinders [7]; for each length and for each type of stiffening two test specimens have been investigated. The experimental buckling loads scatter within the hatched area, whereas the characteristic postbuckling loads have practically no scatter region.

From this figure it can be deduced, that the ratio of the characteristic postbuckling load to the theoretical buckling load depends on the type of stiffening and on the length of the cylinder.

Further it can be seen, that the circumferential wave number of the characteristic postbuckling pattern is only one or two integer numbers smaller than the theoretical wave number n_{cr}, for all lengths of the cylinder within the treated region.

Cylinders with greater length than presented in Fig. 3 buckle with two-tier patterns and thus are beyond the scope of this paper.

3.3. Formulation of the Mathematical Problem

From the results of the tests discussed before and from results of numerous other buckling and postbuckling tests on axially compressed thinwalled, carefully manufactured Mylar cylinders, performed at the DFL in Braunschweig, the following general statements can be derived:

a) The characteristic postbuckling pattern is the same for cylinders of equal nominal dimensions and stiffness properties, even if the scatter of the experimental buckling loads indicates a considerable difference in the pattern of the initial imperfections.

Exceptions are possible for cylinders, for which two postbuckling patterns have nearly the same qualification to appear after buckling, e. g., if in systematic investigations of cylinders with different lengths, a value of the length is approached, at which the characteristic postbuckling pattern changes from one wave number to another.

b) The minimum loads of the postbuckling curves obtained for equal postbuckling patterns at cylinders with the same nominal dimensions and stiffness properties, are almost equal and thus practically independent of the initial imperfections.

c) The characteristic postbuckling load is a load which is safely carried by the cylinder.

d) The characteristic postbuckling loads has been found to be in the order of magnitude of 35% and 55% of the theoretical buckling load for the outside and inside stiffened cylinders, respectively [7], and in the order of 30% of the theoretical buckling load for isotropic cylinders [3].

e) The ratio of the characteristic postbuckling load to the buckling load of the ideal cylinder depends on the geometry and the stiffness properties of the shell.

From the statements a) to e) two important conclusions can be drawn:

The characteristic postbuckling load can be regarded as a basis of reasonable magnitude for the definition of a design load for the cylinder.

The characteristic postbuckling load can be predicted theoretically, since it is little influenced by initial imperfections.

The difficulty of the calculation of the characteristic postbuckling load lies in the determination of the characteristic postbuckling pattern. For this the postpuckling curves, connected with the critical wave number of the ideal cylinder and with adjacent wave numbers have to be calculated until beyond the minimum of each curve, and then the states of equilibrium have to be investigated with regard to their stability.

The highest circumferential wave number, for which stable states of equilibrium can be found, is the wave number of the characteristic postbuckling pattern. The minimum load of the corresponding postbuckling curve is the characteristic postbuckling load of the cylinder.

In the following these calculations will be performed for ideal cylinders. The analysis will be restricted to centrically stiffened cylinders subjected to axisymmetric loading. For the numerical examples it will be presupposed that the postbuckling patterns are periodic one-tier patterns with more than four waves in circumferential direction.

4. Description of the Calculation Procedure

4.1. Survey

The radial deformations of a thinwalled cylinder in the postbuckling region are of finite magnitude. Consequently for the calculation of the postbuckling states of equilibrium a nonlinear shell theory has to be employed.

The application of a nonlinear shell theory yields differential equations, which are so complicated, that no exact explicite solutions have been found until now, without mentioning the simple case of axisymmetric deformations. Approximate solutions can be obtained by means of energy methods. Thereby the unknown functions are described by series with a limited number of free coefficients; these coefficients are determined by the condition of a stationary value of the total potential energy.

The calculated postbuckling states of equilibrium are stable, if the second variation of the total potential energy is positive definite.

4.2. Total Potential Energy

In this paper Donnell's shallow shell theory [9] will be applied to the postbuckling problem. Due to the simplifications of this theory the postbuckling states of stress and deformation can be described by the radial displacement function w and the stress function Φ [10].

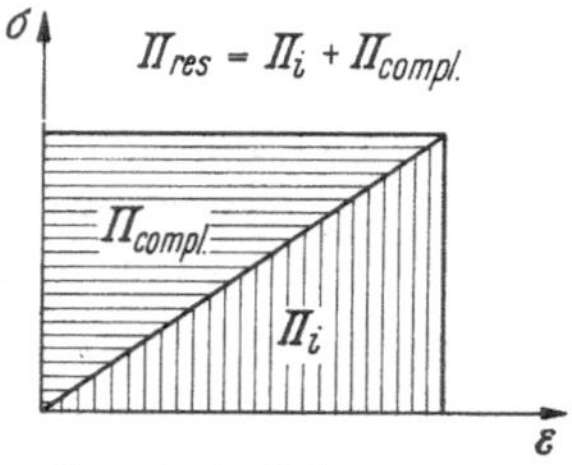

Fig. 4. Potential energy of the internal forces.

The total potential energy is the sum of the potential energy of the internal forces and of the external load:

$$\Pi = \Pi_i + \Pi_e \tag{1}$$

The potential energy Π_i of the internal forces will be expressed as the difference of the resulting and the complementary energy as shown in Fig. 4:

$$\Pi_i = \Pi_{\text{res}} - \Pi_{\text{compl.}} \tag{2}$$

The potential energy Π_e of the external loads is given for the general case that the edges can move and rotate in all directions

$$\Pi_e = -\left[\int_0^{2\pi r}\left(N_x u + N_{xy}\cdot v + R_x w + M_x \frac{\partial w}{\partial x}\right) dy\right]_{x=-l/2}^{x=l/2} + \\ + \int_0^{2\pi r}\int_{-l/2}^{l/2} p w \, dx\, dy. \tag{3}$$

Using the Eqs. (1) until (3) and the definitions and relations compiled in the Appendix, sec. 1.1—1.3, the total potential energy takes the form [11]

$$\Pi = \int_0^{2\pi r}\int_{-l/2}^{l/2}\left\{\frac{\partial^2\Phi}{\partial y^2}\left[\frac{\partial u}{\partial x} + \frac{1}{2}\left(\frac{\partial w}{\partial x}\right)^2\right] + \frac{\partial^2\Phi}{\partial x^2}\left[\frac{\partial v}{\partial y} + \frac{1}{2}\left(\frac{\partial w}{\partial y}\right)^2 - \frac{w}{r}\right] - \right. \\ - \frac{\partial^2\Phi}{\partial x\,\partial y}\left[\frac{\partial u}{\partial y} + \frac{\partial v}{\partial x} + \frac{\partial w}{\partial x}\frac{\partial w}{\partial y}\right] - \\ - \frac{1}{2}A_{11}\left(\frac{\partial^2\Phi}{\partial y^2}\right)^2 - A_{12}\frac{\partial^2\Phi}{\partial x^2}\frac{\partial^2\Phi}{\partial y^2} - \frac{1}{2}A_{22}\left(\frac{\partial^2\Phi}{\partial x^2}\right)^2 - \frac{1}{2}A_{33}\left(\frac{\partial^2\Phi}{\partial x\,\partial y}\right)^2 + \\ + M_x\frac{\partial^2 w}{\partial x^2} + M_y\frac{\partial^2 w}{\partial y^2} + M_{xy}\cdot 2\frac{\partial^2 w}{\partial x\,\partial y} -$$

$$-\frac{1}{2}T_{11}M_x{}^2 - T_{12}M_xM_y - \frac{1}{2}T_{22}M_y{}^2 - \frac{1}{2}T_{33}M_{xy}^2\Big\}\,dx\,dy -$$

$$-\left[\int_0^{2\pi r}\left(\frac{\partial^2\Phi}{\partial y^2}u - \frac{\partial^2\Phi}{\partial x\,\partial y}v + R_x\cdot w + M_x\frac{\partial w}{\partial x}\right)dy\right]_{-l/2}^{l/2} +$$

$$+\int_0^{2\pi r}\int_{-l/2}^{l/2} pw\,dx\,dy. \tag{4}$$

In this form the total potential energy is a function of the displacements u, v, w, the stress function Φ, the moments M_x, M_y and M_{xy} and the radial edge force R_x.

The displacements u and v can be eliminated because of the fact, that the stress function satisfies the inplane equilibrium conditions and because of the choice of proper boundary conditions in the following. The moments will be expressed by the derivatives of the radial displacement w with aid of the relations given in the Appendix, section 1.3. The radial edge force R_x will be eliminated by the choice of proper boundary conditions [12]. Then the variation of the potential energy turns out to be a function of w and Φ only.

4.3. Postbuckling States of Equilibrium

The postbuckling states of equilibrium of the cylinder are determined from the condition of a stationary value of the total potential energy, i. e., from the condition that the first variation of the potential energy vanishes

$$\delta\Pi = 0. \tag{5}$$

The variation of the total potential energy, Eq. (4), is performed, with respect to Φ and w independently, for the purpose of obtaining approximate solutions of the postbuckling problem. This variational principle was first formulated by E. Reissner [13] and leads to the expression

$$\delta\Pi = \int_0^{2\pi r}\int_{-l/2}^{l/2}\Big\{\Big[D_{11}\frac{\partial^4 w}{\partial x^4} + 2(D_{12}+2D_{33})\frac{\partial^4 w}{\partial x^2\,\partial y^2} + D_{22}\frac{\partial^4 w}{\partial y^4} -$$

$$-\frac{\partial^2\Phi}{\partial y^2}\frac{\partial^2 w}{\partial x^2} - \frac{\partial^2\Phi}{\partial x^2}\frac{\partial^2 w}{\partial y^2} + 2\frac{\partial^2\Phi}{\partial x\,\partial y}\frac{\partial^2 w}{\partial x\,\partial y} - \frac{1}{r}\frac{\partial^2\Phi}{\partial x^2} + p\Big]\delta w -$$

$$-\Big[A_{11}\frac{\partial^4\Phi}{\partial y^4} + 2\left(A_{12}+\frac{1}{2}A_{33}\right)\frac{\partial^4\Phi}{\partial x^2\,\partial y^2} + A_{22}\frac{\partial^4\Phi}{\partial x^4} +$$

$$+\frac{\partial^2 w}{\partial x^2}\frac{\partial^2 w}{\partial y^2} - \left(\frac{\partial^2 w}{\partial x\,\partial y}\right)^2 + \frac{1}{r}\frac{\partial^2 w}{\partial x^2}\Big]\delta\Phi_b\Big\}\,dx\,dx +$$

$$+\text{ edge terms.} \tag{6}$$

In the following the analysis is restricted to the subsequent listed boundary conditions.

For the axisymmetric forces and deformations

$$\begin{aligned} u &= \pm \frac{u_0}{2} &\quad \text{or} \quad& N_x = N_{x0} \\ v &= 0 & & \\ w &= 0 &\quad \text{or} \quad& w = a_{00} \\ \frac{\partial w}{\partial x} &= 0 &\quad \text{or} \quad& M_x = 0 \end{aligned} \tag{7}$$

and for the non-axisymmetric forces and deformations

$$\begin{aligned} u &= 0 &\quad \text{or} \quad& N_x = 0 \\ v &= 0 &\quad \text{or} \quad& N_{xy} = 0 \\ w &= 0 &\quad \text{or} \quad& R_x = 0 \\ \frac{\partial w}{\partial x} &= 0 &\quad \text{or} \quad& M_x = 0\,. \end{aligned} \tag{8}$$

The case $u = \pm \frac{u_0}{2}$ refers to a shortening controlled system, whereas the case $N_x = N_{x0}$ refers to a load controlled system.

The case $w = a_{00}$ means that the axisymmetric radial displacement due to the membrane forces N_{x0} and pr is not restrained at the edges; this free radial edge displacement is

$$a_{00} = -(A_{12} N_{x0} + A_{22}\, pr) r\,. \tag{9}$$

For the boundary conditions, Eqs. (7)—(9), the edge terms in the expression for the first variation of the total potential energy, Eq. (6), vanish. The variation $\delta \Phi_b$ contains no variation of N_{x0} and the variation δw contains no variation of a_{00} [12]. Hence, $\delta \Pi$ is the same for load and shortening controlled systems, as well as for systems with the radial edge displacements $w = 0$ or $w = a_{00}$.

The condition $\delta \Pi = 0$ is satisfied for arbitrary variations $\delta \Phi_b$ and δw only if the square brackets connected with $\delta \Phi_b$ and δw, respectively, vanish.

From the requirement that the square bracket connected with $\delta \Phi_b$ must be equal zero, one obtains the compatibility condition

$$A_{11} \frac{\partial^4 \Phi}{\partial y^4} + 2\left(A_{12} + \frac{1}{2} A_{33}\right) \frac{\partial^4 \Phi}{\partial x^2\, \partial y^2} + A_{22} \frac{\partial^4 \Phi}{\partial x^4} + \\ + \frac{\partial^2 w}{\partial x^2} \frac{\partial^2 w}{\partial y^2} - \left(\frac{\partial^2 w}{\partial x\, \partial y}\right)^2 + \frac{1}{r} \frac{\partial^2 w}{\partial x^2} = 0\,. \tag{10}$$

From the requirement, that the square bracket connected with δw must vanish, one obtains the equilibrium condition

$$D_{11}\frac{\partial^4 w}{\partial x^4}+2(D_{12}+2D_{33})\frac{\partial^4 w}{\partial x^2\,\partial y^2}+D_{22}\frac{\partial^4 w}{\partial y^4}-$$
$$-\frac{\partial^2 \Phi}{\partial y^2}\frac{\partial^2 w}{\partial x^2}-\frac{\partial^2 \Phi}{\partial x^2}\frac{\partial^2 w}{\partial y^2}+2\frac{\partial^2 \Phi}{\partial x\,\partial y}\frac{\partial^2 w}{\partial x\,\partial y}-\frac{1}{r}\frac{\partial^2 \Phi}{\partial x^2}+p=0. \quad (11)$$

The Eqs. (10) and (11) form a set of two nonlinear differential equations for the determination of the unknown functions w and Φ. For this set of equations no exact solutions are known, except those, which describe an axisymmetric state of deformation.

Approximate solutions can be obtained by assuming for the unknown functions w and Φ series with free coefficients a_{jm} and $b_{\tilde{j}\tilde{m}}$

$$\Phi=\sum_{\tilde{j}}\sum_{\tilde{m}} b_{\tilde{j}\tilde{m}}\, g_{\tilde{j}}(y)\cdot g_{\tilde{m}}(x)+N_{x0}\frac{y^2}{2}+pr\frac{x^2}{2} \quad (12)$$

$$w=a_{00}+\sum_{j}\sum_{m} a_{jm}\cdot f_j(y)\cdot f_m(x) \quad (13)$$

and by determination of these coefficients from the condition of a stationary value of the total potential energy, Eq. (5).

For actual calculations the series (12) and (13) are limited to a finite number of terms. Consequently, the variations $\delta\Phi$ and δw are no longer arbitrary, but restricted to the variation of the free coefficients a_{jm} and $b_{\tilde{j}\tilde{m}}$

$$\delta\Phi=\delta\Phi_b+\delta N_{x0}\frac{y^2}{2}=\sum_{\tilde{j}}^{\tilde{j}_e}\sum_{\tilde{m}}^{\tilde{m}_e}\delta b_{\tilde{j}\tilde{m}}\, g_{\tilde{j}}(y)\, g_{\tilde{m}}(x)+\delta N_{x0}\frac{y^2}{2} \quad (14)$$

$$\delta w=\sum_{j}^{j_e}\sum_{m}^{m_e}\delta a_{jm} f_j(y)\cdot f_m(x). \quad (15)$$

Introduction of the variations $\delta\Phi_b$ and δw, Eqs. (14) and (15), into Eq. (5) together with Eq. (6) yields the approximate compatibility condition in the form

$$\int_0^{2\pi r}\int_{-l/2}^{l/2}\left\{\left[A_{11}\frac{\partial^4\Phi}{\partial y^4}+2\left(A_{12}+\frac{1}{2}A_{33}\right)\frac{\partial^4\Phi}{\partial x^2\,\partial y^2}+A_{22}\frac{\partial^4\Phi}{\partial x^4}+\right.\right.$$
$$\left.\left.+\frac{\partial^2 w}{\partial x^2}\frac{\partial^2 w}{\partial y^2}-\left(\frac{\partial^2 w}{\partial x\,\partial y}\right)^2+\frac{1}{r}\frac{\partial^2 w}{\partial x^2}\right]\sum_{\tilde{j}}^{\tilde{j}_e}\sum_{\tilde{m}}^{\tilde{m}_e}\delta b_{\tilde{j}\tilde{m}} g_{\tilde{j}}(y)\, g_{\tilde{m}}(x)\right\}dx\,dy=0 \quad (16)$$

and the approximate equilibrium condition in the form

$$\int_0^{2\pi r}\int_{-l/2}^{l/2}\left\{\left[D_{11}\frac{\partial^4 w}{\partial x^4}+2(D_{12}+2D_{33})\frac{\partial^4 w}{\partial x^2\,\partial y^2}+D_{22}\frac{\partial^4 w}{\partial y^4}-\right.\right.$$
$$\left.-\frac{\partial^2\Phi}{\partial y^2}\frac{\partial^2 w}{\partial x^2}-\frac{\partial^2\Phi}{\partial x^2}\frac{\partial^2 w}{\partial y^2}+2\frac{\partial^2\Phi}{\partial x\,\partial y}\frac{\partial^2 w}{\partial x\,\partial y}-\frac{1}{r}\frac{\partial^2\Phi}{\partial x^2}+p\right]\times$$
$$\left.\times\sum_{j}^{j_e}\sum_{m}^{m_e}\delta a_{jm}f_j(y)\,f_m(x)\right\}dx\,dy=0. \tag{17}$$

Equation (16) represents a set of as many differential equations as the series for the stress function Φ, Eq. (12), contains free coefficients $b_{\bar{j}\bar{m}}$. Analogous to this, Eq. (17) represents a set of as many differential equations as the series for the radial displacement w, Eq. (13), contains free coefficients a_{jm}.

For the numerical calculation of the postbuckling states of equilibrium the following procedure is employed: The series for Φ, Eq. (12), and the series for w, Eq. (13), are introduced in the compatibility condition (10) or (16) and the coefficients $b_{\bar{j}\bar{m}}$ are determined as function of the coefficients a_{jm}.

$$b_{\bar{j}\bar{m}}=b_{\bar{j}\bar{m}}(a_{jm}). \tag{18}$$

If Eq. (16) is used, one obtains an approximate relation between $b_{\bar{j}\bar{m}}$ and a_{jm} [11, 14, 15]. If Eq. (10) is used, one obtains the exact relation [1, 16, 17]; in the latter case the series for Φ, Eq. (12), is not arbitrary, but must be choosen in such a way, that Eq. (10) can be solved by equating like terms. In both cases it presents no mathematical difficulties to derive the relation Eq. (18), since the compatibility Eqs. (10) and (16) are linear in $b_{\bar{j}\bar{m}}$.

With aid of Eq. (18) the coefficients $b_{\bar{j}\bar{m}}$ in the equilibrium condition, Eq. (17), will be eliminated. Then Eq. (17) represents a system of nonlinear equations for the coefficients a_{jm}, which can be solved by an iteration procedure.

When the coefficients a_{jm} and $b_{\bar{j}\bar{m}}$ are known, the problem of determination of the state of stress and deformation in the postbuckling region is solved.

The axial shortening can be obtained with aid of the geometric and elastic relations for ε_x given in the Appendix, section 1.2

$$u_0=\frac{1}{2\pi r}\int_0^{2\pi r}\int_{-l/2}^{l/2}\frac{\partial u}{\partial x}\,dx\,dy=A_{11}N_{x0}\,l+$$
$$+\frac{1}{2\pi r}\int_0^{2\pi r}\int_{-l/2}^{l/2}\left[A_{12}\frac{\partial^2\Phi}{\partial x^2}-\frac{1}{2}\left(\frac{\partial w}{\partial x}\right)^2\right]dx\,dy. \tag{19}$$

4.4. Stability Criterion

In order to find the stability criterion the second variation of the total potential energy has to be investigated. The states of equilibrium are stable, if this second variation is positive definite.

Throughout the second variation of the total potential energy one has to preserve the validity of the compatibility condition, Eq. (10) or (16), respectively.

If the compatibility condition is satisfied the first variation of the total potential energy can be written in the simple form

$$\delta \Pi = \sum_j \sum_m \frac{\partial \Pi(w, \Phi, f_j(y)\, f_m(x))}{\partial a_{jm}} \cdot \delta a_{jm}$$

$$= \int_0^{2\pi r} \int_{-l/2}^{l/2} \left\{ \left[D_{11} \frac{\partial^4 w}{\partial x^4} + 2(D_{12} + 2D_{33}) \frac{\partial^4 w}{\partial x^2\, \partial y^2} + D_{22} \frac{\partial^4 w}{\partial y^4} - \right.\right.$$

$$\left. - \frac{\partial^2 \Phi}{\partial y^2} \frac{\partial^2 w}{\partial x^2} - \frac{\partial^2 \Phi}{\partial x^2} \frac{\partial^2 w}{\partial y^2} + 2 \frac{\partial^2 \Phi}{\partial x\, \partial y} \frac{\partial^2 w}{\partial x\, \partial y} - \frac{1}{r} \frac{\partial^2 \Phi}{\partial x^2} + p \right] \times$$

$$\left. \times \sum_j \sum_m \delta a_{jm} f_j(y)\, f_m(x) \right\} dx\, dy. \tag{20}$$

From the condition, that the compatibility condition remains valid throughout the second variation follows that $\delta b_{\bar{j}\bar{m}}$ is no longer an arbitrary variation but depends on the variations δa_{jm}

$$\delta b_{\bar{j}\bar{m}} = \delta b_{\bar{j}\bar{m}}(a_{jm}, \delta a_{jm}). \tag{21}$$

The variation of the axial load is for a load controlled system

$$\delta N_{x0} = 0 \tag{22a}$$

and for a shortening controlled system, according to Eq. (19),

$$\delta N_{x0} = \frac{1}{l \cdot A_{11}} \frac{1}{2\pi r} \int_0^{2\pi r} \int_{-l/2}^{l/2} \left[- A_{12} \frac{\partial^2 (\delta \Phi)}{\partial x^2} + \frac{\partial w}{\partial x} \frac{\partial (\delta w)}{\partial x} \right] dx\, dy. \tag{22b}$$

The second variation of the total potential energy is obtained in two steps: First variation of $\delta \Pi$ in the form of Eq. (20) with respect to both functions w and Φ

$$\delta^2 \Pi = \int_0^{2\pi r} \int_{-l/2}^{l/2} \Big\{ \Big[D_{11} \frac{\partial^4 (\delta w)}{\partial x^4} + 2 (D_{12} + 2 D_{33}) \frac{\partial^4 (\delta w)}{\partial x^2 \partial y^2} + D_{22} \frac{\partial^4 (\delta w)}{\partial y^4} -$$
$$- \frac{\partial^2 (\delta \Phi)}{\partial y^2} \frac{\partial^2 w}{\partial x^2} - \frac{\partial^2 (\delta \Phi)}{\partial x^2} \frac{\partial^2 w}{\partial y^2} + 2 \frac{\partial^2 (\delta \Phi)}{\partial x \partial y} \frac{\partial^2 w}{\partial x \partial y} -$$
$$- \frac{\partial^2 \Phi}{\partial y^2} \frac{\partial^2 (\delta w)}{\partial x^2} - \frac{\partial^2 \Phi}{\partial x^2} \frac{\partial^2 (\delta w)}{\partial y^2} + 2 \frac{\partial^2 \Phi}{\partial x \partial y} \frac{\partial^2 (\delta w)}{\partial x \partial y} -$$
$$- \frac{1}{r} \frac{\partial^2 (\delta \Phi)}{\partial x^2} \Big] \sum_j \sum_m \delta a_{jm} f_j(y) f_m(x) \Big\} \, dx \, dy \qquad (23)$$

and then elimination of $\delta \Phi$ with aid of Eq. (14) together with Eqs. (21) and (22). This procedure results in an expression of the form

$$\delta^2 \Pi = \sum_j \sum_m \sum_i \sum_k \frac{\partial^2 \Pi}{\partial a_{jm} \partial a_{ik}} \delta a_{jm} \delta a_{ik}. \qquad (24)$$

The investigated state of equilibrium is stable, if the second variation $\delta^2 \Pi$ turns out to be positive definite, i.e., if the determinant of the coefficient matrix and all its principle minors are positive.

In the stability investigation it is examined, whether the buckled cylinder has the tendency to transform into another postbuckling pattern. Consequently the series for w and Φ, describing the basic pattern, must be extended by terms, which describe new patterns. These new patterns must be of a principally different type than the basic pattern, otherwise a non positive second variation of the total potential energy would indicate only a transition to an improved postbuckling state of stress and deformation.

In the following chapter the calculation procedure is outlined on the basis of a simple example.

5. Examples

5.1. Approximate Calculation of Postbuckling States of Equilibrium and their Stability

5.1.1. Formulation of the Problem. The procedure for the calculation of the postbuckling states of equilibrium and the evaluation of the stability criterion will be demonstrated with help of rigorously truncated series.

For this purpose we choose an orthotropic cylinder under axial load. First postbuckling states of equilibrium, characterized by a one-tier pattern with n waves in circumferential direction, will be calculated, comp. Fig. 1. Then the stability of these equilibrium states will be examined under the assumption that the system is load controlled.

The postbuckling states of equilibrium are described by the simple series

$$\Phi = \left(b_{0,1} + b_{n,1} \cdot \cos\frac{ny}{r} + b_{2n,1} \cdot \cos\frac{2ny}{r}\right)\cos\frac{\pi x}{l} + N_{x0}\frac{y^2}{2}, \qquad (25)$$

$$w = a_{00} + \left(a_{0,1} + a_{n,1} \cdot \cos\frac{ny}{r}\right)\cos\frac{\pi x}{l}. \qquad (26)$$

For the examination of the stability of these postbuckling states it will be analysed whether the cylinder has the tendency to assume one of the following patterns:

a) one-tier patterns: $n_s = n + 1$, $n_s = n - 1$, $n_s = 1$ } without phase shift

b) two-tiers patterns: $n_s = n$, $n_s \neq n$ } with and without phase shift 90°

c) unbuckled state of stress and deformation.

For the derivation of the formulas we restrict the stability investigation to the question, whether the cylinder tends to assume the two-tiers pattern $n_s = n$ with phase shift 90° or to reassume the unbuckled state. This questioning requires the extension of the series for Φ and w, Eqs. (25) and (26), by the following terms

$$\Phi_s = b_{n,2} \cdot \sin\frac{ny}{r} \cdot \sin\frac{2\pi x}{l}, \qquad (27)$$

$$w_s = a_{n,2} \cdot \sin\frac{ny}{r} \cdot \sin\frac{2\pi x}{l}. \qquad (28)$$

The series, Eqs. (25—28), for the stress function and the radial displacements satisfy the classical boundary conditions

$$\frac{\partial N_x}{\partial y} = 0 \qquad v = 0 \qquad w = a_{00} \qquad M_x = 0 \qquad (29)$$

with a_{00} according to Eq. (9).

5.1.2. Postbuckling States of Equilibrium. The series (25) and (26) allow only rough approximations to the actual state of stress and [illegible] mation in the postbuckling region. Hence, the approximate compatibility condition, Eq. (16), and the approximate equilibrium condition, Eq. (17), must be used in the analysis.

Introduction of the series for the stress function Φ, Eq. (25), and the series for the radial displacement function w, Eq. (26), into the approximate compatibility condition, Eq. (16), (comp. Appendix, sec. 2.1) and integration over the entire surface of the cylinder yields three relations between the coefficients $a_{0,1}$, $a_{n,1}$ and $b_{0,1}$, $b_{n,1}$, $b_{2n,1}$:

$$\begin{aligned} b_{0,1} V_{0,1} \left(\frac{l}{\pi}\right)^2 &= a_{0,1} \frac{1}{r} - a_{n,1}^2 \left(\frac{n}{r}\right)^2 \frac{2}{3\pi}, \\ b_{n,1} V_{n,1} \left(\frac{l}{\pi}\right)^2 &= a_{n,1} \frac{1}{r} - a_{0,1} a_{n,1} \left(\frac{n}{r}\right)^2 \frac{8}{3\pi}, \\ b_{2n,1} V_{2n,1} \left(\frac{l}{\pi}\right)^2 &= \qquad - a_{n,1}^2 \left(\frac{n}{r}\right)^2 \frac{2}{\pi}. \end{aligned} \tag{30}$$

Introduction of the series, Eqs. (25) and (26), into the approximate equilibrium condition, Eq. (17) (comp. Appendix, sec. 2.2) and integration over the entire surface of the cylinder yields two further relations between the coefficients $a_{0,1}$, $a_{n,1}$ and $b_{0,1}$, $b_{n,1}$, $b_{2n,1}$,

$$\begin{aligned} & a_{0,1} B_{0,1} \left(\frac{l}{\pi}\right)^2 + N_{x0} \cdot a_{0,1} + \frac{1}{r} b_{0,1} - a_{n,1} b_{n,1} \left(\frac{n}{r}\right)^2 \frac{4}{3\pi} = 0, \\ & a_{n,1} B_{n,1} \left(\frac{l}{\pi}\right)^2 + N_{x0} \cdot a_{n,1} + \frac{1}{r} \cdot b_{n,1} - (b_{n,1} a_{0,1} + b_{0,1} a_{n,1}) \left(\frac{n}{r}\right)^2 \frac{8}{3\pi} - \\ & \qquad - b_{2n,1} a_{n,1} \left[\left(\frac{n}{r}\right)^2 + \frac{2n}{r} \frac{n}{r}\right] \frac{4}{3\pi} = 0. \end{aligned} \tag{31}$$

The 5 equations, Eqs. (30) and Eqs. (31), are sufficient to determine the 5 free coefficients in the series, Eqs. (25) and (26). The solution of this set of nonlinear equations can be performed by an iteration procedure [15, 17]. With known values for the free coefficients $a_{0,1}$, $a_{n,1}$, $b_{0,1}$, $b_{n,1}$ and $b_{2n,1}$ the state of stress and deformation of the cylinder is known within the limits of accuracy given by the presupposed simple series, Eqs. (25) and (26).

The axial shortening is obtained by introduction of Eqs. (25) and (26) into Eq. (19) (comp. Appendix 2.3) and integration over the entire surface of the cylinder

$$\frac{\Delta l}{l} = A_{11} N_{x0} - A_{12} \cdot b_{0,1} \left(\frac{\pi}{l}\right)^2 \frac{2}{\pi} - \frac{1}{8} \cdot (2 a_{0,1}^2 + a_{n,1}^2) \left(\frac{\pi}{l}\right)^2. \tag{32}$$

The postbuckling states of equilibrium for an orthotropic cylinder, the geometrical and stiffness properties of which are listed in Table 1,

Table 1. *Stiffness coefficients of the plywood cylinders, the postbuckling curves of which are presented in Fig. 5 and Fig. 6*

$A_{11} = 1.313 \cdot 10^{-4}$ cm/kg	$D_{11} = 3.790$ cmkg
$A_{12} = 0$	$D_{12} = 0$
$A_{22} = 2.297 \cdot 10^{-4}$ cm/kg	$D_{22} = 0.268$ cmkg
$A_{33} = 1.127 \cdot 10^{-4}$ cm/kg	$D_{33} = 0.261$ cmkg

are presented in Fig. 5 for a variety of circumferential wave numbers n.

5.1.3. Stability Investigation. As has been mentioned above, in the stability investigation it will be examined, whether the existing state of equilibrium tends to transform to the two-tiers pattern $n_s = n$ with phase shift 90° or to reassume the unbuckled state. For this investigation the sum of the series, Eqs. (25) and (27), and (26) and (28), respectively,

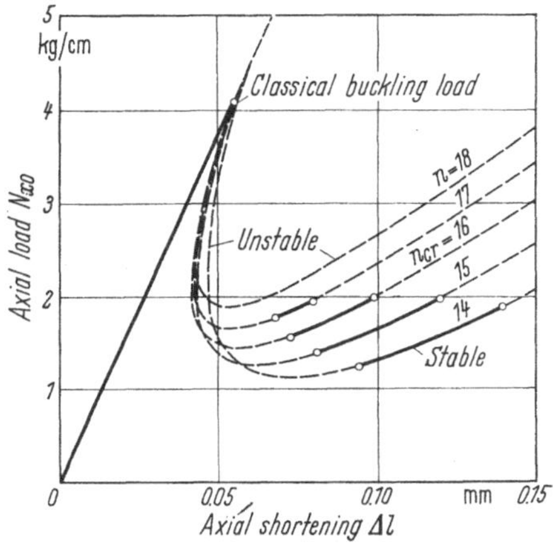

Fig. 5. Theoretical postbuckling curves of an axially loaded longitudinally stiffened plywood cylinder, $l = 100$ mm, $r = 200$ mm, stiffness coefficients listed in Table 1.

$$\Phi_{\text{res}} = \Phi + \Phi_s = \left(b_{0,1} + b_{n,1} \cdot \cos\frac{ny}{r} + b_{2n,1} \cdot \cos\frac{2ny}{r}\right) \cdot \cos\frac{\pi x}{l} + + b_{n,2} \cdot \sin\frac{ny}{r} \cdot \sin\frac{2\pi x}{l} + N_{x0}\frac{y^2}{2} \tag{33}$$

$$w_{\text{res}} = w + w_s = a_{00} + \left(a_{0,1} + a_{n,1} \cdot \cos\frac{ny}{r}\right)\cos\frac{\pi x}{l} + + a_{n,2} \cdot \sin\frac{ny}{r} \cdot \sin\frac{2\pi x}{l} \tag{34}$$

has to be employed. The variation of these two series leads to the following expressions

$$\delta\Phi_{\text{res}} = \left(\delta b_{0,1} + \delta b_{n,1} \cdot \cos\frac{ny}{r} + \delta b_{2n,1} \cdot \cos\frac{2ny}{r}\right) \cdot \cos\frac{\pi x}{l} + + \delta b_{n,2} \cdot \sin\frac{ny}{r} \cdot \sin\frac{2\pi x}{l} + \delta N_{x0}\frac{y^2}{2} \tag{35}$$

$$\delta w_{\text{res}} = \left(\delta a_{0,1} + \delta a_{n,1} \cdot \cos\frac{ny}{r}\right)\cos\frac{\pi x}{l} + \delta a_{n,2} \cdot \sin\frac{ny}{r} \cdot \sin\frac{2\pi x}{l}. \tag{36}$$

Introduction of Eqs. (33) and (34) into the approximate compatibility equation, Eq. (16), and then integration over the whole surface of the cylinder yields the following relations between the free coefficients a_{jm} and $b_{\tilde{j}\tilde{m}}$:

$$b_{0,1} V_{0,1} \left(\frac{l}{\pi}\right)^2 = a_{0,1} \frac{1}{r} - a_{n,1}^2 \left(\frac{n}{r}\right)^2 \frac{2}{3\pi} - a_{n,2}^2 \left(\frac{n}{r}\right)^2 \frac{8}{15\pi}$$

$$b_{n,1} V_{n,1} \left(\frac{l}{\pi}\right)^2 = a_{n,1} \frac{1}{r} - a_{0,1} a_{n,1} \left(\frac{n}{r}\right)^2 \frac{8}{3\pi}$$

$$b_{2n,1} V_{2n,1} \left(\frac{l}{\pi}\right)^2 = \qquad - a_{n,1}^2 \left(\frac{n}{r}\right)^2 \frac{2}{\pi}$$

$$b_{n,2} V_{n,2} \left(\frac{l}{2\pi}\right)^2 = a_{n,2} \frac{1}{r} - a_{0,1} a_{n,2} \left(\frac{n}{r}\right)^2 \frac{8}{15\pi}. \tag{37}$$

By variation of Eq. (37) one obtains

$$\delta b_{0,1} V_{0,1} \left(\frac{l}{\pi}\right)^2 = \delta a_{0,1} \frac{1}{r} - a_{n,1} \cdot \delta a_{n,1} \left(\frac{n}{r}\right)^2 \frac{4}{3\pi} - a_{n,2} \delta a_{n,2} \left(\frac{n}{r}\right)^2 \frac{16}{15\pi}$$

$$\delta b_{n,1} V_{n,1} \left(\frac{l}{\pi}\right)^2 = \delta a_{n,1} \frac{1}{r} - (a_{0,1} \delta a_{n,1} + \delta a_{0,1} \cdot a_{n,1}) \left(\frac{n}{r}\right)^2 \frac{8}{3\pi}$$

$$\delta b_{2n,1} V_{2n,1} \left(\frac{l}{\pi}\right)^2 = \qquad - a_{n,1} \delta a_{n,1} \left(\frac{n}{r}\right)^2 \frac{4}{\pi}$$

$$\delta b_{n,2} V_{n,2} \left(\frac{l}{2\pi}\right)^2 = \delta a_{n,2} \frac{1}{r} - (\delta a_{0,1} a_{n,2} + a_{0,1} \delta a_{n,2}) \left(\frac{n}{r}\right)^2 \frac{8}{15\pi}. \tag{38}$$

The evaluation of the stability criterion for the shortening controlled cylinder requires the variation of the axial load too. Introduction of the series Φ_{res} and w_{res}, Eqs. (33) and (34), into Eq. (22b) (comp. Appendix, section 3.1) and integration over the whole surface of the cylinder yields

$$\delta N_{x0} = \frac{A_{12}}{A_{11}} \delta b_{0,1} \left(\frac{\pi}{l}\right)^2 \frac{2}{\pi} + \frac{1}{4A_{11}} (2a_{0,1} \delta a_{0,1} + a_{n,1} \delta a_{n,1}) \left(\frac{\pi}{l}\right)^2. \tag{39}$$

The second variation of the total potential energy is obtained by introduction of the series, Eqs. (33)—(36), in Eq. (23) (comp. Appendix, sec. 3.2), integration over the whole surface of the cylinder and then elimination of $\delta b_{0,1}$, $\delta b_{n,1}$, $\delta b_{2n,1}$, $\delta b_{n,2}$ and δN_{x0} with aid of the Eqs.

(38) and (39). This procedure leads to an expression of the form

$$\frac{2}{\pi r l} \cdot \delta^2 \Pi = K_{00} \cdot (\delta a_{0,1})^2 + K_{nn} \cdot (\delta a_{n,1})^2 + \\ + 2 K_{n0} \cdot \delta a_{0,1} \cdot \delta a_{n,1} + K_{ss} (\delta a_{n,2})^2 \tag{40}$$

for the case of a load controlled system ($\delta N_{x0} = 0$) with the abbreviations

$$K_{00} = 2\left[B_{0,1} + N_{x0}\left(\frac{\pi}{l}\right)^2 + \left(\frac{\pi}{l}\right)^4 \frac{1}{V_{0,1}}\left(\frac{1}{r}\right)^2\right] + \left(\frac{\pi}{l}\right)^4 \frac{1}{V_{n,1}}\left[\left(\frac{n}{r}\right)^2 \frac{8}{3\pi} a_{n,1}\right]^2$$

$$K_{nn} = B_{n,1} + N_{x0}\left(\frac{\pi}{l}\right)^2 + \left[\left(\frac{\pi}{l}\right)^2\left(\frac{n}{r}\right)^2 \frac{4}{\pi} a_{n,1}\right]^2 \left[\frac{1}{V_{0,1}}\frac{2}{9} + \frac{1}{V_{2n,1}}\right] + \\ + \left(\frac{\pi}{l}\right)^4 \cdot \frac{1}{V_{n,1}}\left[\left(\frac{n}{r}\right)^2 \frac{8}{3\pi} \cdot a_{0,1} - \frac{1}{r}\right]^2 - \left(\frac{\pi}{l}\right)^2\left(\frac{n}{r}\right)^2 \frac{4}{\pi}\left[\frac{2}{3} b_{0,1} + b_{2n,1}\right]$$

$$K_{n0} = \left(\frac{\pi}{l}\right)^4\left(\frac{n}{r}\right)^2 \frac{8}{3\pi} \cdot a_{n,1}\left\{\frac{1}{V_{n,1}}\left[\left(\frac{n}{r}\right)^2 \frac{8}{3\pi} \cdot a_{0,1} - \frac{1}{r}\right] - \frac{1}{V_{0,1}}\frac{1}{r}\right\} - \\ - \left(\frac{\pi}{l}\right)^2\left(\frac{n}{r}\right)^2 \frac{8}{3\pi} b_{n,1}$$

$$K_{ss} = B_{n,2} + N_{x0}\left(\frac{2\pi}{l}\right)^2 + \left(\frac{2\pi}{l}\right)^4 \frac{1}{V_{n,2}}\left[\left(\frac{n}{r}\right)^2 \frac{8}{15\pi} a_{0,1} - \frac{1}{r}\right]^2 - \\ - \left(\frac{2\pi}{l}\right)^2\left(\frac{n}{r}\right)^2 \frac{8}{15\pi} b_{0,1} \tag{41a}$$

and for the case of a shortening controlled system

$$K_{00} = K_{00\,(\delta N_{x0}=0)} + \frac{1}{A_{11}}\left(\frac{\pi}{l}\right)^4 a_{0,1}\left[a_{0,1} + A_{12}\left(\frac{\pi}{l}\right)^2 \frac{1}{V_{0,1}}\frac{4}{\pi}\frac{1}{r}\right]$$

$$K_{nn} = K_{nn(\delta N_{x0}=0)} + \frac{1}{A_{11}}\left(\frac{\pi}{l}\right)^4 a_{n,1}^2\left[\frac{1}{4} - A_{12}\left(\frac{\pi}{l}\right)^2 \frac{1}{V_{0,1}}\left(\frac{n}{r}\right)^2 \frac{8}{3\pi^2}\right]$$

$$K_{n0} = K_{n0\,(\delta N_{x0}=0)} + \frac{1}{A_{11}}\left(\frac{\pi}{l}\right)^4 \frac{a_{n,1}}{2}\left\{a_{0,1} - A_{12}\left(\frac{\pi}{l}\right)^2 \frac{1}{V_{0,1}}\frac{2}{\pi} \times \\ \times \left[\left(\frac{n}{r}\right)^2 \frac{8}{3\pi} a_{0,1} - \frac{1}{r}\right]\right\}$$

$$K_{ss} = K_{ss\,(\delta N_{x0}=0)}\,. \tag{41b}$$

The coefficients $a_{0,1}$, $a_{n,1}$, $b_{0,1}$, $b_{n,1}$ and $b_{2n,1}$ in Eq. (41) are known from the calculation of the postbuckling state of equilibrium, the stability of which is examined.

If the second variation of the total potential energy is written as matrix of a quadratic form

	$\delta a_{0,1}$	$\delta a_{n,1}$	$\delta a_{n,2}$
$\delta a_{0,1}$	K_{00}	K_{n0}	0
$\delta a_{n,1}$	K_{n0}	K_{nn}	0
$\delta a_{n,2}$	0	0	K_{ss}

(42)

the criterion of a stable state of equilibrium is given by the condition that the determinant and all principle minors of the matrix are positive.

5.1.4. Numerical Results. Fig. 5 shows numerical results for a plywood cylinder with the dimensions $l = 100$ mm, $r = 200$ mm, $t = 0{,}7$ mm, the stiffness coefficients of which are listed in Table 1.

The application of the stability criterion to the calculated postbuckling states confirms the fact, known from tests, that the postbuckling states of equilibrium are stable only within a limited region of axial shortening. The states of equilibrium connected with wave numbers n greater than the critical wave number $n_{cr} = 16$ are unstable, with exception of a very short region of the curve $n = n_{cr} + 1 = 17$. From tests, performed on similar cylinders, we conclude that this region also would turn out to be unstable, if in our stability calculations more different patterns had been considered.

The lower limits of the region containing the stable states of equilibrium are given by bifurcation points. At these limits the cylinder tends to assume a pattern with a circumferential wave number reduced by one. We know from tests that at the lower stability limits single buckles jump out successively until the cylinder has reassumed the unbuckled state. The results of our stability investigations seem to confirm this tendency.

However, in the location of these lower limits there exists a fundamental difference between the experimental and theoretical results. In the tests we found in the region of the minimum postbuckling loads stable states of equilibrium, whereas our simple calculations indicate, that the equilibrium states in this region are unstable. This discrepancy is probably due to the different boundary conditions. The edges of the test cylinders were glued into rigid, parallely guided endplates, whereas in our calculations the edges of the cylinder were not restraint against warping and tilting.

The upper stability limits are again given by bifurcation points. At these bifurcation points the cylinder shows the tendency to transform into a two-tiers pattern. The offer of possible two-tiers patterns included circumferential wave numbers ranging from $n_s = n - 5$ to $n_s = n + 5$.

In tests on similar cylinders no transition to a two-tiers pattern has been observed; at the upper stability limits the test cylinder always snapped through into a one-tier pattern with a circumferential wave number reduced by one. It is possible that this transition to another one-tier pattern passes over the intermediate formation of a two-tiers pattern, a phenomenon which is similar to the behavior of a radially loaded curved panel [11]. If this is true the calculations would describe correctly the actual behavior of the shell.

If, as in this example, the stability limits are given by bifurcation points, the location of these stability limits are independent whether the system is load or shortening controlled. (The difference between the two systems becomes apparent only in the case that the lower stability limit is located at very small values of axial shortening: For load controlled systems the stable region of each postbuckling curve cannot extend further than to the minimum load; for shortening controlled systems the stable region cannot extend further than to the minimum shortening of the postbuckling curve.)

It is surprising that the results of the simple calculation discussed above, show so good agreement with the experimental postbuckling behavior, in spite of the fact that the series for w and Φ contain only a very small number of free coefficients and in spite of the fact that the boundary conditions in the tests and calculations are considerably different.

5.2. Exact Calculation of Postbuckling States of Equilibrium

Since the rough approximations, described and discussed in the preceding chapter, do not permit quantitatively correct statements on the behavior of compressed cylinders, we have endeavoured to obtain better approximations to the solution of the postbuckling problem. Thereby we succeeded in finding practically exact solutions for the set of differential equations, Eqs. (10) and (11), describing the postbuckling states of equilibrium; but these calculations were not yet extended to the investigation of the stability.

Fig. 6 shows postbuckling curves, calculated for an orthotropic axially loaded plywood cylinder, having the dimensions $l = 200$ mm, $r = 200$ mm and $t = 0{,}7$ mm. The stiffness coefficients of this cylinder are listed in Table 1.

For the calculation of the postbuckling curves so many terms in the series for the stress function Φ and the radial displacement function w had been considered, that further terms had no influence on the numerical results.

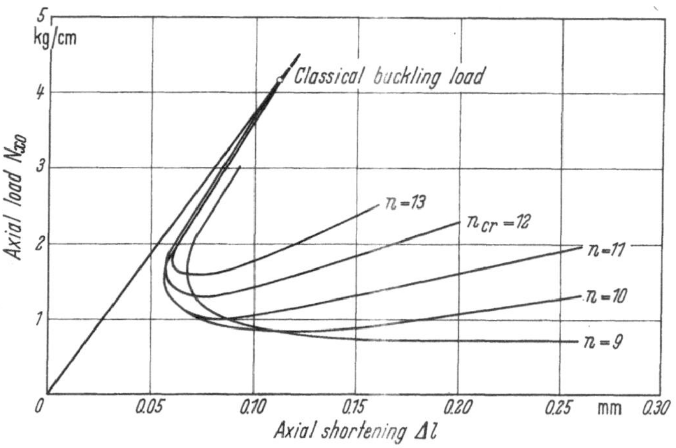

Fig. 6. Theoretical postbuckling curves of an axially loaded longitudinally stiffened plywood cylinder $l = 200$ mm, $r = 200$ mm, stiffness coefficients listed in Table 1.

Comparison of Figs. 2 and 6 shows that for the treated cylinder the experimental and theoretical postbuckling curves have the same character. We have to point out that the two diagrams have been obtained for longitudinally stiffened, axially loaded cylinders, but with different geometry and with different stiffness properties.

The diagram Fig. 2 presenting the experimental results contains only two postbuckling curves; it was not possible to find patterns with more or less waves in circumferential direction. The diagram Fig. 6 showing the theoretical results contains a family of postbuckling curves and it remains the task of the stability investigations to decide which equilibrium states are stable and thus describe the actual postbuckling behavior.

6. Summary and Outlock to Further Investigations

In the first part of this paper it has been deduced from postbuckling tests in the region of elastic deformations that for each thinwalled, compressed circular cylinder, which buckles with elastic deformations, there exists one characteristic postbuckling load, which can certainly be carried by the cylinder. This characteristic postbuckling load is the minimum load of the characteristic postbuckling curve, i. e. the curve, connected with the highest circumferential wave number, for which stable states of equilibrium exist.

This characteristic postbuckling load lies far above the absolutely smallest postbuckling load, the determination of which was for a long time the aim of the investigators, who based their analysis on the postbuckling calculations of von Kármán.

From the experimental investigations it has been inferred that the characteristic postbuckling load is not much influenced by initial imperfections. Thus it seems reasonable to calculate this load, to begin with, for the ideal cylinder.

In the second part of the paper a calculation procedure is outlined, which allows the determination of the characteristic postbuckling load of the ideal cylinder. The application of this procedure is demonstrated on simple examples. The theoretical results do not quite agree with the experimental results, but the comparison indicates already, that with more extensive numerical calculations and some improvements in the tests, sufficient agreement between the theoretical and experimental results might be obtained.

If it turns out, that this procedure yields no satisfying agreement between the theoretical and experimental results, initial imperfections have to be introduced into the analysis. The effect of the initial imperfections can only be, that the wave numbers of the characteristic postbuckling pattern are reduced by small integer numbers. With known characteristic postbuckling pattern, the characteristic postbuckling load can be calculated disregarding the initial imperfections. From the fact, that the initial imperfections can cause the characteristic postbuckling pattern and, hence, the characteristic postbuckling load, to alter only by discrete steps, follows, that it will not be necessary to take into account the exact distribution of the initial imperfections, but it will be sufficient to classify the cylinders according to the amplitudes of their initial imperfections and to calculate the characteristic postbuckling load for each quality class.

After having extended the investigations to postbuckling patterns with two tiers of buckles and after having succeeded to obtain sufficient agreement between theoretical and experimental results it will be possible to draft, with an acceptable amount of calculation work, design curves which supply reliable informations on the load carrying capacity of actual cylinders.

Appendix

1. Donnell's Shallow Shell Theory

1.1 Definition of Stress-Function

$$N_x = \frac{\partial^2 \Phi}{\partial y^2}, \qquad N_{xy} = -\frac{\partial^2 \Phi}{\partial x\, \partial y}, \qquad N_y = \frac{\partial^2 \Phi}{\partial x^2}.$$

1.2 Relation between Inplane Strains, Deformations and Membrane Forces

$$\varepsilon_x = \frac{\partial u}{\partial x} + \frac{1}{2}\left(\frac{\partial w}{\partial x}\right)^2 \qquad = A_{11}\frac{\partial^2 \Phi}{\partial y^2} + A_{12}\frac{\partial^2 \Phi}{\partial x^2},$$

$$\varepsilon_y = \frac{\partial v}{\partial x} + \frac{1}{2}\left(\frac{\partial w}{\partial y}\right)^2 - \frac{w}{r} = A_{12}\frac{\partial^2 \Phi}{\partial y^2} + A_{22}\frac{\partial^2 \Phi}{\partial x^2},$$

$$\gamma_{xy} = \frac{\partial u}{\partial y} + \frac{\partial v}{\partial x} + \frac{\partial w}{\partial x}\frac{\partial w}{\partial y} \qquad = -A_{33}\frac{\partial^2 \Phi}{\partial x\,\partial y}.$$

1.3 Relation between Moments and Changes of Curvature and Twist

$$\frac{\partial^2 w}{\partial x^2} = T_{11} M_x + T_{12} M_y, \qquad M_x = D_{11}\frac{\partial^2 w}{\partial x^2} + D_{12}\frac{\partial^2 w}{\partial y^2},$$

$$\frac{\partial^2 w}{\partial y^2} = T_{12} M_x + T_{22} M_y, \qquad M_y = D_{12}\frac{\partial^2 w}{\partial x^2} + D_{22}\frac{\partial^2 w}{\partial y^2},$$

$$2\frac{\partial^2 w}{\partial x\,\partial y} = T_{33} M_{xy}. \qquad M_{xy} = D_{33}\cdot 2\frac{\partial^2 w}{\partial x\,\partial \gamma}.$$

1.4 Combination of Stiffness Coefficients and Wave Numbers

$$V_{\bar{j},\bar{m}} = A_{11}\left(\frac{\bar{j}n}{r}\right)^4 + 2\left(A_{12} + \frac{1}{2}A_{33}\right)\left(\frac{\bar{m}\pi}{l}\right)^2\left(\frac{\bar{j}n}{r}\right)^2 + A_{22}\left(\frac{\bar{m}\pi}{l}\right)^4,$$

$$B_{j,m} = D_{11}\left(\frac{m\pi}{l}\right)^4 + 2(D_{12} + 2D_{33})\left(\frac{m\pi}{l}\right)^2\left(\frac{jn}{r}\right)^2 + D_{22}\left(\frac{jn}{r}\right)^4.$$

2. Postbuckling States of Equilibrium

2.1 Compatibility Condition

$$\int_0^{2\pi r}\int_{-l/2}^{l/2}\Bigg[\left(b_{0,1}V_{0,1} + b_{n,1}V_{n,1}\cdot\cos\frac{ny}{r} + b_{2n,1}V_{2n,1}\cdot\cos\frac{2ny}{r}\right)\cos\frac{\pi x}{l} +$$

$$+\left(a_{0,1} + a_{n,1}\cdot\cos\frac{ny}{r}\right)a_{n,1}\left(\frac{\pi}{l}\right)^2\left(\frac{n}{r}\right)^2\cdot\cos\frac{ny}{r}\cos^2\frac{\pi x}{l} -$$

$$- a_{n,1}^2\cdot\left(\frac{n}{r}\right)^2\cdot\left(\frac{\pi}{l}\right)^2\cdot\sin^2\frac{ny}{r}\cdot\sin^2\frac{\pi x}{l} -$$

$$-\frac{1}{r}\left(a_{0,1} + a_{n,1}\cdot\cos\frac{ny}{r}\right)\left(\frac{\pi}{l}\right)^2\cdot\cos\frac{\pi x}{l}\Bigg]\times$$

$$\times\left[\delta b_{0,1} + \delta b_{n,1}\cdot\cos\frac{ny}{r} + \delta b_{2n,1}\cdot\cos\frac{2ny}{r}\right]\cos\frac{\pi x}{l}\,dx\,dy = 0.$$

2.2 Equilibrium Condition

$$\int_0^{2\pi r}\int_{-l/2}^{l/2}\Bigg\{\left[a_{0,1}\cdot B_{0,1}+a_{n,1}\cdot B_{n,1}\cdot\cos\frac{ny}{r}\right]\cos\frac{\pi x}{l}-$$
$$-\left[b_{n,1}\left(\frac{n}{r}\right)^2\cdot\cos\frac{ny}{r}+b_{2n,1}\left(\frac{2n}{r}\right)^2\cos\frac{2ny}{r}\right]\left[a_{0,1}+a_{n,1}\cdot\cos\frac{ny}{r}\right]\times$$
$$\times\left(\frac{\pi}{l}\right)^2\cos^2\frac{\pi x}{l}-\left[b_{0,1}+b_{n,1}\cdot\cos\frac{ny}{r}+b_{2n,1}\cdot\cos\frac{2ny}{r}\right]\times$$
$$\times a_{n,1}\left(\frac{\pi}{l}\right)^2\left(\frac{n}{r}\right)^2\cdot\cos\frac{ny}{r}\cos^2\frac{\pi x}{l}+$$
$$+2\left[b_{n,1}\cdot\frac{n}{r}\cdot\sin\frac{ny}{r}+b_{2n,1}\cdot\frac{2n}{r}\cdot\sin\frac{2ny}{r}\right]\times$$
$$\times a_{n,1}\frac{n}{r}\cdot\left(\frac{\pi}{l}\right)^2\cdot\sin\frac{ny}{r}\sin^2\frac{\pi x}{l}+$$
$$+\frac{1}{r}\left(b_{0,1}+b_{n,1}\cdot\cos\frac{ny}{r}+b_{2n,1}\cos\frac{2ny}{r}\right)\left(\frac{\pi}{l}\right)^2\cdot\cos\frac{\pi x}{l}+$$
$$+N_{x0}\left[a_{0,1}+a_{n,1}\cdot\cos\frac{ny}{r}\right]\left(\frac{\pi}{l}\right)^2\cdot\cos\frac{\pi x}{l}\Bigg\}\times$$
$$\times\left[\delta a_{0,1}+\delta a_{n,1}\cdot\cos\frac{ny}{r}\right]\cdot\cos\frac{\pi x}{l}\,dx\,dy=0.$$

2.3 Axial Shortening

$$\frac{\Delta l}{l}=A_{11}N_{x0}+\frac{1}{2\pi r l}\int_0^{2\pi r}\int_{-l/2}^{l/2}\left[-A_{12}\left(b_{0,1}+b_{n,1}\cdot\cos\frac{ny}{r}+b_{2n,1}\cdot\cos\frac{2ny}{r}\right)\times\right.$$
$$\left.\times\left(\frac{\pi}{l}\right)^2\cdot\cos\frac{\pi x}{l}-\frac{1}{2}\left(a_{0,1}+a_{n,1}\cdot\cos\frac{ny}{r}\right)^2\left(\frac{\pi}{l}\right)^2\cdot\sin^2\frac{\pi x}{l}\right]dx\,dy.$$

3. Stability Criterion

3.1 Variation of the Axial Load

$$\delta N_{x0}=\frac{1}{A_{11}\cdot l}\frac{1}{2\pi r}\int_0^{2\pi r}\int_{-l/2}^{l/2}\left[+A_{12}\left(\delta b_0+\delta b_{n,1}\cdot\cos\frac{ny}{r}+\delta b_{2n,1}\cdot\cos\frac{2ny}{r}\right)\times\right.$$
$$\times\left(\frac{\pi}{l}\right)^2\cdot\cos\frac{\pi x}{l}+\left(a_{0,1}+a_{n,1}\cdot\cos\frac{ny}{r}\right)\times$$
$$\left.\times\left(\delta a_{0,1}+\delta a_{n,1}\cdot\cos\frac{ny}{r}\right)\left(\frac{\pi}{l}\right)^2\cdot\sin^2\frac{\pi x}{l}\right]dx\,dy.$$

3.2 Second Variation of the Total Potential Energy

$$\delta^2 \Pi = \int\int \Big\{\Big[\Big(\delta a_{0,1} B_{0,1} + \delta a_{n,1} B_{n,1} \cdot \cos\frac{ny}{r}\Big)\cos\frac{\pi x}{l} +$$

$$+ a_{n,2} B_{n,2} \sin\frac{ny}{r} \sin\frac{2\pi x}{l}\Big] -$$

$$- \Big\langle\Big[\delta b_{n,1}\Big(\frac{n}{r}\Big)^2 \cos\frac{ny}{r} + \delta b_{2n,1}\Big(\frac{2n}{r}\Big)^2 \cos\frac{2ny}{r}\Big]\cos\frac{\pi x}{l} + \delta b_{n,2}\Big(\frac{n}{r}\Big)^2 \times$$

$$\times \sin\frac{ny}{r} \cdot \sin\frac{2\pi x}{l} - \delta N_{x0}\Big\rangle\Big[a_{0,1} + a_{n,1} \cdot \cos\frac{ny}{r}\Big]\Big(\frac{\pi}{l}\Big)^2 \cos\frac{\pi x}{l} -$$

$$- \Big\langle\Big[\delta b_{0,1} + \delta b_{n,1} \cdot \cos\frac{ny}{r} + \delta b_{2n,1} \cdot \cos\frac{2ny}{r}\Big]\Big(\frac{\pi}{l}\Big)^2 \cdot \cos\frac{\pi x}{l} +$$

$$+ \delta b_{n,2}\Big(\frac{2\pi}{l}\Big)^2 \cdot \sin\frac{ny}{r} \cdot \sin\frac{2\pi x}{l}\Big\rangle a_{n,1}\Big(\frac{n}{r}\Big)^2 \cdot \cos\frac{ny}{r} \cdot \cos\frac{\pi x}{l} +$$

$$+ 2\Big\langle\Big[\delta b_{n,1}\frac{n}{r} \cdot \sin\frac{ny}{r} + \delta b_{2n,1}\frac{2n}{r} \cdot \sin\frac{2ny}{r}\Big]\frac{\pi}{l} \cdot \sin\frac{\pi x}{l} +$$

$$+ \delta b_{n,2}\frac{n}{r}\frac{2\pi}{l} \cdot \cos\frac{ny}{r} \cdot \cos\frac{2\pi x}{l}\Big\rangle a_{n,1}\frac{n}{r}\frac{\pi}{l} \cdot \sin\frac{ny}{r} \cdot \sin\frac{\pi x}{l} -$$

$$- \Big\langle\Big[b_{n,1}\Big(\frac{n}{r}\Big)^2 \cdot \cos\frac{ny}{r} + b_{2n,1}\Big(\frac{2n}{r}\Big)^2 \cdot \cos\frac{2ny}{r}\Big]\cos\frac{\pi x}{l} + b_{n,2}\Big(\frac{n}{r}\Big)^2 \times$$

$$\times \sin\frac{ny}{r} \cdot \sin\frac{2\pi x}{l} - N_{x0}\Big\rangle\Big[\Big(\delta a_{0,1} + \delta a_{n,1} \cdot \cos\frac{ny}{r}\Big)\Big(\frac{\pi}{l}\Big)^2 \times$$

$$\times \cos\frac{\pi x}{l} + \delta a_{n,2} \cdot \Big(\frac{2\pi}{l}\Big)^2 \sin\frac{ny}{r} \cdot \sin\frac{2\pi x}{l}\Big] -$$

$$- \Big\langle\Big[b_{0,1} + b_{n,1} \cdot \cos\frac{ny}{r} + b_{2n,1} \cdot \cos\frac{2ny}{r}\Big]\Big(\frac{\pi}{l}\Big)^2 \cdot \cos\frac{\pi x}{l} +$$

$$+ b_{n,2}\Big(\frac{2\pi}{l}\Big)^2 \cdot \sin\frac{ny}{r} \cdot \sin\frac{2\pi x}{l}\Big\rangle\Big[\delta a_{n,1} \cdot \Big(\frac{n}{r}\Big)^2 \cdot \cos\frac{ny}{r} \cdot \cos\frac{\pi x}{l} +$$

$$+ \delta a_{n,2}\Big(\frac{n}{r}\Big)^2 \cdot \sin\frac{ny}{r} \cdot \sin\frac{2\pi x}{l}\Big] +$$

$$+ 2\Big\langle\Big[b_{n,1}\frac{n}{r} \cdot \sin\frac{ny}{r} + b_{2n,1} \cdot \frac{2n}{r} \cdot \sin\frac{2ny}{r}\Big]\frac{\pi}{l} \cdot \sin\frac{\pi x}{l} +$$

$$+ b_{n,2}\frac{n}{r}\frac{2\pi}{l} \cdot \cos\frac{ny}{r} \cdot \cos\frac{2\pi x}{l}\Big\rangle\Big[\delta a_{n,1} \cdot \frac{n}{r}\frac{\pi}{l} \cdot \sin\frac{ny}{r} \cdot \sin\frac{\pi x}{l} +$$

$$+ \delta a_{n,2}\frac{n}{r}\frac{2\pi}{l} \cdot \cos\frac{ny}{r} \cdot \cos\frac{2\pi x}{l}\Big] +$$

$$+ \frac{1}{r} \left\langle \left[\delta b_{0,1} + \delta b_{n,1} \cdot \cos \frac{ny}{r} + \delta b_{2n,1} \cdot \cos \frac{2ny}{r} \right] \left(\frac{\pi}{l} \right)^2 \cdot \cos \frac{\pi x}{l} + \right.$$

$$\left. + \delta b_{n,2} \left(\frac{2\pi}{l} \right)^2 \cdot \sin \frac{ny}{r} \cdot \sin \frac{2\pi x}{l} \right\rangle \Big\} \times$$

$$\times \left[\left(\delta a_{0,1} + \delta a_{n,1} \cdot \cos \frac{ny}{r} \right) \cos \frac{\pi x}{l} + \delta a_{n,2} \cdot \sin \frac{ny}{r} \cdot \sin \frac{2\pi x}{l} \right] \Big\} dx\, dy.$$

References

1. von Kármán, Th., L. G. Dunn and H.-S. Tsien: The buckling of thin cylindrical shells under axial compression. J. Aeron. Sci. **8**, No. 8, 303—312 (1941).
2. Hoff, N. J., W. R. Madson and J. Mayers: The postbuckling equilibrium of axially compressed circular cylindrical shells. Stanford Univ., Dept. Aeron. Astron. SUDAER No. 221, Feb. 1965, 25 + 7 pp.
3. Thielemann, W. F., and M. E. Esslinger: On the postbuckling behavior of thin-walled axially compressed circular cylinder of finite length. Proc. 70th Anniv. Symp. on the Theory of Shells to Honor Lloyd Hamilton Donnell, ed. by D. Muster, Univ. of Houston, Houston, Texas, 1967, pp. 433—479.
4. Koiter, W. T.: Over de stabiliteit van het elastisch evenwicht. Diss. Techn. Hogeschool Delft 1945, Amsterdam: Paris 1945, 233 S.
5. Gerard, G.: Compressive stability of orthotropic cylinders. J. Aerospace Sci. **29**, No. 10, 1171—1179 (1962).
6. Appel, H.: Buckling modes of orthotropic circular cylinders under axial compression for various combinations of stiffness parameters. Deutsche Luft- u. Raumfahrt, Forsch. Ber. DLR FB 65—47, München: ZLDI 1965, pp. 29.
7. Garkisch, H. D.: Experimentelle Untersuchung des Beulverhaltens von Kreiszylinderschalen mit exzentrischen Längsversteifungen. Deutsche Luft- u. Raumfahrt, Forsch. Ber. DLR FB 67—75, München: ZLDI 1967, 20 + 28 S
8. Geier, B.: Das Beulverhalten versteifter Zylinderschalen. Teil 1: Differentialgleichungen. Z. Flugwiss. **14**, Nr. 7, 306—323, 1966.
9. Donnell, L. H.: A new theory for the buckling of thin cylinders under axial compression and bending. Trans. ASME Vol. 56, Nov. 1934, pp. 795—806.
10. Thielemann, W. F.: New developments in the nonlinear theories of the buckling of thin cylindrical shells. Aeronautics and Astronautics, Proc. Durand Cent. Conf., Stanford Univ. 1959 Oxford/London/New York/Paris: Pergamon Press 1960, pp. 76—119.
11. Wolmir, A. S.: Biegsame Platten und Schalen. Berlin: VEB Verl. f. Bauwesen 1962.
12. Esslinger, M.: Stabilitätsrechnung für dünnwandige Kreiszylinderschalen im Nachbeulbereich. Deutsche Luft- u. Raumfahrt, Forsch. Ber. DLR FB 67, 70, München: ZLDI 1967.
13. Reissner, E.: On a variational problem in elasticity. J. Math. Phys. 90—95 (1950).
14. Cicala, P.: The effect of initial deformations on the behaviour of a cylindrical shell under axial compression. Quart. Appl. Math. **9**, No. 3, 273—293 (1951).
15. Esslinger, M.: Nachbeulrechnung dünnwandiger Kreiszylinderschalen. Deutsche Luft- u. Raumfahrt, Forsch. Ber. DLR FB 67—25, München: ZLDI März 1967, 77 S.

16. Thielemann, W., W. Schnell und G. Fischer: Beul- und Nachbeulverhalten orthotroper Kreiszylinderschalen unter Axial- und Innendruck. (DVL-Ber. 153) Z. Flugwiss. 8, Nr. 10/11, 284—293 (1960).
17. Meyer-Piening, H. R.: Zur Berechnung von Nachbeullasten dünnwandiger Kreiszylinder endlicher Länge. WGLR-DGRR-Jahrestag. Okt. 1967, Karlsruhe, Vortrag Nr. 21, Vorabdruck d. Deutsch. Forsch. Anst. f. Luft- u. Raumfahrt, Braunschweig 1967, 23 S.

Discussion

A. van der Neut: First of all I might compliment Professor Thielemann on his carefully carried-out tests. Professor Thielemann advocated the use of the post-buckling load carrying capacity in the design of shell structures. His motivation was that this has been accepted in the case of flat plates. I am not confident that this philosophy is acceptable. One should consider that the folds are much sharper in shell buckling than in plate buckling, also due to the snap-through effect. This means that bending stresses are far more severe and easily cause permanent deformation.

As an engineer I am a little surprised that Professor Thielemann, who claimed to look upon his problem as an engineer as well, payed so much attention to the post-buckling behaviour.

My conclusion is that the engineer is interested only in the primary buckling load, and here imperfection sensitivity is of primary importance.

W. Thielemann: I agree of course that the problem of the sensitivity to initial imperfections is a very important one. On the other hand, I hope I have made clear, that the primary buckling load may deviate from the theoretical buckling load even down to zero if you have one local buckle, which is similar to one which will be seen afterwards in the post-buckling region.

It is of course unusual to have such large buckles, but if you have them, there will be no primary buckling load at all. Nevertheless the shell is able to carry a considerable load.

J. Singer: Was the anisotropy of mylar taken into account in the interpretation of the test results?

W. Thielemann: We have taken into account in all the calculations that mylar is not an isotropic material.

N. J. Hoff: In the case of hydrostatic pressure it might be acceptable to design circular cylindrical shells for the minimal postbuckling load but such a procedure would lead to far too conservative design when the load is axial compression because the difference is very large between the maximal load carried in accordance with Koiter's theory and the minimal load emphasized by investigators following in the footsteps of von Kármán and Tsien. In this respect I would like to note that my earlier conclusion, referred to by Professor Thielemann, namely that the minimization procedure used by these investigators leads to a vanishing minimal load as the number of terms retained in the truncated series is increased, was accompanied by the statement that the thickness-to-radius ratio also approaches zero. This means simply that the particular approximate procedure results in a trivial solution and certainly not that the post-buckling load of real cylindrical shells is zero.

I would like to add that NARASIMHAN's calculations mentioned in my remarks to Professor BUDIANSKY's paper indicate that the length-to-diameter ratio has little effect on the maximal stress a compressed circular cylindrical shell can carry. On the other hand, the number of buckles around the circumference depends very much on this ratio. It is of interest to note that the number of buckles predicted by the large-displacement theory for the post-buckled state differs little from the number of buckles obtained from FLÜGGE's solution of his small-displacement equations published in 1933.

J. W. HUTCHINSON: The small deflection restriction on shallow-shell theory, which is implied by its usual derivation, can be removed if a proper interpretation is made of the dependent variables appearing in these equations, see KOITER, W. T., "On the nonlinear theory of thin elastic shells", Koninkl. Nederl. Akademie van Weten., Series B. 69, No. 1, 1966.

W. THIELEMANN: It will be interesting to evaluate this improved nonlinear shell theory numerically.

Geometrical Methods in the Non-Linear Theory of Shells

By

A. V. Pogorelov

Physical-Technical Institute of Low Temperatures, Charkov, USSR

1. The fundamental problem of the theory of shells is to determine the deformations of the shell as well as the stresses in the material under the action of external loads. A well-known method for the solution of this problem is based upon the variational principle of LAGRANGE. If the load acting on the shell is conservative, this variational principle consists of the following.

A shell subjected to a load q among all possible forms F satisfying the condition of clamped edges takes the form for which the functional

$$W = U(F) - A_q(F)$$

is stationary. This means its variation is $\delta W = 0$. Here, U is the strain energy of the shell and A the work done by the load q.

Let u, v, w denote the components of the deflection of the points of the middle surface of the shell during its deformation. The integrand in the strain energy U has a rather complex expression. It contains the functions u, v, w and their derivatives to the second order. The Euler-Lagrange equation for the functional W is a system of three differential equations of the forth order for the functions u, v, w. It is clear that the solution of the fundamental problem of the theory of shells in this way is rather difficult if not hopeless at all.

In a number of cases it is possible to assume that the deflection of the points of the middle surface and the change of its normal curvatures will be rather small. In this assumption the functional W may be simplified if we confine ourselves to its quadratic part. The corresponding system of the Euler-Lagrange equations will be linear. The solution of the fundamental problem of the theory of shells in the assumption of the above smallness of deformation is the object of the linear theory.

If the deflections of the shell are not accompanied by considerable changes of normal curvatures, then in the expression of the strain energy U the part, which is caused by the bending of the shell, may be omitted. It is known that the theory of shells which also includes this simplifying assumption is called momentless theory.

We shall consider the elastic states of the shell which are characterized by considerable changes in the initial form. In this case the linearity of the problem as well as the momentless theory are not permissible. However, I am going to show that it is the assumption of the great changes of the shell form during the deformation that makes possible a new approach to the solution of the problem based on simple geometrical ideas.

2. The fact is that the assumption of the considerable changes of the shell form involves an important conclusion about the character of deformation. Namely, the form of the middle surface during such deformation is always like one of the forms of its isometric transformation. Indeed, for the basic construction materials, metals and their alloys, the moduli of elasticity are about 10^5 to 10^6 kg/cm^2 but the limit of elasticity is 10^2 to 10^3 kg/cm^2. Thus the maximum relative elastic deformation is of the order of magnitude 10^{-3}. From this it follows that in spite of the great changes of the shell form the inner metric of its middle surface practically does not change (the change is of the order of magnitude of 0.1%). The deformation of the surface without the change of its inner metric is geometrical bending. Naturally, the deformation with a small change of the inner metric results in forms that are like isometric transformations. That is why in the solution for the variational problem for the functional $W = U - A$ we shall take the forms which are like isometric transformations. The solution is also made easier by some specific features of isometric transformations near which is the form we look for.

The thing is that because of the clamped edges the middle surface does not usually allow regular geometrical bendings. We say that the shell is geometrically rigid. The geometrically non-rigid shell would receive the load acting on it by bending, but because of the small thickness of the shell its rigidity to bending is negligible. That is why any shell constructed correctly must be geometrically rigid. As the middle surface does not allow regular bending our isometric transformations must belong to a larger class of surfaces with a break of smoothness along some lines. These singularities of isometric transformations give a key to the solution of the main problem.

3. The solution of the variational problem for the functional $W = U - A$ we shall devide into two stages. At first the isometric transformation is fixed and the functional is considered on the forms which are like this isometric transformation. At this stage the solution of the problem is obtained in compact form with very common assumptions about the shell surface and its isometric transformation. As a result of

this solution the functional W will depend only on isometric transformations. Accordingly, the common variational principle takes the following geometrical form (variational principle A).

The considerable deformation of the elastic shell under the given load is like the form of isometric transformation of the initial surface that makes the functional $W = U(F) - A(F)$ stationary.

This functional depends on isometric transformations of the middle surface of the shell. The term $U(F)$ (energy of deformation) is expressed in the following way:

$$U(\tilde{F}) = \frac{E\delta^3}{24(1-\nu^2)} \iint_F (\varDelta\varkappa_1{}^2 + \varDelta\varkappa_2{}^2 + 2\nu\varDelta\varkappa_1\varDelta\varkappa_2)\, d\sigma + \\ + cE\delta^{5/2}\int_\gamma \frac{\alpha^{5/2}}{\varrho^{1/2}}\, dS_\gamma + \frac{E\delta^3}{12(1-\nu^2)}\int_\gamma \alpha(-2\varkappa + \varkappa_e + \varkappa_i)\, dS_\gamma .$$

Here $\varDelta\varkappa_1$ and $\varDelta\varkappa_2$ are the extremal changes of the normal curvatures during the deformation from the initial form to the isometric transformation $\tilde{F}$; 2α is the angle between the tangential planes of the surface F along the ridge γ; ϱ is the radius of curvature of the curve $\tilde{\gamma}$; $\varkappa_e$ and $\varkappa_i$ are the normal curvatures of the surface $\tilde{F}$ in the direction normal to the ridge γ, $\varkappa$ is the normal curvature F_0 in the corresponding direction; δ is the thickness of the shell; E is the modulus of elasticity; ν is the Poission's ratio and c is the constant $\simeq 0.19$. In the first term of the expression U the integration is over the area of the surface, in the two others over the arc of the ridge γ.

The term $A(F)$ of the functional W is the work done by the external load during the deformation of the shell into the F form and is calculated in the usual way.

4. The application of the variational principle A we shall demonstrate by the example of determining elastic states caused by the loss of stability of strictly covex shell submitted to a uniform external pressure. The shell is supposed to be sufficiently sloping and the edges clamped.

According to the principle A the determining of the elastic state of the shell at considerable buckling reduces to the variational problem for the functional W defined on the isometric transformations of the initial form. Because of the clamped edge and the strict convexity of the shell, the purely geometrical consideration shows that each isometric transformation of the middle surface with a break of smoothness along the convex curve γ reduces to the specular reflection of some segment of the shell in the plane of its base.

Assume that the region of buckling is small, or more generally, that the surface of the initial shell over the region of buckling is approximated well by the adjoining paraboloid in the centre of the buckling. Then the corresponding integration for the strain energy at the specular buckling gives the following simple expression.

$$U = \pi c E (2h)^{3/2} \delta^{5/2} (\varkappa_1 + \varkappa_2) .$$

Here $2h$ is the flexure in the centre of the buckling; $\varkappa_1$ and $\varkappa_2$ are the main curvatures of the middle surface, δ is the thickness of the shell, E is the modulus of elasticity and $c \simeq 0.19$.

In the same assumption for the work A done by the external pressure p during deformation of the shell we obtain the following expression

$$A = \frac{2\pi h^2 p}{\sqrt{\varkappa_1 \varkappa_2}} .$$

From the stationarity of the functional W in the state of elastic equilibrium of the shell we find the dependence of the pressure p received by the shell from the flexure $2h$ in the centre of the buckling

$$p = \frac{3}{2} c E (\varkappa_1 + \varkappa_2) \sqrt{\varkappa_1 \varkappa_2} \frac{\delta^{5/2}}{\sqrt{2h}} .$$

From this equation it is seen that the pressure (p) received by the shell decreases with the increase of deformation ($2h$) and it shows the instability of the post-critical deformations under external pressure.

The minimal pressure (lower critical load) received by the shell is determined by this expression if $2h$ is substituted by the maximum geometrically possible flexure. For example, for the spherical segment of the height h, the radius of curvature R and the thickness δ the following expression is obtained for the value of the lower critical pressure

$$p_i = 3 c E \left(\frac{\delta}{R}\right)^2 \sqrt{\frac{\delta}{2h}} .$$

This expression gives a somewhat smaller value because while solving the variational problem for the functional at the first stage we did not take into account the effect of the clamped edge on the deformation of the shell near the ridge of the isometric transformation. In a more accurate discussion the coefficient c in the expression of the strain energy is no longer constant (0.19). It depends on the dimensionless parameter that characterizes the proximity of the ridge of isometric transformation to the clamped edge of the shell. Accordingly, for the value of the lower critical pressure on the spherical segment we obtain the following more precise expression

$$p_i = \frac{3 c E \left(\frac{\delta}{R}\right)^2 \sqrt{\frac{\delta}{2h}}}{1 - 1{,}85 \sqrt{\frac{\delta}{2h}}} .$$

The value p_i given by this expression agrees well with the results of experimental investigations ($2h/\delta = 10-30$).

5. Let us give two more examples which show the effectiveness of the application of the geometrical principle A to the study of elastic states of stricktly convex shells during great deformations. Here we mean the loading of the shell by a concentrated force normal to the shell surface and combined loading by the concentrated force and the external pressure. Omitting the proof that is not much different from the above mentioned, let us formulate the results obtained.

The concentrated force acting normal to the shell surface results in the flexure $2h$ that is connected with the value of the force f by the following equation

$$f = \frac{3\pi c}{2} E \delta^{5/2} (\varkappa_1 + \varkappa_2) \sqrt{2h}.$$

Here $\varkappa_1$ and $\varkappa_2$ are the normal curvatures of the shell at the point where the force is applied, δ is the thickness of the shell, E is the modulus of elasticity and $c \simeq 0.19$. From this equation we see that with the increase of deformation ($2h$) the load (f) received by the shell increases. And this shows the stability of the equilibrium states of the shell under the concentrated load. For the case of the spherical segment of the radius R we have accordingly

$$2h = \frac{R^2 f^2}{9\pi^2 c^2 E^2 \delta^5}.$$

The experimental data agree well with this expression.

Let us take a sloping stricktly convex shell with the clamped edge, loaded by the concentrated force f and the uniform outer pressure which is smaller than the upper critical value. I mean the pressure at which the shell loses its stability. The elastic states of the shell here are determined as in the above cases. This discussion shows that at the given external pressure p and the small value of the force f the elastic states are stable but at a considerable f they are unstable. The critical combination of the p and f values separating the stable and unstable states is defined by the equation

$$pf = \frac{9}{16} \pi c^2 E^2 \delta^5 (\varkappa_1 + \varkappa_2)^2 \sqrt{\varkappa_1 \varkappa_2}.$$

Here $\varkappa_1$, $\varkappa_2$, E, δ and c have the former values. The application of this equation is restricted by the natural condition, namely, the pressure p must be considerably smaller than the upper critical value.

6. The main difficulty we have while solving the variational problem for the functional W consists in the analytical definition of the isometric

transformations of the middle surface of the shell. In the above examples these isometric transformations are very simple. That is why the solution of the corresponding problems is obtained in compact form. Very often the class of the possible isometric transformations is greatly reduced by the visual considerations of the form of the expected shell deformations. Let us give an example.

The experiments show that the cylindrical shell loaded on the edges by a uniformly distributed contracting stress loses stability and begins to buckle. This buckling is characterized by a definite periodicity both along the length and in the circular direction. Taking this periodicity of the structure of the deformed shell into consideration the corresponding isometric transformations are found by means of pureley geometrical discussion. It appears that they make possible the simple geometrical description, which consists in the following.

Let us take a regular prism with an even number of sides (2n) and on one of its side faces α_1 let us draw a periodical curve γ_1. Let us reflect specularly the curve γ_1 in the plane β passing through the side face α_1 and the prism axis. In this way we get a curve γ_2 lying in the side face α_2 adjoining α_1. Now let us construct a similar curve γ_3 in the face α_3 adjoining α_2 and so on. Thus, in each face α_i a curve γ_i will be constructed. Now through the curves γ_1 and γ_2 let us draw a cylindrical surface Z_{12} with the formative normals of the plane β. Similarly, we shall construct the surfaces Z_{23} and so on. The surfaces Z_{12}, Z_{23}, etc. constitute a tubular surface Z that is isometric to the cylinder with the break of smoothness and the formation of the ridges along the γ_i curves.

Considering the elastic states resulting from the loss of stability of the cylindrical shell at axial contraction we shall variate the functional W on the isometric transformations of "Z" type. As a result of it W will depend on the curve γ_1 taken on the prism face α_1 and two integer parameters which characterize the periodicity of the form of the deformed shell. The integer parameters are defined by the character of the stability loss, and the curve γ is defined by the stationarity of the functional W.

The investigation of the post-critical elastic states of the cylindrical shell in the way mentioned results in a definite dependence of the value of the axial stress received by the shell from deformation (ends draw together). For the magnitude of the minimum load received (i. e., for the lower critical load) we obtain the following simple expression:

$$p_i = 0.16 E \frac{\delta}{R}.$$

This expression agrees well with the data of experimental investigation ($R/\delta = 500-1300$).

Similar concepts have been applied for the investigation of the post-critical elastic states of cylindrical shells loaded by a uniform external pressure and twisting.

7. The geometrical method of investigation of the postcritical elastic states of stricktly convex shells may be applied to the study of the initial stage of post-critical deformation, in particular, to the determining of the upper critical load, i. e., the load at which the shell loses its stability. This is possible because the load received by the shell at the moment of the loss of stability is stationary. As a result of it at considerable buckling the load received by the shell does not change much. If the deformation is great we are under conditions of the application of the concepts given above. Such sort of discussion brings us to a certain variational principle "B" in which the infinitely small bendings take the place of isometric transformations.

References

Pogorelov, A. V.: Geometrical methods in the non-linear theory of elastic shells (in Russian), Moscow 1967.

Discussion

W. T. Koiter: My first question is how the second term in the energy equation, involving the constant C, has been derived. In addition to this question I should like to draw Professor Pogorelov's attention to similar previous work by Yoshimura and Kirste for cylindrical shells, and by Ashwell (at the first IUTAM-Symposium on thin shells) for spherical shells.

A. V. Pogorelov: The constant C in the expression of the deformation energy is associated with the minimum of some functional. An exhaustive answer to your question may be found in my publication referred to in my paper. See chapter II.

I know of the existence of the papers by Kirste, Yoshimura and Ashwell. Unfortunately, they were not published in sources accessible to me and therefore their contents are not known to me. According to the titles of these papers they appear to deal with particular cases, while my work is concerned with the general principles.

T. Brøndum-Nielsen: In your lecture you presented a formula for the critical combination of a uniformly distributed load p and a concentrated load f. According to this formula, f will approach infinity as p approaches zero and vice versa. Consequently the formula seems applicable only for values of p and f within certain limits.

A. V. Pogorelov: As I mentioned in my lecture, the allowable range of the formula determining the initial combination of p and f is restricted by some natural conditions. That is, the pressure p must not exceed the critical value and the deformation caused by the concentrated force f is to be within the limits of what is geometrically allowed.

Bending of a Cylindrical Shell Subject to Axial Loading

By

A. Kildegaard

Technical University of Denmark, Copenhagen, Denmark

1. Introduction

The present paper contains a solution of the equations for the semi-infinite circular cylindrical shell subject to an axial end-load in equilibrium. Forces, moments and displacements are computed using the Novozhilov equations of thin shells. A result of this investigation is the fact that the deformation cannot be considered as localized near the end as might perhaps be concluded from the St. Venant principle. A careful experimental investigation has also been carried out, and the results of it are found to be in complete agreement with theory.

There have been several earlier investigations of this subject of which the work by N. J. Hoff [1] seems most relevant. However, in that paper the simplified Donell equations were used, which does not seem to be quite satisfactory in all cases.

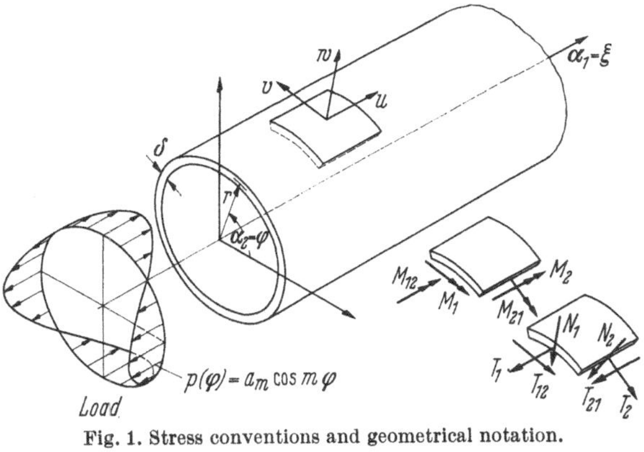

Fig. 1. Stress conventions and geometrical notation.

2. Basic Equations

In Fig. 1 a semi-infinite circular cylindrical shell with radius r and thickness δ is shown with a coordinate system ξ, φ, where ξ is the non-dimensional axial coordinate, and φ the angle in circumferential direction.

The load which acts upon the shell is an axial end-load in equilibrium, $p = a_m \cos m\varphi$, where a_m is a constant and m an integer. In order to have equilibrium, one must assume $m \geqq 2$. Displacements, forces and moments are taken positiv as shown in Fig. 1. Following the notations of NOVOZHILOV [2], the membrane force is denoted by T and the transversal force per unit length by N.

Also following NOVOZHILOV, let $\tilde{T}_1$, $\tilde{T}_2$, and $\tilde{S}$[1] be the complex forces, defined by the following equations

$$\tilde{T}_1 = T_1 - i\frac{2b^2}{r}\frac{M_2 - \mu M_1}{1-\mu^2} \tag{2.1}$$

$$\tilde{T}_2 = T_2 - i\frac{2b^2}{r}\frac{M_1 - \mu M_2}{1-\mu^2} \tag{2.2}$$

$$\tilde{S} = S + i\frac{2b^2}{r}\frac{H}{1-\mu} \tag{2.3}$$

where μ is Poisson's ratio and

$$4b^4 = 12(1-\mu^2)\left(\frac{r}{\delta}\right)^2. \tag{2.4}$$

The quantities S and H are given by

$$S = T_{12} - \frac{M_{21}}{r} = T_{21} \tag{2.5}$$

$$H = (M_{12} + M_{21})/2. \tag{2.6}$$

Following NOVOZHILOV we introduce the auxiliary function

$$\tilde{T} = \tilde{T}_1 + \tilde{T}_2 \tag{2.7}$$

whereby the shell equations are reduced to the single complex equation

$$\Delta\Delta\tilde{T} + \frac{\partial^2\tilde{T}}{\partial\varphi^2} + i\,2b^2\frac{\partial^2\tilde{T}}{\partial\xi^2} = 0. \tag{2.8}$$

In terms of the auxiliary function $\tilde{T}$ the complex forces $\tilde{T}_1$, $\tilde{T}_2$ and $\tilde{S}$ and the displacements $\tilde{u}$, $\tilde{v}$ and $\tilde{w}$ are obtained from the following relations

$$\tilde{T}_1 = \tilde{T} - i\frac{\Delta\tilde{T}}{2b^2} \tag{2.9}$$

$$\tilde{T}_2 = i\frac{\Delta\tilde{T}}{2b^2} \tag{2.10}$$

$$\frac{\partial\tilde{S}}{\partial\xi} = -\frac{i}{2b^2}\left\{\frac{\partial\Delta\tilde{T}}{\partial\varphi} + \frac{\partial\tilde{T}}{\partial\varphi}\right\} \tag{2.11}$$

---

[1] ~ (thilda) denotes a complex quantity.

$$\frac{\partial \tilde{S}}{\partial \varphi} = -\frac{\partial \tilde{T}}{\partial \xi} + i\frac{1}{2b^2}\frac{\partial}{\partial \xi}(\Delta \tilde{T}) \tag{2.12}$$

$$\frac{\partial \tilde{u}}{\partial \xi} = \frac{r}{E\delta}\left\{\tilde{T} - i\frac{1+\mu}{2b^2}\Delta \tilde{T}\right\} \tag{2.13}$$

$$\frac{\partial^2 \tilde{v}}{\partial \xi^2} = -\frac{r}{E\delta}\left\{\left(1 + i\frac{1+\mu}{b^2}\right)\frac{\partial \tilde{T}}{\partial \varphi} + i\frac{1+\mu}{2b^2}\frac{\partial}{\partial \varphi}(\Delta \tilde{T})\right\} \tag{2.14}$$

$$\frac{\partial^2 \tilde{w}}{\partial \xi^2} = \frac{r}{E\delta}\left\{\Delta \tilde{T} + i\frac{1+\mu}{2b^2}\frac{\partial^2 \tilde{T}}{\partial \varphi^2}\right\}. \tag{2.15}$$

From the last three equations the physical displacements can be found as the real part of corresponding complex displacement, i. e.

$$u = Re(\tilde{u}), \quad v = Re(\tilde{v}) \quad \text{and} \quad w = Re(\tilde{w}). \tag{2.16}$$

3. Solution of the Shell Equation

The self-equilibrating axial load applied to the semi-infinite shell is given by

$$p = a_m \cos m\varphi. \tag{3.1}$$

The solution of (2.8) is found in the form

$$\tilde{T} = \tilde{C} e^{\lambda \xi} \cos m\varphi \tag{3.2}$$

where $\tilde{C}$ and λ are complex numbers.

By inserting (3.2) into the shell equation (2.8) we find the following bi-quadratic equation for λ

$$\lambda^4 - 2(m^2 - i b^2)\lambda^2 + (m^4 - m^2) = 0. \tag{3.3}$$

The four complex roots of the characteristic equation (3.3) are given by

$$\lambda_i = \pm\sqrt{m^2 - i b^2 \pm \sqrt{m^2 - b^4 - 2 i b^2 m^2}} \quad i = 1, 2, 3, 4. \tag{3.4}$$

Hence one can write the solution of (2.8) in the following form

$$\tilde{T} = \sum_{i=1}^{4} \tilde{C}_i e^{\lambda_i \xi} \cos m\varphi. \tag{3.5}$$

The characteristic equation has two roots with positiv real parts. Since a bounded behaviour is required in the whole region, the corresponding coefficients C_i must be equal to zero.

The solution is therefore given by

$$\tilde{T} = (\tilde{C}_1 e^{\lambda_1 \xi} + \tilde{C}_2 e^{\lambda_2 \xi}) \cos m\varphi \tag{3.6}$$

where $\lambda_1 = \lambda_{1R} + i\lambda_{1J}$ and $\lambda_2 = \lambda_{2R} + i\lambda_{2J}$ have negativ real parts such that $\lambda_{1R} < \lambda_{2R} < 0$.

There are four boundary conditions for $\xi = 0$, from which the complex constants $\tilde{C}_1$ and $\tilde{C}_2$ can be found. These are

$$\begin{aligned} T_1 &= a_m \cos m\varphi \\ M_1 &= 0 \\ S + \frac{2H}{r} &= 0 \\ 2\frac{\partial H}{\partial \varphi} + \frac{\partial M_1}{\partial \xi} &= 0 . \end{aligned} \tag{3.7}$$

By substituting $\tilde{T}$ from (3.6) into (2.9), (2.10), (2.11) and (2.12) the complex forces $\tilde{T}_1$, $\tilde{T}_2$ and $\tilde{S}$ are obtained as functions of the coordinates ξ and φ. Separating the real and imaginary part of $\tilde{T}_1$, $\tilde{T}_2$ and $\tilde{S}$ and using Eqs. (2.1), (2.2) and (2.3) the forces and moments take the following form

$$\begin{aligned} T_1 &= \{e^{\lambda_{1R}\xi}(B_{1,1}\cos\lambda_{1J}\xi + B_{2,1}\sin\lambda_{1J}\xi + \\ &\quad + e^{\lambda_{2R}\xi}(B_{1,2}\cos\lambda_{2J}\xi + B_{2,2}\sin\lambda_{2J}\xi)\}\cos m\varphi \\ T_2 &= \{e^{\lambda_{1R}\xi}(B_{3,1}\cos\lambda_{1J}\xi + B_{4,1}\sin\lambda_{1J}\xi) + \\ &\quad + e^{\lambda_{2R}\xi}(B_{3,2}\cos\lambda_{2J}\xi + B_{4,2}\sin\lambda_{2J}\xi)\}\cos m\varphi \\ M_1 &= r\{e^{\lambda_{1R}\xi}(B_{5,1}\cos\lambda_{1J}\xi + B_{6,1}\sin\lambda_{1J}\xi) + \\ &\quad + e^{\lambda_{2R}\xi}(B_{5,2}\cos\lambda_{2J}\xi + B_{6,2}\sin\lambda_{2J}\xi)\}\cos m\varphi \\ M_2 &= r\{e^{\lambda_{1R}\xi}(B_{7,1}\cos\lambda_{1J}\xi + B_{8,1}\sin\lambda_{1J}\xi) + \\ &\quad + e^{\lambda_{2R}\xi}(B_{7,2}\cos\lambda_{2J}\xi + B_{8,2}\sin\lambda_{2J}\xi)\}\cos m\varphi \\ S &= \{e^{\lambda_{1R}\xi}(B_{9,1}\cos\lambda_{1J}\xi + B_{10,1}\sin\lambda_{1J}\xi) + \\ &\quad + e^{\lambda_{2R}\xi}(B_{9,2}\cos\lambda_{2J}\xi + B_{10,2}\sin\lambda_{2J}\xi)\}\sin m\varphi \\ H &= r\{e^{\lambda_{1R}\xi}(B_{11,1}\cos\lambda_{1J}\xi + B_{12,1}\sin\lambda_{1J}\xi) + \\ &\quad + e^{\lambda_{2R}\xi}(B_{11,2}\cos\lambda_{2J}\xi + B_{12,2}\sin\lambda_{2J}\xi)\}\sin m\varphi . \end{aligned} \tag{3.8}$$

Correspondingly, we obtain from (2.13), (2.14) and (2.15) the displacements u, v and w

$$\begin{aligned} u &= \frac{r}{E\delta}\{e^{\lambda_{1R}\xi}(B_{13,1}\cos\lambda_{1J}\xi + B_{14,1}\sin\lambda_{1J}\xi) + \\ &\quad + e^{\lambda_{2R}\xi}(B_{13,2}\cos\lambda_{2J}\xi + B_{14,2}\sin\lambda_{2J}\xi)\}\cos m\varphi \end{aligned}$$

$$v = \frac{r}{E\delta}\{e^{\lambda_{1R}\xi}(B_{15,1}\cos\lambda_{1J}\xi + B_{16,1}\sin\lambda_{1J}\xi) + \\ + e^{\lambda_{2R}\xi}(B_{15,2}\cos\lambda_{2J}\xi + B_{16,2}\sin\lambda_{2J}\xi)\}\sin m\varphi$$

$$w = \frac{r}{E\delta}\{e^{\lambda_{1R}\xi}(B_{17,1}\cos\lambda_{1J}\xi + B_{18,1}\sin\lambda_{1J}\xi) + \\ + e^{\lambda_{2R}\xi}(B_{17,2}\cos\lambda_{2J}\xi + B_{18,2}\sin\lambda_{2J}\xi)\}\cos m\varphi. \tag{3.9}$$

For the coefficients $B_{i,j}$ we have

$$\begin{aligned}
B_{1,1} &= C_{1R} - C_{4,1}/2b^2 \\
B_{2,1} &= C_{1J} + G_{3,1}/2b^2 \\
B_{3,1} &= G_{4,1}/2b^2 \\
B_{4,1} &= -G_{3,1}/2b^2 \\
B_{5,1} &= -(\mu C_{1J} + (1-\mu)\cdot G_{3,1}/2b^2)/2b^2 \\
B_{6,1} &= -(\mu C_{1R} + (1-\mu)\, G_{4,1}/2b^2)/2b^2 \\
B_{7,1} &= -(C_{1J} - (1-\mu)\cdot G_{3,1}/2b^2)/2b^2 \\
B_{8,1} &= -(C_{1R} - (1-\mu)\, G_{4,1}/2b^2)/2b^2 \\
B_{9,1} &= (-G_{1,1} + G_{6,1}/2b^2)/m \\
B_{10,1} &= (-G_{2,1} - G_{5,1}/2b^2)/m \\
B_{11,1} &= (1-\mu)\,(G_{2,1} + G_{5,1}/2b^2)/2b^2 m \\
B_{12,1} &= (1-\mu)\,(-G_{1,1} + G_{6,1}/2b^2)/2b^2 m \\
B_{13,1} &= \{G_{7,1}(1 + (1+\mu)\,\gamma_1/2b^2) + G_{8,1}\,(1+\mu)\,\beta_1/2b^2\}/(\lambda_{1R}^2 + \lambda_{1J}^2) \\
B_{14,1} &= \{G_{7,1}(1+\mu)\,\beta_1/2b^2 - G_{8,1}(1 + (1+\mu)\,\gamma_1/2b^2)\}/(\lambda_{1R}^2 + \lambda_{1J}^2) \\
B_{15,1} &= m\,\{G_{9,1}(1 - (1+\mu)\cdot\gamma_1/2b^2) - \\
&\quad - G_{10,1}\cdot(1+\mu)\,(\beta_1 + 2)/2b^2\}/(\lambda_{1R}^2 + \lambda_{1J}^2)^2 \\
B_{16,1} &= m\{-G_{10,1}(1 - (1+\mu)\,\gamma_1/2b^2) - \\
&\quad - G_{9,1}(1+\mu)\,(\beta_1 + 2)/2b^2\}/(\lambda_{1R}^2 + \lambda_{1J}^2)^2 \\
B_{17,1} &= \{1 - m^2(\beta_1 + m^2 + (1+\mu)\cdot\gamma_1/2b^2)/(\lambda_{1R}^2 + \lambda_{1J}^2)^2\}\,C_{1R} + \\
&\quad + m^2 C_{1J}\,\{(1+\mu)\,(\beta_1 + m^2)/2b^2 - \gamma_1\}/(\lambda_{1R}^2 + \lambda_{1J}^2)^2 \\
B_{18,1} &= m^2 C_{1R}\,\{(1+\mu)\,(\beta_1 + m^2)/2b^2 - \gamma_1\}/(\lambda_{1R}^2 + \lambda_{1J}^2)^2 - \\
&\quad - \{1 - m^2(\beta_1 + m^2 + (1+\mu)\,\gamma_1/2b^2)/(\lambda_{1R}^2 + \lambda_{1J}^2)^2\}\,C_{1J}
\end{aligned} \tag{3.10}$$

where

$$
\begin{aligned}
G_{1,1} &= C_{1R}\lambda_{1R} - C_{1J}\lambda_{1J}\\
G_{2,1} &= -C_{1R}\lambda_{1J} - C_{1J}\lambda_{1R}\\
G_{3,1} &= C_{1R}\beta_1 - C_{1J}\cdot\gamma_1\\
G_{4,1} &= -C_{1R}\gamma_1 - C_{1J}\beta_1\\
G_{5,1} &= C_{1R}(-\gamma_1\lambda_{1J} + \beta_1\cdot\lambda_{1R}) - C_{1J}(\beta_1\lambda_{1J} + \gamma_1\lambda_{1R})\\
G_{6,1} &= C_{1R}(-\gamma_1\lambda_{1R} - \beta_1\cdot\lambda_{1J}) + C_{1J}(-\beta_1\lambda_{1R} + \gamma_1\lambda_{1J})\\
G_{7,1} &= C_{1R}\cdot\lambda_{1R} + C_{1J}\cdot\lambda_{1J}\\
G_{8,1} &= C_{1J}\lambda_{1R} - C_{1R}\lambda_{1J}\\
G_{9,1} &= (\lambda_{1R}^2 - \lambda_{1J}^2)\,C_{1R} + 2\lambda_{1R}\lambda_{1J}C_{1J}\\
G_{10,1} &= (\lambda_{1R}^2 - \lambda_{1J}^2)\,C_{1J} - 2\lambda_{1R}\lambda_{1J}\cdot C_{1R}
\end{aligned}
\tag{3.11}
$$

and

$$
\begin{aligned}
\beta_1 &= \lambda_{1R}^2 - \lambda_{1J}^2 - m^2\\
\gamma_1 &= 2\lambda_{1R}\lambda_{1J}\,.
\end{aligned}
\tag{3.12}
$$

The coefficients $B_{i,2}$ ($i = 1, 2, \ldots, 18$) are found by changing index 1 to 2 in (3.10), (3.11) and (3.12). Inserting the values of T_1, M_1, S and H from Eq. (3.8) in (3.7), we get the following system of four equations for the four coefficients C_{1R}, C_{1J}, C_{2R} and C_{2J}

$$
A_{i,j}\begin{Bmatrix} C_{1R}\\ C_{1J}\\ C_{2R}\\ C_{2J}\end{Bmatrix} = \begin{Bmatrix} a_m\\ 0\\ 0\\ 0\end{Bmatrix}
\tag{3.13}
$$

where $A_{i,j}$ is the coefficient matrix depending upon m, μ, b, λ_{1R}, λ_{1J}, λ_{2R} and λ_{2J}.

The elements of the coefficient matrix $A_{i,j}$ can be written as

$$
\begin{aligned}
A_{1,1} &= 1 + \gamma_1/2b^2\\
A_{1,2} &= b_1/2b^2\\
A_{1,3} &= 1 + \gamma_2/2b^2\\
A_{1,4} &= \beta_2/2b^2\\
A_{2,1} &= (1-\mu)\cdot\beta_1/2b^2\\
A_{2,2} &= \mu - (1-\mu)\gamma_1/2b^2\\
A_{2,3} &= (1-\mu)\,\beta_2/2b^2\\
A_{2,4} &= \mu - (1-\mu)\,\gamma_2/2b^2
\end{aligned}
$$

$$A_{3,1} = -\lambda_{1R} + (-\gamma_1\lambda_{1R} - \beta_1\lambda_{1J})/2b^2 + (1-\mu)\{-\lambda_{1J} + (-\gamma_1\lambda_{1J} + \beta_1\lambda_{1R})/2b^2\}/b^2$$

$$A_{3,2} = \lambda_{1J} + (-\beta_1\lambda_{1R} + \gamma_1\lambda_{1J})/2b^2 + (1-\mu)\{-\lambda_{1R} - (\beta_1\lambda_{1J} + \gamma_1\lambda_{1R})/2b^2\}/b^2$$

$$A_{3,3} = -\lambda_{2R} + (-\gamma_2\lambda_{2R} - \beta_2\lambda_{2J})/2b^2 + (1-\mu)\{-\lambda_{2R} + (-\gamma_2\lambda_{2J} + \beta_2\lambda_{2R})/2b^2\}/b^2$$

$$A_{3,4} = \lambda_{2J} + (-\beta_2\lambda_{2R} + \gamma_2\lambda_{2J})/2b^2 + (1-\mu)\{-\lambda_{2R} - (\beta_2\lambda_{2J} + \gamma_2\lambda_{2R})/2b^2\}/b^2$$

$$A_{4,1} = (1-\mu)(-\beta_1\lambda_{1R} + \gamma_1\lambda_{1J})/2b^2 + (2-\mu)\lambda_{1J}$$

$$A_{4,2} = (1-\mu)(\beta_1\lambda_{1J} + \gamma_1\lambda_{1R})/2b^2 + (2-\mu)\lambda_{1R}$$

$$A_{4,3} = (1-\mu)(-\beta_2\lambda_{2R} + \gamma_2\lambda_{2J})/2b^2 + (2-\mu)\lambda_{2J}$$

$$A_{4,4} = (1-\mu)(\beta_2\lambda_{2J} + \gamma_2\lambda_{2R})/2b^2 + (2-\mu)\lambda_{2R} \tag{3.14}$$

The results of a numerical calculation of the roots of the characteristic equation (3.3) are for different values of m and b given in Table 1. The coefficients $B_{i,j}$ in Eq. (3.10) are for the end-load $p = 1 \cdot \cos m\varphi$ and Poisson's ratio $\mu = 0.3$ given in Table 2.

Table 1. *Values of the Roots λ_{1R}, λ_{1J}, λ_{2R} and λ_{2J} of Eq.* (3.3)

$b = 6$

m	2	6	10	14	18	22
λ_{1R}	− 6.3477	− 9.1762	−13.0487	−17.0267	−21.0209	−25.0190
λ_{1J}	5.6843	4.4413	3.8767	3.6218	3.4792	3.3884
λ_{2R}	− 0.3029	− 3.1341	− 7.0067	−10.9849	−14.9791	−18.9772
λ_{2J}	− 0.2712	− 1.5169	− 2.0817	− 2.3366	− 2.4792	− 2.5701

$b = 10$

m	2	6	10	14	18	22
λ_{1R}	−10.2034	−11.9844	−15.2919	−19.1154	−23.0588	−27.0361
λ_{1J}	9.8036	8.5733	7.4196	6.7602	6.3728	6.1227
λ_{2R}	− 0.1765	− 1.9592	− 5.2668	− 9.0904	−13.0338	−17.0111
λ_{2J}	− 0.1696	− 1.4016	− 2.5554	− 3.2148	− 3.6022	− 3.8524

$b = 18$

m	2	6	10	14	18	22
λ_{1R}	−18.1117	−19.0481	−21.0485	−24.0043	−27.5241	−31.2994
λ_{1J}	17.8895	17.0599	15.7198	14.3940	13.3674	12.6266
λ_{2R}	− 0.0969	− 1.0341	− 3.0346	− 5.9904	− 9.5102	−13.2855
λ_{2J}	− 0.0956	− 0.9261	− 2.2663	− 3.5921	− 4.6187	− 5.3595

Table 2. *Values of the Coefficients* $B_{i,j}$ *in Eq.* (3.8) *for* $b = 6$

	$m = 2$		$m = 6$		$m = 10$	
$i \backslash j$	1	2	1	2	1	2
1	−0.0108	1.0108	−0.1234	1.1234	−0.3958	1.3958
2	0.0042	−0.8283	0.3753	−0.4801	0.7613	−0.8003
3	0.0947	0.0401	1.0707	−0.1070	1.3848	−0.3910
4	0.1874	0.0599	−0.3797	0.4111	−0.7677	0.7794
5	0.0026	−0.0026	−0.0037	0.0037	−0.0075	0.0075
6	−0.0013	−0.0048	−0.0144	−0.0032	−0.0176	−0.0004
7	0.0008	−0.0113	0.0036	−0.0050	0.0074	−0.0079
8	−0.0002	−0.0142	−0.0027	−0.0152	−0.0003	−0.0178
9	−0.0462	0.0407	−0.4664	0.4654	−0.8116	0.8114
10	−0.0173	−0.2625	0.4826	−0.5348	0.8400	−0.8513
11	0.0002	0.0026	−0.0047	0.0052	−0.0082	0.0083
12	−0.0004	0.0004	−0.0045	0.0045	−0.0079	0.0079
13	0.0075	−3.2188	0.0183	−0.3742	0.0364	−0.2387
14	0.0015	−0.0881	−0.0622	0.0114	−0.0868	0.0767
15	0.0125	10.1934	0.0597	0.1022	0.0955	−0.0200
16	−0.0045	11.9634	−0.1250	0.4712	−0.1292	0.2006
17	0.0730	−20.6499	0.7496	−1.0570	0.5490	−0.6102
18	0.1952	−23.6184	0.2575	−2.2722	0.2960	−0.9861

	$m = 14$		$m = 18$		$m = 22$	
$i \backslash j$	1	2	1	2	1	2
1	−0.7027	1,7027	−1.0196	2.0196	−1.3408	2.3408
2	1.1205	−1.1478	1.4702	−1.4940	1.8157	−1.8382
3	1.6975	−0.7000	2.0163	−1.0176	2.3384	−1.3392
4	−1.1264	1.1346	−1.4760	1.4832	−1.8216	1.8283
5	−0.0110	0.0110	−0.0144	0.0144	−0.0177	0.0177
6	−0.0206	0.0026	−0.0238	0.0057	−0.0269	0.0088
7	0.0109	−0.0112	0.0143	−0.0146	0.0176	−0.0179
8	0.0027	−0.0207	0.0058	−0.0238	0.0089	−0.0269
9	−1.1445	1.1444	−1.4749	1.4749	−1.8045	1.8045
10	1.1810	−1.1848	1.5199	−1.5215	1.8584	−1.8591
11	−0.0115	0.0115	−0.0148	0.0148	−0.0181	0.0181
12	−0.0111	0.0111	−0.0143	0.0143	−0.0175	0.0175
13	0.0507	−0.1942	0.0606	−0.1719	0.0676	−0.1586
14	−0.0964	0.0942	−0.1010	0.1010	−0.1036	0.1043
15	0.1068	−0.0551	0.1107	−0.0711	0.1122	−0.0800
16	−0.1238	0.1487	−0.1198	0.1310	−0.1172	0.1230
17	0.4134	−0.4401	0.3289	−0.3441	0.2723	−0.2822
18	0.2701	−0.6028	0.2394	−0.4260	0.2130	−0.3259

Table 2. *Values of the Coefficients* $B_{i,j}$ *in Eq.* (3.8) *for* $b = 10$
(Continued)

	$m = 2$		$m = 6$		$m = 10$	
$i \backslash j$	1	2	1	2	1	2
1	−0.0051	1.0051	−0.0267	1.0267	−0.1296	1.1296
2	−0.0029	−0.9284	0.1387	−0.3620	0.3826	−0.4425
3	−0.1384	0.0178	0.8436	0.0027	1.0999	−0.1203
4	0.2587	0.0207	−0.1136	0.1806	0.3850	0.4030
5	0.0013	−0.0013	−0.0004	0.0004	−0.0014	0.0014
6	0.0007	−0.0016	−0.0042	−0.0016	−0.0053	−0.0011
7	0.0004	−0.0046	0.0005	−0.0015	0.0013	−0.0016
8	0.0002	−0.0051	−0.0011	−0.0051	−0.0010	−0.0055
9	−0.0114	0.0100	−0.2515	0.2507	−0.4821	0.4819
10	−0.0399	−0.1672	0.2389	−0.3580	0.4888	−0.5217
11	0.0001	0.0006	−0.0008	0.0013	−0.0017	0.0018
12	−0.0000	0.0000	−0.0009	0.0009	−0.0017	0.0017
13	0.0021	−5.5896	0.0086	−0.4469	0.0115	−0.2212
14	0.0059	−0.0763	−0.0206	−0.1073	−0.0382	−0.0003
15	0.0058	31.1901	0.0136	0.3049	0.0374	0.0460
16	0.0034	33.2940	−0.0512	1.0218	−0.0763	0.2805
17	−0.1484	−62.6638	0.7700	−2.1349	0.7648	−0.9188
18	0.2527	−66.2887	0.1522	−5.8414	0.2633	−2.2693

	$m = 14$		$m = 18$		$m = 22$	
$i \backslash j$	1	2	1	2	1	2
1	−0.2875	1.2875	−0.4651	1.4651	−0.6510	1.6510
2	0.6139	−0.6407	0.8333	−0.8499	1.0464	−1.0589
3	1.2784	−0.2838	1.4608	−0.4631	1.6484	−0.6497
4	−0.6172	0.6253	−0.8362	0.8412	−1.0489	1.0527
5	−0.0022	0.0022	−0.0029	0.0029	−0.0037	0.0037
6	−0.0060	−0.0005	−0.0066	0.0001	−0.0073	0.0008
7	0.0021	−0.0023	0.0029	−0.0030	0.0037	−0.0037
8	−0.0005	−0.0060	0.0001	−0.0066	0.0008	−0.0073
9	−0.6889	0.6888	−0.8908	0.8908	−1.0912	1.0912
10	0.6993	−0.7117	0.9028	−0.9086	1.1047	−1.1079
11	−0.0024	0.0025	−0.0032	0.0032	−0.0039	0.0039
12	−0.0024	0.0024	−0.0031	0.0031	−0.0038	0.0038
13	0.0181	−0.1629	0.0243	−0.1360	0.0295	−0.1206
14	−0.0482	0.0335	−0.0537	0.0470	−0.0570	0.0535
15	0.0527	0.0001	0.0603	−0.0204	0.0641	−0.0318
16	−0.0785	0.1519	−0.0765	0.1107	−0.0744	0.0929
17	0.6264	−0.6708	0.5148	−0.5352	0.4330	−0.4449
18	0.2973	−1.3096	0.2911	−0.8972	0.2732	−0.6742

Table 2. *Values of the Coefficients* $B_{i,j}$ *in Eq.* (3.8) *for* $b = 18$ (Continued)

	$m = 2$		$m = 6$		$m = 10$	
$i \backslash j$	1	2	1	2	1	2
1	−0.0018	1.0018	−0.0076	1.0076	−0.0215	1.0215
2	−0.0015	−0.9774	0.0251	−0.5102	0.1271	−0.2873
3	−0.2477	0.0060	0.4674	0.0219	0.8833	−0.0019
4	0.2874	0.0062	0.0874	0.0582	−0.1057	0.1538
5	0.0004	−0.0004	0.0001	−0.0001	−0.0001	0.0001
6	0.0004	−0.0005	−0.0007	−0.0005	−0.0014	−0.0005
7	0.0001	−0.0015	0.0001	−0.0008	0.0001	−0.0004
8	0.0001	−0.0015	−0.0002	−0.0016	−0.0004	−0.0016
9	−0.0021	0.0018	−0.0953	0.0949	−0.2450	0.2448
10	−0.0297	−0.0952	0.0581	−0.2435	0.2339	−0.3187
11	0.0000	0.0001	−0.0001	0.0003	−0.0003	0.0003
12	−0.0000	−0.0000	−0.0001	0.0001	−0.0003	0.0003
13	0.0004	−10.2850	0.0043	−0.7908	0.0051	−0.2689
14	0.0045	−0.0397	−0.0038	−0.1979	−0.0114	−0.0909
15	0.0020	105.8769	0.0031	1.5387	0.0066	0.1596
16	0.0018	107.8665	−0.0095	3.1386	−0.0284	0.6919
17	−0.2512	−212.0473	0.4510	−9.5123	0.8237	−1.9042
18	0.2843	−215.4327	0.1369	−18.6204	0.1404	−6.6788

	$m = 14$		$m = 18$		$m = 22$	
$i \backslash j$	1	2	1	2	1	2
1	−0.0643	1.0643	−0.1330	1.1330	−0.2173	1.2173
2	0.2558	−0.3203	0.3874	−0.4195	0.5155	−0.5342
3	1.0231	−0.0544	1.1173	−0.1279	1.2100	−0.2145
4	−0.2535	0.2728	−0.3886	0.3982	−0.5172	0.5228
5	−0.0003	0.0003	−0.0004	0.0004	−0.0006	0.0006
6	−0.0015	−0.0004	−0.0017	−0.0003	−0.0018	−0.0002
7	0.0003	−0.0004	0.0004	−0.0005	0.0006	−0.0006
8	−0.0004	−0.0016	−0.0003	−0.0017	−0.0002	−0.0018
9	−0.3733	0.3732	−0.4910	0.4910	−0.6050	0.6049
10	0.3725	−0.4101	0.4936	−0.5124	0.6087	−0.6192
11	−0.0004	0.0004	−0.0005	0.0006	−0.0007	0.0007
12	−0.0004	0.0004	−0.0005	0.0005	−0.0007	0.0007
13	0.0053	−0.1623	0.0066	−0.1219	0.0085	−0.1010
14	−0.0170	−0.0302	−0.0215	−0.0025	−0.0249	0.0113
15	0.0140	0.0502	0.0213	0.0212	0.0266	0.0065
16	−0.0389	0.2787	−0.0429	0.1552	−0.0438	0.1051
17	0.8457	−1.0765	0.7746	−0.8492	0.6901	−0.7221
18	0.2139	−3.5324	0.2672	−2.2692	0.2921	−1.6296

4. Experimental Results

In order to verify the theory, a careful experimental investigation has been carried out in which the normal deflection was mesured. A cylindrical shell was welded together from rolled sections of a thin steel plate. The dimensions of the shell were the following: radius 118 mm, thickness 1.91 mm and length 3000 mm (= 25.4. radius), so that the associated number b was equal to 10.10.

Fig. 2. Test setup showing load device and supports of the shell.

At one end of the shell four concentrated, equally spaced, axial forces are applied. Two opposite forces are pulling, and the other two pushing. The other edge of the shell is free. During the experiment the shell was placed with its axis vertically. The load device and the supports of the shell are shown on Fig. 2.

The load device consist of a screw which through a strain gauge dynamometer acts upon a horizontal beam. Each end of the beam is connected with a flexible wire for transfering the pulling forces to the shell. The weight of the shell and the reactions from the pulling forces in the wires is taken up by two columns.

As the shell is much more sensitive for radial forces than for the axial forces for which the experiment is carried out all possible care was taken to avoid obliquity in the load. In order to secure parallelity of the wires its supports on the beam were made adjustabel. The columns were supported on a roller with its centre in the point of contact of the column and the shell so that horizontal forces could not be transmitted to the shell.

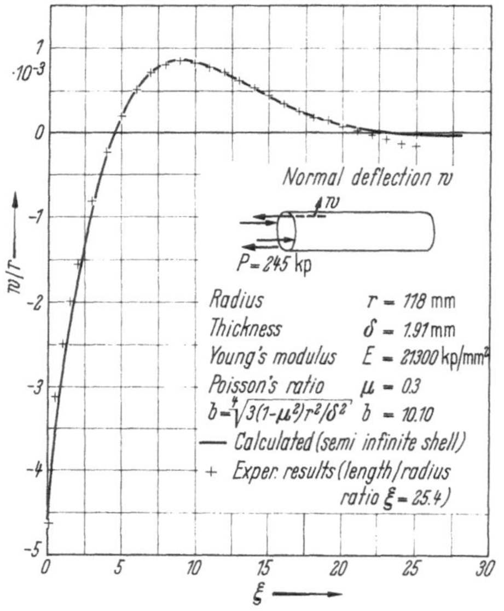

Fig. 3. Normal deflections along the generator.

The normal deflection of the shell along the circumference was mesured by means of 12 equally spaced dial gauges mounted in a ring, see Fig. 2. By placing the ring at different positions along the shell, and for every position making a load cycle, the normal deflections along 12 generators of the shell were mesured. As a result of the mesurements the normal deflections along one generator are shown on Fig. 3.

5. Discussion

The theoretical solution for the semi-infinite cylindrical shell subject to the same load as used in the experimental investigation is found by expanding the four forces, each of magnitude P, in a Fourier series

$$p(\varphi) = \frac{4P}{\pi r} \sum_{m=2,6,10,\ldots}^{\infty} \cos m\varphi . \tag{5.1}$$

The corresponding normal deflection is calculated and shown together with the experimental results on Fig. 3. The figure shows a very good agreement between theory and experiment. In some distance from the edge (about 3 radii) only the first term in the Fourier series is significant, specially so since the $m = 4$ term is abscent. The deviation of the experimental results from the theoretical curve for $\xi > 20$ can be attributed to the finite length of the shell.

From (3.8) and (3.9) it is seen, that the axial decay-length ξ^* is determined by the real parts λ_{1R} and λ_{2R} of the roots of the characteristic equation (3.3).

Fig. 4 shows in a typical case how λ_{1R} and λ_{2R} varies with the circumferential wave number m. At some distance from the edge of the shell, only the term with the factor $e^{-\lambda_{2R}\xi}$ in Eq. (3.8) and (3.9) is significant, thus the acial decaylength is mainly determined by λ_{2R}. As $-\lambda_{2R}$ increases with m the axial decay-length ξ^* decrease with increasing m.

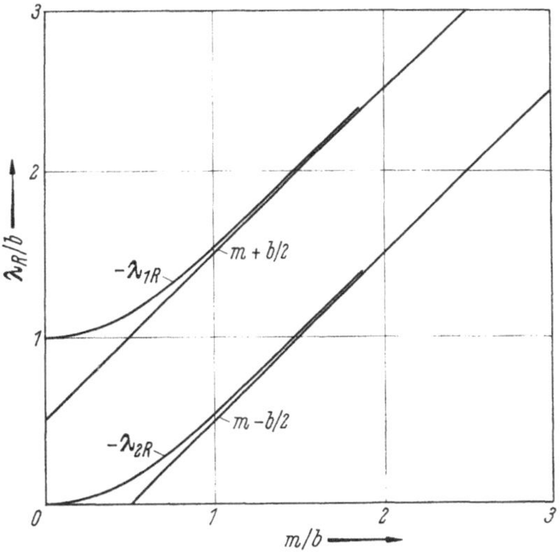

Fig. 4. Real parts of the roots of the characteristic equation.

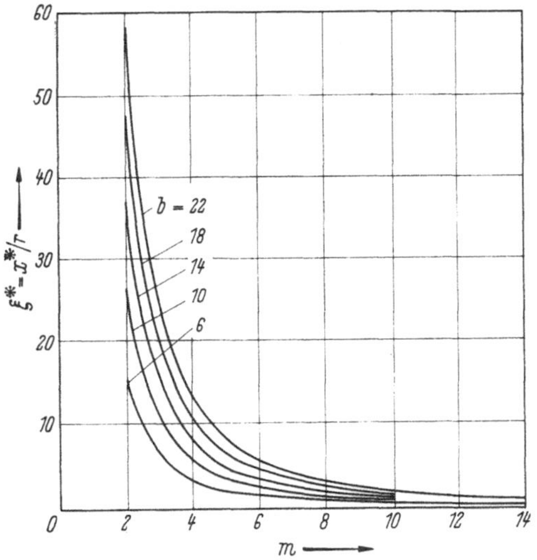

Fig. 5. Axial decay-length ξ^*.

Defining for example ξ^* as the distance from the edge where the responce is damped to 1% of its original value, ξ^* is determined by $e^{-\lambda_{2R}\xi^*} = 0.01$. Fig. 5 shows how ξ^* varies with m for different values of b.

For low values of the circumferencial wave number m, it appears that the effect of a self-equilibrating end load, propagates far down the length of the shell. The effect can only be considered as localized near the end for sufficiently large values of m ($m > b$). Hence St. Venant's principle must be applied with caution to thin shells.

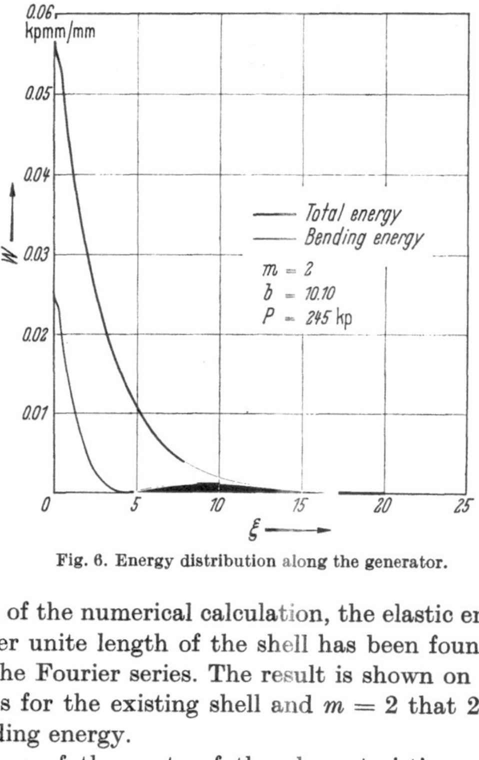

Fig. 6. Energy distribution along the generator.

As a result of the numerical calculation, the elastic energy of bending and tension per unite length of the shell has been found using just the first term in the Fourier series. The result is shown on Fig. 6. Furthermore one finds for the existing shell and $m = 2$ that 27% of the total energy is bending energy.

If the values of the roots of the characteristic equation (3.3) are compared with the corresponding roots of the Donnell equations [1] it is found that for small values of the circumferencial wave number m ($m \ll b$) the latter roots are in error by about 10%. This is a well known result [3], [4]. On the other hand, for small values of m Vlasov's semi-membrane theory (see [2]) shows to be in very good agreement with the precent one of NOVOZHILOV, while for larger values of m ($m > b$) the semi-membrane theory is very much in error.

References

1. HOFF, N. J.: Boundary-value problems of the thin-walled circular cylinder. J. Appl. Mech., Trans. ASME, **76**, 343–350 (1954).
2. NOVOZHILOV, V. V.: The Theory of Thin Shells, Groningen: Noordhoff 1959.

3. KEMPNER, J.: Remarks on Donnell's Equations, J. Appl. Mech., Trans. ASME, 77, 117–118 (1955).
4. HOFF, N. J.: The accuracy of Donnell's Equations. J. Appl. Mech. Trans. ASME, 77, 329–334 (1955).

Discussion

A. VAN DER NEUT: There exists a relation between this problem and the former presentation by Professor POGORELOV. By "isometric" (inextensional) deformation the loaded edge can obtain displacements in the direction of the load system. Therefore, if the shell thickness goes to zero the shell cannot resist this end load. If the shell thickness is small the load can be sustained but the deformation is close to inextensional where the "wave length" is infinite and one gets very slowly decaying deformations. This means that thin shells are not appropriate as a structure for this load system and one has to apply reinforcement by rings.

A. L. GOL'DENWEIZER: I would like to ask the author if he is familiar with the papers of Professor VLASOV in which the subject of decay of the stress-strain state in circular cylindrical shells, subjected to self-equilibrated edge loads is treated. These papers appeared about 1934–37.

V. KRUPKA: Using the semi-bending theory the same problem was solved by V. Z. VLASOV. Since your solution is more accurate the relation between semi-bending theory and your theory will be interesting. It seems that especially results at a greater distance from the points of application of the forces should be very similar.

A. KILDEGAARD: VLASOV's semi-membrane theory is for low circumferential wave numbers, i. e. for $m \ll b$, in good agreement with the present theory of NOVOZHILOV. For large wave numbers, i. e. for $m > b$, the semi-membrane theory is very much in error.

W. FLÜGGE: The paper illustrates the well-known fact that St. Venant's principle is not a principle — counterexamples show that it sometimes does not hold. All what the principle really states, is, that the fundamental equations of elasticity are elliptic.

The Theory of Symmetrically Loaded Weakmoment Shells of Revolution, Made of Differentmodulus Material

By

S. A. Ambartsumyan

Academy of Sciences of the Armenian SSR, Erevan, USSR

In this paper we consider, in certain aspects, the theory of symmetrically loaded weak moment shells of revolution, made of different modulus material [1, 2].

1. Let us discuss a shell whose median surface is a surface of revolution with axis z.

The location of any point M of the shell will be defined by the Gaussian coordinates: φ, the angle, which is the azimuth of the plane extended trough the point M and the z-axis, the meridian arc s, and normal, to the median surface the rectilinear coordinate γ [3, 4].

The material of the shell is such that at pure tension in any direction it has the modulus of elasticity E^+, and at pure compression in any direction it has the modulus of elasticity E^-. The Poisson ratios are ν^+-characterizing a transverse contraction at tension, and ν^--characterizing a transverse expansion at compression.

It is assumed that at simultaneous tension and compression in different principal directions, the modulus of elasticity remains E^+ and E^- respectively.

The shell is assumed to be loaded axisymmetrically, and the surfaces conditions are

$$\tau_{s\gamma} = \frac{p}{2}, \quad \sigma_\gamma = \frac{q}{2} \qquad \text{by} \qquad \gamma = \pm \frac{h}{2}. \tag{1.1}$$

Finally, "weak moment" shells we call those shells, or their parts, in which the normal stresses from bending moments (M_s, M_φ) do not surpass the normal stresses from tangential forces (T_s, T_φ). That means, that the normal stresses along the shell thickness do not change their sign.

2. The theory proposed is based on the following assumptions: the normal stresses σ_s and σ_φ along the shell thickness vary according to the linear law,

the distances along the normal (γ) between two points of the shell remain unchanged after deformation,

the relationship establishing the connection between shearing stress $\tau_{s\gamma}$ and shear deformation $e_{s\gamma}$, is taken from the conditions of pure elastic shear,

the normal stress σ_γ may be disregarded in comparison with other stresses.

3. Since a symmetrically loaded shell of revolution will remain a body of revolution, during deformation its internal forces and displacements will not be functions of the angular coordinate φ.

Then for differential elements of shells of revolution we have the following [3—5] geometrical relations:

$$e_\gamma = \frac{\partial u_\gamma}{\partial \gamma}, \qquad e_s = \frac{1}{H_1}\left(\frac{\partial u_s}{\partial s} + k_1 u_\gamma\right)$$

$$e_\varphi = \frac{1}{H_2}\left(u_\gamma \cos\vartheta - u_s \sin\vartheta\right)$$

$$e_{s\gamma} = H_1 \frac{\partial}{\partial\gamma}\left(\frac{1}{H_1} u_s\right) + \frac{1}{H_1}\frac{\partial u_\gamma}{\partial s} \tag{3.1}$$

and the following equilibrium equations:

$$\frac{\partial}{\partial s}(H_2\sigma_s) + \sigma_\varphi H_1 \sin\vartheta + \frac{1}{H_1}\frac{\partial}{\partial\gamma}(H_1^2 H_2 \tau_{s\gamma}) = 0$$

$$\frac{\partial}{\partial\gamma}(H_1 H_2 \sigma_\gamma) - \sigma_s k_1 H_2 - \sigma_\varphi r k_2 H_1 + \frac{\partial}{\partial s}(H_2 \tau_{s\gamma}) = 0 \tag{3.2}$$

where $e_i(s, \gamma)$ — are the components of the deformation tensor, $\sigma_i(s, \gamma)$, $\tau_{s\gamma}$ — are the components of the stress tensor, $u_i(s, \gamma)$ — are the components of displacements of any point of the shell, $R_i(s)$ — are the principal radii of curvature of the median surface of the shell, $r(s)$ — the distance from point M to the axis of revolution z, $\vartheta(s)$ — the angle between the tangent to the meridian and the axis z, and where $H_1 = 1 + \gamma/R_1$, $H_2 = r(1 + \gamma/R_2)$ — are the Lame coefficients ($i = s, \varphi, \gamma$).

In view of the axial symmetry it is obvious, that $\tau_{\varphi s} = 0$, and $\tau_{\varphi\gamma} = 0$. Consequently the normal stress σ_φ acting on the planes $\varphi = \text{const}$ is a principal stress, so the direction φ, becomes principal too. The other two principal directions α and β in every point of the shell are to be found in the plane of the corresponding meridian.

The principal stresses σ_α, σ_β are only dependent on stresses $\sigma_s, \sigma_\gamma, \tau_{s\gamma}$, the principal deformations e_α, e_β, are only dependent on deformations e_s, $e_{s\gamma}$, and could be determined by means of the usual / formulae for plane problems.

Using the fundamental hypothesis for principal stresses and deformations we obtain

$$\sigma_\alpha = \frac{1}{2}\left(\sigma_s + \sqrt{\sigma_s^2 + 4\tau_{s\gamma}^2}\right), \qquad \sigma_\beta = \frac{1}{2}\left(\sigma_s - \sqrt{\sigma_s^2 + 4\tau_{s\gamma}^2}\right)$$

$$e_\alpha = \frac{1}{2}\left(e_s + \sqrt{e_s^2 + e_{s\gamma}^2}\right), \qquad e_\beta = \frac{1}{2}\left(e_s - \sqrt{e_s^2 + e_{s\gamma}^2}\right). \tag{3.3}$$

Without imposing any restrictions, the principal stress σ_β will have minus sign, that is σ_β is a compressive stress, thus $\sigma_\alpha > 0$, $\sigma_\beta < 0$.

The relative disposition of principal directions φ, α, β, with the coordinates s, φ, γ at the particular point should be determined by means of the direction cosines $l_j = \cos(\alpha, i)$, $m_j = \cos(\beta, i)$, $n_j = \cos(\varphi, i)$, $(j = 1, 2, 3;\ i = s, \varphi, \gamma)$, which are determined by means of the following known formulae [1, 2],

	α	β	φ
s	l_1	m_1	0
φ	0	0	1
γ	l_3	m_3	0

$$m_1^2 = \frac{1}{1+k^2}, \qquad m_3^2 = \frac{k^2}{1+k^2}$$

$$m_1 m_3 = \frac{k}{1+k^2} \tag{3.4}$$

$$k = -\frac{\sigma_s + \sqrt{\sigma_s^2 + 4\tau_{s\gamma}^2}}{2\tau_{s\gamma}} = -\frac{e_s + \sqrt{e_s^2 + e_{s\gamma}^2}}{e_{s\gamma}}.$$

In case of a stress state having axial symmetry for shells of revolution, the generalized law of elasticity in the system of coordinates S, φ, γ, is to be written as follows

$$e_s = a_{11}\sigma_s + a_{12}\sigma_\varphi + (a_{22} - a_{11}) m_1^2 \sigma_\beta$$

$$e_\varphi = a_{33}\sigma_\varphi + a_{12}\sigma_s$$

$$e_{s\gamma} = (a_{11} + a_{22} - 2a_{12})\tau_{s\gamma}, \quad e_{s\gamma} = G_i^{-1}\tau_{s\gamma} \tag{3.5}$$

or

$$\sigma_s = c_{33} e_s - c_{12} e_\varphi - c_{33}(a_{22} - a_{11}) m_1^2 \sigma_\beta$$

$$\sigma_\varphi = c_{11} e_\varphi - c_{12} e_s + c_{12}(a_{22} - a_{11}) m_1^2 \sigma_\beta$$

$$\tau_{s\gamma} = G_i e_{s\gamma}, \qquad c_{ik} = \frac{a_{ik}}{a_{11}a_{33} - a_{12}^2}. \tag{3.6}$$

We take $\sigma_\alpha > 0$, $\sigma_\beta < 0$, with the result that the coefficients of elasticity a_{ik} have the following values

$$a_{11} = \frac{1}{E^+}, \qquad a_{22} = \frac{1}{E^-}, \qquad a_{12} = -\frac{\nu^+}{E^+} = -\frac{\nu^-}{E^-}. \tag{3.7}$$

For the coefficient of elasticity a_{33} according to the sign of principal stress σ_φ, we have: $a_{33} = 1/E^+$ if $\sigma_\varphi > 0$, or $a_{33} = 1/E^-$ if $\sigma_\varphi < 0$.

In virtue of (3.7), for shear modulus of differentmodulus materials by pure shear we have [1]

$$G_i = \frac{E^+E^-}{E^+(1+\nu^-) + E^-(1+\nu^+)}. \tag{3.8}$$

Using the relationships (3.3) we obtain from the generalized law of elasticity in the principal directions φ, α, β [1, 2, 6]

$$m_1{}^2\sigma_\beta = \frac{b_{22}e_s + b_{23}e_\varphi}{2} - \frac{(b_{22} - b_{12})e_{s\gamma}^2 + 2e_s(b_{22}e_s + b_{23}e_\varphi)}{4\sqrt{e_s{}^2 + e_{s\gamma}^2}} \tag{3.9}$$

where

$$b_{22} = \frac{a_{11}a_{33} - a_{12}^2}{\Omega}, \qquad b_{12} = -\frac{a_{12}(a_{33} - a_{12})}{\Omega}, \qquad b_{23} = -\frac{a_{12}(a_{11} - a_{12})}{\Omega}$$

$$\Omega = a_{11}a_{22}a_{33} - a_{12}^2(a_{11} + a_{22} + a_{33} - 2a_{12}). \tag{3.10}$$

4. Solving the first equilibrium equation (3.2) for shearing stress $\tau_{s\gamma}$, and considering linear law for normal stresses σ_s, σ_φ, we obtain

$$\tau_{s\gamma} = p\frac{\gamma}{h} + \frac{1}{2}\left(\frac{h^2}{4} - \gamma^2\right)\varphi(s) \tag{4.1}$$

where $\varphi(s)$ — is the unknown sought function.

By virtue of the fundamental hypothesis, from (3.1) with (3.5) we obtain

$$e_\gamma = \frac{\partial u_\gamma}{\partial\gamma} = 0, \qquad u_\gamma = u_\gamma(s) = w(s)$$

$$H_1\frac{\partial}{\partial\gamma}\left(\frac{1}{H_1}u_s\right) + \frac{1}{H_1}\frac{dw}{ds} = \frac{1}{G_i}\left[p\frac{\gamma}{h} + \frac{1}{2}\left(\frac{h^2}{4} - \gamma^2\right)\varphi\right]. \tag{4.2}$$

Integrating the latter equation for γ over the limits from zero to γ and considering that with $\gamma = 0$ $u_s = u(s)$, with exactness the linear law for σ_s and σ_φ, we obtain the following values for the tangential displacement of any point of the shell

$$u_s = u - \gamma\left(W - \frac{h^2}{8G_i}\varphi\right), \qquad W = \frac{dw}{ds} - \frac{u}{R_1} \tag{4.3}$$

where $u(s)$ and $w(s)$ — are the sought displacements of the median surface of the shell.

Inserting the values of u_γ and u_s into relationships (3.1), we obtain for the deformation components

$$e_s = \varepsilon_1 + \gamma \varkappa_1, \qquad e_\varphi = \varepsilon_2 + \gamma \varkappa_2$$

$$\varepsilon_1 = \frac{du}{ds} + \frac{w}{R_1}, \qquad \varepsilon_2 = \frac{1}{r}\,(w \cos\vartheta - u \sin\vartheta)$$

$$\varkappa_1 = -\frac{dW}{ds} + \frac{h^2}{8G_i}\frac{d\varphi}{ds}, \qquad \varkappa_2 = \left(W - \frac{h^2}{8G_i}\varphi\right)\frac{\sin\vartheta}{r}. \tag{4.4}$$

This relationships are identical with the corresponding relationships of the theory of shell of revolution without the hypothesis of nondeformable normals [4, 7].

5. Because the normal stresses σ_s and σ_φ along the shell thickness vary according to the linear law, nonlinear terms in the formulae of stresses σ_s and σ_φ, probably distribute in the series for γ, and from this series must be taken only the first two, linear by γ, terms.

Then for stresses σ_s, σ_φ and $\tau_{s\gamma}$ we obtain

$$\sigma_s = \frac{A_1}{h}\varepsilon_1 - \frac{A_2}{h}\varepsilon_2 + c_{33}(a_{22} - a_{11})\,\Phi + {}$$
$$+ \gamma\left[\frac{A_1}{h}\varkappa_1 - \frac{A_2}{h}\varkappa_2 + c_{33}(a_{22} - a_{11})\,\Psi\right.$$

$$\sigma_\varphi = \frac{A_3}{h}\varepsilon_2 - \frac{A_4}{h}\varepsilon_1 - c_{12}(a_{22} - a_{11})\,\Phi + {}$$
$$+ \gamma\left[\frac{A_3}{h}\varkappa_2 - \frac{A_4}{h}x_1 - c_{12}(a_{22} - a_{11})\,\Psi\right.$$

$$\tau_{s\gamma} = p\,\frac{\gamma}{h} + \frac{1}{2}\left(\frac{h^2}{4} - \gamma^2\right)\varphi \tag{5.1}$$

where for nonlinear functions $\Phi(s)$ and $\Psi(s)$ we have

$$\Phi(s) = \frac{1}{4}\left[(b_{22} - b_{12})\left(\frac{h^2}{8G_i}\right)^2\varphi^2 + 2\varepsilon_1(b_{22}\varepsilon_1 + b_{23}\varepsilon_2)\right] \times$$
$$\times \left[\left(\frac{h^2}{8G_i}\right)^2\varphi^2 + \varepsilon_1{}^2\right]^{-1/2}$$

$$\Psi(s) = \frac{1}{4}\left\{(b_{22} - b_{12})\left(\frac{h^2}{8G_i}\right)^3\frac{p}{G_i h}\varphi^3 + \right.$$
$$+ [(3b_{22} + b_{12})\varepsilon_1\varkappa_1 + 2b_{23}(\varepsilon_1\varkappa_2 + \varepsilon_2\varkappa_1)]\left(\frac{h^2}{8G_i}\right)^2\varphi^2 -$$

$$- 2\varepsilon_1(b_{12}\varepsilon_1 + b_{23}\varepsilon_2)\frac{h^2}{8G_i}\frac{p}{G_i h}\varphi + 2b_{22}\varepsilon_1{}^3\varkappa_1 +$$

$$+ 2b_{23}\varepsilon_1{}^3\varkappa_2\Big\}\left[\left(\frac{h^2}{8G_i}\right)^2\varphi^2 + \varepsilon_1{}^2\right]^{-3/2}. \tag{5.2}$$

We established the laws for change in stresses directed along the thickness of the shell; however, as in the theory of classic homogeneous shells, instead of the stresses, it is convenient to introduce their statically equivalent internal forces and moments.

From the conditions of static equivalence for internal tangential (T_s, T_φ) and transverse (N) forces, as well as for bending moments (M_s, M_φ), we have

$$\begin{aligned}
T_s &= A_1\varepsilon_1 - A_2\varepsilon_2 + B_{33}(a_{22} - a_{11})\Phi\\
T_\varphi &= A_3\varepsilon_2 - A_y\varepsilon_1 - B_{12}(a_{22} - a_{11})\Phi\\
M_s &= K_1\varkappa_1 - K_2\varkappa_2 + D_{33}(a_{22} - a_{11})\Psi\\
M_\varphi &= K_3\varkappa_2 - K_y\varkappa_1 - D_{12}(a_{22} - a_{11})\Psi\\
N &= \frac{h^3}{12}\varphi
\end{aligned} \tag{5.3}$$

where

$$\begin{aligned}
A_1 &= B_{33} - B_{33}(a_{22} - a_{11})\frac{b_{22}}{2}, & A_2 &= B_{12} + B_{33}(a_{22} - a_{11})\frac{b_{23}}{2}\\
A_3 &= B_{11} + B_{12}(a_{22} - a_{11})\frac{b_{23}}{2}, & A_y &= B_{12} - B_{12}(a_{22} - a_{11})\frac{b_{22}}{2}\\
K_1 &= D_{33} - D_{33}(a_{22} - a_{11})\frac{b_{22}}{2}, & K_2 &= D_{12} + D_{33}(a_{22} - a_{11})\frac{b_{23}}{2}\\
K_3 &= D_{11} + D_{12}(a_{22} - a_{11})\frac{b_{23}}{2}, & K_y &= D_{12} - D_{12}(a_{22} - a_{11})\frac{b_{22}}{2}\\
B_{ik} &= hc_{ik}, & D_{ik} &= \frac{h^3}{12}c_{ik}.
\end{aligned} \tag{5.4}$$

6. The equilibrium equations of the classic theory of symmetrically loaded shells of revolution remain unchanged in the case of differentmodulus shells, and have the following form [3—5]

$$\begin{aligned}
&\frac{d}{ds}(rT_s) + T_\varphi\sin\vartheta + \frac{r}{R_1}N = -rp\\
&\frac{d}{ds}(rN) - r\left(\frac{T_s}{R_1} + \frac{T_\varphi}{R_2}\right) = -rq\\
&\frac{d}{ds}(rM_s) + M_\varphi\sin\vartheta - rN = 0.
\end{aligned} \tag{6.1}$$

Inserting the values of the internal forces and moments from (5.3) into equilibrium equations (6.1), we obtain the following system of three nonlinear differential equations in terms of the three sought functions $u(s)$, $w(s)$, $\varphi(s)$

$$
\begin{aligned}
& A_1 \frac{d^2 u}{ds^2} - (A_1 - A_2 + A_4) \frac{\sin\vartheta}{r} \frac{du}{ds} + \left(A_2 \frac{1}{R_1 R_2} - A_3 \frac{\sin^2\vartheta}{r^2}\right) u + \\
& \quad + \left(A_1 \frac{1}{R_1} - A_2 \frac{1}{R_2}\right) \frac{dw}{ds} + \left\{A_1 \frac{d}{ds}\left(\frac{1}{R_1}\right) - \right. \\
& \quad \left. - \left[(A_1 - A_2 + A_4) \frac{1}{R_1} - A_3 \frac{1}{R_2}\right] \frac{\sin\vartheta}{r}\right\} w + \frac{1}{R_1} \frac{h^3}{12} \varphi + \\
& \quad + (a_{22} - a_{11}) \left[B_{33} \frac{d\Phi}{ds} - (B_{33} + B_{12}) \frac{\sin\vartheta}{r} \Phi\right] = -p \\
& \frac{h^3}{12} \frac{d\varphi}{ds} - \frac{h^3}{12} \frac{\sin\vartheta}{r} \varphi - \left(A_1 \frac{1}{R_1} - A_4 \frac{1}{R_2}\right) \frac{du}{ds} - \left(A_2 \frac{1}{R_1} - A_3 \frac{1}{R_2}\right) \times \\
& \quad \times \frac{\sin\vartheta}{r} u - \left[A_1 \frac{1}{R_1^2} - (A_2 + A_4) \frac{1}{R_1 R_2} + A_3 \frac{1}{R_2^2}\right] w - \\
& \quad - (a_{22} - a_{11}) \left(B_{33} \frac{1}{R_1} - B_{12} \frac{1}{R_2}\right) \Phi = -q \qquad (6.2) \\
& K_1 \frac{d^3 w}{ds^3} + (K_2 - K_1 - K_4) \frac{\sin\vartheta}{r} \frac{d^2 w}{ds^2} + \left(K_2 \frac{1}{R_1 R_2} - K_3 \frac{\sin^2\vartheta}{r^2}\right) \frac{dw}{ds} - \\
& \quad - K_1 \frac{1}{R_1} \frac{d^2 u}{ds^2} - \left[2 K_1 \frac{d}{ds}\left(\frac{1}{R_1}\right) + (K_2 - K_1 - K_4) \frac{1}{R_1} \frac{\sin\vartheta}{r}\right] \frac{du}{ds} - \\
& \quad - \left[K_1 \frac{d^2}{ds^2}\left(\frac{1}{R_1}\right) + (K_2 - K_1 - K_4) \frac{d}{ds}\left(\frac{1}{R_1}\right) \frac{\sin\vartheta}{r} + K_2 \frac{1}{R_1 R_2} - \right. \\
& \quad \left. - K_3 \frac{1}{R_1} \frac{\sin^2\vartheta}{r^2}\right] u - K_1 \frac{h^2}{8G_i} \frac{d^2\varphi}{ds^2} - (K_2 - K_1 - K_4) \frac{\sin\vartheta}{r} \times \\
& \quad \times \frac{h^2}{8G_i} \frac{d\varphi}{ds} - \left(K_2 \frac{1}{R_1 R_2} - K_3 \frac{\sin^2\vartheta}{r^2}\right) \frac{h^2}{8G_i} \varphi + \frac{h^3}{12} \varphi - \\
& \quad - (a_{22} - a_{11}) \left[D_{33} \frac{d\Psi}{ds} - (D_{33} + D_{12}) \frac{\sin\vartheta}{r} \Psi\right] = 0 .
\end{aligned}
$$

The latter terms of these equations are very complicated nonlinear functions of the sought functions u, w, φ.

As usual, in solving specific edge problems the applicable boundary conditions are imposed on the differential equations of solutions (6.2) of the shell. The boundary conditions are determined by the usual methods [4, 7]

— free edge; $M_s = 0, \quad T_s = 0, \quad N = 0$

— free supported edge; $M_s = 0, \quad T_s = 0, \quad w = 0$

— hinge — fixed edge; $M_s = 0, \quad u = 0, \quad w = 0$

— fixed edge; $\frac{dw}{ds} - \frac{h^2}{8G_i}\varphi = 0, \quad u = 0, \quad w = 0.$

Equations (6.2) and boundary conditions constitute a complete system of three nonlinear differential equations in terms of the three sought functions u, w, φ. Having the values of the sought functions, by means of formulae (5.1) — (5.4), it is not difficult to find the values of the stresses, forces and moments.

Analyzing the formulae of stresses, moments, forces and differential equations of solution, we can see, that their structure differ from the structure of corresponding formulae and equations of the classical theory of shell of revolution, only in having terms containing nonlinear functions Φ and Ψ. Of course in this case the "classic parts" of this formulae and equations by their containing will not coincide with the corresponding formulae and equations of the classic theory, because the ratios of formulae and equations given here, have differentmodulus containing.

An important peculiarity of this differential equations and calculating formulae is that all nonlinear terms contain a multiplier $(a_{22} - a_{11})$, and when having classic elastic material $(E^+ = E^- = E, \ \nu^+ = \nu^- = \nu)$ become zero.

We must also point, that the multiplier with all nonlinear terms, after certain transformations become into a small parameter $(a_{22} - a_{11})/(a_{22} + a_{11})$.

7. For illustration let us discuss an example. A closed circular shell (radius of curvature R, length L), supports a sinusarly distributed, normally applied surface load of maximum intensity q_0. The shell boundaries are defiend by two transverse sections perpendicular to the axis of revolution Z. One of the ends $(s = 0)$ of the shell is hinge-fixed and the other end $(s = l)$ is free supported but loaded uniformly distributed axial forces of intensity.

Let the shell be made on a differentmodulus material, and for simplification $\nu^+ = \nu^- = 0$.

For this shell we have

$$R_1 = \infty, \qquad R_2 = r = R, \qquad \vartheta = 0, \qquad p = 0, \qquad q = q_0 \sin\frac{\pi s}{l}$$

$$a_{11} = \frac{1}{E^+}. \qquad a_{22} = \frac{1}{E^-}, \qquad a_{12} = a_{21} = 0$$

$$c_{11} = \frac{1}{a_{33}}, \qquad c_{33} = E^+, \qquad c_{12} = c_{21} = 0$$

$$b_{11} = E^+, \qquad b_{22} = E^-, \qquad b_{33} = \frac{1}{a_{33}}$$

$$b_{12} = b_{13} = b_{23} = 0, \qquad G_i = \frac{E^+ E^-}{E^+ + E^-}. \tag{7.1}$$

The differential equations system of solution takes the following form

$$\frac{d^2 u}{d s^2} + 2\mu \frac{d}{d s} \frac{\left(\frac{h^2}{8 G_i}\right)^2 \varphi^2 + 2 \left(\frac{d u}{d s}\right)^2}{4 \sqrt{\left(\frac{h^2}{8 G_i}\right)^2 \varphi^2 + \left(\frac{d u}{d s}\right)^2}} = 0$$

$$\frac{d\varphi}{d s} - \frac{12}{a_{33}} \frac{1}{h^2 R^2} w = - \frac{12}{h^3} q_0 \sin \frac{\pi s}{l}$$

$$\frac{d^3 w}{d s^3} - \frac{h^2}{8 G_i} \frac{d^2 \varphi}{d s^2} + \frac{2}{E^+ + E^-} \varphi -$$

$$- 2\mu \frac{d}{d s} \frac{\left[\left(\frac{h^2}{8 G_i}\right)^2 \varphi^2 + 2 \left(\frac{d u}{d s}\right)^2\right] \left(\frac{d u}{d s}\right) \left(\frac{h^2}{8 G_i} \frac{d\varphi}{d s} - \frac{d^2 w}{d s^2}\right)}{4 \sqrt{\left[\left(\frac{h^2}{8 G_i}\right)^2 \varphi^2 + \left(\frac{d u}{d s}\right)^2\right]^3}} = 0. \tag{7.2}$$

where with nonlinear terms we have small parameter

$$\mu = \frac{a_{22} - a_{11}}{a_{22} + a_{11}} = \frac{E^+ - E^-}{E^+ + E^-}. \tag{7.3}$$

Then for internal forces, moments and stresses we have

$$T_s = h \frac{E^+ + E^-}{2} \left[\frac{d u}{d s} + \mu \frac{\left(\frac{h^2}{8 G_i}\right)^2 \varphi^2 + 2 \left(\frac{d u}{d s}\right)^2}{2 \sqrt{\left(\frac{h^2}{8 G_i}\right)^2 \varphi^2 + \left(\frac{d u}{d s}\right)^2}} \right]$$

$$T_\varphi = \frac{h}{a_{33}} \frac{w}{R}, \qquad M_\varphi = 0, \qquad \sigma_\varphi = \frac{1}{a_{33}} \frac{w}{R}$$

$$M_s = \frac{h^3}{12} \frac{E^+ + E^-}{2} \left[\frac{h^2}{8 G_i} \frac{d\varphi}{d s} - \frac{d^2 w}{d s^2} + \right.$$

$$\left. + \mu \frac{\left[3 \left(\frac{h^2}{8 G_i}\right)^2 \varphi^2 + 2 \left(\frac{d u}{d s}\right)^2\right] \left(\frac{d u}{d s}\right) \left(\frac{h^2}{8 G_i} \frac{d\varphi}{d s} - \frac{d^2 w}{d s^2}\right)}{2 \sqrt{\left[\left(\frac{h^2}{8 G_i}\right)^2 \varphi^2 + \left(\frac{d u}{d s}\right)^2\right]^3}} \right.$$

$$\sigma_s = \frac{E^+ + E^-}{2}\left\{\frac{du}{ds} + \gamma\left(\frac{h^2}{8G_i}\frac{d\varphi}{ds} - \frac{d^2w}{ds^2}\right) + \right.$$

$$+ \mu\left[\frac{\left(\frac{h^2}{8G_i}\right)^2\varphi^2 + 2\left(\frac{du}{ds}\right)^2}{2\sqrt{\left(\frac{h^2}{8G_i}\right)^2\varphi^2 + \left(\frac{du}{ds}\right)^2}} + \right.$$

$$\left.\left. + \gamma\frac{\left[3\left(\frac{h^2}{8G_i}\right)^2\varphi^2\left(\frac{du}{ds}\right) + 2\left(\frac{du}{ds}\right)^3\right]\left(\frac{h^2}{8G_i}\frac{d\varphi}{ds} - \frac{d^2w}{ds^2}\right)}{2\sqrt{\left[\left(\frac{h^2}{8G_i}\right)^2\varphi^2 + \left(\frac{du}{ds}\right)^2\right]^3}}\right]\right\}. \tag{7.4}$$

For boundary conditions we have

$$\begin{aligned} &u = 0, && M_s = 0, && w = 0 && \text{by} && s = 0 \\ &T_s = T^*, && M_s = 0, && w = 0 && \text{by} && s = l. \end{aligned} \tag{7.5}$$

The principal method of obtaining the solution of this problem will be considered to be the small parameter method, with small parameter μ.

The solution of the system will be represented in the form

$$\begin{aligned} u &= u_0(s) + \mu u_1(s) + \mu^2 u_2(s) + \dots \\ w &= w_0(s) + \mu w_1(s) + \mu^2 w_2(s) + \dots \\ \varphi &= \varphi_0(s) + \mu\varphi_1(s) + \mu^2\varphi_2(s) + \dots . \end{aligned} \tag{7.6}$$

The boundary conditons, in the small parameters method, is rewritten in the following form

$$\begin{aligned} &u_0 = 0, && M_{s,0} = 0, && w_0 = 0 && \text{by} && s = 0 \\ &T_{s,0} = T^*, && M_{s,0} = 0, && w_0 = 0 && \text{by} && s = l \\ &u_j = 0, && M_{s,j} = 0, && w_j = 0 && \text{by} && s = 0 \\ &T_{s,j} = 0, && M_{s,j} = 0, && w_j = 0 && \text{by} && s = l \\ & && (j = 1, 2, 3, \dots). \end{aligned} \tag{7.7}$$

Without going into details, limiting ourselves to the third approximation, we permit the final results

$$w = -\frac{12 q_0 \Delta_0}{h^3 \Delta}\left\{1 - \mu\frac{2\lambda^4}{(E^+ + E^-)\Delta\Delta_0} + \right.$$

$$\left. + \mu^2\frac{\lambda^6 A}{\Delta\Delta_0}\left[1 + \frac{2\lambda^2}{\Delta(E^+ + E^-)\Delta}\right]\right\}\sin\lambda s$$

$$\varphi = -\frac{12 q_0 \lambda^3}{h^3 \Delta}\left[1 + \mu \frac{24 E^-}{(E^+ + E^-)\Delta R^2 h^2} - \right.$$
$$\left. - \mu^2 \frac{12 \lambda^2 A E^-}{\Delta R^2 h^2}\left[1 + \frac{2\lambda^2}{A(E^+ + E^-)\Delta}\right]\right\} \cos \lambda s$$

$$u = \frac{2T^*}{h(E^+ + E^-)}(1 - \mu + \mu^2)s, \qquad T_s = T^*$$

$$M_s = -\frac{q_0 \lambda^2}{\Delta}\left\{1 + \mu\left[1 - \frac{12 A \lambda^2 E^-}{\Delta R^2 h^2} - \frac{\lambda^4}{\Delta}\right] + \right.$$
$$+ \mu^2 \frac{A(E^+ + E^-)\lambda^2}{2}\left[1 + \frac{2\lambda^2}{\Delta(E^+ + E^-)\Delta}\right] \times$$
$$\left. \times \left[1 - \frac{12 A \lambda^2 E^-}{\Delta R^2 h^2} - \frac{\lambda^4}{\Delta}\right]\right\} \sin \lambda s$$

$$T = \frac{h}{R} E^- w, \qquad N = \frac{h^3}{12}\varphi$$

$$\Delta = \lambda^4 + \frac{12 E^-}{h^2 R^2}\Delta_0, \qquad \Delta_0 = A\lambda^2 + \frac{2}{E^+ + E^-}$$

$$A = \frac{h^2}{8 G_i} = \frac{h^2}{8}\frac{E^+ + E^-}{E^+ E^-}, \qquad \lambda = \frac{\pi}{l}. \tag{7.8}$$

Because $w < 0$, then $T_\varphi < 0$, $\sigma_\varphi < 0$, therefore we took $a_{33} = 1/E^-$.

8. Let us discuss a numerical example.

Let $E^+ = 1.5E$, $E^- = E$, $\mu = 0.2$, $R = l$, $h = 0.2R$.

Then we obtain

$$w = -3.6549 \frac{q_0 l}{E}(1 - 0.0488 + 0.0036) \sin \frac{\pi s}{l} = -3.4897 \frac{q_0 l}{E} \sin \frac{\pi s}{l}$$

$$\varphi = -128.4 \frac{q_0}{l^2}(1 + 0.1326 - 0.0099) \cos \frac{\pi s}{l} = -144.0 \frac{q_0}{l^2} \cos \frac{\pi s}{l}$$

$$u = 4 \frac{T^*}{El}(1 - 0.2 + 0.04)s = 3.36 \frac{T^*}{El} s$$

$$M_s = -0.0273 q_0 l^2 (1 + 0.1326 + 0.0099) \sin \frac{\pi s}{l}$$
$$= -0.0312 q_0 l^2 \sin \frac{\pi s}{l}, \qquad T_s = T^*$$

$$T_\varphi = -0.6919 q_0 l \sin \frac{\pi s}{l}, \qquad N = -0.069 q_0 l \cos \frac{\pi s}{l}.$$

If we have classic material, $E^{+} = 1.5E$, $E^{-} = 1.5E$, $\mu = 0$, $R = l$, $h = 0.2R$, we obtain

$$w = -2.5129 \frac{q_0 l}{E} \sin \frac{\pi s}{l}, \qquad \varphi = -108.9 \frac{q_0}{l^2} \cos \frac{\pi s}{l}$$

$$u = 3.333 \frac{T^*}{El} s, \qquad M_s = -0.0231 q_0 l^2 \sin \frac{\pi s}{l}$$

$$T_\varphi = -0.1119 q_0 l \sin \frac{\pi s}{l}, \qquad N = -0.0726 q_0 l \cos \frac{\pi s}{l}$$

or if $E^{+} = E$, $E^{-} = E$, $\mu = 0$, $R = l$, $h = 0.2R$, we have

$$w = -3.8593 \frac{q_0 l}{E} \sin \frac{\pi s}{l}, \qquad \varphi = -108.9 \frac{q_0}{l^2} \cos \frac{\pi s}{l}$$

$$u = 4.999 \frac{T^*}{El} s, \qquad M_s = -0.0231 q_0 l^2 \sin \frac{\pi s}{l}$$

$$T_\varphi = -0.7719 q_0 l \sin \frac{\pi s}{l}, \qquad N = -0.0126 q_0 l \cos \frac{\pi s}{l}.$$

In examining the results of calculation, we note that the values of the displacements, stresses, forces and moments as determined from the classical theory of shells of revolution, without the hypotheses of nondeformable normals, may differ substantially from the corresponding values as found from the proposed theories.

References

1. Ambartsumyan, S. A., and A. A. Khachatryan: Basic equation of the theory of elasticity for materials with different response to tension and compression (in Russian), Engineering Journal, MTT, No. 2, 1966.
2. Ambartsumyan, S. A.: Equations of the plane problem of the theory of elasticity for materials with different moduli (in Russian), Bull. Acad. Sci. Armenian SSR, Mechanics, XII, No. 2, 1966.
3. Lurje, A. I.: Statics of thin elastic shells (in Russian), Gostehizdat, 1947.
4. Ambartsumyan, S. A.: Theory of anisotropic shells (in Russian), Fizmatgiz, 1961.
5. Gol'denweizer, A. L.: Theory of elastic thin shells (in Russian), Gostehizdat, 1953.
6. Ambartsumyan, S. A.: Axial-Symmetric problem of a circular cylindrical shell made of a material with different response to tension and compression (in Russian), Bull. Acad. Sci. USSR, Mechanics.
7. Ambartsumyan, S. A.: A contribution to the general theory of anisotropic shells (in Russian), PMM, XII, No. 2, 1958.

On Incremental Collapse of Shells under Cyclic Loading

By

A. Sawczuk

University of Grenoble, France, and The Institute for Basic Technical Research, Warsaw, Poland

Abstract. Certain consequences of KOITER's "inadaptation theorem" for elastic plastic structures subjected to cyclic multiparameter loadings, are discussed. Incremental collapse of shells is studied in terms of generalized variables. A method for finding interaction surfaces of load multipliers at incremental collapse is formulated. Examples are considered.

1. Introduction

An elastic-perfectly plastic structure subjected to a loading prescribed except for a single magnitude parameter μ, collapses into a mechanism of instantaneous motion at the uniquely specified yield-point value μ_0 of the load multiplier. The elastic unloading, $\mu < \mu_0$, checks the motion and results at $\mu = 0$ in a state of residual stress in the structure. This self-equilibrated system of stresses has no influence on the collapse load multiplier of another loading program. A sequence of loading cycles, however, may result either in low cycle fatigue or in development of irreversible deformations by increments, long before the load of a specific cycle attains the intensity associated with an instantaneous collapse. Whenever deformations can increase by fractions an elastic-perfectly plastic structure will not shake down to the prescribed loading program. It becomes unserviceable after a finite number of cycles because of excessive deformations (cf. PRAGER [1]).

For a given loading program (multiparametric) it is therefore necessary to establish the domain $f(\mu_1, \ldots) = \text{const}$ of safe multipliers μ_1, $\mu_2, \ldots, \mu_\alpha$ for which the incremental collapse can not occur. Still, however, a structure safe against the incremental collapse is not necessarily safe against the low cycle fatigue, the last one being generally move restrictive.

The first studies of incremental collapse concerned frames subjected to non-proportional concentrated loads (NEAL and SYMONDS [2]),

HODGE [3]). The general theorem of inadaptation of elastic-plastic continua is due to KOITER [4]. His generalisation of the upper bound theorem of limit analysis specifies the necessary condition of incremental collapse in terms of the energy dissipation. The theorem was not, however, applied to structures until recently GOKHFELD [5] made use of it in an approximate analysis of adaptation of plates and shells to cycles of pressure and temperature. Other studies on shake-down, not restricted to frames, concerned generalisations and modifications of Melan's theorem of adaptation, [6], to plates and shells (KÖNIG [7, 8]).

In the present note the incremental collapse is studied in terms of generalized variables. Certain consequences of Koiter's theorem for piece-wise linear yield loci are discussed. Domains of safe load multipliers are established on examples of two-parameter loading of cylindrical shells.

2. Preliminaries

The generalized stresses Q_i and the generalized strain rates $\dot{q}_i$, $i = 1, \ldots, n$, are related to the Cartesian components of the field quantities by the following requirement

$$\int_A Q_i \dot{q}_i \, dA = \int_V \sigma_{\alpha\beta} \dot{\varepsilon}_{\alpha\beta} \, dV, \tag{1}$$

but otherwise arbitralily defined. In (1) A stands for the volume in the intrinsic space of a structure, V is the actual volume and $\alpha, \beta = 1, 2, 3$.

At plastic deformations the generalized stresses are interrelated by a homogeneous function of degree N in Q_i

$$\Phi(Q_i) - k^N = 0, \qquad \dot{\Phi} = 0, \tag{2}$$

and which represents a closed and convex yield surface in n-dimensional space of generalized stresses (Fig. 1). The stress states outside the yield locus can not be equilibrated. The surface (2) is assumed to be the potential surface (or a generalized potential surface) for the plastic strain rates $\dot{q}_i^P$, hence for the regular regimes

$$\dot{q}_i^P = \nu(\partial\Phi/\partial Q_i), \qquad \nu \geq 0. \tag{3}$$

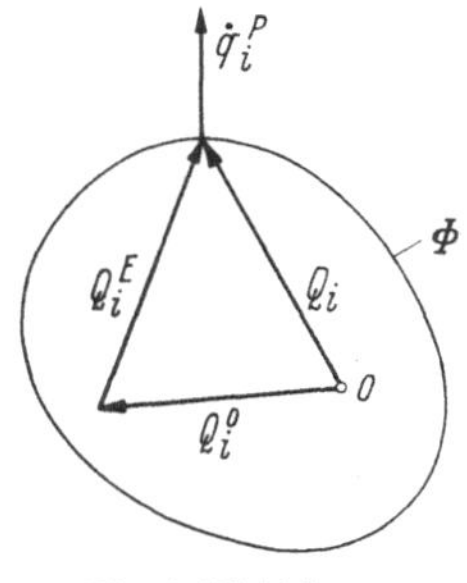

Fig. 1. Yield locus and generalized variables.

The stresses Q_i satisfying (2) are decomposed into the residual part Q_i^0 and the "elastic" one Q_i^E, which would appear if the structure were to respond elastically to the stress state Q_i, thus:

$$Q_i = Q_i^E + Q_i^0. \tag{4}$$

It has to be noticed that the requirement (1) leaves unspecified certain residual stresses in V. Therefore, whatever is the adopted definition of Q_i (i. e. the transformation rule $\sigma_{\alpha\beta} \rightarrow Q_i$), the Q_i^0 are not related to residual stresses $\sigma^R_{\alpha\beta}$ by the rule connecting $\sigma_{\alpha\beta}$ and Q_i. In other words the present approach excludes from the analysis the partially plastic cross-sections, and leaves outside its scope the problem of plastic fatigue (cyclic collapse), which has to be considered in terms of the physical field quantities. In what follows the deformations and displacements are assumed to remain small and the loading cycles sufficiently slow to consider only the quasi-static response.

A process of plastic deformation can be described in terms of the work absorbed by the actually considered structure. For a plastic point at any instant of plastic flow $\dot{q}_i^P \neq 0$, and the dissipation function — a homogeneous function of degree one in the strain rates — is necessarily positive, $D = Q_i \dot{q}_i^P > 0$. Furthermore, in view of (2) and (3) the following identities hold

$$D(\dot{q}_i^P) = Q_i \dot{q}_i^P = \frac{\partial D}{\partial \dot{q}_i^P} \dot{q}_i^P = \nu \frac{\partial \Phi}{\partial Q_i} Q_i = \nu N k^N. \tag{5}$$

In a loading cycle within $t_1 \leq t \leq t_2$, and producing $\dot{q}_i^P \neq 0$, compatible with the kinematical boundary restraints, the increase of plastic deformation, [4]

$$\Delta q_i^P = \int_{t_1}^{t_2} \dot{q}_i^P dt \tag{6}$$

is also kinematically admissible. Therefore, whenever the stress field is independant of time, $Q_i \equiv \bar{Q}_i$, the plastic work on a cycle is

$$W^P = \int_{t_1}^{t_2} Q_i \dot{q}_i^P dt = \bar{Q}_i \Delta q_i^P. \tag{7}$$

3. Koiter's Inadaptation Theorem

The inadaptation theorem states that an elastic perfectly plastic structure will not adapt itself to a given n-parameter loading $P_\alpha(x, t) = \mu_k(t) p_\alpha^k(x)$, if there exists in a time interval $t_1 < t < t_2$, a kinematically admissible cycle of plastic strain rates such that

$$\int_{t_1}^{t_2} dt \int_S P_\alpha \dot{u}_\alpha dS > \int_{t_1}^{t_2} dt \int_A D(\dot{q}_i^P) dA. \tag{8}$$

In (8) $\dot{u}_\alpha$ denotes a kinematically admissible velocity field and in the definition of P_α, μ_k is the load multipliers and p_α^k stands for the unit load distribution functions. For simplicity we assume $P_\alpha(x, t_1) = P_\alpha(x, t_2) = 0$.

In view of the properties of the dissipation function as recalled in (5), the right hand side of (8) can be integrated on a cycle $(0, t)$, say, yielding the following expression:

$$\int_0^t dt \int D(\dot{q}_i^P)\, dA = \int_A dA \int_0^t \frac{\partial D}{\partial \dot{q}_i^P} \dot{q}_i^P dt = \int_A D(\Delta q_i^P)\, dA. \tag{9}$$

The plastic work is therefore uniquely specified by the plastic strain increments (6).

The principe of virtual work applied to the considered loading cycle gives the relation:

$$\int_0^t dt \int_S P_\alpha \dot{u}_\alpha dS = \int_0^t dt \int_A Q_i \dot{q}_i dA. \tag{10}$$

In view of the definition (4) the right hand side of (10) can be rewritten as follows

$$\int_0^t dt \int_A Q_i \dot{q}_i dA = \int_0^t dt \int_A Q_i^E \dot{q}_i dA, \tag{11}$$

because the term containing $Q_i^0 \dot{q}_i$ vanishes when integrated over the volume, since Q_i^0 constitutes a self-equilibrated system and $\dot{q}_i$ is kinematically admissible. Moreover, the strain rate $\dot{q}_i$ can be decomposed into the plastic and the elastic parts, $\dot{q}_i = \dot{q}_i^P + q_i^0$, the elastic one being associated with the residual stress Q_i^0. For a closed, kinematically admissible plastic strain cycle it is

$$Q_i^0(0) = Q_i^0(t), \tag{12}$$

the relation (11) takes the form:

$$\int_0^t dt \int_A Q_i \dot{q}_i dA = \int_0^t dt \left(\int_A Q_i^E \dot{q}_i^0 dA + \int_A Q_i \dot{q}_i^P dA \right). \tag{13}$$

The first term on the right hand side represents the change in the elastic energy produced by a virtual change $\dot{q}_i^0$ in the strain rates such that $\dot{q}_i^0$ are related to the residual stresses Q_i by an appropriate elastic law. Since for plastic states

$$\int_A Q_i^0 \dot{q}_i^0 dA = 0, \tag{14}$$

it can be concluded that the first term in (13) vanishes. Moreover, whenever plastic deformation accurs during the considered loading cycle the stress Q_i^E attains the value $\overline{Q}_i^E$ belonging to the envolope of all possible elastic stresses which could appear if the structure were elastic during the considered loading cycle. This envelope, maximum and minimum values of the components Q_i^E, is independent of time. Therefore, utilising the (7) and (12), the formula (8) takes eventually the following form:

$$\int_0^t dt \int_S P_\alpha \dot{u}_\alpha dS = \int_A \bar{Q}_i^E \Delta q_i^P dA > \int_A D(\Delta q_i^P) dA. \tag{15}$$

The relation (15) in terms of the variables in a real three-dimensional Cartesian space was first obtained by Gokhfeld [5], by a slightly different reasoning.

The essential point of the above given form of Koiter's theorem is that the expression is independent of time. For a given loading program it allows to obtain an interaction surface of incremental collapse $f(\mu_1 \ldots) = f_{00}$ in an n-dimensional space of load multipliers μ_k, similarly as it is the case in the limit analysis (cf. [9, 10]). Any interaction surface obtained from (15) constitutes, by virtue of the original theorem, an upper bound to the shakedown loading program.

For certain piece-wise linear yield loci the velocity field of plastic motion at the collapse is the same for a large variety of loadings. It is so whenever the differential equations for the displacement rates can be reduced to a stress-free form. The velocity field is then of the type:

$$\dot{u}_\alpha(x, t) = u_0(t) f_\alpha(x) + C(t). \tag{16}$$

For such cases the interaction surface of incremental collapse can easily be expressed in terms of the respective collapse load multipliers, i. e. in terms of the limit analysis solutions for particular loads acting independently on a rigid-perfectly plastic structure.

4. Applications

We consider first a short elastic-plastic cylindrical shell of geometry and loading as shown in Fig. 2 the loads are cyclic and vary slowly with time according to unspecified laws. As the criterion of yielding we take the limited interaction yield locus as shown in Fig. 3, M_0 and N_0 being respectively the axial yield moment and the yield hoop force of the shell cross-section. It is known that only the stress profiles AB and CD, the end points included, are kinematically admissible at the collapse state (cf. [3]).

To have appropriate reference values for the considered incremental collapse we recall first the limit analysis solution for the case when both loads increase proportionally to a single parameter. Since for the uniform pressure the stress profile is represented by the point H and for the ring of force there must be a stress discontinuity somewhere on the length, the stress profile at the combined (but one parameter) loading is HEFGK. There is a jump in N_φ at $x = x_0$, namely $N_\varphi] = 2N_0$. Integration of the cylindrical shell equilibrium equation for the indicated stress

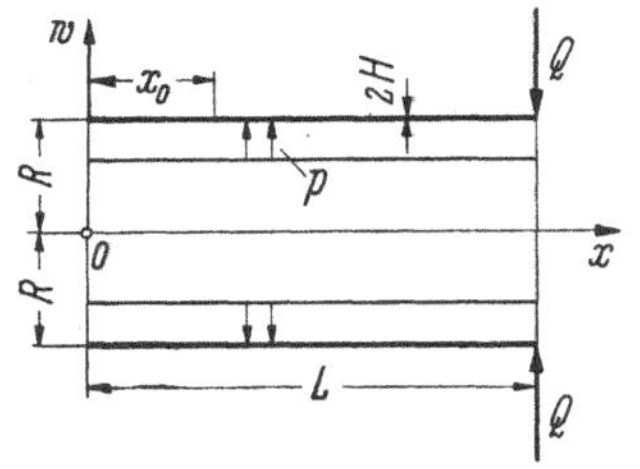

Fig. 2. Loading of a short cylindrical shell.

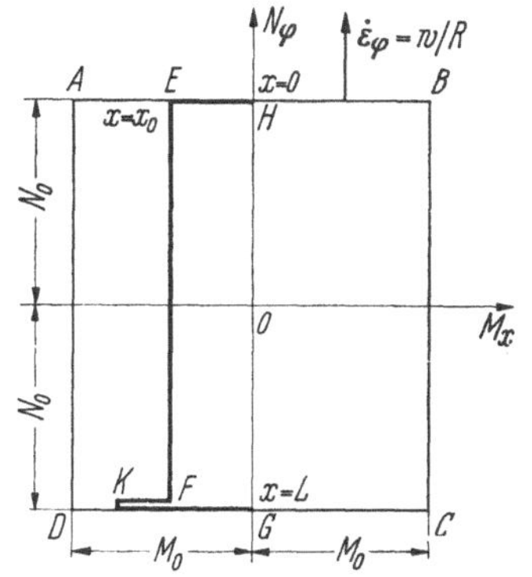

Fig. 3. Stress profile.

profile, subject to the stress boundary conditions and the continuity requirements $M_x] = 0$, $(dM/dx)] = 0$ at $x = x_0$ eventually yields

$$\xi_0 \equiv \frac{x_0}{L} = 1 - \sqrt{\frac{1 - p_0}{2}} \tag{17}$$

for the stress discontinuity position and:

$$\frac{p}{p_0} - 1 + \sqrt{2\left(1 - \frac{p}{p_0}\right)} - \frac{Q}{Q_0}\left(\sqrt{2} - 1\right) = 0 \tag{18}$$

as the interaction curve for one parameter loadings. The yield point loads for individually acting loadings are respectively:

$$p_0 = \frac{N_0}{R}, \qquad Q_0 = \frac{N_0 L}{R}\left(\sqrt{2} - 1\right). \tag{19}$$

The solution (18) is valid until the stress point K remains within the yield rectangle $ABCD$. In the most unfavorable conditions, $(p = 0)$ such requirement leads to the following limitation for the shell length $L \leq L_{cr} = \sqrt{\dfrac{2M_0R}{N_0\left(3 - 2\sqrt{2}\right)}}$. Further analysis of incremental collapse we restrict to the shells satisfying the above requirement.

To obtain for incremental collapse an interaction curve analogous to (18) refering to one parameter loading we have to derive first the envelope of elastic states $\bar{Q}_i^E$ under the prescribed loads. Having this done the formula (15) will furnish the required result.

In the elastic response the relevant stress resultants are (cf. [11])

$$M_x = +Dw'', \quad N_\varphi = +4D\beta^4 Rw \tag{20}$$

where D stands for the rigidity and:

$$\beta^4 = \frac{2HE}{4R^2D} = \frac{3}{4}\,\frac{1-\nu^2}{H^2R^2}, \tag{21}$$

ν being the Poisson ratio (assumed in further numerical examples to be zero) and $2H$ is the thickness of the wall (cf. Fig. 2). Since we have restricted the analysis to short shells we expand the general integral of the elastic shell equation:

$$w = C_1 \sin\beta x \operatorname{sh}\beta x + C_2 \sin\beta x \operatorname{ch}\beta x + C_3 \cos\beta x \operatorname{sh}\beta x + C_4 \cos\beta x \operatorname{ch}\beta x \tag{22}$$

into a power series and retain the terms up to the fourth order. This allows us to keep the numerical side of the problem simple, without loosing any feature of the incremental collapse analysis.

For the case of ring of forces the stress boundary conditions are:

$$w''(0) = w'''(0) = w''(\beta L) = 0, \qquad w'''(\beta L) = -Q/D. \tag{23}$$

After evaluation of constants for the relevant generalized stresses we obtain eventually:

$$M_x = -QL\xi^2(1-\xi), \qquad N_\varphi = \frac{2QR}{L}(1-3\xi), \tag{24}$$

where $\xi = x/L$.

Since the loads p and Q vary independently within the limits $(0, p)$ and $(0, Q)$ respectively (and in opposite directions), the envelope of the elastic hoop force is:

$$\bar{N}^E_{\max} = \begin{cases} pR + \dfrac{2QR}{L}(1-3\xi), & 0 \le \xi \le \dfrac{1}{3}, \\ pR, & \dfrac{1}{3} \le \xi \le 1, \end{cases} \tag{25}$$

$$\bar{N}^E_{\min} = \begin{cases} 0, & 0 \le \xi \le \dfrac{1}{3}, \\ \dfrac{2QR}{L}(1-3\xi), & \dfrac{1}{3} \le \xi \le 1. \end{cases} \tag{26}$$

For the considered yield locus the only possible mechanism of plastic motion compatible with the associated flow law is such that $\dot{W}'' = 0$. Thus the velocity field for a short shell (entirely plastic) is linear, and the radial displacement increment is of the form:

$$\Delta \dot{W} = A_0(t)x + B(t), \qquad (27)$$

where A_0 and B are unknown. For the illustration sake the envelope and the displacement rates are sketched in Fig. 4a, b respectively.

We shall now proceed with application of the theorem (15), bearing in mind that in the considered case:

$$\Delta q^P = \Delta W/R,$$
$$D(\Delta q^P) = N_0 \Delta W/R. \qquad (28)$$

The right hand side term of the inequality (15) is found to take the following value:

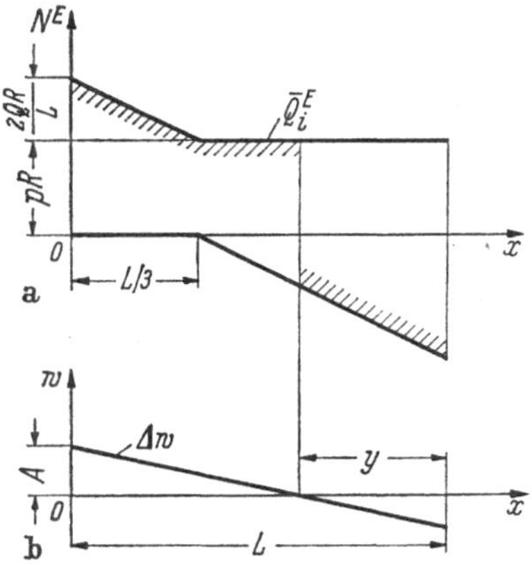

Fig. 4. Envelope of elastic hoop force and the deflection increments.

$$\int_A D(\Delta q^P)\,dA = \int_0^L \frac{\Delta W}{R} N_0\,dx = \frac{N_0 A_0}{2R} \frac{(L-y)^2 + y^2}{L-y} \qquad (29)$$

for $0 \leq y \leq \frac{2}{3} L$. The left hand side is easily determined using the visualisation of (24)—(26) as given in Fig. 4. We obtain eventually:

$$\int_A \bar{Q}_i^E \Delta q_i^P\,dA = \frac{pA_0}{2R}(L-y) + \frac{QA_0}{L-y}\left[\frac{8L-9y}{27} + \frac{y^2}{L}\left(2 - \frac{y}{L}\right)\right]. \qquad (30)$$

The expressions (29) and (30) relate to the unit of the shell circumference.

To derive the interaction curve of incremental collapse we take the equality sign in (15) and use the collapse load values for the individual load as given in (19), we arrive at the following expression:

$$F = \frac{p}{p_0} \frac{(1-\eta)^2}{(1-\eta)^2 + \eta^2} + 2\left(\sqrt{2} - 1\right) \frac{Q}{Q_0} \frac{8 - 9\eta + 27\eta^2(2-\eta)}{27[(1-\eta)^2 + \eta^2]} = 1, \qquad (31)$$

where $\eta = y/L$, $0 \leq \eta \leq \frac{2}{3}$. This family of straight lines in the plane of load multipliers has the envelope, which is the interaction curve we are looking for. If the displacement rate field were independent of the parameter η the respective interaction curve would be directly that of (31). This is the case if the collapse mode is considered such that $\eta < 0$, where eventually one obtains a straight line, belonging to the family (31), however. In the considered case collapse modes other than (27) being impossible, the envelope of (31), obtained eliminating η among $F = 1$ and $dF/d\eta = 0$ specifies the load domain safe against the incremental collapse, $\eta \leq \sqrt{2}/2$.

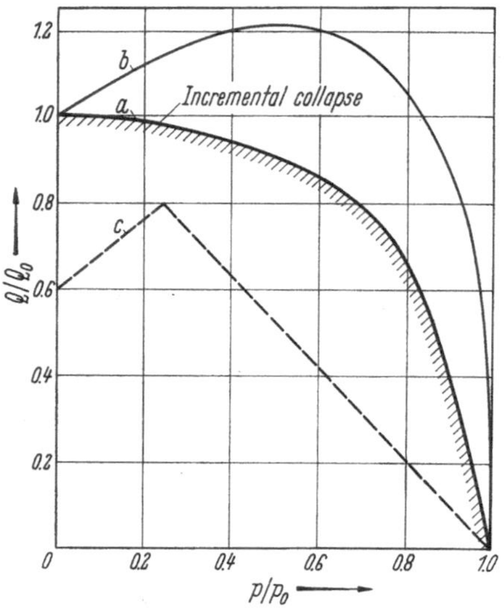

Fig. 5. Interaction curves. *a*) incremental collapse, *b*) limit analysis, *c*) purely elastic response.

This envelope is shown in Fig. 5, together with the collapse load interaction curve given by (18). The differences between these two curves are important. The incremental collapse curve is well approximated by the straight lines:

$$\frac{p}{p_0} + \frac{16}{27}\left(\sqrt{2} - 1\right)\frac{Q}{Q_0} = 1,$$

$$\frac{1}{5}\frac{p}{p_0} + \frac{Q}{Q_0} = 1. \qquad (32)$$

Broken line in Fig. 5 sketches the boundary of purely elastic response for the stress resultants (24).

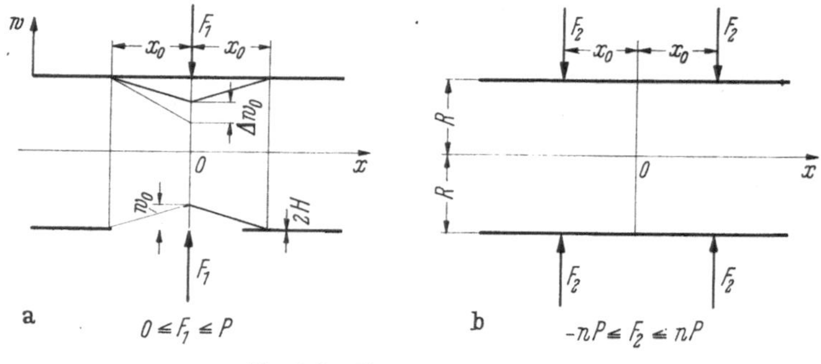

Fig. 6. Loading of a long cylinder.

As the second example we consider a long cylindrical shell loaded by a two parameter system as indicated in Fig. 6. The loads vary independently within the limits:

$$0 \leq F_1 \leq P \qquad \text{and} \qquad -nP \leq F_2 \leq nP, \qquad n < 1.$$

For the load F_1 only acting, the collapse ring load and the collapse mode are respectively (cf. [3]):

$$P_0 = 4\sqrt{\frac{M_0 N_0}{R}}, \qquad x_0 = 2\sqrt{\frac{M_0 R}{N_0}}. \tag{33}$$

In the present example we want to determine, the load intensity P such at which the shell will deform in the collapse mode $\dot{w} = w_0(t)\, f(x)$. In other words we want to determine the load intensities for which the central force will continue to sink even if the collapse load as computed for one parameter loads is not attained (cf. Fig. 6a).

For the yield locus of Fig. 3 the collapse mode:

$$\Delta W = \Delta W_0 (1 - x/x_0) \tag{34}$$

correspond to the yield point load P_0 and is of the required type. The stress profile corresponding to the mode (34) is that represented by the line CD on the yield locus shown in Fig. 3. The stress points C and D of the stress profile are associated with the yield hinges at $x = 0$ and $x = \pm x_0$.

The right hand side of (15) is immediately obtained from a formula analogous to (29). Accounting for the dissipation at the hinges, the resulting expression (per unit of the circumference) is:

$$2\int_A D(\Delta q)\, dA = 2\int_0^{x_0} \frac{\Delta W}{R} N_0\, dx + 4 M_0 \frac{\Delta W_0}{x_0} = 4\Delta W_0 \sqrt{\frac{M_0 N_0}{R}}. \tag{35}$$

To calculate the left hand side of the relation (15) we have to establish first the envelope of the extremum elastic states. Using appropriately the elastic solution (cf. [11])

$$M_x = \frac{P}{4\beta} e^{-\beta x} (\cos\beta x - \sin\beta z), \quad N_\varphi = \frac{P R \beta}{2} e^{-\beta x} (\sin\beta x + \cos\beta x), \tag{36}$$

we evaluate the stress resultants for the loads F_1 and F_2. The resulting diagrams, together with their envelopes are shown in Fig. 7. After an integration analogous to that of Eq. 30 one obtains eventually:

$$\int_A \bar{Q}_i^E \Delta q\, dA = P\Delta w_0 + \frac{nP}{\sqrt{6}} \Delta W_0 \Big[2 + 4 e^{-\sqrt{3/2}} \Big(\sin\sqrt{\tfrac{3}{2}} - \cos\sqrt{\tfrac{3}{2}}\Big) + \\ + \sqrt{2}\, e^{-\frac{3\pi}{4}} \Big(1 - \frac{3\pi}{4}\Big) + e^{-\sqrt{6}} \big(\cos\sqrt{6} - \sin\sqrt{6}\big)\Big] = (1 + 0.985 n)\, P\, \Delta W_0. \tag{37}$$

Hence results (35) and (37) yield the relation:

$$P = \frac{4}{1 + kn} \sqrt{\frac{M_0 N_0}{R}}, \tag{38}$$

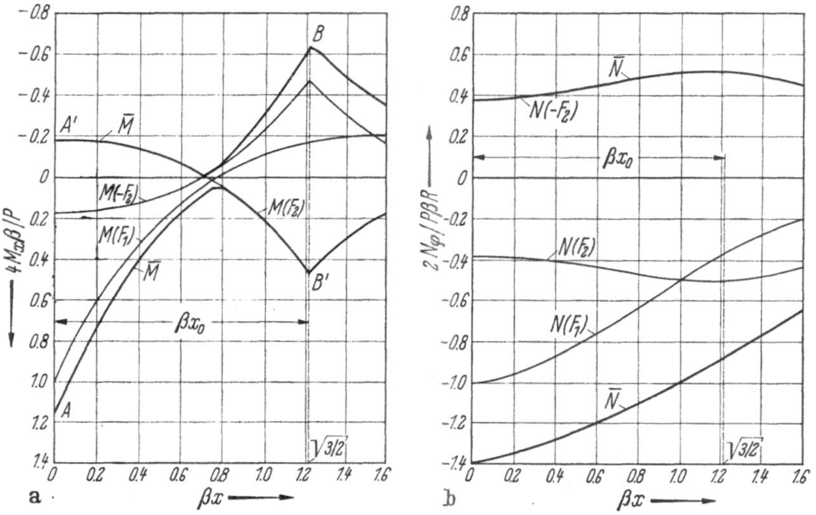

Fig. 7. Envelope of elastic stress resultants.

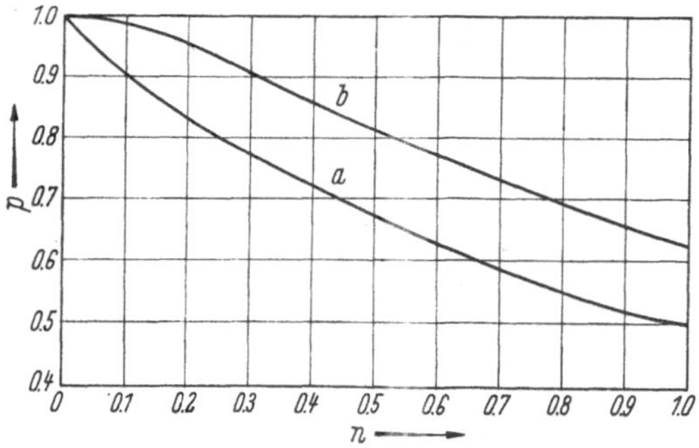

Fig. 8. Load intensity. *a*) **incremental collapse at cyclic loading,** *b*) **yield point load.**

where $k = 0{,}985$, thus practically $k = 1$ might be taken. The load of incremental collapse (38) is to be compared with the result (33), or with the result pertaing to the case when F_1 and F_2 increase proportionally. In this that circumstance an upper bound to the collapse load is:

$$P_0 = 4\, \frac{2n + \sqrt{4n^2 + (1 + 2n)^2}}{(1 + 2n)^2} \sqrt{\frac{N_0 M_0}{R}}\,. \tag{39}$$

In Fig. 8 the results (38) and (39) (in a dimensionless form) are plotted against the load ratio n.

From the presented examples it is seen that whenever the elastic solutions for particular loads are known, the domain safe against an incremental collapse in a specific collapse mode, can be established from (15).

References

1. PRAGER, W.: Problem types in the theory of perfectly plastic materials. J. Aeronaut Sci., **16**, 337—341 (1948).
2. NEAL, B. G., and P. S. SYMONDS: A method for calculating the failure loads for a framed structure subjected to fluctuating loads. J. Inst. Civil Engrs. (London) **35**, 186—197 (1950).
3. HODGE, P. G.: Plastic analysis of structures, New York: McGraw-Hill 1959.
4. KOITER, W. T.: A new general theorem on shake-down of elastic-plastic structures. Proc. Koninkl. Ned. Akad. Wettenschap. **B 59**, 24—34, 1956.
5. GOKHFELD, D. A.: Some problems of the adaptation theory for shells and plates (in Russian). Proc. 6th. Union Confer. Shells (Baku 1966), Moscow: Nauka 1966 pp. 284—291.
6. MELAN, E.: Theorie statisch unbestimmter Systeme aus ideal plastischem Baustoff. Sitz. Ber. Akad. Wiss. Wien, Abt IIa, **145**, 195—218, 1936.
7. KÖNIG, J. A.: Theory of shake-down of elastic-plastic structure. Arch. Mech. Stos **18** (1966).
8. KÖNIG, J. A.: Shake-down theory of plates and shells. Arch. Mech. Stos (in press).
9. SAWCZUK, A., and TH. JAEGER: Grenztragfähigkeitstheorie der Platten, Berlin/Göttingen/Heidelberg: Springer 1963.
10. HODGE, P. G., and CHANG-KUEI SUN: On plastic interaction curves. DOMITT Rep. 1—34, Chicago 1967.
11. TIMOSHENKO, S., and S. WOINOWSKY-KRIEGER: Theory of plates and shells, New York: McGraw-Hill 1959.

Discussion

W. T. KOITER: Professor SAWCZUK's application of the "inadaptation" theorem is indeed of great interest. I may perhaps add that the performance of an integration with respect to time of the general theorem presupposes that no reversal of plastic strain rates should occur, but this assumption seems reasonable in the loading conditions considered. I also wonder, and here I am linking up with Dr. LECKIE's remarks, whether a lower bound may perhaps also be obtained in the present examples by means of MELAN's theorem. Finally, I would appreciate the detailed references to other recent applications of both shakedown theorems.

F. A. LECKIE: I agree with the reactions of Professor SAWCZUK. I was pretty disillusioned with elasticity theory, when it came to the practical application to shells, because I found that the people, who are making shells are far and far ahead of the people, who are using elastic theory. But it does seem to me that there is a lot of hope in using the results of the elastic theory together with the plasticity theorems. We at Cambridge have been using the shake-down theorem a great deal in the solution of shell problems. We get lower bounds and we use self equilibrating sets of stress, so that in a sphere-cylinder intersection for example, one has edge solutions, which are elastic. One can optimize these using linear programming and get a lower bound on a shake-down value. This also has value on situations, which are completely beyond the computer even.

So we have been working on lower bounds, which may be a little unconservative, because one tends to get hardening, and hardening induces sort of "incremental shake-down".

A. Sawczuk: In relation to the comments by Professors Koiter and Leckie I may only remark further on the integration of the general theorem with respect to time. Passing from the components of the stress and strain rate tensors to a set of generalized variables we may loose control of certain strains, still controlling the displacements. At an incremental collapse expressed in terms of actually chosen generalized variables, there can appear such zones, where cyclic changes of strain rates take place and still integration is possible for the given set of generalized variables. Moreover, whenever the mechanism of incremental motion of a structure remains unchanged during the actual loading cycle — as it is the case for certain linear yield loci — the integration can also be performed. Thus there exist the cases where effective applications of Koiter's theorem — in the manner exposed by Gokhfeld and in the present note — can yield explicit upper bounds to the shakedown cycle and complement the techniques commented on by Professor Leckie.

The references cited and those discussed in Koiter's "General Theorems for Elastic-Plastic Solids" (Progress in Solid Mechanics, **1**, Amsterdam: North-Holland 1960) I may append by the following:

References

1. Rozenblum, V. I.: On the shakedown theorem for elastic-plastic bodies (in Russian) Izv. ANSSSR, OTN, No. 6, 47—53 (1958).
2. Rozenblum, V. I.: On adaptation of uneven heated elastic-plastic solids (in Russian), Prikl. Mekh. Tekh. Fiz. (PMTF), No. 5, 98—101 (1965).
3. Gokhfeld, D. A.: Carrying capacity of turbine disces under unsteady operating conditions (in Russian), Mashinovedenie, No. 6, 61—68 (1965).
4. Gokhfeld, D. A., and M. M. Kononov: Investigation of the temperature fields and thermal stresses in the rotor of a centripetal gas turbine (in Russian), in Teplovye Napryazhenya, Part 3, Naukova Dumka, Kiev 1965, 224—232.
5. Leckie, F. A.: Shakedown pressures for flush cylindersphere shell interactions. J. Mech. Engng. Sci. **7**, 367—371 (1965).
6. Leckie, F. A., and R. K. Penny: Shakedown loads for radial nozzles in spherical pressure vessels, Int. J. Solids Structures, **3**, 743—755 (1967).
7. Hodge, P. G., and A. J. Kalinowski: Shakedown interaction curve for a circular arch, DOMIIT Rep. 1—36, Chicago 1967.

Rheological States of Geometrically Nonlinear Rotational Membranes

By

Z. Bychawski and **W. Olszak**
Polish Academy of Sciences, Warsaw, Poland

Summary. The present paper is concerned with a rheological theory of geometrically non-linear deformation processes of rotational shells in the membrane state under internal pressure.

A rotational shell is assumed to be of small and constant thickness and made of an isotropic, homogeneous and incompressible material which, in general, behaves non-linearly and shows combined instantaneous (elasto-plastic) and time-dependent (viscous) properties.

The constitutive equation of the membrane material is given in an operational form which corresponds to a non-linear integral equation. This equation may be reduced to a differential form.

The basic set of membrane equations consists of the compatibility condition and the equation of equilibrium for a deformed state.

A general method of solution is given for the above system of equations. In principle, it consists in assuming the solutions for the stress functions and the displacement in the form of double power series containing a small physical parameter. In particular, it is shown that the creep problem may be solved on the proposed analogy.

The theory and the method of solution are illustrated by some results obtained for a spherical shell.

1. Introduction

Thin rotational shells find a very expanded application in modern engineering constructions, especially, in the conditions of high loads and at elevated temperatures. It is evident that the analysis of shells based on the classical concepts of geometrical and physical linearity and time-independence is no more sufficient is such cases.

The complexity of conditions in which the shells may resist to loads suggests a new approach in investigating the problems of deformation and stress analysis in shells. This approach considers the shell-behavior as a deformation process, during which the nonlinearities involved — as a consequence of both geometrical and physical properties of the shell — are taken into account.

From the geometrical point of view a characteristic feature of very thin shells is their relatively small resistance to bending and twisting moments. Under sufficiently high loads and for displacements several times greater than the shell thickness, it is observed that the influence of moments is usually negligible. Thus, in the state of a geometrically nonlinear deformation a very thin shell may be considered to be a membrane.

The assumption about the membrane state shells enables us to neglect all moments of internal forces in deriving the fundamental shell equations.

The process of shell deformation may also have a considerable influence on the geometrical properties of shells due to strains and displacements increasing in time, which may attain significant values. This fact restrains the validity of the geometrical assumptions to certain definite time-interval only.

From the physical point of view an appropriate constitutive equation of the shell material should be assumed, in order to represent qualitatively and quantitatively the physical properties of the deformation process. It is clear that such a constitutive equation must be nonlinear if we do wish to obtain an adequate picture of the shell material behavior in advanced states of deformation.

The general problems of physically and geometrically nonlinear rotational membranes in a combined deformation state have been studied by BYCHAWSKI [1]. Some solutions to elasto-plastic and creep deformation problems of a nonlinear shallow spherical membrane have been given by BYCHAWSKI and KOPECKI [2]. A detailed analysis of some nonlinear viscoelastic membranes of revolution has been made by KOPECKI [4].

The present paper is concerned with a general theory of physically and geometrically nonlinear processes of rotational membranes and various problems and aspects of this theory.

We shall study rotational membranes of small and constant thickness which deform under the normal internal load varying, in general, with the surface coordinates and time. We assume that in the time-interval considered, the strain tensor and strain rate tensor are small quantities, the rotation angles are very small and only the normal component of displacement vector has a finite value.

Further, we assume that the material of the membrane is isotropic, homogeneous and incompressible, and behaves, in general, nonlinearly exhibiting combined elasto-plastic and viscous properties. Such assumptions seem to be appropriate and necessary when investigating intensive extensions of geometrically nonlinear membranes.

2. Geometrical Relations

We consider the geometry of a nonlinear process of deformation of a rotational surface, and assume that the undeformed surface is generated by the revolution of a plane regular curve which does not imply any singularities.

We postulate certain initial conditions related to the state of the surface at initial instant, and assume that the membrane behavior is characterized by an instantaneous response to the applied load. According to the assumed properties of the membrane material, we may have the following possibilities for $t = t_0$, where t_0 is the initial instant:

a) the initial state of a membrane is caused by the application of a distributed load $p = p(t)$ such that at $t = t_0$, $p(t_0) = 0$. Therefore, the initial state of the surface is neutral;

b) the initial state of a membrane is caused by an instantaneous response (in general, elasto-plastic) to the load $p(t_0) = p_0$ which may depend on the surface coordinates;

c) the initial state of a membrane is considered at $t = \bar{t}$, where $\bar{t}$ is an intermediate instant from the interval $[t_0, t]$, i. e., the deformation of the membrane surface is caused by a combined elasto-plastic and viscous response.

In the last case, the initial conditions are related to a certain intermediate state of the surface and express in an instantaneous manner the effects occured in the time-interval $(-\infty, \bar{t}]$.

Further, we assume that the changes of the membrane thickness are neglibible and the stress and deformation states are completely described by the corresponding states of the membrane middle surface at every time-instant.

We fix on the membrane middle surface a set of nondimensional Gaussian coordinates Θ_α $(\alpha = 1,2)$, the lines of which coincide with the lines of main curvatures of the undeformed surface.

Let r_0 be the radius vector of an arbitrary point on the undeformed surface S_0

$$\boldsymbol{r}_0 = \boldsymbol{r}_0(\Theta_1, \Theta_2). \tag{2.1}$$

During the deformation process the point is carried to the new position on the deformed surface S given by the radius vector

$$\boldsymbol{r} = \boldsymbol{r}(\Theta_1, \Theta_2, t) = \boldsymbol{r}_0 + \boldsymbol{u}, \tag{2.2}$$

at arbitrary instant t. Here $\boldsymbol{u}$ is the displacement vector.

The vector $\boldsymbol{u}$ may be presented in the form

$$\boldsymbol{u} = u^\alpha \boldsymbol{a}_\alpha + w \boldsymbol{a}_3, \tag{2.3}$$

where $\boldsymbol{a}_\alpha$ are basic vectors of the surface S_0, w is the displacement component normal to S_0 and $\boldsymbol{a}_3$ the vector of the unit length normal to S_0.

The basic vectors of the deformed surface S relative to the coordinates on S_0 are found from Eqs. (2.2) and (2.3)

$$\boldsymbol{A}_\alpha = \boldsymbol{r}_{0,\alpha} + \boldsymbol{u}_{,\alpha} = \alpha_\alpha{}^\varrho \boldsymbol{a}_\varrho + \beta_\alpha \boldsymbol{a}_3,$$

where the coefficients are

$$\alpha_\alpha{}^\varrho = \delta_\alpha{}^\varrho + u^\varrho|_\alpha - b_\alpha{}^\varrho w, \qquad \beta_\alpha = w_\alpha + b_\alpha{}^\varrho u_\varrho. \tag{2.4}$$

Denoting by

$$a_{\alpha\beta} = \boldsymbol{a}_\alpha \cdot \boldsymbol{a}_\beta, \quad A_{\alpha\beta} = \boldsymbol{A}_\alpha \boldsymbol{A}_\beta, \tag{2.5}$$

the corresponding metric tensors of S_0 and S, we define now the deformation tensor

$$\gamma_{\alpha\beta} = \frac{1}{2}\,(A_{\alpha\beta} - a_{\alpha\beta}). \tag{2.6}$$

Introducing Eq. (2.5) into Eq. (2.6) we find

$$\gamma_{\alpha\beta} = \frac{1}{2}\,[u_\alpha|_\beta + u_\beta|_\alpha - 2b_{\alpha\beta} w + (w_{,\alpha} + b_\alpha{}^\mu u_\mu)(w_{,\beta} + b_\beta{}^\varrho u_\varrho) - \\ - (u^\mu|_\alpha - b_\alpha{}^\mu w) b_\beta{}^\varrho a_{\varrho\mu} w]. \tag{2.7}$$

The above expression may be considerably simplified, if we have in mind our assumptions concerning the magnitude of strains and displacements. Thus, for nondimensional quantities we can write the inequality

$$\overline{u}_\alpha \ll \overline{w} \ll 1. \tag{2.8}$$

By neglecting the corresponding terms in Eq. (2.7) in connection with Eq. (2.8), we now obtain

$$\gamma_{\alpha\beta} = \frac{1}{2}\,(u_\alpha|_\beta + u_\beta|_\alpha - 2b_{\alpha\beta} w + w_{,\alpha} w_{,\beta} + b_{\alpha\varrho} b_\beta{}^\varrho w w). \tag{2.9}$$

The coefficients b appearing in Eqs. (2.7) and (2.9) are components of the surface tensor of the second fundamental form.

We consider now the extensions $e_{\alpha\alpha}$ of a line element ds along the curvilinear coordinates Θ_α

$$e_{\alpha\alpha} = (ds_\alpha - ds_{0\alpha})/ds_{0\alpha}, \tag{2.10}$$

where

$$ds_\alpha = \sqrt{A_{\alpha\alpha}}\, d\Theta, \qquad ds_{0\alpha} = \sqrt{a_{\alpha\alpha}}\, d\Theta. \tag{2.11}$$

The above definition of extensions is characteristic for large deformations. Since we assume that strains are small quantities, we find

$$e_{\alpha\alpha} = \sqrt{A_{\alpha\alpha}/a_{\alpha\alpha}} - 1 = \sqrt{1 + 2\varepsilon_{\alpha\alpha}} - 1 \approx \varepsilon_{\alpha\alpha}. \tag{2.12}$$

It should be noticed that on the basis of Eq. (2.2) all quantities describing the deformation process of the surface S are functions of time. Accordingly, if we put $t = t_0$, all these quantities will describe the instantaneous response of the membrane as shown in b) or the combined response as in c).

3. Equations of Equilibrium

It is known from the theory of large deformations that the equations of equilibrium have the same form as in the classical theory, if only all quantities are related to the coordinates on the deformed surface S. From the general equation of equilibrium known from the membrane theory we find the following relations:

$$n^{\alpha\beta}\|_\alpha = 0, \tag{3.1}$$

$$n^{\alpha\beta} B_{\alpha\beta} + p = 0, \tag{3.2}$$

where n are stress resultants and B coefficients of the second fundamental form of the deformed surface S.

We now wish to express Eqs. [3.1) and (3.2) in terms of quantities relative to the undeformed surface S_0.

According to our simplifications made in paragraph 2, we can now write the approximate expressions for the basic vectors of the surface S in the form (see Eq. (2.4) and (2.4a))

$$\boldsymbol{A}_\alpha = \boldsymbol{a}_\alpha + w_{,\alpha} \boldsymbol{a}_3, \tag{3.3}$$

and then we find the approximate value of the vector normal to S

$$\boldsymbol{A}_3 = \boldsymbol{A}_1 \times \boldsymbol{A}_2 = \boldsymbol{a}_3 - w_{,\alpha} \boldsymbol{a}^\alpha. \tag{3.4}$$

On the basis of Eqs. (3.3) and (3.4) we calculate the components of the tensor B related with the second fundamental form of the surface S. Thus, we obtain approximately (and sufficiently for our purposes) the components of tensor B

$$\beta_{\alpha\beta} = \boldsymbol{A}_3 \cdot \boldsymbol{A}_{\alpha,\beta} = b_{\alpha\beta} + w|_{\alpha\beta}, \tag{3.5}$$

which now are expressed by the quantities related to the undeformed surface S_0.

Since we have assumed that the deformation tensor is small (see Eq. (2.9)), we may calculate the stress resultants n with respect to the undeformed surface. Thus, for $a_{\alpha\beta} \gg \gamma_{\alpha\beta}$ we simply find

$$n^{\alpha\beta}\|_{\alpha} = n^{\alpha\beta}|_{\alpha}. \tag{3.6}$$

The relation between the physical stress resultants $n_{\alpha\beta}$ and $n^{\alpha\beta}$ is

$$n_{\alpha\beta} = \sqrt{a_{\beta\beta}/a^{\alpha\alpha}}\, u^{\alpha\beta}, \tag{3.7}$$

and since our system of coordinates is orthogonal, these components are equal to each other.

4. Physical Relations

According to our assumptions about the geometrical properties of the deformation process of the membrane we now may write the resultants of internal forces expressed by the physical components of stress σ

$$n_{\alpha\beta} = h\,\sigma_{\alpha\beta} = h\left(s_{\alpha\beta} + \frac{1}{3}\,\delta_{\alpha\beta}\,\sigma_{\mu\mu}\right), \tag{4.1}$$

where h denotes the thickness of the membrane and by s the components of the stress deviator.

The constitutive equation for a complete nonlinear viscoelastic material, which we shall use here, has been proposed by BYCHAWSKI and FOX [4]. It has been shown that the constitutive equation for hereditary process includes as particular cases some previous theories of creep of metals and nonmetallic materials, and, therefore, may be of use, especially, in investigating the non-steady states.

The constitutive equation is derived on the basis of a generalized superposition principle which may be expressed in the form

$$e_{\alpha\beta}(t) = \int_{-\infty}^{t} d\,e_{\alpha\beta}(t), \tag{4.2}$$

written for the infinitesimal strain tensor components e. The principle is founded on the integrability in the sense of Stieltjes.

The Eq. (4.2) may be reduced to a simpler form containing the Riemann integral if certain conditions concerning the functions appearing under integral are satisfied. Thus, for our purposes we have

$$e_{\alpha\beta}(t) = s_{\alpha\beta}(t)\,\Phi[s(t)] - \int_{t_0}^{t} s_{\alpha\beta}(t)\cdot \partial_t \tilde{H}\,[t, \tau, s(\tau)]\cdot d\tau, \tag{4.3}$$

where $\partial_\tau = \partial/\partial_\tau$. Here, $\tilde{H}$ is the generalized creep function

$$\tilde{H}[t, \tau, s(\tau)] = C(t - \tau)\,\Theta(t^* - \tau) + H[t, \tau, s(\tau)] \cdot \Theta(\tau - t^*), \quad (4.4)$$

t^* denoting the time instant at which the effective stress

$$s(\tau) = \frac{3}{2}\,[s_{\alpha\beta}(\tau)\, s_{\alpha\beta}(\tau)]^{1/2}, \quad (4.5)$$

attains its limit value, C standing for the creep function of the linear range, H for the creep function of the nonlinear range and Θ for the Heaviside distribution.

Equation (4.3) is the constitutive relation for a complete nonlinear viscoelastic material in which the first part represents the nonlinear instantaneous response expressed by the function Φ and the second part is related to the time-dependent deformation.

Equation (4.3) may be reduced to a differential form if some special assumptions are made with respect to the form of the generalized creep function. For example, if we assume that this function satisfies the condition

$$\partial_\tau \tilde{H}[t, \tau, s(\tau)] = F[s(\tau)]\, \partial_\tau C(t - \tau), \quad (4.6)$$

where

$$F[s(\tau)] = \Theta(t^* - \tau) + \Theta(\tau - t^*)\, F[s(\tau)], \quad (4.7)$$

then Eq. (4.3), as shown in [2], reduces to the form

$$L\{e_{\alpha\beta}(t) - s_{\alpha\beta}(t)\, \Phi[s(t)]\} = -F[s(t)] \cdot s_{\alpha\beta}(t), \quad (4.8)$$

where L is a linear differential operator with constant coefficients

$$L = a_k \partial_t{}^k. \quad (4.9)$$

If we put into Eqs. (4.8) and (4.3)

$$C(t - \tau) = \lambda(t - \tau), \qquad \lambda = \text{const.}, \quad (4.10)$$

then we obtain the constitutive equation of the Odqvist creep theory

$$e_{\alpha\beta}(t) = s_{\alpha\beta}(t)\, \Phi[s(t)] + \lambda \int_{t_0}^{t} F[s(\tau)]\, s_{\alpha\beta}(\tau)\, d\tau. \quad (4.11)$$

The same equation may also be obtained from the differential form (4.8) if the coefficients of the operator L are put

$$a_0 = 0, \qquad a_1 = 1/\lambda, \qquad a_k = 0, \qquad (k = 2, 3, \ldots). \quad (4.12)$$

We shall now write the constitutive Eq. (4.3) in the operational form

$$e_{\alpha\beta} = N s_{\alpha\beta}, \tag{4.13}$$

where N is the nonlinear integral operator, the form of which is clearly seen from Eq. (4.3).

Consider now the initial conditions as in paragraph 2. In the case a) the strain tensor disappears. In the case b) by putting $t = t_0$ into Eq. (4.3), the integral disappears and we obtain

$$e^0_{\alpha\beta} = N_0 s^0_{\alpha\beta}, \qquad e^0_{\alpha\beta} = e_{\alpha\beta}(t_0), \qquad s^0_{\alpha\beta} = s_{\alpha\beta}(t_0), \tag{4.14}$$

where N_0 is the operator of the instantaneous deformation, the form of which is clearly seen from Eq. (4.3).

According to c), Eq. (4.3) gives the operational relation

$$e^*_{\alpha\beta} = N s^*_{\alpha\beta}, \quad e^*_{\alpha\beta} = e_{\alpha\beta}(\bar{t}), \quad s^*_{\alpha\beta} = s_{\alpha\beta}(\bar{t}). \tag{4.15}$$

Let us consider the particular case of the theory as expressed by Eq. (4.1) which we now write in the operational form

$$e_{\alpha\beta} = \overline{N} s_{\alpha\beta} \tag{4.16}$$

where N is the nonlinear operator of Eq. (4.11). Equation (4.16) may also be written in a differential form which is directly obtained by differentiation with respect to time

$$\dot{e}_{\alpha\beta} = \dot{\overline{N}} s_{\alpha\beta}. \tag{4.17}$$

Here, as is usually done, we assume the nonlinear functions Φ and F in the form (for $\lambda = 1$)

$$\Phi = \frac{3}{2} D_i s^{m-1}, \qquad F = \frac{3}{2} D_c s^{n-1}, \tag{4.18}$$

where D_i, D_c and m, n are physical constants (m, n denoting odd natural numbers, $m, n \geq 1$).

If according to Eq. (4.14) we consider the instantaneous state at $t = t_0$, then the corresponding operator of instantaneous deformation becomes

$$e^0_{\alpha\beta} = \overline{N}_0 s^0_{\alpha\beta}, \tag{4.19}$$

or alternatively

$$\dot{e}^0_{\alpha\beta} = \dot{\overline{N}} s^0_{\alpha\beta}.$$

It should be noticed that since the membrane material is assumed to be incompressible, the components of strain or strain rate tensor must satisfy the conditions

$$e_{kk} = 0, \qquad \dot{e}_{kk} = 0, \qquad (k = 1, 2, 3). \tag{4.20}$$

However, the influence of the transversal deformation and the thickness change are disregarded in our theory.

5. Combined Deformation State and Stress State

Let us allow the curvilinear system of coordinates Θ_α to coincide with the orthogonal system $\Theta_1 = \xi$, $\Theta_2 = \varphi$ on the surface S_0, where ξ is the arc length measured along the meridian from a fixed point and φ is the angle relative to parallels. If ϱ is the distance of a point on S_0 from the axis of symmetry and ξ its coordinate measured on this axis, then ϱ, φ, ξ are cylindrical coordinates of the surface.

The square of a linear element ds_0 on S_0 may be written as follows:

$$ds_0{}^2 = \boldsymbol{dr}_0 \cdot \boldsymbol{dr}_0 = d\xi^2 + \varrho^2\, d\varphi^2, \tag{5.1}$$

where $d\boldsymbol{r}_0$ is the differential of the radius vector (2.1). From Eq. (5.1) we find

$$a_{11} = 1, \qquad a_{22} = \varrho^2, \qquad a_{\alpha\beta} = 0, \qquad (\alpha \neq \beta). \tag{5.2}$$

Further, we assume the full symmetry of the deformed state of the surface under a constant internal pressure p.

The condition of equilibrium (3.1) expressed by means of Eq. (4.1) takes the form

$$d_\xi(\varrho\sigma_1) = \sigma_2 d_\xi \varrho, \tag{5.3}$$

and the condition (3.2) becomes

$$\sigma_1(k_1 - d_\xi{}^2 w) + \sigma_2\left(k_2 - \frac{1}{\varrho}\, d_\xi w \cdot d_\xi \varrho\right) = \frac{p}{h}, \tag{5.4}$$

where in order to simplify the notations we reduce the double subscript for stresses.

Since an axially symmetric deformation process is assumed, the displacement component u_2 disappears, and by putting $u_1 = u$ we obtain on the basis of Eq. (2.9) and Eq. (2.12) (when reducing the double subscript for strains)

$$\begin{aligned} e_1 &= d_\xi u + \frac{1}{2}\, d_\xi w \cdot d_\xi w + k_1 w + \frac{1}{2}\,(k_1 w)^2, \\ e_2 &= \frac{1}{\varrho}\, u d_\xi \varrho + k_2 w + \frac{1}{2}\,(k_2 w)^2. \end{aligned} \tag{5.5}$$

Here, and in Eq. (5.4), k_1, k_2 are the main curvatures of the surface.

The relations for main strains (5.5) are not independent. By eliminating the displacement u we obtain the condition of compatibility of deformations in the form

$$\varrho d_\xi \varrho \cdot d_\xi e_2 + (e_2 - e_1)(d_\xi \varrho)^2 - \varrho e_2 d_\xi^2 \varrho = -\frac{1}{2}(d_\xi w)^2 (d_\xi \varrho)^2 + \\ + \frac{1}{2} d_\xi [\varrho k_2 w(2 + k_2 w)] d_\xi \varrho - \frac{1}{2} k_1 w(2 + k_1 w)(d_\xi \varrho)^2 - \\ - \frac{1}{2} \varrho k_2 w(2 + k_2 w) d_\xi^2 \varrho. \tag{5.6}$$

It should be noticed that Eqs. [5.5) are derived without any restrictions concerning the magnitudes of the main curvatures. If, for example, the membrane may be assumed to be shallow, then the last terms in Eqs. (5.5) containing the squares of curvatures may be neglected. Further, in this case we may put $d\varrho = d\xi$, and in consequence Eq. (5.6) simplifies considerably.

Thus, for a shallow membrane we find the following set of equations:

$$d_\varrho(\varrho \sigma_1) = \sigma_2, \tag{5.7}$$

$$\sigma_1(k_1 - d_\varrho^2 w) + \sigma_2\left(k_2 - \frac{1}{\varrho} d_\varrho w\right) = \frac{p}{h}, \tag{5.8}$$

$$\varrho d_\varrho e_2 + e_2 - e_1 = -\frac{1}{2}(d_\varrho w)^2 + d_\varrho(\varrho k_2 w) - k_1 w. \tag{5.9}$$

Further, we introduce the following substitutions:

$$r = \varrho^2/R^2, \qquad \sigma_1 = \frac{1}{r} D z, \qquad D = p/c, \qquad c = h/R, \\ \overline{w} = w/h, \qquad \overline{k}_1 = R k_1, \qquad \overline{k}_2 = R k_2, \tag{5.10}$$

and represent the stress deviator components in the form

$$s_1 = \frac{1}{3} D\left(3\frac{z}{r} - 2 d_r z\right), \qquad s_2 = \frac{1}{3} D\left(4 d_r z - 3\frac{z}{r}\right). \tag{5.11}$$

By introducing the quantities (5.10), we satisfy Eq. (5.7), and Eq. (5.8) becomes

$$\frac{z}{r}\left[\overline{k}_1 - 4c\sqrt{r}\, d_r\left(\sqrt{r}\, d_r \overline{w}\right)\right] + \left(2 d_r z - \frac{z}{r}\right)(k_2 - 2c\, d_r \overline{w}) = 1. \tag{5.12}$$

We now make use of the physical relations (4.13) and according to Eq. (511) we have

$$e_1 = N s_1 = N\left[\frac{1}{3} D\left(3\frac{z}{r} - 2 d_r z\right)\right],$$
$$e_2 = N s_2 = N\left[\frac{1}{3} D\left(4 d_r z - 3\frac{z}{r}\right)\right]. \tag{5.13}$$

On the basis of Eq. (5.13) the condition (5.9) may be now written

$$2\sqrt{r}\, d_r\left(\sqrt{r}\, N s_2\right) - N s_1 = -2c^2 r (d_r \overline{w})^2 + 2c\sqrt{r}\, d_r\left(\sqrt{r}\, \bar{k}_2 \overline{w}\right) - c\bar{k}_1 \overline{w}. \tag{5.14}$$

The set of Eqs. (5.12) and (5.14) is a system of two equations with respect to two unknown functions: the nondimensional stress function z and the nondimensional displacement $\overline{w}$. Thus, a combined deformation process of the membrane is fully described by this system. There are no objections in using the more complex system of Eqs. (5.4) and (5.6), and the methods applied for a shallow membrane may also have application for a more general case.

There are certain alternatives in presenting the set of Eqs. (5.12) and (5.14) according to the meaning of the physical operator N. For example, by putting N_0 into Eq. (5.14) instead of N, and denoting by z_0 and w_0 the corresponding stress and deflection functions of the instantaneous state, we obtain from Eqs. (5.12) and (5.14) a set of equations for the instantaneous problem of the membrane.

By differentiating Eq. (5.14) with respect to time and by putting $\dot{\overline{N}}$ instead of $\dot{N}$, we obtain, together with Eq. (5.12), a set of equations expressed by the strain rates according to Eq. (4.17).

6. Pure Creep and Analogy

It has been found by Bychawski [7, 8] that there exists an analogy between the set of equations for the instantaneous problem and creep problem of nonlinear flat membranes. It has been pointed out that this analogy exists also in the case of a spherical membrane [2], if Eq. (4.16) together with Eq. (418) are used separately for the instantaneous and creep deformations. The analogy has also been found for the special case of Eq. (4.13).

The operator (4.17) may be presented in the form

$$\dot{\overline{N}} = \dot{\overline{N}}_i + \dot{\overline{N}}_c, \tag{6.1}$$

where $\overline{N}_i$ is the instantaneous part of the operator (6.1) and $\dot{\overline{N}}$ is the operator of creep deformation.

We shall show that the application of the analogy is also possible in the case of general rotational membranes.

Assume that the operators $\dot{\overline{N}}_i$ and $\dot{\overline{N}}_c$ are expressed by the functions (4.18), respectively, and are separately introduced into Eq. (5.14) when differentiated with respect to time. Then we obtain two equations which are formally analogous. The first equation containing $\dot{\overline{N}}_i$ may be at once intergrated giving the form (5.14) with the substitution of $\overline{N}_i$ instead of N and $z = z_i$, $\overline{w}_i = \overline{w}_i$, $\bar{k}_\alpha = \bar{k}_\alpha{}^i$.

Further, we put (the index c is related to creep)

$$z_c = \Psi(t) z_c{}^0, \qquad \overline{w}_c = \eta(t) w_c{}^0, \qquad \bar{k}^c = \Psi(t) k^{c0}. \tag{6.2}$$

Substituting the values (6.2) into Eq. (5.14) when diffenrentiated with respect to time, we find the possibility of representing $\overline{N}_0$ in the form

$$\overline{N}_c = [\Psi(t)]^n \, \dot{\overline{N}}_{c0}, \tag{6.3}$$

where $\dot{\overline{N}}_{c0}$ is a time-independent operator of creep deformation. Here, we notice that on the basis of Eq. (4.18) we have

$$\dot{\overline{N}}_{c0} = \overline{N}_i, \tag{6.4}$$

for $D = D_i$ and $n = m$.

It may easily be shown that by substituting Eq. (6.2) into the set of Eqs. (5.12) and (5.14) we obtain a set of time-dependent equations with respect to two unknown functions $\Psi(t)$ and $\eta(t)$; this can be solved in a simple way.

On the other hand, we find that the time-independent set of equations from the latter case is formally analogous to the set obtained by application of the operator $\overline{N}_i$.

7. Method of Solution

The system of Eqs. (5.12) and (5.14) is related to a combined instantaneous and time-dependent deformation of the membrane. In general, the proposed method of solution consists in assuming the resolving stress function z and the reflection $\overline{w}$ in the form of double power series

$$z = z(r, t) = \sum_{i,j=0}^{\infty} z_{ij} \delta^j r^{i+1}, \qquad \overline{w} = \overline{w}(r, t) = \sum_{i,j=0}^{\infty} \overline{w}_{ij} \delta^j r^{i+1}, \tag{7.1}$$

where δ is a small parameter which has a physical meaning. By substituting Eqs. (7.1) into the set of Eqs. (5.12) and (5.14), we obtain two recurrent systems of integral (or differential) equations with respect to time for the unknown coefficients z_{ij} and $\overline{w}_{ij}$ as functions of time.

Since the system of Eqs. (5.12) and (5.14) is highly nonlinear then, in order to avoid raising to the powers of the double series, it is convenient to apply at first the single power series

$$z = \sum_{v=0}^{\infty} z_v r^{v+1}, \qquad z_v = \sum_{q=0}^{\infty} z_{vq}(t)\,\delta^q,$$

$$\overline{w} = \sum_{v=0}^{\infty} w_v r^{v+1}, \qquad \overline{w}_v = \sum_{q=0}^{\infty} w_{vq}(t)\,\delta^q. \tag{7.2}$$

It is found that the basic solution for the method given here is the solution of the instantaneous problem for the functions z_0 and $\overline{w}_0$. All time-dependent coefficients of Eqs. (7.1) or (7.2) are expressed by the coefficients of the development

$$z_0 = \sum_{v=0}^{\infty} z_{v0} r^{v+1}, \qquad \overline{w}_0 = \sum_{v=0}^{\infty} \overline{w}_{v0} r^{v+1}, \tag{7.3}$$

which are found from the boundary conditions for $t = t_0$.

The solution of a combined state of the membrane obtained on the basis of the concept of an instantaneous perturbated state can practically be valid for comparatively short time-intervals. This conclusion follows from the fact that the perturbated development may, in principle, describe the initial state of the deformation process only.

8. Boundary Conditions

In order to simplify our considerations we shall assume that the boundaries are rigid and the membrane is hinged at the boundary. Then, at an arbitrary instant t, we have

a) the displacement or displacement rate on the boundary

$$u = 0, \qquad \dot{u} = 0, \qquad \text{for} \qquad r = 1, \tag{8.1}$$

b) the deflection or deflection rate on the boundary

$$\overline{w} = 0, \qquad \dot{\overline{w}} = 0, \qquad \text{for} \qquad r = 1, \tag{8.2}$$

c) the stresses at the center point of the membrane are equal to each other

$$\sigma_1 = \sigma_2, \qquad \text{for} \qquad r = 0. \tag{8.3}$$

It should be noticed that the first coefficient of the developments 7(.1) are found from the boundary conditions (8.1) and (8.2) for $t = t_0$ and may be calculated with the desired degree of accuracy from the algebraic equations.

9. Results

Some results obtained for a spherical shell by using the presented method of solution are shown in the figures.

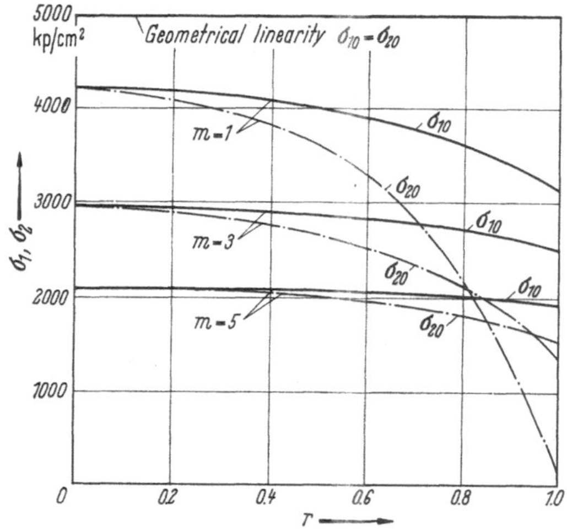

Fig. 1.

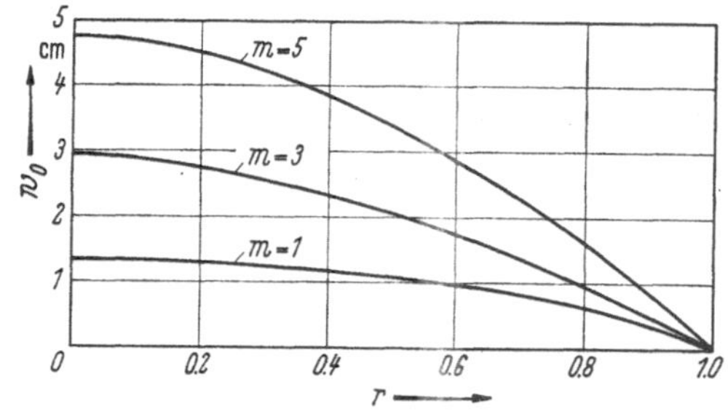

Fig. 2.

Fig. 1 shows the stress distribution, and Fig. 2 the deflection for an instantaneous state of the membrane characterized by the different coefficients of nonlinearity n.

In Figs. 3 and 4 are shown, respectively, the stress distribution and the deflection in the presence of a stabilizing creep.

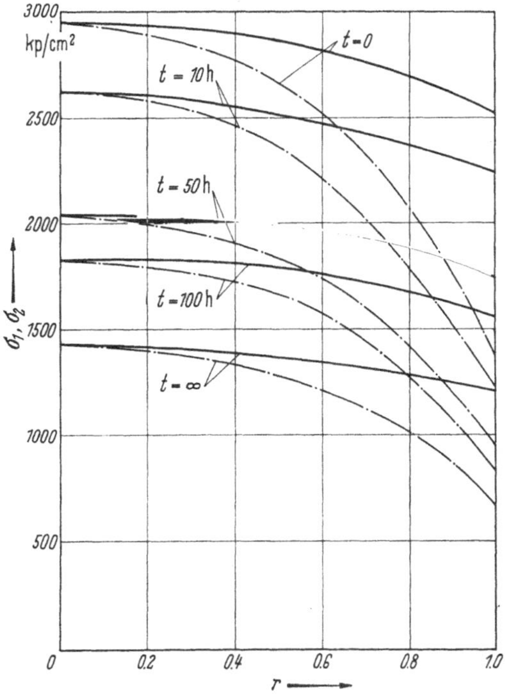

Fig. 3.

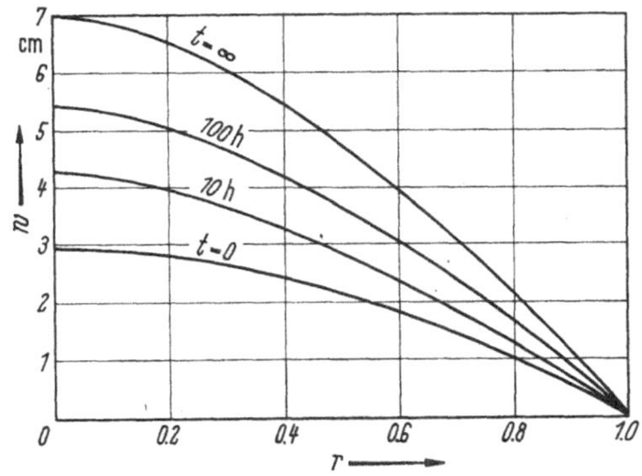

Fig. 4.

The Figs. 5 and 6 represent the solutions for the stress distribution and the deflection in the case of a combined state of deformation.

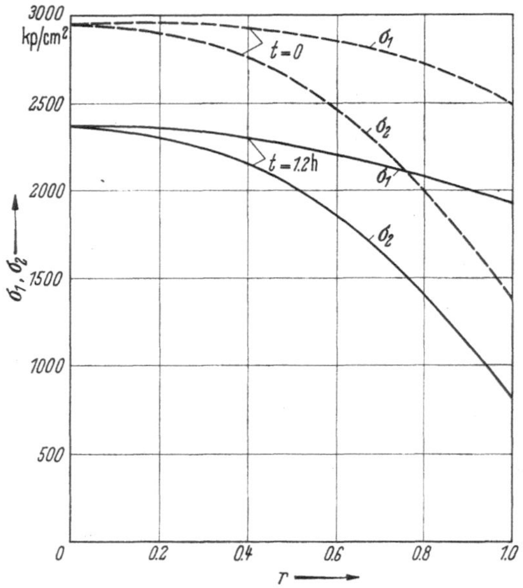

Fig. 5.

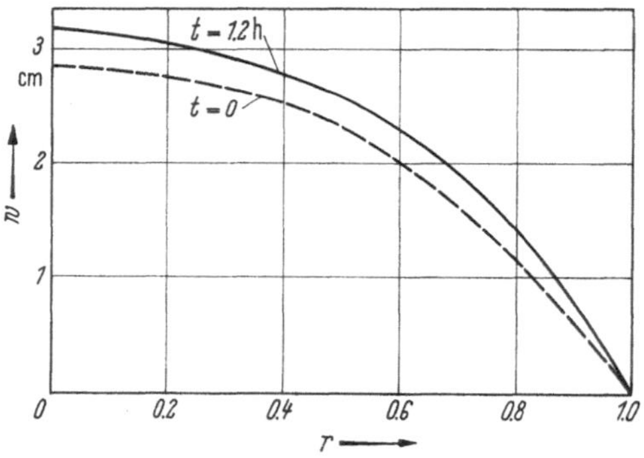

Fig. 6.

References

1. BYCHAWSKI, Z.: Combined instantaneous and creep deformation of rotational shells in a nonlinear membrane state, Southeastern Conference on Theoretical and Applied Mechanics, Auburn 1966.
2. BYCHAWSKI, Z., and H. KOPECKI: A spherical shell under elastic-plastic deformation and creep (in Polish), Rozprawy Inzynierskie, **15**, 2 (1967).
3. OLSZAK, W.: Les critères de transition en élasto-visco-plasticité, Bull. Acad. Polon. Sci., 1, **14**, 1966.

4. Kopecki, H.: Rheological problems of nonlinear deformations of rotational shells in membrane state (in Polish), Dissertation, Cracow 1967.
5. Bychawski, Z., and A. Fox: The constitutive equation for a complete nonlinear viscoelastic material, Acta Mechanica, Vienna (in print).
6. Olszak, W.: Sur la théorie des phénomènes visco-plastiques (in print), Milano.
7. Bychawski, Z.: On the applicability of the elastic analogy in the range of geometrically nonlinear theory of creep of circular membranes (in Polish), Rozprawy Inzynierskie, **13**, **3**, 1965.
8. Bychawski, Z.: Elastic analogue in the general case of a geometrically nonlinear membrane subjected to creep, Archiwum Mechaniki Stosowanej, **17**, 4 (1965).

Discussion

F. Odqvist: Your constitutive equations, connecting tensors of total creep deformation and stress have analytical form. Presumably you must have some additional condition to secure irreversibility of creep deformation?

Z. Bychawski: Your remark is very fundamental from the theoretical and practical point of view. As yet, no assumptions have been made in our theory as regards the problem of irreversibility. However, there are potential possibilities to secure irreversibility of creep deformation which are contained in the general form of the generalized creep function. We intend to consider this problem in a separate paper.

W. Nachbar: Was the assumption concerning very small rotations verified in the numerical calculations?

Z. Bychawski: There is an assumption about very small rotations of the membrane elements in our theory. However, it seems that there is no necessity of a verification of this assumption in the numerical calculations. The method of solution for a general case of a combined instantaneous and time-dependent deformation is founded on the basis of the perturbation method. The creep precess gives then only a small deviation from the instantaneous state for which our assumption is satisfied.

Plastic Instability of a Spherical Shell

By

F. A. Leckie

University Engineering Laboratories, Cambridge, England

Introduction

The problem which is to be studied is that of a hemispherical shell subjected to an inward radial force applied through a rigid boss (Fig. 1). This problem has been studied by ASHWELL for shells which suffer large

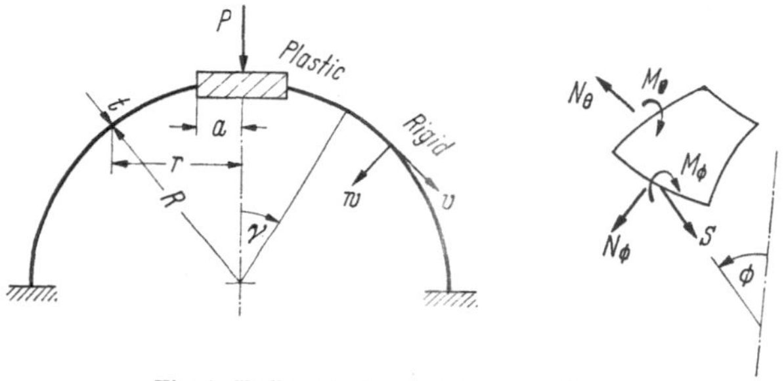

Fig. 1. Shell geometry and stress resultants.

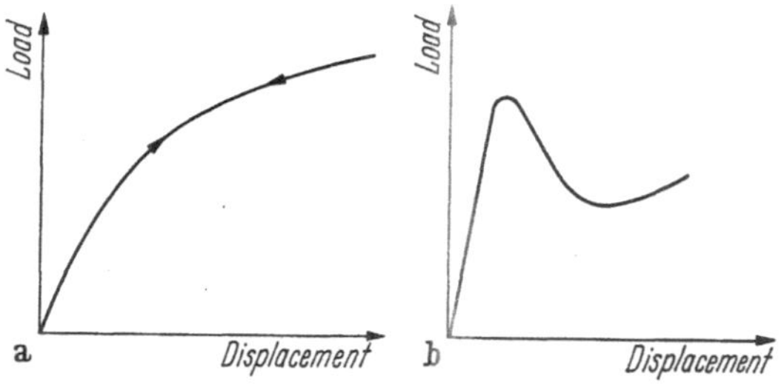

Fig. 2. a) Load-displacement diagram for elastic shell; b) Load-displacement diagram for plastic shell.

displacements but which remain elastic. The theory of Ashwells predicts a load displacement curve which is monotonically increasing (Fig. 2a), and this prediction has been confirmed experimentally by ASHWELL and

EVAN-IWANOWSKI [2]. In both series of experiments the shells tested were quite thin, the radius to thickness ratio being in excess of 200.

Experiments performed by the author on metallic shells with radius to thickness ratio of approximately 100, revealed a completely different behavior with a snap through action (Fig. 2b) characteristic of many elastic shell instabilities [3]. In these tests the shells suffered considerable plastic deformation and it seems probable that the plastic deformations account for this difference in behavior.

Since plastic deformations do occur, it would seem reasonable to perform a rigid-plastic analysis. The analysis is done in the same spirit as these analysis which make use of KOITER's [4] theory of elastic instability. In the elastic analysis, the buckling load and the corresponding slope of the post buckling curve are determined. In a similar way, the rigid-plastic analysis determines the yield load and the corresponding slope of the load deflection curve.

2. Rigid Plastic Analysis

Equilibrium Equations and Kinematics. The sign convention for the stress resultants acting on the shell are shown in Fig. 1. In dimensionless form the equilibrium equations are [5];

$$\frac{\partial}{\partial \varphi}(n_\varphi \sin\varphi) - n_\Theta \cos\varphi = \Delta \sin\varphi$$

$$s \cos\varphi \sin\varphi + n_\varphi \sin^2\varphi + \bar{p} h = 0$$

$$\frac{\partial}{\partial \varphi}(m_\varphi \sin\varphi) - m_\Theta \cos\varphi = \frac{s}{h} \sin\varphi \qquad (1\text{a—c})$$

where

$$\bar{p} = \frac{P}{2\Pi M_0} \qquad M_0 = \frac{\sigma_0 t^2}{4} \qquad N_0 = \sigma_0 t$$

$$n_\varphi = \frac{N_\varphi}{N_0} \qquad n_\Theta = \frac{N_\Theta}{N_0} \qquad m_\varphi = \frac{M_\varphi}{M_0} \qquad m_\Theta = \frac{M_\Theta}{M_0}$$

$$s = \frac{S}{N_0} \qquad h = \frac{M_0}{R N_0} = \frac{t}{4R}$$

and

σ_0 is the yield stress of the material in simple tension.

The components of displacement are V and W as shown in Fig. 1. Introducing

$$\dot{v} = \frac{\dot{V}}{R} \qquad \text{and} \qquad \dot{w} = \frac{\dot{W}}{R}$$

the strain and curvature rates are;

$$\dot{e}_\Theta = \dot{v}\cot\varphi - \dot{w} \qquad \dot{e}_\varphi = \frac{\partial \dot{v}}{\partial\varphi} - w$$

$$\dot{\varkappa}_\Theta = -h\cot\varphi\left(\dot{v} + \frac{\partial\dot{w}}{\partial\varphi}\right) \qquad \dot{\varkappa}_\varphi = -h\frac{\partial}{\partial\varphi}\left(\dot{v} + \frac{\partial\dot{w}}{\partial\varphi}\right) \quad (2\text{a}-\text{d})$$

where the dot represents partial derivations w. r. t. time.

It is noted in passing that $\overline{p} = 1$ is the yield load for a plate subjected to a concentrated load, and that the geometric parameter h is small compared to unity.

The Yield Surface. Because of its simplicity, the two-moment limited interaction yield surface proposed by Hodge [6] is to be used. The limited interaction yield surface is obtained by assuming that interaction accurs between direct forces n_φ and n_Θ, and between moments m_φ and m_Θ, but that no interaction occurs between direct forces and moments. The projections of the yield surface on the force and moment planes are shown in Fig. 3.

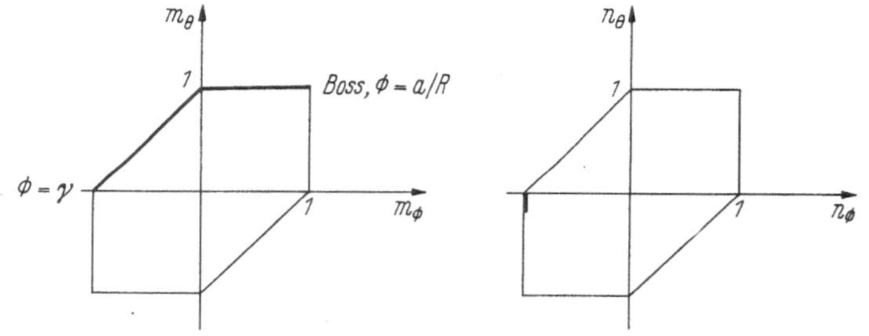

Fig. 3. Yield surfaces.

It has been shown by Hodge [6] that for plastic deformation to take place, it is necessary that the stresses lie on the boundary of the projection of the yield surface on the force plane, although they may lie within the boundary of the projection on the moment plane. However, when $n_\varphi = \pm 1$, it is readily shown that plastic deformation can only occur when the stresses also lie on the boundary of the projection on the moment plane.

Conditions at a Rigid-Plastic Boundary. The experimental work indicates that when the shell is subjected to a radial load only a small local region becomes plastic. Consequently, the position ($\varphi = \gamma$) of the boundary, separating the plastic from the rigid region, is one of the unknowns of the problem.

At a point in the plastic zone a small angle $\Delta\varphi$ from the boundary let the velocities be

$$\dot{v} = \dot{v}' \qquad w = \dot{w}' = 0 \qquad \frac{\partial \dot{w}}{\partial \varphi} = \frac{\partial \dot{w}'}{\partial \varphi}$$

while at the rigid region

$$\dot{v} = \dot{w} = \frac{\partial \dot{w}}{\partial \varphi} = 0 .$$

Following HODGE [5] the average strain rate vector components at $\varphi = \gamma$ are,

$$\dot{e}_\Theta = \dot{v}' \cot\gamma \qquad\qquad \dot{e}_\varphi = -\frac{\dot{v}'}{\Delta\varphi}$$

$$\dot{\varkappa}_\Theta = -h \cot\gamma \left(\dot{v}' + \frac{\partial \dot{w}'}{\partial \varphi}\right) \qquad \dot{\varkappa}_\varphi = \frac{h}{\Delta\varphi}\left(\dot{v}' + \frac{\partial \dot{w}'}{\partial \varphi}\right)$$

so that as $\Delta\varphi \to 0$ the strain rate vector $\dot{q}$ becomes

$$\begin{aligned} \dot{q} &= \left\{\dot{e}_\Theta, \dot{e}_\varphi, \dot{\varkappa}_\Theta, \dot{\varkappa}_\varphi\right\} \\ &= \frac{1}{\Delta\varphi}\left\{\dot{v}' \cot\gamma\, \Delta\varphi,\ -\dot{v}',\ -h \cot\gamma \left(\dot{v}' + \frac{\partial \dot{w}'}{\partial \varphi}\right) \Delta\varphi,\ h\left(\dot{v}' + \frac{\partial \dot{w}'}{\partial \varphi}\right)\right\}. \end{aligned}$$

Hence as $\Delta\varphi \to 0$ the direction of the strain rate vector is defined by

$$\dot{e}_\Theta = 0 \quad \text{and} \quad \dot{\varkappa}_\Theta = 0$$

and it follows on account of the normality rule that

$$n_\varphi = \pm 1 \quad \text{and} \quad m_\varphi = \pm 1 .$$

In order to determine the position γ of the rigid plastic boundary, certain assumptions are made about the stress distribution in the plastic region. It will be found that these assumptions are in agreement with the solution determined in the next section.

It would seem reasonable to expect that in the plastic region at $\varphi = \gamma$,

$$n_\varphi = -1 \quad m_\varphi = -1 \quad \text{and} \quad m_\Theta = 0 .$$

If plastic yielding is to be avoided in the rigid region, it follows that n_φ and m_φ must get no smaller and consequently

$$\frac{\partial n_\varphi}{\partial \varphi} \geq 0 \qquad \frac{\partial m_\varphi}{\partial \varphi} \geq 0$$

in the rigid region at $\varphi = \gamma$.

From equations (1b) and (1c)

$$\sin\varphi\frac{\partial m_\varphi}{\partial\varphi}+\cos\varphi(m_\varphi-m_\Theta)=\frac{s}{h}\sin\varphi=-\frac{1}{\cos\varphi}\left[\overline{p}+n_\varphi\frac{\sin^2\varphi}{h}\right].$$

Assuming that γ is small, and using the fact that at $\varphi=\gamma$

$$m_\varphi=-1 \quad \text{and} \quad n_\varphi=-1$$

$$\gamma\frac{\partial m_\varphi}{\partial\varphi}=1+m_\Theta+\frac{\gamma^2}{h}-\overline{p}.$$

At the edge of the plastic region it has been assumed that

$$m_\varphi=-1 \quad \text{and} \quad m_\Theta=0$$

so that

$$\gamma\frac{\partial m_\varphi}{\partial\varphi}\bigg|_P=-(\overline{p}-1)+\frac{\gamma^2}{h}$$

and since in the plastic region m_φ decreases from $+1$ to -1 as φ increases from α to γ it seems reasonable to except that $\frac{\partial m_\varphi}{\partial\varphi}\leq 0$ throughout this region.

In the rigid region at $\varphi=\gamma$

$$\gamma\frac{\partial m_\varphi}{\partial\varphi}\bigg|_R=-(\overline{p}-1)+\frac{\gamma^2}{h}+m_{\Theta R}=\gamma\frac{\partial m_\varphi}{\partial\varphi}\bigg|_P+m_{\Theta R}.$$

In crossing over the plastic-rigid boundary discontiniuty in m_Θ is permissable, but for yielding to be avoided $m_{\Theta R}$ must be -ive. Consequently, in order to ensure the condition

$$\frac{\partial m_\varphi}{\partial\varphi}\bigg|_R\geq 0$$

it follows that,

$$\frac{\partial m_\varphi}{\partial\varphi}\bigg|_P=0 \quad \text{and} \quad m_{\Theta R}=0.$$

The condition defining the plastic-rigid boundary is therefore

$$\frac{\partial m_\varphi}{\partial\varphi}\bigg|_P=\frac{\partial m_\varphi}{\partial\varphi}\bigg|_R=0$$

or

$$\frac{\gamma^2}{h}=\overline{p}-1. \tag{3a—b}$$

A similar argument applied to n_φ produces a less severe criterion.

The Stress Fields. In this analysis it is assumed that φ is so small that φ^2 can be neglected by comparison with unity. It seems reasonable to expect that $n_\varphi = -1$ throughout the plastic region. As a result of the discussion on the yield surface, it follows that the stresses must then lie on the boundary of the projection of the yield surface on the moment plane. Consequently, it would seem reasonable to use the solution of the corresponding plate problem [5] as a guide. Pursuing this line leads naturally to the following assumed stress distribution:

Region 1 for $\alpha < \varphi < \beta$

$$m_\Theta = 1 \quad \text{and} \quad n_\varphi = -1 .$$

Substituting in equations (1) and integrating yields

$$\begin{aligned} n_\varphi &= -1 & n_\Theta &= (-1 + \overline{p}h)(1 + \varphi^2/2) \\ m_\Theta &= 1 & m_\varphi &= (1 - \overline{p}) + \frac{\varphi^2}{3h} + \frac{A}{\varphi} . \end{aligned} \tag{4}$$

Region 2 for $\beta < \varphi < \gamma$

$$m_\Theta - m_\varphi = 1 \quad \text{and} \quad n_\varphi = -1$$

which yield on integration of equations (1)

$$\begin{aligned} m_\varphi &= (1 - \overline{p}) \ln \varphi + \varphi^2/2h + B \qquad m_\Theta = 1 + m_\varphi \\ n_\varphi &= -1 \qquad n_\Theta = (-1 + \overline{p}h)(1 + \varphi^2/2) . \end{aligned} \tag{5}$$

A and B are constants of integration. Using the following conditions

$$\begin{aligned} \text{at} \quad \varphi &= \alpha & m_\varphi &= 1 \\ \varphi &= \beta & m_\varphi &= 0 \\ \varphi &= \gamma & m_\varphi &= -1 \end{aligned}$$

yields the equations

$$\begin{aligned} 1 &= (1 - \overline{p}) + \frac{\alpha^2}{3h} + \frac{A}{\alpha} \\ 0 &= (1 - \overline{p}) + \frac{\beta^2}{3h} + \frac{A}{\beta} \\ 0 &= (1 - \overline{p}) \ln \beta + \frac{\beta^2}{2h} + B \\ -1 &= (1 - \overline{p}) \ln \gamma + \frac{\gamma^2}{2h} + B \end{aligned} \tag{6}$$

which together with the previous condition

$$\frac{\gamma^2}{h} = (\overline{p} - 1) \tag{3 b}$$

provide the five equations for the five unknowns $\overline{p}$, β, γ, A and B.

If we let

$$\gamma = x\beta \quad \text{and} \quad \gamma = z\alpha \tag{7}$$

then equations (6) can be solved to give

$$\overline{p} = \frac{\ln x + 1/2x^2 + 1/2}{\ln x + 1/2x^2 - 1/2}$$

and

$$\frac{1}{y} = (\overline{p} - 1)\left(\frac{1}{x} - \frac{1}{y}\right) + \frac{\overline{p} - 1}{3}\left(\frac{1}{y^3} - \frac{1}{x^3}\right). \tag{8a—b}$$

Choosing a value for x it is then possible to obtain the corresponding values of $\overline{p}$, y, γ^2/h and α^2/h. The results of this calculation are shown in graphical form in Fig. 4, in which $\overline{p}$ and $\acute{y}$ are plotted against $\varrho = \frac{a}{R}\sqrt{\frac{R}{T}}$.

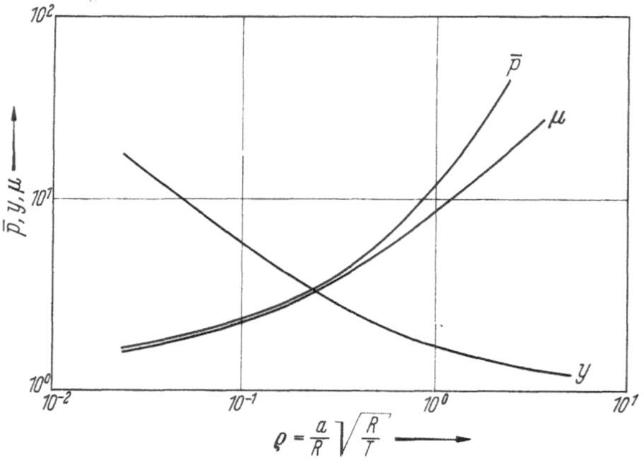

Fig. 4. Variation of $\overline{p}$, y and μ with p.

It is not difficult to show that the assumed stress field satisfies the assumptions made in the section on Conditions at a Rigid-Plastic Boundary, but it still remains to check that the corresponding velocity field is suitable. The velocity field is found by using the normality rule in conjunction with equations (2).

Region 1

From the normality rule

$$\dot{e}_\Theta = 0 \quad \text{and} \quad \dot{\varkappa}_\varphi = 0$$

and integrating the corresponding equation (2a) and (2c) gives

$$\dot{w} = C\varphi + D \qquad \dot{v} = C\varphi^2 + D\varphi$$

$$\dot{\chi} = \frac{\partial \dot{w}}{\partial \varphi} + \dot{v} = C.$$

Region 2

From the normality rule

$$\dot{e}_\Theta = 0 \quad \text{and} \quad \dot{\varkappa}_\varphi = -\dot{\varkappa}_\Theta$$

and integrating equations (2) gives

$$\dot{w} = E \ln\varphi + F \quad \dot{v} = E\varphi \ln\varphi + F\varphi \quad \dot{\chi} = E/\varphi.$$

The constants C, D, E and F are obtained by applying the conditions at

$\varphi = \alpha$	$\dot{w} = \dot{w}_0$ (given)
$\varphi = \beta$	$\dot{w}$ and $\dot{\chi}$ are compatible
$\varphi = \gamma$	$\dot{w} = 0$

(Satisfying compatibility of $\dot{w}$ and $\dot{\chi}$ at β automatically satisfied compatibility of $\dot{v}$).

Solving the resulting equations gives

$$C = -\frac{\dot{w}_0}{\beta[1 + \ln\gamma/\beta - \alpha/\beta]} \qquad D = \frac{\dot{w}_0(1 + \ln\gamma/\beta)}{[1 + \ln\gamma/\beta - \alpha/\beta]}$$

$$E = -\frac{\dot{w}_0}{[1 + \ln\gamma/\beta - \alpha/\beta]} \qquad F = \frac{\dot{w}_0 \ln\gamma}{[1 + \ln\gamma/\beta - \alpha/\beta]}. \tag{9}$$

Using these results it is not difficult to show that $\dot{\varkappa}_\Theta$ is everywhere positive and $\dot{e}_\varphi$ everywhere negative. Hence the velocity fields is everywhere compatible and consistent with the normality rule.

The Stress Distribution in the Rigid Region.

Before the present solution is complete a stress field must be found within the rigid region which nowhere violates yield.

Satisfactory stress distributions have been found for the ranges of $\bar{p}$ of (1) $0 < \bar{p} < 3.87$ and (2) $3.87 < \bar{p} < 16.2$. In order to save space only brief details of the stress distribution are given.

Range 1 ($0 < \overline{p} < 3.87$).

For this range three different stress regions have been assumed. These are

(i) $$\gamma < \varphi < \psi .$$

Assume $n_\Theta - n_\varphi = 1 - \varepsilon$, where ε is small compared to unity, and $m_\Theta = 0$. This angle ψ is defined by the condition $\varphi = \psi$ when $n_\varphi = -1/2$.

(ii) $$\psi < \varphi < \delta .$$

Assume $n_\Theta = 0$ and $m_\Theta = 0$. The angle δ is defined by the condition $\varphi = \delta$ when $s = 0$.

(iii) $$\delta < \varphi < \pi/2 .$$

Assume $s = 0$ and $m_\Theta = 0$.

After integration of the equilibrium equation (1) the stress profiles obtained are similar to those illustrated in Fig. 5a.

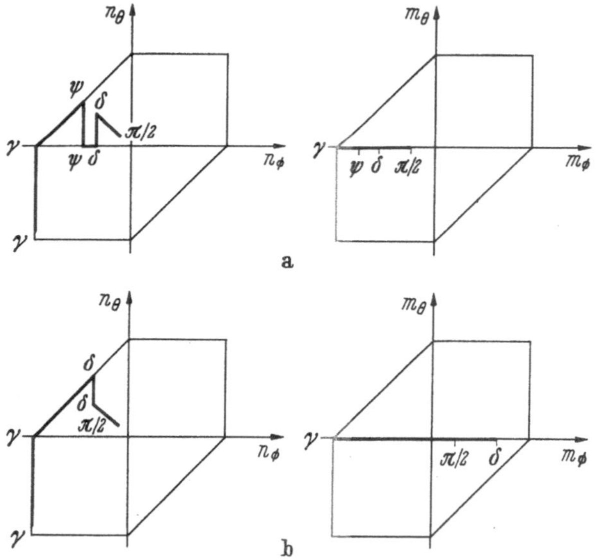

Fig. 5. Suitable stress profiles in the rigid region.

Range 2 ($3.87 < p < 16.2$).

For this range two stress regions are defined

(i) $$\gamma < \varphi < \delta .$$

Assume $n_\Theta - n_\Theta = 1 - \varepsilon$ and $m_\varphi = 0$ and $\varphi = \delta$ when $s = 0$

(ii) $$\delta < \varphi < \pi/2\,.$$

Assume $s = 0$ and $m_\Theta = 0$.

A typical stress profile is shown in Fig. 5b and it can be seen that for this range the limiting value of $\overline{p}$ is obtained when $m_\varphi = 1$ at $\varphi = \delta$.

Further improvements in raising the value of $\overline{p}$ should be possible by assuming other stress fields. Actually $\overline{p} = 16.2$ probably represents a reasonable limit of the present analysis in which it is assumed that φ^2 can be neglected by comparison with unity.

Assuming that $\gamma^2 = 1/50$ represents a reasonable maximum for φ^2 and recalling the relation of Eq. (3a)

$$\frac{\gamma^2}{h} = \overline{p} - 1$$

we readily find that

$$\frac{R}{t} < 190$$

which represents an extremely thin shell. Because of this, no further effort has been made to find higher values of $\overline{p}$.

3. The Slope of the Load/Deflection Curve at Yield

Having found the velocity and stress fields it is now possible to obtain the slope of the load-deflection curve at yield by making use of the integral relation obtained by BATTERMAN (7). This relation is obtained in a manner similar to the virtual work equation, except that the equilibrium equations are replaced by the corresponding rate equilibrium equation (i. e. the partial derivature w. r. t. time of the equilibrium equation) and the displacements by velocities.

In terms of the present convention, the integral relation is

$$\left[\dot{w}\left(\frac{rs}{R}\right)^{\cdot} + \dot{v}\left(\frac{rn_\varphi}{R}\right)^{\cdot} - \dot{\chi}h\left(\frac{rm_\varphi}{R}\right)^{\cdot} + \varphi\dot{\chi}(\dot{w}n_\varphi - \dot{v}s)\right]_\alpha^\gamma$$

$$= \int_\alpha^\gamma \left((n_\varphi + n_\Theta)\dot{e}_\varphi \dot{e}_\Theta + m_\Theta(\dot{\varkappa}_\Theta \dot{e}_\varphi + h\dot{\chi}^2) + m_\varphi \dot{\varkappa}_\varphi \dot{e}_\Theta + n_\varphi \dot{\chi}^2 - 2s\dot{e}_\varphi \dot{\chi}\right)\varphi\, d\varphi \tag{10}$$

where r is the radius indicated in Fig. 1.

With the stress and velocity fields available is it not a difficult matter to compute the R. H. S. of Eq. (10), but the computations involved in the L. H. S. require some explanation.

We note that

$$(rs)^{\cdot} = \dot{r}s + r\dot{s} \quad \text{etc.}$$

It is not difficult to obtain the relation (7)

$$\frac{\dot{r}}{r} = \dot{e}_\Theta$$

and since $\dot{e}_\Theta = 0$ throughout the shell, it follows that $\dot{r} = 0$.

If we now consider the moment at the junction of the plastic and rigid regions we have that the total increment of moment at the junction is

$$dm_\varphi = \frac{\partial m_\varphi}{\partial t} dt + \frac{\partial m_\varphi}{\partial \varphi} d\varphi .$$

Now $dm_\varphi = 0$ since $m_\varphi = -1$ at $\varphi = \gamma$. Also it was argued that $\frac{\partial m_\varphi}{\partial \varphi} = 0$ [cf. Eq. (3a)].

Consequently

$$\frac{\partial m_\varphi}{\partial t} = 0 .$$

Finally multiplication of Eq. 1(b) by R gives

$$s r \cos\varphi + n_\varphi r \sin\varphi + \bar{p} h R = 0$$

and performing the partial derivation w. r. t. time, and noting that $\dot{r} = 0$ we obtain

$$\dot{s} r \cos\varphi - s r \sin\varphi \dot{\varphi} + \dot{n}_\varphi r \sin\varphi + n_\varphi r \cos\varphi \dot{\varphi} + \dot{\bar{p}} h R = 0$$

from which s can be calculated.

With the evaluation of $\dot{s}$, all the information is now available for the computation of the integral which eventually yields

$$\dot{w}_0 \dot{\bar{p}} h = -\left(E^2 \ln \gamma/\beta + \frac{c^2}{2} (\beta^2 - \alpha^2)\right).$$

Substituting the values for E and C (Eq. 9) and using the relation of Eq. 7 yields the result,

$$\frac{d\bar{p}}{d\left(\frac{W}{t}\right)} = -\frac{4\left(\frac{1}{2}\left(1 - \frac{x^2}{y^2}\right) + \ln y\right)}{\left(1 + \ln x - \frac{x}{y}\right)^2} = -\mu . \tag{11}$$

A plot of μ is shown in Fig. 4.

4. Discussion on the Results of the Rigid Plastic Analysis

The most significant result of this rigid plastic analysis is that the slope of the load deflection curve at the collapse load is negative, and that the magnitudes of $\overline{p}$ and μ are approximately the same. This means the slope is such that for a displacement $W/t = 1$, the load is approximately zero (Fig. 6). Because the slope of the rigid plastic equilibrium path is so large the initial elastic deformation shell causes considerable modification. In fact, if it is assumed that the shell were initially purely elastic and then purely plastic, the load displacement curve would have the form illustrated in Fig. 6. Because of the effects of interaction between the elastic and plastic deformations, the actual curves are certain to be rounded off. Further analysis would be required to determine the precise shape of the equilibrium path, and it may be that, provided the general shape is correct, such detail is more readily obtained from experiment.

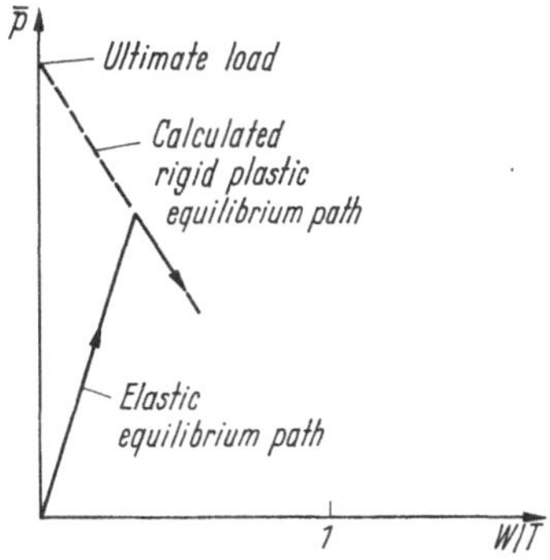

Fig. 6. Approximate equilibrium path.

The results shown in Fig. 4 also indicate that for a constant value of R/T, the yield load rises with the increase of the boss size parameter a/R. However, as $\overline{p}$ increases the value of y decreases which means that as the boss size increases, the region of plastic deformation becomes more localized to an area in the vicinity of the boss.

5. Experimental Programme

The snap through behavior was initially observed on 12″ diameter shells spun from aluminum sheets, but for controlled experiments these shells were unsatisfactory from two points of view. Firstly, the geometry was very variable, and secondly, since soft aluminum was used for spinning, it was extremely difficult to decide upon the appropriate value of the yield stress. As a result, it was decided to manufacture the shells from the solid. The material finally chosen was an aluminum alloy which has the stress strain curve shown in Fig. 7. It was decided to use a value of 41,000 lb/sq. in for the yield stress σ_0.

On the face of it would seem an extravagence to manufacture shells from the solid, but the method to be described has been developed to the stage that it is a simple, routine process and one which leads to reliable quantitative results and hence the need for fewer shells. The shells were manufactured in a milling machine, the solid metal being mounted on a base which is slowly rotated about the vertical axis OA (Fig. 8)). A fly cutter is mounted on the inclined axis OB which rotates quickly. The combined motion causes a spherical surface to be generated in the metal. The fly cutter is then advanced until the desired profile is obtained.

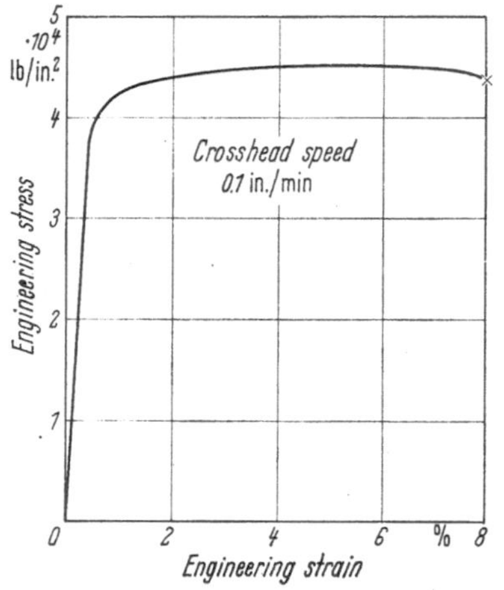

Fig. 7. Stress strain curve for the shell material.

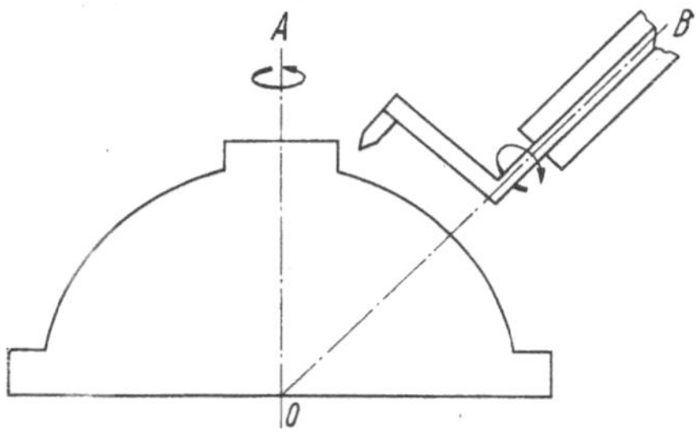

Fig. 8. Arrangement of the miller.

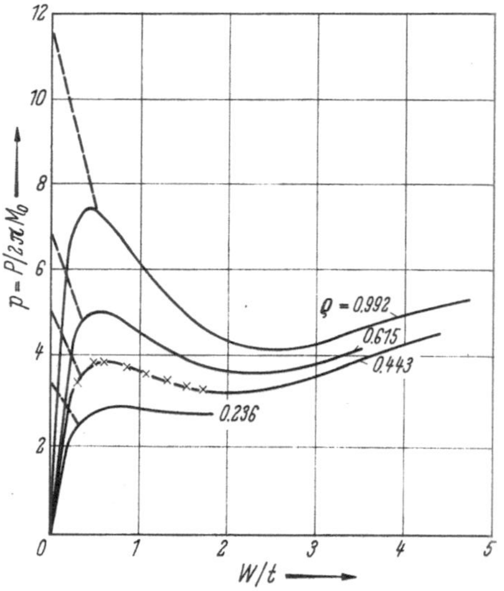

Fig. 9. Load-displacement results for boss-loaded shells.

In the present case, shells could be manufactured with a very small tolerance on thickness — better than 2% — and most of this was towards the shell base.

Five shells were made in this manner. They were of the same nominal radius and thickness 3 in. and 0.05 in. respectively. Four different boss sizes were chosen, two of the shells having the same boss size so that the repeatability could be checked. The shells were loaded through a screw driven cross head, and the load was measured by means of a load cell. The cross head displacement was also measured and in plotting the load/displacement curves due allowance was made for the load cell flexibility.

The load displacement curves obtained from the test are shown in Fig. 9. Almost perfect repeatability was obtained for the two shells with the same boss size, the crosses in Fig. 9 indicating the duplicate test.

The load displacement curves obtained from the tests all have the same characteristic shape. After an initial linear response, the curves continue to rise with decreasing slope until a maximum load is obtained, when the deflection is about half the shell thickness. The load then starts to fall and reaches a minimum value when the displacement is approximately twice the shell thickness. Thereafter, the load again rises. While the character of the curves remains the same. the load variation become more pronounced with increased values of ϱ. The shape of the curves indicate that if these shells were subjected to dead loading then snap through action would occur when the load reaches its maximum value. For the larger values of ϱ the snap through action would be very severe and causing displacements many times the shell thickness. Because of this behavior, any attempt to calculate the load-displacement curve for static loading would be somewhat wasteful.

The predictions of the rigid plastic theory for the values of the ultimate load and for the initial shape of the load displacement are also shown in Fig. 9. As predicted previously in the section on the rigid plastic analysis, the values of the ultimate load are in excess of the maximum load achieved by the shell. Better results are obtained, however, if following the procedure shown in Fig. 6, the maximum load is assumed to be given by the intersection of the pure elastic and rigid plastic response. This procedure gives values in excess of the true maximum load for $\varrho = .992$, while for $\varrho = .236$, the predicted value is slightly less than the experimental (Fig. 9). In order to determine the maximum load more exactly, it is clear that some sophisticated analysis would be required, but the predictions of the simple elastic, rigid-plastic interaction, are sufficiently close to the experimental values to believe that this simplified mechanism approximates the maximum load rather well.

6. The Effect of Imperfections

Shell imperfections appear to play an equivalent role in the rigid plastic analysis that "major" and "minor" imperfections (8) play in elastic instability.

It is clear that imperfections which have the same form as the plastic deformation shall cause a reduction in the collapse load of $\mu\, \delta/t$ where δ is the magnitude of the imperfection at the boss. Such an imperfection can be described as a "major imperfection". The rigid plastic equilibrium path for a shell with such an imperfection is obtained by displacing the origin by an amount δ/t (Fig. 10). It might be expected that the actual equilibrium path would be modified as illustrated in Fig. 10. This speculation has not been checked experimentally.

The effect of another type of imperfection was studied by completing a rigid plastic analysis for a spherical shell with a small dimple of reverse curvature R (Fig. 11). The analysis revealed that the collapse

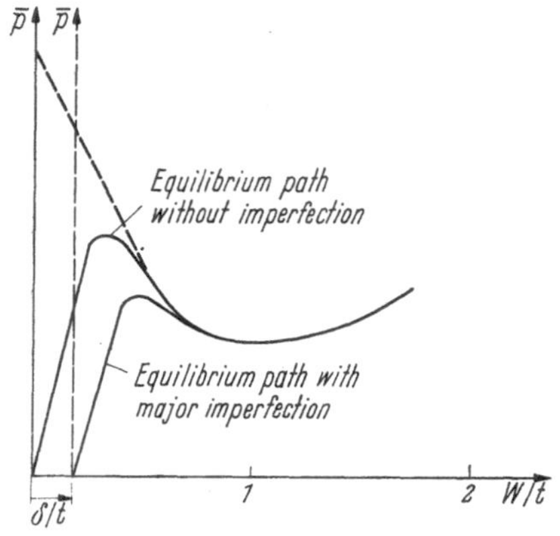

Fig. 10. Effect of major imperfection.

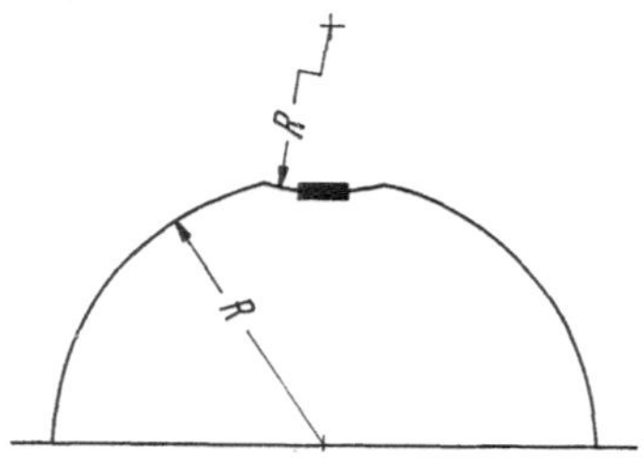

Fig. 11. Spherical shell with minor imperfection.

load and the slope at collapse are unaffected provided the radius of the dimple is less than γr. Such an imperfection can be regarded as a "minor" imperfection.

7. Conclusions

It has been demonstrated by theory and experiment that when a metal shell is loaded by a load applied through a rigid boss, snap-through action can be expected, and that at the end of snap-through the extent of the permanent displacement is likely to be several times the shell thickness. Unfortunately, because of this action, designs based on the ultimate load concept are likely to be quite unsafe. Under these circumstances a small displacement elastic analysis would be more appropriate. Such an analysis would be used to predict the load at which yielding would first occur. The loads so calculated would be sufficiently conservative to offset the possibility of catastrophic snap-through.

References

1. ASHWELL, D. G.: On the large deflection of a spherical shell with an inward pointing load, Proc. of the IUTAM Symposium on The Theory of Thin Elastic Shells, Delft, 1959.
2. EVAN-IWANOWSKI, R. M., H. S. CHENG and T. C. LOO: Experimental investigation of deformations and stability of spherical shells subjected to a concentrated load at the apex, Proc. of the 4th US Cong. of Appl. Mech., 1962.

3. BUDIANSKY, B., and J. W. HUTCHINSON: A survey of some buckling problems, AIAA Journal, **4**, No. 9 (September 1966).
4. KOITER, W. T.: Elastic stability and post-buckling behavior, Non-linear Problems, ed. R. E. Langer, Madison, Wisconsin: University of Wisconsin Press 1963.
5. HODGE, P. G., Jr.: Limit analysis of rotationally symmetric plates and shells, Prentice-Hall 1963.
6. HODGE, P. L., Jr.: Yield conditions for rotationally symmetric shells under axisymmetric loading, J. Appl. Mech., **27**, 323—331 (1960).
7. BATTERMAN, S. C.: Load-deformation behavior of shells of revolution, Proc. ASCE Eng. Mech. Div., December, 1964.
8. ROORDA, J.: The buckling behavior of imperfect structural systems, J. Mech. and Phys. of Solids, **13** (1965).

Discussion

A. N. SHERBOURNE: Professor LECKIE is to be congratulated on providing a simple engineering approach to the actual load carrying capacity of a shell loaded centrally through a rigid boss.

The philosophy bounds the real behaviour by an elastic loading line and a rigid-plastic unloading line which is specified only as to ultimate load and initial negative slope. In establishing the unloading characteristics, uniqueness is satisfied through the upper and lower bound approaches involving virtual work and static equilibrium respectively. In establishing the lower bound however, a boundary condition ($m_\varphi = n_\varphi = -1$) is used at the rigid/plastic interface. Is this valid *i*) in view of the uncoupling of the moment and axial force yield surfaces und *ii*) in view of the ohysical inability of a section to sustain simultaneously the fully plastic values of moment and axial force.

The experiments are also premised upon ductility of the material. Is this correct-in view of work hardening arizing from machining operations?

F. LECKIE: I agree with this. All I can say to this is that I probably should take more complicated yield conditions. I think my reason was, that I was not inclined to make it more complicated. Secondly, I was in the learning process, when I was doing this analysis and thirdly, it seemed to me, that I was assuming a shell that was stronger than the actual one, and that the predictions of snap through would be conservative.

C. R. STEELE: I also thank you for a very enjoyable presentation. Your result that the plastic region becomes small for large boss size is of particular interest to me. I have briefly looked at the possibilities of an elastic-plastic analysis and found a similar result for such boundaries in the "steep" region. The small size of the plastic zone, in which all the nonlinearities occur, provides a possibility for a simplifying approximation since the curvature is large but the deflection is nearly constant in this zone. Thus for at least a class of elastic-plastic shell problems it appears that a reasonably simple solution can be obtained.

Plastic Analysis of Shallow Spherical Shells at Moderately Large Deflections

By

M. Duszek

Polish Academy of Sciences, Warsaw, Poland

Abstract. Post yield behaviour of a shallow, spherical, rigid-plastic shell is investigated. Employing the appropriate set of shell equations relating to moderately large tranverse deflection, the load-deflection relationship is established and the transition from bending to purely membrane response is discussed. The analysis discloses significant differences between the actual load carrying capacity and that yielded by the limit analysis theory. Particular examples are considered.

1. Introduction

The load-carrying capacity of axi-symmetric rigid, perfectly plastic shells was widely studied [1—7]. All these analyses, however, involve an assumption that deflections of shells remain small as compared with the shell thickness. The plastic motion which commences at the yield point load is assumed to continue without any change in the load intensity. On the other hand, it is well known that experimentally recorded values of the load-carrying capacity of shells may differ considerably from those computed assuming that changes in the shell geometry are negligible.

The plastic analysis of circular plates [8—12] and cylindrical shells [13—16] at large deflections shows that the plastic deformation process leads to important changes in the stress field. The membrane forces develop and effectively produce changes in the actual carrying capacity.

In the present paper, the incremental theory of plasticity for infinitesimal strains is used. The solutions sought for should satisfy equilibrium equations, strain-displacement relations accounting for moderately large displacements, a yield condition and physical relations which for the incremental theory of plasticity are those of the associated flow rule.

2. Formulation of the Problem

Consider a shallow spherical shell loaded by a uniformly distributed internal pressure P. The shell of radius A is of constant thickness $2H$ (Fig. 1).

We attempt to obtain the expressions specifying the interdependence between the applied load and the "prevailing" displacement.

It is assumed that the shell material is rigid, perfectly plastic, obeying the Tresca yield conditions and the associated flow rule. The yield surface employed is that of uniform shells.

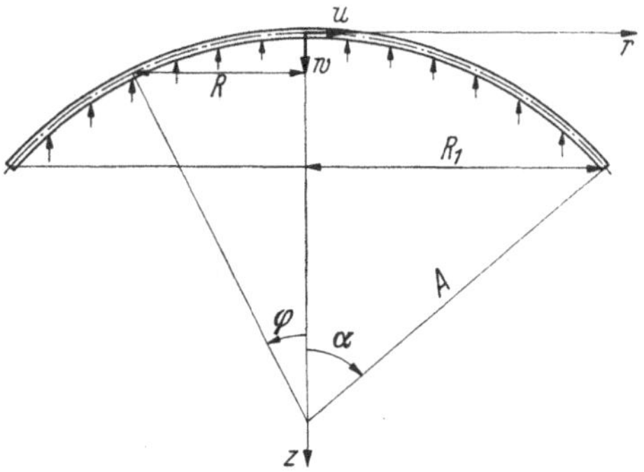

Fig. 1. The geometry of a shallow spherical shell.

For further convenience we introduce the following dimensionless quantities:

$$n_\varphi = \frac{N_\varphi}{N_0}, \quad n_\Theta = \frac{N_\Theta}{N_0}, \quad m_\varphi = \frac{M_\varphi}{M_0}, \quad m_\Theta = \frac{M_\Theta}{M_0}, \quad p = \frac{PA}{N_0},$$

$$w = \frac{W}{A}, \quad u = \frac{U}{A}, \quad h = \frac{H}{2A}, \quad c = \sin\alpha = \frac{R_1}{A}, \quad r = \sin\varphi = \frac{R}{A},$$

$$\varkappa_\varphi = \frac{1}{2} H K_\varphi, \quad \varkappa_\Theta = \frac{1}{2} H K_\Theta, \quad d_{\text{ext}} = \frac{D_{\text{ext}}}{2\pi N_0 A^2}, \quad d_{\text{int}} = \frac{D_{\text{int}}}{2\pi N_0 A^2} \tag{2.1}$$

where $N_0 = 2\sigma_0 H$ and $M_0 = \sigma_0 H^2$ are respectively the fully plastic membrane force and the yield moment for a uni-axial state of stress, and σ_0 denotes the yield stress in tension, K_φ, K_Θ stand for the dimensional rates of curvature and D is the dissipation. The meaning of remaining quantities is evident from Fig. 1.

3. Strain-Displacement Relations

In a derivation of kinematical relations we follow the analysis presented by DONELL [17] who obtained a general form of relations between strains and displacements, applicable to large strains (of the order of unity) and large deflections (of the order of a radius of curvature of shells). Much attention has been devoted to the analysis of thin shells under small strains (small in comparison with unity) and moderately large transverse deflections (normal to the shell middle-surface) but small displacement in the remaining directions.

By moderately large deflections we understand displacement of the order of shell thickness yet small compared to dimensions such as the radius of curvature or the length of a displacement wave. Further simplifications are obtained for shells which remain axially-symmetric throughout the deformation process. In particular for spherical shells the strain-displacement relations have the form

$$\varepsilon_\varphi = u_\varphi' - w_n + \frac{1}{2}(w_n')^2 - (w_n'' + w_n)\frac{z}{A}$$

$$\varepsilon_\Theta = u_\varphi \operatorname{ctg}\varphi - w_n - (w_n + w_n' \operatorname{ctg}\varphi)\frac{z}{A} \tag{3.1}$$

where ε_φ and ε_Θ denote principal components of the Almansi strain tensor at an arbitrary point located at a distance z from the middle surface, u_φ and w_n are respectively displacements in the meridional and the normal directions and a prime denotes differentiation with respect to φ.

The simplifications were introduced on the basis of classification according to the "order of magnitude", shown in Table 1, based essentially on the analysis due to Donell [17], (where $\varepsilon \ll 1$)

Table 1

Order of magnitude	1	$\varepsilon^{1/2}$	ε	ε^2	ε^3
Classified quantities	$\frac{H}{W}$	c	h, w, w', w''	u, u', u''	$w'u$, hu'

In the equations employed in the present paper the terms up to the order ε^2 have been retained. The above classification applies to thin shells for moderately large vertical deflections but small tangential deflections and small strains. These assumptions are valid for the cases of technical interest.

Expressing for shallow shells the components of the displacement vector in a cylindrical coordinate system (r, z, Θ) and neglecting terms of the order higher than ε^2, the strain-displacement relations for the shell middle surface take the following form:

$$\lambda_\varphi = u_r' + w_z' r + \frac{1}{2}(w_z')^2, \qquad \varkappa_\varphi = -h w_z'',$$
$$\lambda_\Theta = \frac{u_r}{r}, \qquad \varkappa_\Theta = -\frac{h}{r} w_z'. \tag{3.2}$$

Now a prime indicates differentiation with respect to the radius. As $\frac{d}{d\varphi} = \cos\varphi \frac{d}{dr}$, and since on the account that for shallow shells $\cos\varphi \approx 1$, we eventually have $\frac{d}{d\varphi} \approx \frac{d}{dr}$. The differentiation of (3.2) with respect to time provides the following expressions for the strain rates:

$$\dot\lambda_\varphi = \dot u_r' + \dot w_z' r + w_z' \dot w_z', \qquad \dot\varkappa_\varphi = -h \dot w_z'',$$
$$\dot\lambda_\Theta = \frac{\dot u_r}{r}, \qquad \dot\varkappa_\Theta = -\frac{h \dot w_r'}{r}. \tag{3.3}$$

In the limiting case $\alpha \to 0$ and $A \sin\alpha \to R_1$, equations (3.2) describe the kinematics of a circular plate radius R_1

$$\lambda_\varphi = u_r' + \frac{1}{2}(w_z')^2, \qquad \varkappa_\varphi = -h w_z'',$$
$$\lambda_\Theta = \frac{u_r}{r}, \qquad \varkappa_\Theta = -\frac{h w_z'}{r}. \tag{3.4}$$

Equations (3.4) coincide with the strain-displacement relations used in the theory of thin circular plates at moderately large deflections [9—12].

4. Equation of Equilibrium

Equations of equilibrium for spherical shells at large deflections could be derived in a similar way as it is usually done, i. e. by considering equilibrium of a deformed shell element in the undeformed coordinate system and neglecting the terms of higher order. However the above procedure might lead to the results inconsistent with the principle of virtual work on which proofs of limit analysis theorems are based. Therefore, we shall derive equations of equilibrium starting from the latter principle

$$D_{\text{ext}} = D_{\text{int}}$$

where D_{ext} and D_{int} denote respectively the rate of dissipation of energy due to external and internal forces (cf. [6]). According to the definition (2.1) the above equation in dimensionless quantities reads

$$d_{\text{ext}} = d_{\text{int}} \tag{4.1}$$

where explicitly

$$d_{\text{int}} = \int (n_\Theta \dot{\lambda}_\Theta + n_\varphi \dot{\lambda}_\varphi + m_\Theta \dot{\varkappa}_\Theta + m_\varphi \dot{\varkappa}_\varphi)\, r\, dr \tag{4.2}$$

$$d_{\text{ext}} = \int p \dot{w}_z r\, dr + [\overline{n}_\varphi \dot{u}_\varphi + \overline{q}_\varphi \dot{w}_n + \overline{m}_\varphi \dot{\psi}] r \tag{4.3}$$

and $\dot{\psi} = -h\dot{w}_n'$. Quantities $\overline{n}_\varphi$, $\overline{q}_\varphi$, $\overline{m}_\varphi$ denote dimensionless forces on the edge of the shell. Substitution of (3.3) into (4.2) results in

$$d_{\text{int}} = \int [n_\Theta \dot{u}_r + n_\varphi r\, (\dot{u}_r' + \dot{w}_z' r + w_z' \dot{w}_z') - h m_\Theta \dot{w}_z' - h m_\varphi r \dot{w}_z''] \, dr.$$

Integrating by parts and defining

$$h[(r m_\varphi)' - m_\Theta] = r q_\varphi, \tag{4.4}$$

we obtain the following expression for the dimensionless rate of dissipation of internal energy:

$$\begin{aligned} d_{\text{int}} = \int \{ [n_\Theta - (n_\varphi r)'] \, \dot{u}_r - [(n_\varphi r)' \, (r + w_z') + n_\varphi r\, (1 + w_z'') + \\ + (r q_\varphi)'] \, \dot{w}_z \} \, dr + n_\varphi r \dot{u}_r + [n_\varphi r (r + w_z') + r q_\varphi] \, \dot{w}_z - \\ - h m_\varphi r \dot{w}_z' . \end{aligned} \tag{4.5}$$

Substituting (4.3) and (4.5) into (4.1) we get

$$\begin{aligned} \int \{ [n_\Theta - (n_\varphi r)'] \, \dot{u}_r - [(n_\varphi r)' \, (r + w_z') + n_\varphi r\, (1 + w_z'') + \\ + (q_\varphi r)' + r p] \dot{w}_z \} \, dr + \{ (n_\varphi - \overline{n}_\varphi) \, r \dot{u}_r + [(n_\varphi - \overline{n}_\varphi) \, (r + w_z') + \\ + (q_\varphi - \overline{q}_\varphi) \, r \dot{w}_z + (m_\varphi - \overline{m}_\varphi) \, r \dot{\psi}] = 0. \end{aligned} \tag{4.6}$$

For the Eq. (4.6) to be satisfied for all continuous values of $\dot{u}_r$ and $\dot{w}_z$, the terms in brackets appearing in the integrand should vanish,

$$\begin{aligned} n_\Theta - (n_\varphi r)' &= 0 \\ (r q_\varphi)' + \big(r n_\varphi \, (r + w_z')\big)' + r p &= 0 \end{aligned} \tag{4.7}$$

and n_φ, q_φ and m_φ should be continuous on the shell edge. Eqs. (4.4) and (4.7) constitute the sought system of equilibrium equations.

Eliminating the shearing force, we finally obtain

$$\begin{aligned} (r n_\varphi)' - n_\Theta &= 0 \\ h \big((r m_\varphi)' - m_\Theta\big)' + \big(r n_\varphi \, (r + w_z')\big)' + r p &= 0 . \end{aligned} \tag{4.8}$$

A limit transition to the plate with radius R_1 is obtained by putting $\alpha \to 0$ and $A \sin\alpha \to R_1$. Then Eqs. (4.8) yield

$$\begin{aligned} (r n_\varphi)' - n_\Theta &= 0 \\ h \big((r m_\varphi)' - m_\Theta\big)' + (r n_\varphi w_z')' + r q^* &= 0 \end{aligned} \tag{4.9}$$

where dimensionless quantities r and w are related to the plate radius R_1 and q^* denotes $q^* = \frac{p R_1^2}{6 \sigma_0 H^2}$. Eqs. (4.9) are identical with the most commonly used equations of equilibrium for circular plates at moderately large deflections [9—12].

5. Yield Surface

The general form of the Tresca yield condition for uniform shells, derived by Onat and Prager [1] in terms of three parameters

$$p = -\frac{\dot{\lambda}_\varphi}{4\dot{\varkappa}_\varphi}, \qquad r = -\frac{\dot{\lambda}_\Theta}{4\dot{\varkappa}_\Theta}, \qquad q = -\frac{1}{4}\,\frac{\dot{\lambda}_\varphi + \dot{\lambda}_\Theta}{\dot{\varkappa}_\varphi + \dot{\varkappa}_\Theta} \tag{5.1}$$

is presented in Tables 2 and 3.

Table 2

Mean para-meter	Hyper surface	Generalized stress			
		n_φ	n_Θ	m_φ	m_Θ
p	$G_\varphi^\pm$	$\mp(p+q)$	$\mp(q-r)$	$\pm 1 \mp 2(p^2+q^2)$	$\pm 2(r^2-q^2)$
q	$G_{\Theta\varphi}^\pm$	$\mp(p+q)$	$\mp(q+r)$	$\pm 1 \mp 2(p^2+q^2)$	$\pm 1 \mp 2(q^2+r^2)$
r	$G_\Theta^\pm$	$\mp(q-r)$	$\mp(q+r)$	$\pm 2(p^2-q^2)$	$\pm 1 \mp 2(q^2+r^2)$

Table 3

	Hyper-surface	Yield condition	Strain rates
for $q=r$	$H_\Theta^\pm$	$m_\Theta = \pm(1-n_\Theta^2)$	$\dot{\lambda}_\Theta : \dot{\varkappa}_\Theta = \pm 2 n_\Theta, \ \dot{\lambda}_\varphi = \dot{\varkappa}_\varphi = 0$
for $p=q$	$H_\varphi^\pm$	$m_\varphi = \pm(1-n_\varphi^2)$	$\dot{\lambda}_\varphi : \dot{\varkappa}_\varphi = \pm 2 n_\varphi, \ \dot{\lambda}_\Theta = \dot{\varkappa}_\Theta = 0$
for $p=r$	$H_{\Theta\varphi}^\pm$	$m_\Theta - m_\varphi = \pm[1-(n_\Theta - n_\varphi)^2]$	$\dot{\lambda}_\Theta : \dot{\varkappa}_\Theta = \pm 2(n_\Theta - n_\varphi), \ \dot{\lambda}_\varphi = -\dot{\lambda}_\Theta, \ \dot{\varkappa}_\varphi = -\dot{\varkappa}_\Theta$

The generalized strain rate vector is expressed as

$$q = [\dot{\lambda}_\varphi, \dot{\lambda}_\Theta, \dot{\varkappa}_\varphi, \dot{\varkappa}_\Theta] = \mu[-4p(q-r), -4r(p-q), q-r, p-q]. \tag{5.2}$$

If parameters p, q and r are distinct and lie within the interval $(-1/2, +1/2)$ then the intermediate parameter is that contained between remaining parameters. If one of the parameters is greater than $1/2$ or smaller than $1/2$, its value should be replaced respectively by $1/2$ or $-1/2$.

A stress profile represented by an apex of the Tresca hexagon corresponds to distinct values of p, q and r, and then Table 2 applies. When arbitrary two parameters are equal then the stress profile is represented by a side of the Tresca hexagon.

The yield hypersurface is described parametrically by the system of six relations (Table 2) representing three pairs of non-cylindrical smooth hypersurfaces and the system of six square relations (Table 3) representing three pairs of parabolic hypercylinders (cf. Fig. 2 for intersection with $n_\Theta = 0$).

6. Solution

Consider a shallow spherical cap with a simply supported edge restricted from any motion, and subjected to pressure uniformly distributed over the plane.

Let us assume that the centre of the shell is in a pure membrane state thus $m_\varphi = m_\Theta = 0$ and $n_\varphi = n_\Theta = 1$.

A justification of this assumption will be presented later on. Such a state can cover a certain zone of the radius $(0 \leq r \leq \xi)$. We suppose that this zone propagates in the course of plastic deformations. By an analysis

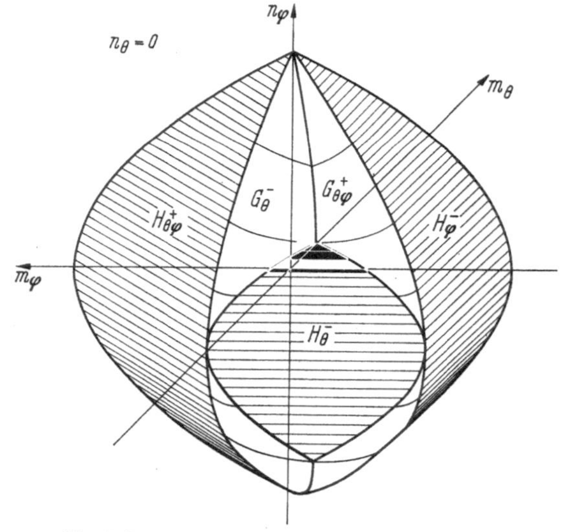

Fig. 2. Intersections of the yield surface for $n_\Theta = 0$.

of statically admissible states of stress and a discussion of the load carrying capacity solutions we may expect the stress profile corresponding to the remaining part of the shell to lie on the hypersurface H_Θ^-, Fig. 2. We will be looking for a solution corresponding to the yield hypersurface consisting of hypercylinder H_Θ^-. For the considered problem the kinematical and statical boundary conditions and the continuity requirements on the boundary between two zones are the following:

$$\text{for } r = 0: \quad w_z = w_0, \quad w_z' = 0, \quad n_\varphi = n_\Theta, \quad m_\varphi = m_\Theta \tag{6.1}$$

$$\text{for } r = \xi: \quad w_z] = 0, \quad w_z'] = 0, \quad u_r] = 0, \quad n_\varphi] = 0, \quad m_\varphi] = 0 \tag{6.2}$$

$$\text{for } r = c: \quad w_z = 0, \quad u_r = 0, \quad m_\varphi = 0 \tag{6.3}$$

where $\beta]$ denotes a jump in the quantity β. As for the initial conditions at the instant of initial yielding we assume

$$t = 0: \quad u = w = 0. \tag{6.4}$$

The adopted procedure of solution consists of integration of the equilibrium equations for the assumed stress profile and of checking the statical admissibility of the obtained stress resultant field.

I. Central zone $0 \leq r \leq \xi$. According to the introduced assumption, the membrane state in the central zone is specified by

$$n_\varphi = n_\Theta = 1, \quad m_\varphi = m_\Theta = 0. \tag{6.5}$$

Substituting (6.5) to the equation of equilibrium (4.8), we obtain

$$(r^2 + r w_z')' + r p = 0.$$

After evaluation of the integration constants from the boundary conditions $(6.1)_1$ and $(6.1)_2$, the following formula on the shells deflection in the central zone is obtained

$$w_z = -\frac{r^2}{2}\left(1 + \frac{p}{2}\right) + w_0. \tag{6.6}$$

II. Boundary zone $\xi \leq r \leq c$. The state of stress in the boundary zone is supposed to be represented by the hypersurface H_Θ^- described by the equation

$$n_\Theta^2 - m_\Theta = 1. \tag{6.7}$$

The extent of validity of these hypothesis will be discussed later. For the yield equation (6.7) the associated flow rule yields

$$\dot\lambda_\varphi = 0, \quad \varkappa_\varphi = 0, \quad \dot\lambda_\Theta : \dot\varkappa_\Theta = -2 n_\Theta. \tag{6.8}$$

Substituting (3.3) into (6.8), we get

$$\dot w_z'' = 0$$

$$\dot u_r' + \dot w_z' r + w_z' \dot w_z' = 0$$

$$\frac{\dot u_r}{h \dot w_z'} = 2 n_\Theta. \tag{6.9}$$

The relations (6.7) and (6.9) furnish a system of four equations for the unknowns w_z, u_r, n_Θ and m_Θ. This system has an elementary solution. Using the boundary and the initial conditions (6.3), (6.4) as well as the continuity requirements in evaluation of integration constants the solution is eventually found to be

$$w_z = \left(1 + \frac{p}{2}\right)(c - r)\xi, \tag{6.10}$$

$$u = \frac{r\xi}{2}\left(1 + \frac{p}{2}\right)\left[r - \xi\left(1 + \frac{p}{2}\right)\right] - \frac{c\xi}{2}\left(1 + \frac{p}{2}\right)\left[c - \xi\left(1 + \frac{p}{2}\right)\right], \tag{6.11}$$

$$n_\Theta = \frac{1}{2h}\left[\frac{1}{2}(c^2 - r^2) + \xi(r - c)\left(1 + \frac{p}{2}\right)\right], \tag{6.12}$$

$$m_\Theta = \frac{1}{4h^2}\left[\frac{1}{2}(c^2 - r^2) + \xi(r - c)\left(1 + \frac{p}{2}\right)\right]^2 - 1. \tag{6.13}$$

The remaining functions n_φ and m_φ are determined from the equation of equilibrium. Substituting (6.10), (6.12) and (6.13) into (4.8), integrating and using the continuity condition, we obtain

$$\begin{aligned} n_\varphi = &-\frac{1}{2h}\left[\frac{r^2}{6} - \frac{c^2}{2} + \left(1 + \frac{p}{2}\right)\left(c - \frac{r}{2}\right)\xi\right] + \\ &+ \frac{\xi}{2hr}\left[\frac{\xi^2}{6} - \frac{c^2}{2} + \left(1 + \frac{p}{2}\right)\left(c - \frac{\xi}{2}\right)\xi\right] + \frac{\xi}{r}, \end{aligned} \tag{6.14}$$

$$\begin{aligned} m_\varphi = &\frac{7}{240h^2}r^4 - \frac{7}{48h^2}\xi r^3\left(1 + \frac{p}{2}\right) + \\ &+ \left\{\frac{\xi^2}{6h^2}\left(1 + \frac{p}{2}\right)^2 + \frac{1}{4h^2}\left[c\xi\left(1 + \frac{p}{2}\right) - \frac{c^2}{2}\right] - \frac{p}{6h}\right\}r^2 + \\ &+ \left\{\frac{-\xi}{2h} - \frac{\xi}{4h^2}\left[\frac{\xi^2}{6} - \frac{c^2}{2} + \left(1 + \frac{p}{2}\right)\left(c - \frac{\xi}{2}\right)\xi\right] + \right. \\ &\left. + \frac{\xi c}{2h^2}\left(1 + \frac{p}{2}\right)\left[-\xi\left(1 + \frac{p}{2}\right) + \frac{c}{2}\right]\right\}r + \\ &+ \left\{\frac{\xi^2}{h}\left(1 + \frac{p}{2}\right) + \frac{\xi^2}{2h^2}\left(1 + \frac{p}{2}\right)\left[\frac{\xi^2}{6} - \frac{c^2}{2} + \left(1 + \frac{p}{2}\right)\left(c - \frac{\xi}{2}\right)\xi\right] + \right. \\ &\left. + \frac{c^2}{4h^2}\left[-\xi\left(1 + \frac{p}{2}\right) + \frac{c}{2}\right]^2 - 1\right\} + \\ &+ \left\{\frac{\xi^4}{h^2}\left[\frac{1}{80} - \frac{1}{16}\left(1 + \frac{p}{2}\right) + \frac{1}{12}\left(1 + \frac{p}{2}\right)^2\right] - \right. \\ &\left. - \frac{\xi^2 p}{h}\left[\frac{1}{12} + \frac{1}{2p}\left(1 + \frac{p}{2}\right)\right] - \left[\frac{c^2}{4h} - \frac{c\xi}{2h}\left(1 + \frac{p}{2}\right)\right]^2 + 1\right\}\frac{\xi}{r}. \end{aligned} \tag{6.15}$$

So far boundary conditions $(6.2)_1$ and $(6.3)_5$ have not been utilized. These conditions allow to find the central shell deflection

$$w_0 = \left(1 + \frac{p}{2}\right)\left(c - \frac{\xi}{2}\right)\xi \tag{6.16}$$

and to obtain an expression for the radius of the momentless zone

$$\begin{aligned}
&\frac{7}{240\,h^2}\,c^4 - \frac{7}{48\,h^2}\,\xi c^3\left(1 + \frac{p}{2}\right) + \\
&+ \Big\{\frac{\xi^2}{6\,h^2}\left(1 + \frac{p}{2}\right)^2 + \frac{1}{4\,h^2}\left[c\,\xi\left(1 + \frac{p}{2}\right) - \frac{c^2}{2}\right] - \frac{p}{6\,h}\Big\}\,c^2 + \\
&+ \Big\{-\frac{\xi}{2h} - \frac{\xi}{4\,h^2}\left[\frac{\xi^2}{6} - \frac{c^2}{2} + \left(1 + \frac{p}{2}\right)\left(c - \frac{\xi}{2}\right)\xi\right] + \\
&+ \frac{\xi c}{2\,h^2}\left(1 + \frac{p}{2}\right)\left[-\,\xi\left(1 + \frac{p}{2}\right) + \frac{c}{2}\right]\Big\}\,c + \\
&+ \Big\{\frac{\xi^2}{h}\left(1 + \frac{p}{2}\right) + \frac{\xi^2}{2\,h^2}\left(1 + \frac{p}{2}\right)\left[\frac{\xi^2}{6} - \frac{c^2}{2} + \left(1 + \frac{p}{2}\right)\left(c - \frac{\xi}{2}\right)\xi\right] + \\
&+ \frac{c^2}{4\,h^2}\left[-\,\xi\left(1 + \frac{p}{2}\right) + \frac{c}{2}\right]^2 - 1\Big\} + \\
&+ \Big\{\frac{\xi^4}{h^2}\left[\frac{1}{80} - \frac{1}{16}\left(1 + \frac{p}{2}\right) + \frac{1}{12}\left(1 + \frac{p}{2}\right)^2\right] - \\
&- \frac{\xi^2 p}{h}\left[\frac{1}{12} + \frac{1}{2p}\left(1 + \frac{p}{2}\right)\right] - \left[\frac{c^2}{4h} - \frac{c\,\xi}{2h}\left(1 + \frac{p}{2}\right)\right]^2 + 1\Big\}\,\frac{\xi}{c} = 0\,.
\end{aligned} \tag{6.17}$$

The above equations represent a parametric form of the sought dependence between the applied pressure and the central deflection of the shell.

7. Discussion of Results

An illustration of the above solution for two particular cases of geometry of the shells $c = 0.1$, $h = 0.005$ and $c = 0.2$, $h = 0.01$ is shown in Fig. 3 (full line).

A purely membrane solution, readily available from the analysis of equation of equilibrium (4.8) provides the following load-deflection relation

$$p_m = 2\left(\frac{2\,w_0}{c^2} - 1\right). \tag{7.1}$$

This solution is given in Fig. 3 as straight broken lines. The results plotted clearly show a gradual transition from a bending solution to the membrane solution where purely membrane state is reached asymptotically for $w \to \infty$. However, it should be remembered that the solution for thin shallow caps represented by the full line is valid for deflections of the order of shell thickness. For larger deflections, both equilibrium equations and geometrical relations become not sufficiently accurate.

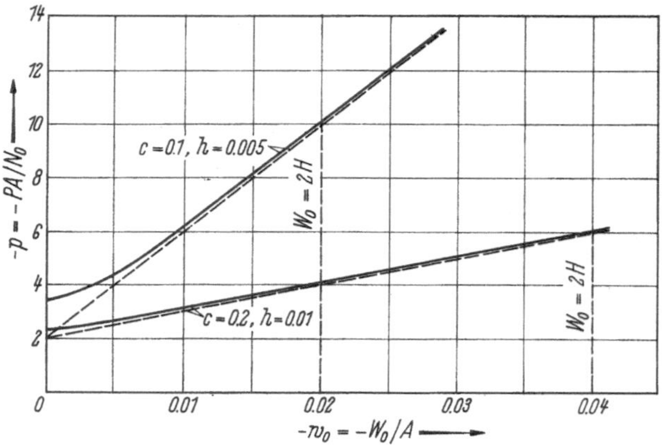

Fig. 3. Deflection of the shell center versus applied load.

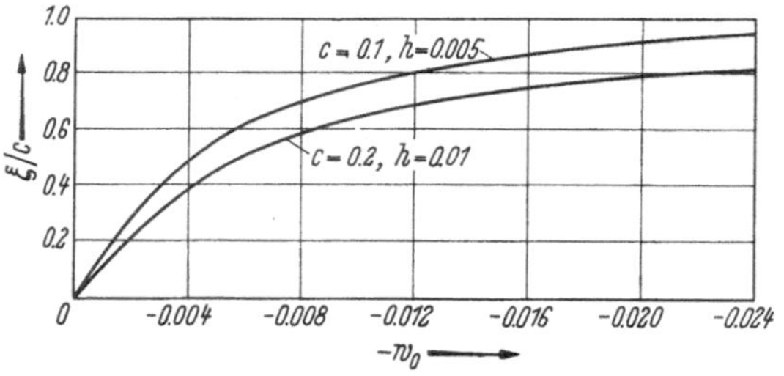

Fig. 4. Propagation of the membrane zone with increasing deflection.

The propagation of the membrane zone as a function of deflection $w_0 = w_0(\xi)$, computed from (6.16) and (6.17), is shown in Fig. 4.

At the instant of yielding ($w_0 = 0$) the membrane zone reduces to the shell centre and then propagates towards support with increasing load. The value of the initial load-carrying capacity of a shell can be computed from (6.16) and (6.17) and is found to be

$$p_y = -\left(6\,\frac{h}{c^2} + 0.2\,\frac{c^2}{h}\right). \tag{7.2}$$

The above relation is plotted in Fig. 5 for two values of the parameter h, namely for $h = 0.005$ and $h = 0.01$. The solutions obtained are valid for relatively small values of c, characteristic for shallow shells, full line in Fig. 5. However, analysing the presented diagrams one can observe that for larger angles the limit load tends to the value corresponding to the membrane state solution $p_y \to 2$ (a horizontal line in Fig. 5). Very similar results to those shown in Fig. 5 were obtained by MRÓZ and XU-BING-JE [4] on the basis of the same yield condition but linearized geometrical

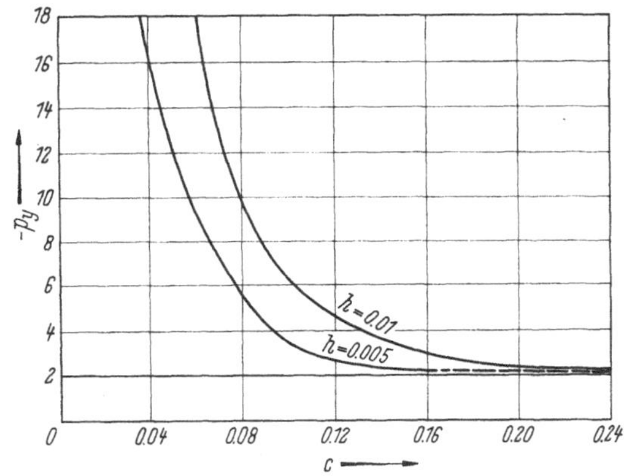

Fig. 5. Initial load-carrying capacity as a function of the shell dimensions.

relations and equilibrium equations as commonly used in the theory of thin spherical shells at small deflections.

Equations (6.12)—(6.15) show that the values of generalized stresses on the edge $r = c$ are $n_\Theta = 0$, $m_\Theta = 1$, $m_\varphi = 0$, $n_\varphi \neq 0$. Hence, this stress field exceeds the limit hypersurface of Fig. 2. Therefore, the obtained formulae provide only an upper estimate of the complete solution at the instant of yielding and an approximate solution for advanced plastic deformations. This solution can be, however, treated as an exact one for a suitably altered yield surface, for example a small outward parallel shifting of hypersurfaces $H^-_{\Theta\varphi}$ and G^-_Θ satisfies this requirement.

The complete solution would involve additional zones corresponding to the relevant equations of the yield surfaces. However attempts of this type cannot be more successful than the attempts to find the closed complete solution to the limit analysis problem, where the exact solution is still not known. Assuming $c \to 0$ and $A\,c \to R_1$ a limit transition to the plate with radius R_1 is obtained. In terms of the notation used in [9]

$$q^* = \frac{P R_1}{6\,\sigma_0 H^2}, \qquad \delta = \frac{W_0}{2H}, \qquad \xi^* = \frac{\xi}{c} \tag{7.3}$$

Eqs. (6.16) and (6.17) in the limiting case yield (Fig. 6, full line)

$$\delta = \frac{3}{4} q^* \xi^* \left(1 - \frac{\xi^*}{2}\right),$$

$$\frac{3}{4} q^{*2} \xi^{*2} (\xi^{*3} - 3\xi^{*2} + 3\xi^* - 1) + q^* (-2\xi^{*3} + 3\xi^{*2} - 1) - 1 + \xi^* = 0. \tag{7.4}$$

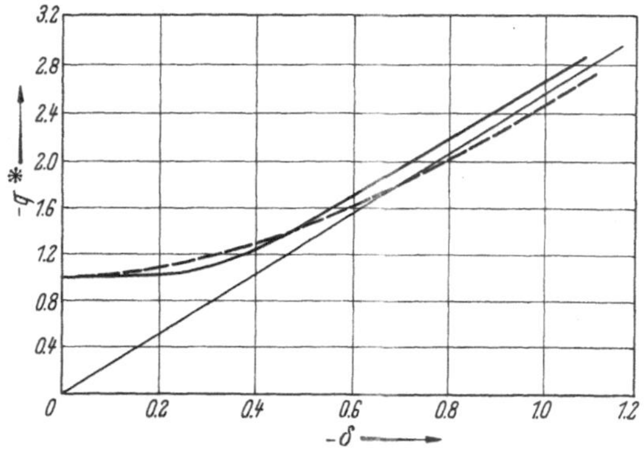

Fig. 6. Comparison of solutions for circular plates at large deflection.

The membrane solution for plates as obtained from (7.1), (Fig. 6, straight line) is

$$q^* = \frac{8}{3} \delta. \tag{7.5}$$

The broken line in Fig. 6 corresponds to the solution for circular plates at moderately large deflections (cf. [9]).

8. Conclusions

The curves of Fig. 4 indicate that the central membrane zone propagates very rapidly with increasing deflections. The thinner the shell the quicker the membrane zones propagate. At $h = 0.01$ and $h = 0.005$ for deflections equal to the shell thickness, the central zone extends over the major part of the shell, $\xi = -0.90c$ i. e. only a narrow region adjacent to the support is subjected to bending. The plot of the limit load versus the deflection w_0 tends asymptotically to the membrane solution, as it was shown in Fig. 3. For the shell deflections equal to the wall thickness, the difference between the present solution p and the membrane solution p_m equals approximately one per cent as far as the value of the load-carrying capacity is concerned. Hence it can be concluded that the pure membrane state in the shell is practically reached at deflections of the order of the wall thickness.

The above solution confirms the results obtained when analysing rigid perfectly plastic plates and cylindrical shells, [9—16].

A general conclusion can now be stated that changes in geometry constitute an essential factor governing the response of structures within a plastic regime leading to the so called "geometrical strengthening". The contrary situation can also arise when the effect of "weakening" may be noticed [18]. Similarly as for plates [12] "unstable" load-deflection relations are then obtained.

In both cases, however, accounting for the deformed shape of a rigid-perfectly structure yields the load values differing significantly from the classical limit analysis solutions.

From the solutions for a circular plate with restrained edges it follows that the purely membrane zone at the center appears at $W_0/H = 0.5$ (cf. [9], [12]). In the present approach we required $m_\varphi = m_\Theta = 0$ for $r = 0$ at $w_0 = 0$. To confirm the correctness of such an assumption we pass to the limit with the stress field (6.12) to (6.15) at $c \to 0$ and $Ac \to R_1$, when $\xi = 0$. One eventually obtains $n_\Theta = 0$, $m_\Theta = 1$, $n_\varphi = 0$, $m_\varphi = r^2 - 1$ for $r \neq 0$.

Accounting for the stress discontinuities allowed by the equilibrium equations it can easily be seen that the resulting stresses coincide with those associated with the yield point load for a Tresca plate, except the point $r = 0$. The singularity occurs on the set of measure zero and therefore correctness of the solution is not influenced. This fact justifies our choice of the stress profile for a shell at $r = 0$, allowing to solve the problem otherwise untreatable.

References

1. ONAT, E. T., and W. PRAGER: Limit analysis of shells of revolution, Proc. Roy. Netherl. Acad. Sci., Ser. B, 57, 1954 pp. 534—548.
2. HODGE, P. G.: The collapse load of a spherical cap, Proc. Fourth Midw. Conf. Solid Mech., Austin, 1959 pp. 108—126.
3. DRUCKER, D. C., and R. T. SHIELD: Limit analysis of symmetrically loaded thin shells of revolution, J. Appl. Mech., 1, 26, 61—68 (1959).
4. MRÓZ, Z., and XU BING-JE: The load-carrying capacities of symmetrically loaded spherical shells, Arch. Mech. Stos., 2, 15, 245—266 (1963).
5. HODGE, P. G.: Plastic analysis of structures, New York: McGraw-Hill 1959.
6. HODGE, P. G.: Limit analysis of rotationally symmetric plates and shells, En lewood Cliffs, N. J.: Prentice-Hall 1963.
7. FEINBERG, S. M.: Plastic flow of a shallow shell as an axisymmetric problem (in Russian), PMM 4, 21, 544—549 (1957).
8. ONAT, E. T., and R. M. HAYTHORNTHWAITE: The load-carrying capacity of circular plates at large deflections, J. Appl. Mech., 23, 49—55 (1956).
9. LEPIK, J. R.: Plastic flow of thin circular plates of a rigid-plastic material (in Russian), Bull. Acad. Sci. USSR, Mech. and Mech. Eng., 2, 78—87 (1960).

10. LEPIK, J. R.: Large deflection of rigid-plastic circular plates clamped at the boundary (in Russian), Conf. Theory of Plates and Shells, Publ. State Univ. of Kazan, 1961.
11. SAWCZUK, A., and L. WINNICKI: Plastic behaviour of simply supported concrete plates at moderately large deflections, Int. J. Solids Struct., **1**, 97–111 (1965).
12. LEPIK, J. R.: A contribution to the axial-symmetric bending of thin circular rigid-plastic plates (in Russian), Mech. of Solids, **4**, 104–110 (1966).
13. DUSZEK, M.: Effect of geometry changes on the carrying capacity of cylindrical shells, Bull. Ac. Pol. Sci., **4**, 13, 1965.
14. DUŞZEK, M.: Plastic analysis of cylindrical shells subjected to large deflections, Arch. Mech. Stos., **5**, 18, 599–614 (1966).
15. LEPIK, J. R.: Large deflections of a rigid-plastic cylindrical shell under internal pressure (in Russian), The 6th All-Union Conference on the Theory of Shells and Plates, Baku, 1966.
16. LEPIK, U.: Large deflections of rigid-plastic cylindrical shells under axial tension and external pressure, Nucl. Eng. Design, **4**, 29–38 (1966).
17. DONNELL, H. L.: General thin shell displacement-strain relations, Proc. 4th US Nat. Congr. Appl. Mech., 1962 pp. 529–536.
18. SHABILII, O. N.: Large deflections of a rigid-plastic shallow spherical shell (in Russian), Notes of the Summer-school on "Physically and Geometrically Nonlinear Problems of the Theory of Plates and Shells", Tartu, 1966.

Discussion

W. T. KOITER: I have listened with interest to the results of Mrs. DUSZEK presented by Professor SAWCZUK. I may add perhaps that the first investigation on the effect of geometric nonlinearities on the carrying capacity of rigid-plastic structures were carried out, as far as I am aware, by HAYTHORNTHWAITE and others at Brown University. Thinking aloud, I believe Dr. LECKIE's paper this morning, and Mrs. DUSZEK's paper this afternoon, indicate the need, and its possible fullfilment, of a general theory of initial post-yielding behaviour of rigid-plastic structures along similar lines as the theory of initial elastic post-buckling behaviour.

721/8/69